Seedling Ecology and Evolution

Seedlings are highly sensitive to their environment. After seeds, seedlings typically suffer the highest mortality rate of any life history stage. This book provides a thoughtful and comprehensive review by leading researchers of the interconnected topics that constitute seedling ecology and ecophysiology, focusing on how and why seedlings are successful. It considers the importance of seedlings in plant communities; environmental factors with special impact on seedlings; the morphological and physiological diversity of seedlings, including mycorrhizae; the relationship of the seedling with other life stages; seedling evolution; and seedlings in human-altered ecosystems, including deserts, tropical rainforests, and habitat-restoration projects. The diversity of seedlings is portrayed by specialized groups, such as orchids, bromeliads, and parasitic and carnivorous plants. This important text sets the stage for future research and is valuable to graduate students and researchers in plant ecology, botany, agriculture, and conservation.

The editors are well known for their work in soil seed-bank ecology. *Mary Allessio Leck*, Emeritus Professor of Biology, Rider University, has worked on seed ecology of tidal freshwater wetland species, and on wetland education for urban youth; *V. Thomas Parker*, Professor of Biology, San Francisco State University, on tidal wetland, chaparral, and mycorrhizal ecology, and *Arctostaphylos* evolution; and *Robert L. Simpson*, Professor of Biology and Environmental Science, University of Michigan – Dearborn, on freshwater wetland ecology and the natural history of Michigan.

Seedling Ecology and Evolution

Editors

Mary Allessio Leck
Emeritus Professor of Biology, Rider University, USA

V. Thomas Parker
Professor of Biology, San Francisco State University, USA

Robert L. Simpson
Professor of Biology and Environmental Science, University of Michigan – Dearborn, USA

CAMBRIDGE
UNIVERSITY PRESS

Shaftesbury Road, Cambridge CB2 8EA, United Kingdom

One Liberty Plaza, 20th Floor, New York, NY 10006, USA

477 Williamstown Road, Port Melbourne, VIC 3207, Australia

314–321, 3rd Floor, Plot 3, Splendor Forum, Jasola District Centre, New Delhi – 110025, India

103 Penang Road, #05–06/07, Visioncrest Commercial, Singapore 238467

Cambridge University Press is part of Cambridge University Press & Assessment, a department of the University of Cambridge.

We share the University's mission to contribute to society through the pursuit of education, learning and research at the highest international levels of excellence.

www.cambridge.org
Information on this title: www.cambridge.org/9780521694667

First published 2008

A catalogue record for this publication is available from the British Library

ISBN   978-0-521-87305-5   Hardback
ISBN   978-0-521-69466-7   Paperback

Contents

Contributors   page xi
Foreword by Peter J. Grubb   xiii
Preface   xvii
Acknowledgments   xix

Part I | Introduction

Chapter 1 | Why seedlings?   3

*Mary Allessio Leck, Robert L. Simpson, and
V. Thomas Parker*

1.1 Seedlings as part of a plant's life cycle   3
1.2 Vulnerabilities and bottlenecks   6
1.3 Making it: filters, safe sites, and establishment   7
1.4 Seedlings: a primer   8
1.5 What seedlings can tell us   10
1.6 The scope of *Seedling Ecology and Evolution*   12

Part II | Seedling diversity

Chapter 2 | Seedling natural history   17

Mary Allessio Leck and Heather A. Outred

2.1 Introduction   17
2.2 The seedling stage and fate of seedlings   17
2.3 Seedling types   20
2.4 Seedling diversity – morphology   24
2.5 Seedling diversity – ecophysiology   31
2.6 Vivipary and seedling equivalents   33
2.7 Longevity   36
2.8 Dispersal   37
2.9 Environmental filters and safe sites   38
2.10 Summary   41
2.11 Acknowledgments   41
     Appendixes   42

Chapter 3 | Specialized seedling strategies I: seedlings in
stressful environments   56

José M. Facelli

3.1 Introduction   56
3.2 Seedling establishment in dense shade   60
3.3 Effects of litter on seedling establishment   64
3.4 Seedling establishment in dry environments   65
3.5 Seedling establishment in saline environments   68

3.6 Seedling establishment in cold environments   70
3.7 Physical damage to seedlings   73
3.8 Interactive effects of stress agents and biological
interactions   75
3.9 Overview: adaptations of seedlings to stressful
environments   76
3.10 Acknowledgments   78

Chapter 4 | Specialized seedling strategies II:
orchids, bromeliads, carnivorous plants,
and parasites   79

*Dennis F. Whigham, Melissa K. McCormick,
and John P. O'Neill*

4.1 Introduction   79
4.2 Epiphytic and terrestrial orchids   80
4.3 Bromeliads   85
4.4 Carnivorous plants   88
4.5 Parasitic plants   91
4.6 Summary and future needs   99
4.7 Acknowledgments   100

Part III | **Seedling morphology, evolution,
and physiology**

Chapter 5 | Embryo morphology and seedling evolution   103

Karl J. Niklas

5.1 Introduction   103
5.2 Embryophyte phylogeny   106
5.3 Embryo structure   108
5.4 Embryophyte embryogenesis   112
5.5 Phylogenetic patterns   122
5.6 Concluding remarks   128

Chapter 6 | Regeneration ecology of early angiosperm
seeds and seedlings: integrating inferences
from extant basal lineages and fossils   130

Taylor S. Feild

6.1 Introduction   130
6.2 Previous views of ancestral angiosperm ecology and
seed/seedling morphology   132
6.3 The phylogenetic revolution: inferences on early
angiosperm regeneration ecology from extant basal
angiosperms   135
6.4 Functional biology of basal angiosperm seeds   136
6.5 Functional biology of basal angiosperm seedlings   142
6.6 Outlook and recommendations for future research   146
6.7 Acknowledgments   149

Chapter 7 | Physiological and morphological changes during early seedling growth: roles of phytohormones   150

Elizabeth J. Farnsworth

7.1   Introduction: phytohormones, molecular biology, and the "real world" of early seedling ecology   150
7.2   Seedling responses to light   156
7.3   Seedling responses to temperature   159
7.4   Seedling responses to water   161
7.5   Seedling responses to nutrients   163
7.6   Insights and common themes   166
7.7   Summary   170
7.8   Acknowledgments   171

Chapter 8 | Seedling ecophysiology: strategies toward achievement of positive net carbon balance   172

Kaoru Kitajima and Jonathan A. Myers

8.1   Introduction   172
8.2   Seed reserve utilization   173
8.3   Ontogenetic trajectories of seedling carbon balance   175
8.4   Species differences in inherent relative growth rate (RGR)   177
8.5   Opportunistic versus conservative strategies   178
8.6   Carbohydrate reserves   182
8.7   Phenotypic plasticity   185
8.8   Concluding remarks   187
8.9   Acknowledgments   188

Chapter 9 | The role of symbioses in seedling establishment and survival   189

Thomas R. Horton and Marcel G. A. van der Heijden

9.1   Introduction   189
9.2   Ectomycorrhizal fungi and seedling establishment   199
9.3   Arbuscular mycorrhizal fungi and seedling establishment   207
9.4   Other plant symbionts and seedling establishment   209
9.5   Conclusions   212
9.6   Acknowledgments   213

Part IV | Life history implications

Chapter 10 | The seedling as part of a plant's life history strategy   217

Angela T. Moles and Michelle R. Leishman

10.1   Introduction   217

10.2 The trade-off between offspring production and
      seedling survival   218
10.3 Understanding seed and seedling ecology as parts of
      a plant's life history strategy   227
10.4 Correlations between seed and seedling strategy and
      other aspects of plant ecology   229
10.5 Seed and seedling strategies within species   232
10.6 Implications of a holistic understanding of plant life
      history strategies   235
10.7 Conclusions   237

Chapter 11 | Seedling recruitment and population
                 ecology   239

Ove Eriksson and Johan Ehrlén

11.1 Introduction   239
11.2 The causes of seedling mortality   241
11.3 Recruitment limitation   245
11.4 Seedling recruitment and population dynamics   248
11.5 Genetic structure and selection in seedling
      populations   251
11.6 Concluding remarks   254
11.7 Acknowledgments   254

Chapter 12 | Seedling communities   255

Jon E. Keeley and Phillip J. van Mantgem

12.1 Introduction   255
12.2 Internal drivers   255
12.3 External drivers affecting seedling communities   265
12.4 Seedling community assembly rules   269
12.5 Conclusions   273

Chapter 13 | Spatial variation in seedling emergence
                 and establishment – functional groups
                 among and within habitats?   274

Johannes Kollmann

13.1 Introduction   274
13.2 Description of the seedling stage   276
13.3 Definition of spatial scales   279
13.4 Microhabitat effects on seedling dynamics   280
13.5 Habitat effects on seedling dynamics   282
13.6 Landscape effects on seedling dynamics   285
13.7 Region effects on seedling dynamics   286
13.8 Biome effects on seedling dynamics: seed size and
      seedling survival   287
13.9 Synthesis   289
13.10 Acknowledgments   292

Part V | Applications

Chapter 14 | Does seedling ecology matter for
biological invasions?   295

Laura A. Hyatt

14.1   Introduction   295
14.2   Invasive seedlings   296
14.3   Invasive effects on native seedlings   303
14.4   Conclusions   305
14.5   Acknowledgments   306

Chapter 15 | The role of seedlings in the dynamics of
dryland ecosystems – their response to
and involvement in dryland heterogeneity,
degradation, and restoration   307

Bertrand Boeken

15.1   Introduction   307
15.2   Importance of the seedling stage   308
15.3   Seedlings and spatial heterogeneity of drylands   312
15.4   Seedlings and dryland system degradation   322
15.5   Conclusions   330

Chapter 16 | Anthropogenic disturbance in tropical
forests: toward a functional understanding
of seedling responses   332

James W. Dalling and David F. R. P. Burslem

16.1   Introduction   332
16.2   Significance of the seedling stage for forest
management   334
16.3   Effects of human disturbances on seedling
regeneration   335
16.4   Application of seedling functional ecology to
tropical forest management and restoration   341
16.5   Future directions   349
16.6   Acknowledgments   351

Chapter 17 | Seedling establishment in restored
ecosystems   352

Susan Galatowitsch

17.1   Introduction   352
17.2   Selecting initial community composition for
restoration   353
17.3   Creating safe sites to promote seedling establishment   359
17.4   Managing biotic interactions that affect seedling
survival and growth   364

17.5 Mimicking the effects of disturbances in restoration   368
17.6 Conclusions   369

Part VI | Synthesis

Chapter 18 | The seedling in an ecological and evolutionary context   373

V. Thomas Parker, Robert L. Simpson, and Mary Allessio Leck

18.1 Introduction   373
18.2 Dispersal, seed bank dynamics, and seedling banks   374
18.3 Dynamics of individual seedlings   376
18.4 Seedlings in heterogeneous environments   378
18.5 Alternative strategies   381
18.6 Conclusions   385

References   391
Index   501

Contributors

Bertrand Boeken
Ben-Gurion University of the Negev
The Wyler Department of Dryland Agriculture
Jacob Blaustein Institutes Sede Boker Campus
    for Desert Research
Ben-Gurion, Israel

David F. R. P. Burslem
University of Aberdeen
Department of Plant and Soil Science
Aberdeen, Scotland, UK

James W. Dalling
University of Illinois, Urbana-Champaign
Department of Plant Biology
Urbana, Illinois, USA

Johan Ehrlén
Stockholm University
Department of Botany
Stockholm, Sweden

Ove Eriksson
Stockholm University
Department of Botany
Stockholm, Sweden

José M. Facelli
The University of Adelaide
Discipline of Ecology and Evolutionary Biology
School of Earth and Environmental Sciences
Adelaide, Australia

Elizabeth J. Farnsworth
New England Wild Flower Society
Framingham, Massachusetts, USA

Taylor S. Feild
The University of Tennessee
Ecology and Evolutionary Biology
Knoxville, Tennessee, USA

Susan Galatowitsch
University of Minnesota
Department of Horticultural Science
St. Paul, Minnesota, USA

Peter J. Grubb
University of Cambridge
Department of Plant Sciences
Cambridge, UK

Thomas R. Horton
State University of New York
College of Environmental Science and
    Forestry
Syracuse, New York, USA

Laura A. Hyatt
Rider University
Biology Department
Lawrenceville, New Jersey, USA

Jon E. Keeley
United States Geological Survey
Sequoia and Kings Canyon Field Station
Three Rivers, California, USA
and
University of California – Los Angeles
Department of Ecology and Evolutionary
    Biology
Los Angeles, California, USA

Kaoru Kitajima
University of Florida
Department of Botany
Gainesville, Florida, USA
and
Smithsonian Tropical Research Institute
Apartado, Balboa, Panama

Johannes Kollmann
University of Copenhagen
Department of Agriculture and Ecology
Frederiksberg C, Denmark

Mary Allessio Leck
Rider University
Biology Department
Lawrenceville, New Jersey, USA

Michelle R. Leishman
Macquarie University
Department of Biological Sciences
Sydney, Australia

Melissa K. McCormick
Smithsonian Environmental
    Research Center
Edgewater, Maryland, USA

Angela T. Moles
University of New South Wales
School of Biological, Earth, and
    Environmental Sciences
Sydney, Australia

Jonathan A. Myers
Louisiana State University
Department of Biological Sciences
Division of Systematics, Ecology, and
    Evolution
Baton Rouge, Louisiana, USA

Karl J. Niklas
Cornell University
Department of Plant Biology
Ithaca, New York, USA

John P. O'Neill
Smithsonian Environmental Research Center
Edgewater, Maryland, USA

Heather A. Outred
Massey University
College of Science
Institute of Molecular Biosciences
Palmerston North, New Zealand

V. Thomas Parker
San Francisco State University
Department of Biology
San Francisco, California, USA

Robert L. Simpson
The University of Michigan – Dearborn
Department of Natural Sciences
Dearborn, Michigan, USA

Marcel G. A. van der Heijden
Agroscope Reckenholz-Tanikon Research
    Station ART
Zurich, Switzerland

Phillip J. van Mantgem
United States Geological Survey
Sequoia and Kings Canyon Field
    Station
Three Rivers, California, USA

Dennis F. Whigham
Smithsonian Environmental Research Center
Edgewater, Maryland, USA

Foreword

The properties of seedlings are potentially important to all plant ecologists, whether they be interested chiefly in understanding seminatural indigenous vegetation, invasive plants, or the problems of restoration. In seminatural vegetation, seedling properties may determine the climatic regions occupied on a continental scale and the habitats occupied within a landscape, the ability of one species to coexist with another in a community, and the abundance of one species relative to another at a given time and place. The requirements of seedlings often determine the sites in which potentially invasive species can succeed and whether a given approach to restoration of seminatural vegetation is effective.

During the last 40 years, there has been a steady increase in the amount of research by ecologists on the properties of seedlings as opposed to those of mature plants. Great pioneers such as F. E. Clements and E. J. Salisbury appreciated the importance of studying seedlings, although papers on experimental studies on seedlings were uncommon before the 1960s. Several factors have driven the increase in work on seedlings. Here I emphasize seven.

First, there has been a desire to seek generalizations about seedlings. For example, how does relative growth rate vary with the mass of reserves in the seed, and how does it differ at a given seed-reserve mass between plants of different growth forms (such as tree vs. herb), or species from different kinds of habitat (where the vegetation shows high and low productivity, respectively)? For the mechanistically minded, the key questions become (1) how do seedlings of species with smaller seeds have higher relative growth rates, and (2) how do species of different functional types have different relative growth rates at a given seed-reserve mass? Of course, the answers to these questions have turned out to be related to our increased understanding of the ecophysiology of the vegetative organs of the adult plant, at least of the leaves – there still is much to learn regarding stems and roots.

Second, there has been a realization that differences among species with regard to the requirements of juveniles may play a significant role in making possible long-term coexistence of species in communities. Within a community, the conditions vary more at the scale of the juvenile than of the adult, and juveniles are generally less tolerant of adverse conditions. Here, we are concerned not only with the seedling as defined in a very narrow sense, but also with plants in their first few weeks, months, years, or decades of life – depending on the type of vegetation.

Third, it seemed at one time that a seed number–seedling survival trade-off had considerable potential in explaining the coexistence of species that differ appreciably in seed size but have very similar requirements for regeneration. In this event, most researchers have concluded that the trade-off by itself is not enough to explain the coexistence of the full range of seed sizes, either where greater survival results from greater competitive ability or where it results from greater tolerance of hazards during establishment.

Fourth, there has been a greatly increased appreciation that seedlings, more often than not, are in symbiosis with a type of micro-organism, most commonly with at least one arbuscular mycorrhizal fungus. Gradually, plant ecologists have come to realize that in one community, some plant species are more dependent on a symbiont than in others, and that symbionts of a given type can have inhibitory as well as stimulatory effects. There have been parallel advances in our knowledge of the seedlings of plants that are partially or wholly par-asitic. There remains open the question of how much specialization exists in the relationship between plant species and their symbionts – a question that can now be tackled more satisfactorily as a result of the development of molecular techniques.

Fifth, the development of molecular biology has greatly increased the potential for advances in understanding the physiology of seedlings – particularly their tolerances of shade, drought, low nutri-ent supply, and excess salt. The same goes for our understanding of seedling development, including the part played by phytohormones.

Sixth, there has been a revolution in our thinking about the kinds of seeds of the most primitive angiosperms and the habitats in which they functioned. Also, there has been renewed attention to the earliest true seeds of gymnosperms and the analogous seed-like structures of certain tree lycophytes.

Seventh, in the last two decades, there has been a surge of inter-est in the long-standing problem of why some species are much more invasive than others and in the related issue of how to restore vege-tation at degraded sites. Some of us feel that it is difficult to extract generalizations in these areas, and, in many cases, the key species are idiosyncratic in their requirements. Nevertheless, the great practical importance of the problems makes it imperative that they be tackled by some of the ablest ecologists. Every stage in a plant's life cycle must be considered, but, in many cases, the seedling stage will turn out to be of critical importance.

With this background, we may welcome a new book that covers the whole range of issues I have outlined. An especially attractive feature of the book is that a good many of the schools of thought that have dominated developments in thinking are represented among the authors and, more specifically, that many of the authors have been among those who have taken leading roles in plant ecology in the last two decades.

Studies on seedlings, despite real advances, are still at an imma-ture stage, and there remain significant disagreements. I cannot accept all of the assertions in this book and, indeed, I have argued in print with some of the authors. However, for me, this does not detract from the value of the book. I strongly recommend it to all those who seek thoughtful, up-to-date reviews of the wide range of interconnected topics that constitute seedling ecology and ecophysiology.

Peter J. Grubb
Department of Plant Sciences
University of Cambridge
June 2007

Preface

Interest in developing this multiauthored book grew from our work with seeds and seed-bank ecology. While seed production and seed-bank dynamics are critical stages, what happens to seedlings is also fundamental to explaining field observations of vegetation dynamics and recruitment. Although several recent books discuss seedlings, indicating their importance to plant regeneration (Fenner, 2000) and to seed ecology (Fenner & Thompson, 2005), only one, Swaine (1996), focuses on seedling ecology; it, however, deals exclusively with tropical forest seedlings and is now more than 10 years old. A fourth volume, Forget *et al.* (2005), is primarily about seed predation and dispersal. *Seedling Ecology and Evolution* will complement these works and provide a more all-encompassing discussion. Moreover, it bridges the life-cycle gap following seeds (e.g. Baskin & Baskin, 1998) and seed banks (e.g. Leck *et al.*, 1989). Additional information about regeneration strategies may be found in Harper (1977), Grubb (1977, 1998), and Grime (2001).

We acknowledge the importance of understanding seedling biology in agriculture and horticulture; however, seedlings are well studied in these settings, whereas in natural systems, seedlings are less studied, and the literature is more diffuse. This book explores seedling adaptations and constraints to regeneration in natural and disturbed systems, where a better understanding of seedlings would stimulate study and development of theory regarding this dynamic and often neglected part of the plant life cycle.

After seeds, seedlings typically suffer the highest mortality rate of any life history stage and, therefore, are important in the selection and evolution of species. Seedlings appear to be a "bottleneck" in plant establishment because they are particularly sensitive to the vagaries of the environment. Our purpose is to explore their ecology and evolution and, in the process, bring a diverse literature together for the first time – examining the diverse morphologies and physiologies of seedlings; environmental factors that impact seedlings; driving factors in the evolution of seedlings, including phylogenetic and ecophysiological constraints; seedlings in plant community dynamics, especially how they relate to species and community sustainability; seedling strategies and syndromes, including seedling banks; and the impact of human-generated perturbations, such as invasive species, desertification, and habitat fragmentation and restoration. To accomplish this, contributors were invited to explore a range of topics that are gathered in the book as follows:

- Part I – Introduction. Chapter 1 provides a review of seedling structure, as well as an introduction to the seedling stage of the seed plant life cycle.
- Part II – Seedling diversity. Chapters 2–4 consider aspects of seedling natural history, strategies in stressful habitats where shade,

drought, inundation, and other stressors affect establishment, and strategies of highly specialized plants, including epiphytes, orchids, and parasites.

- Part III – Seedling morphology, evolution, and physiology. Chapters 5–9 examine seedling evolution in the context of embryo evolution and the rise of angiosperm ecological diversity, as well as seedling morphological and developmental changes, phytohormones, maintenance of carbon balance, and the role of symbioses in establishment and survival.
- Part IV – Life history implications. Chapters 10–13 examine the trade-offs of the seedling stage with other stages, and seedlings in population and community contexts, as well as functional groups among and within habitats.
- Part V – Applications. Chapters 14–17 examine seedlings as the advancing front for biological invasions, in deteriorating ecosystems (e.g. deserts), in systems in which they are used for system maintenance (forests), and for restoration.
- Part VI – Synthesis. Chapter 18 considers the multiple perspectives presented by the chapters of this book, presents overarching seedling strategies, and summarizes areas for future study.

References

Baskin, C. C. & Baskin, J. M. (1998). *Seed Ecology, Biogeography, and Evolution of Dormancy and Germination.* San Diego: Academic Press.

Fenner, M., ed. (2000). *Seeds: the Ecology of Regeneration in Plant Communities.* Wallingford: CAB International.

Fenner, M. and Thompson, K. (2005). *The Ecology of Seeds.* Cambridge: Cambridge University Press.

Forget, P.-M., Lambert, J. E., Hulme, P. E., & Vander Wall, S. B., ed. (2005). *Seed Fate: Predation, Dispersal and Seedling Establishment.* Wallingford: CAB International.

Grime, J. P. (2001). *Plant Strategies, Vegetation Processes, and Ecosystem Processes.* Chichester: John Wiley & Sons.

Grubb, P. J. (1977). The maintenance of species-richness in plant communities: the importance of the regeneration niche. *Biological Reviews*, **52**, 107–45.

Grubb, P. J. (1998). A reassessment of the strategies of plants which cope with shortages of resources. *Perspectives in Plant Ecology, Evolution and Systematics*, **1**, 3–31.

Harper, J. L. (1977). *Population Biology of Plants.* London: Academic Press.

Leck, M. A., Parker, V. T., & Simpson, R. L., ed. (1989). *Ecology of Soil Seed Banks.* San Diego: Academic Press.

Swaine, M. D., ed. (1996). *Ecology of Tropical Forest Seedlings*, Man and the Biosphere Series, Vol. 17. Carnforth: UNESCO & The Parthenon Publishing Group.

Acknowledgments

E-mail facilitated interaction with contributors and reviewers from around the world. We are grateful to those who reviewed and improved chapters: Lubomir Adamec, Institute of Botany – Trebon, Czech Republic; Mitch Aide, University of Puerto Rico, USA; Christopher Baraloto, University of Florida, USA; Carol Baskin, University of Kentucky, USA; Margaret Brock, wetland botanist, Tasmania, Australia; Hans Cornelissen, Vrije Universiteit, The Netherlands; Saarad DeWalt, Clemson University, USA; Ian Dickie, Land Care Research, NZ; Joan Ehrenfeld, Rutgers University, USA; Wayne Ferren, Maser Consulting, New Jersey, USA; Lorena Gomez-Aparicio, Universidad de Granada, Spain; Norma Good, botanist, New Jersey, USA; James Grace, United States Geological Survey, Wetlands Center, Louisiana, USA; Denise Hardesty, CSIRO Atherton, Australia; Colleen Hatfield, California State University – Chico, USA; Jose Hierro, Universidad Nacional de La Pampa, Argentina; Patricia Holmes, Cape Ecological Services, South Africa; Enrique Jurado, Universidad Autónoma de Nuevo León, Mexico; Anwar Maun, University of Western Ontario, Canada; Dan Metcalfe, CSIRO Atherton, Australia; Susan Mopper, University of Louisiana, USA; Kazuhide Nara, University of Tokyo, Japan; Susan Schwinning, Texas State University – San Marcos, USA; Anna Sher, University of Denver, USA; John N. Thompson, University of California – Santa Cruz, USA; Larry Tieszen, United States Geological Survey, South Dakota, USA; Barry Tomlinson, Harvard Forest, USA; Eric von Wettberg, University of California – Davis, USA; Michael Walters, Michigan State University, USA; Michael Williams, Butte College, USA; Amy Zanne, National Evolutionary Synthesis Center, North Carolina, USA; and Jess Zimmerman, University of Puerto Rico, USA.

We thank our colleagues, too many to mention, who contributed to the development of our ideas as this book evolved. We are especially grateful to the contributors who willingly devoted their time and creative energies to this book, and for their good humor in meeting deadlines and responding to our numerous queries. We also acknowledge the many others whose work has contributed to our understanding of seedling biology.

Our special thanks go to Jacqueline Garget of Cambridge University Press and Eleanor Umali of Aptara, who shepherded this book to completion; to Marian and Brewster Young, who lent their home in Monterey, California, for a work retreat; and especially to our spouses, Charles F. Leck, Alison Sanders, and Penelope Simpson, for their enthusiastic and enduring support of this project. Finally, we acknowledge the inspiration of particular seedlings, including *Impatiens capensis* (all); *Bidens laevis* and *Polygonum bistortoides* (Leck); *Ambrosia trifida*, *Typha* spp., and *Zizania aquatica* (Simpson); and *Arctostaphylos canescens* and *Grindelia stricta* var. *angustifolia* (Parker).

Part I

Introduction

Chapter 1

Why seedlings?

Mary Allessio Leck, Robert L. Simpson, and V. Thomas Parker

It was, as it were, a little green star with many rays, half an inch in diameter, lifted an inch and a half above the ground on a slender stem. What a feeble beginning for so long-lived a tree! By the next year it will be a star of greater magnitude, and in a few years, if not disturbed, these seedlings will alter the face of Nature here.

Henry D. Thoreau (1993), writing in approximately 1862 about Pinus rigida (Pinaceae).

1.1 | Seedlings as part of a plant's life cycle

The seedling, the young spermatophyte plant following germination, is but one stage in the continuum of a seed plant's life cycle. For ecological purposes, discussion on the life cycle (illustrated in Fig. 1.1) focuses on the processes involved in replacing the adult and/or colonizing new habitats. A reproductive adult plant produces seeds that, once dispersed, become part of the seed bank (Parker *et al.*, 1989; Simpson *et al.*, 1989). Then, following germination, a seedling faces unpredictable environments and is limited by its particular genetic constraints. However, if successful, it survives to adulthood and reproduction.

Seedlings are highly vulnerable, subject to varied abiotic and biotic factors that affect growth and establishment. Their adversities, although variable in severity – depending on habitat and seedling form – include drought, flooding, herbivory, and lack of resources, such as mycorrhizal associates and light. The probability of a seed producing a successful, established plant is usually quite small (e.g. Simpson *et al.*, 1985; Leck & Simpson, 1994; Bazzaz, 1996; Kitajima,

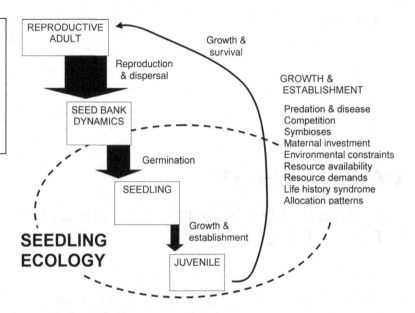

Fig. 1.1 Seed plant life cycle continuum. Shown are factors influencing seedling growth and establishment. The change in thickness of arrows from stage to stage suggests amount of attrition. The dotted line encompasses seedling ecology and indicates the scope of this book.

2007). The seedling stage, therefore, is a bottleneck, and selection pressure is assumed to be high (e.g. Grubb, 1977; Harper, 1977; Fenner & Thompson, 2005; Kitajima, 2007). The successful survival of some seedlings of every species is, ultimately, critical because it underlies the development and sustainability of plant communities.

For an individual plant, changes occur in morphological detail as well as in its reproductive ability, as it passes from "seed, to seedling, juvenile, immature, virginile, reproductive (young), mature (old), subsenile, and senile" stages (Bell, 1991, p. 324). Duration of a particular stage varies with species. Seeds of *Salix* spp. (Salicaceae), for example, germinate within 12–24 hours of dispersal (Young & Young, 1992). Those of other species [e.g. *Verbascum* sp. (Scrophulariaceae)] persist in the soil for more than a century (Telewski & Zeevart, 2002). Similarly, some seedlings, such as those of desert annuals that produce seeds within a few weeks of germination, are not seedlings for long. Conifer seedlings of forests of British Columbia (Canada), in contrast, may be held in the seedling stage for more than 150 years – until light conditions are suitable for continued development (Antos *et al.*, 2005).

Sometimes stages are skipped. For some viviparous species, such as mangroves (e.g. *Bruguiera* spp. Rhizophoraceae; Burger, 1972) and seagrasses (*Thalassia* spp. Hydrocharitaceae; Sculthorpe, 1967), embryo development is continuous and it is not held inactive and dormant within the seed (see Chapter 2). In other cases, plants may proceed directly from seedling to flowering stage. *Chenopodium rubrum* (Chenopodiaceae), *Pharbitis (Ipomoea) nil* (Convolvulaceae), and *Xanthium strumarium* (Asteraceae) have photoperiod sensitive cotyledons, flowering as seedlings following short-day inductive photoperiods; *C. rubrum* produces flowers within six days of initiation of imbibition (Downs & Hellmers, 1975). This precocious behavior is also seen in certain wetland species, including *Lindernia dubia* (Scrophulariaceae) (Leck, pers.

obs.), *Limosella australis* (Scrophulariaceae), and *Myriophyllum variifolium* (Haloragidaceae) (Brock, pers. comm.), that flower within weeks of germination during soil seed-bank experiments. These examples illustrate the plasticity found among plants in the seedling stage of their life cycles.

To become a seedling, the seed must first germinate, often distinguished by the protrusion of the radicle through the seed coat (see Chapter 2). Depending on the species, this process is regulated by dormancy mechanisms interacting with availability of water, quantity and quality of light, (alternating) temperature, levels of oxygen, and/or, in some cases, an external supply of nutrients (see Baskin & Baskin, 1998; Fenner & Thompson, 2005). In a community context, germination is a facet of seed-bank dynamics that are important because what happens to seed banks influences seedlings (Parker *et al.*, 1989; Simpson *et al.*, 1989). Seed banks may be transient or persistent. The relative transience or persistence is related in some habitats to disturbance regime and to seed size, with transient seed-bank species having larger seeds (e.g. Grime, 1989; Leck & Brock, 2000; but see Leishman & Westoby, 1998). Seeds of transient species are present in the soil for <1 year, short-term persistent for >1 year but <5 years, and long-term persistent for >5 years (Fenner & Thompson, 2005). Maintenance of seed banks can involve various mechanisms, including physical, physiological, morphological, and morphophysiological dormancy, and the dormancy level can cycle between dormant and nondormant states (e.g. Baskin & Baskin, 1998). In temperate areas, the larger-seeded transient seed-bank species may germinate at low temperatures (5 °C) and do not require light; their earlier spring germination means that they are in place before later (and smaller) germinators appear (Thompson & Grime, 1979; Leck & Simpson, 1993). Seedling establishment, generally considered to be the process during which a germinated seed achieves independence from maternal reserves (e.g. Fenner & Thompson, 2005), is favored by early germination at least in systems where the environment is predictable. Moreover, seedling establishment requirements of small-seeded persistent species would appear to be different from seedlings of large-seeded transient seed-bank species.

Successful negotiation of stages may vary with species. For example, in tidal freshwater wetland annuals, 91% of *Polygonum punctatum* (Polygonaceae) seeds overwintered to germinate and grow to seedlinghood but less than 1% of *Ambrosia trifida* (Asteraceae) did so (Leck & Simpson, 1994). Survivorship of seedlings varied with species and with location relative to a tidal stream channel (Parker & Leck, 1985). In this tidal freshwater wetland with predictable hydrology and dominated by both annuals and perennials, the later germinating perennials, like *Typha latifolia* (Typhaceae), were not observed to survive in study plots (Parker & Leck, 1985; Leck *et al.*, 1989b; Leck & Simpson, 1995).

Some components of the life cycle are discrete. For example, the seed is an entity, comprised of an embryo, typically with maternally supplied nutrient reserves, within a maternally derived seed coat (e.g.

Baskin & Baskin, 1998). In contrast, the seedling may be more arbitrarily delimited. Although the seedling stage has a defined start when the radicle emerges from the seed coat, its end point is along a growth continuum and is more difficult to recognize (Chapter 2). Furthermore, when a seedling is a seedling may depend on the focus of the viewer. The morphologist considers morphological changes, whereas the physiologist emphasizes the attainment of independence from seed reserves. However, in the case of orchids and parasites, dependence is transferred from maternal resources, if present, to hosts or to mycorrhizal fungi. Thus, it is likely that no one definition, except possibly – *the young spermatophyte plant, following germination* – covers all seedlings.

1.2 | Vulnerabilities and bottlenecks

At each stage of its life cycle, the plant's success is limited by an assortment of intrinsic and extrinsic factors that are to some degree driven by chance. Because of their small size, seedlings have greater susceptibility to resource limitations and other factors that affect establishment and growth (Fig. 1.1). In addition to resource limitations related to size, seedlings may be vulnerable because of low levels of morphological and physiological defenses. Cotyledons of *Toxicodendron pubescens* (Anacardiaceae) suffer herbivory whereas its leaves do not (Miller & Miller, 2005). However, in the case of *Quercus alba*, sprouting causes seeds that are favored by squirrels (*Sciurus carolinensis*) to become less digestible and, under some circumstances, squirrels bite out the embryo, preventing these changes (Steele & Koprowski, 2001). Seeds may present physical barriers that limit granivory; for example, the burs of *Xanthium strumarium* (Asteraceae) are never eaten (N. Good, pers. comm.).

Vulnerability can vary with habitat. As a generalization, dormancy-breaking mechanisms have evolved to increase the probability that germination occurs in a safe site – when and where the likelihood of survival is greatest (e.g. Grubb, 1977; Harper, 1977; Baskin & Baskin, 1998; Fenner & Thompson, 2005). Although a species actually may be able to germinate or live in a range of habitats, it may not find all habitats equally suitable because primary stresses vary. For example, shade-adapted seedlings can slowly acclimate to sunny locations, but survive best inside a forest despite being subject to high levels of herbivory and pathogens; in large gaps, establishment is prevented by competition with fast growing species (see Chapter 8). *Impatiens capensis* (Balsaminaceae) in temperate woodland habitats is more susceptible to white-tailed deer (*Odocoileus virginianus*) herbivory than when growing in a tidal freshwater wetland (Leck, Parker, & Simpson, pers. obs.). Moreover, although germination may occur over a wider range of conditions (along an inundation gradient in a tidal channel), establishment conditions may be narrower (Parker & Leck, 1985; Leck & Simpson, 1994).

Vulnerability varies with phylogeny within a given habitat. Some taxa are more susceptible to an environmental constraint than others. Seedlings of maples (*Acer* spp. Aceraceae) succumb to frost heaving to a greater extent than seedlings of *Carya tomentosa* (Juglandaceae) in successional old fields (New Jersey, USA) (Myster, 1993). Similarly, small seedlings of *Bidens laevis* (Asteraceae), a tidal freshwater wetland dominant, have less predictable establishment than the co-occurring, larger seedlings of *Impatiens capensis* or *Polygonum arifolium* (Leck & Simpson, 1995). In temperate old fields, early spring germinants of *Ambrosia trifida* and *Polygonum* species can tolerate low night temperatures, whereas seedlings of *Abutilon theophrastii* (Malvaceae) and *Ipomoea* cannot; variation in burial depth, resultings in varied emergence time, reduces the intensity of selection (Bazzaz, 1996).

Regardless of the cause of vulnerability, small size, limitation in ability to acclimate, habitat suitability, phylogeny, or other constraints, the seedling stage faces hurdles that are exacerbated by stochastic events. Collectively, these contribute to the significance of the seedling stage as a bottleneck in a species' life history. Selection at the seedling stage may produce seedling specialists or generalists. Examples of specialists are the bulb- and corm-forming seedlings of Australian desert perennials (Pate & Dixon, 1982). These specialized seedlings have the ability, because they possess contractile roots and hypocotyls, to place the apical growing point and a storage bulb, corm, or rhizome well below the soil surface, where they can avoid drought and heat during their first growth season. Another group of specialists are diminutive, woody, microstilt Australian perennials that produce heavily lignified, adventitious stilt roots, which allow the plant to survive desiccation and reduce prolonged soil surface heat stress (Pate, 1989). *Ambrosia trifida*, an example of generalist seedlings, can be found in tidal freshwater marshes or as weeds in agricultural fields. Communities may or may not have high seedling competitive ability, depending on the intensity of competition (Lamb & Cahill, 2006). Seedlings may also be conservative or opportunistic in their use of resources (see Chapter 8) or fugitive or stress tolerators (Shipley *et al.*, 1989).

1.3 | Making it: filters, safe sites, and establishment

The idea of the seedling serving as a bottleneck in a species' life history necessitates considering how the individual survives from the seed bank to establishment and, ultimately, to an adult (Fig. 1.2). A location that assures seedling success and that has all the necessary resources for survival may be termed a *safe site* (e.g. Harper, 1977; Fenner & Thompson, 2005). Safe site requirements vary with species, genotype, and time with functionality related to all the factors that influence establishment and growth (Fig. 1.2).

In a particular habitat, the safe-site filters can vary spatially and temporally, resulting in zonation or in cyclic changes in vegetation.

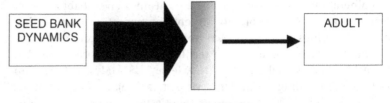

Fig. 1.2 The importance of safe site filters affecting seedling populations. The thickness of the arrows suggests the impact of the filter(s) on seedling establishment to adult.

Safe site filter

In wetlands, for example, water (inundation) can sort species across depth gradients or produce different communities, depending on drawdown/flooding patterns determined by rainfall (e.g. van der Valk, 1981; Leck & Brock, 2000). In a desert, small-scale habitat differences, such as those caused by porcupine digging or a nurse tree (see Chapter 15), can provide safe sites not available nearby.

The array of factors that can result in a seedling's failure to achieve establishment may act together or separately, and are environmental filters (Fig. 1.2). These factors may be abiotic (light, temperature, drought) and/or biotic in nature (competition, availability of microbial symbionts), and may have varying spatial or temporal impact. Seedling attrition may be huge. Moreover, the behavior of the seedling following germination can be intimately tied to seed characteristics, including dispersibility, size, and dormancy. These characteristics are controlled by genetic as well as by environmental factors during development, maturation, and storage (Gutterman, 1993). Individual traits, such as seed mass, can influence susceptibility of seedlings to drought, depth of burial from which seedlings may emerge, range of microsites suitable for seedling establishment (via gap detection mechanisms; e.g. Dalling, 2005), and tolerances to herbivory (Hoshizaki *et al.*, 1997). Agents of burial, whether biotic (e.g. dung beetles; Andresen & Feer, 2005) or abiotic (e.g. soil cracks; Harper, 1977; Bonnis & Lepart, 1994), influence seed position in the soil. In addition to seed dormancy mechanisms (e.g. Baskin & Baskin, 1998), exogenous influences, such as availability of a dispersal vector or disturbance, that place the seed and, thus, the seedling in a position where chances for survival are optimal, cannot be underestimated.

1.4 | Seedlings: a primer

Seedling organs include the *radicle* (or primary root), *cotyledons* (seed leaves), stem, leaves, buds, and surface appendages, such as hairs (Lubbock, 1892; Burger, 1972; de Vogel, 1980). Seed plants include gymnosperms with two to many cotyledons, dicots that typically possess two cotyledons, and monocots with one structure designated as a cotyledon (Fig. 1.3). In each group, major seedling distinctions are based on the position of the cotyledons during germination. When the cotyledons rise above the soil surface, the seedling has *epigeal* germination, and if the cotyledons remain at or below the soil surface,

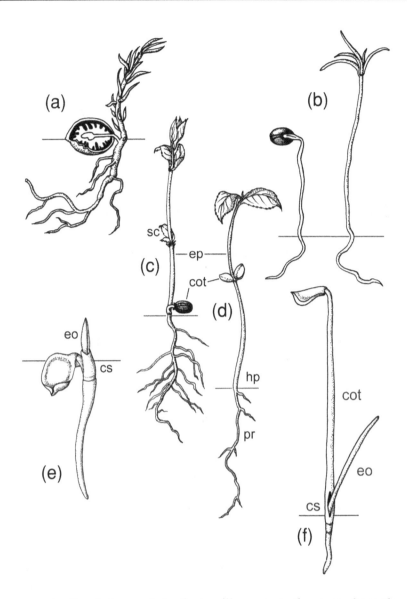

Fig. 1.3 Representative seedlings of gymnosperms, dicots, and monocots illustrating hypogeal and epigeal germination types. Gymnosperms are (a) *Torreya myristica* (hypogeal), with part of the seed removed revealing endosperm and one of two cotyledons, and (b) *Taxus baccata* (epigeal) (Taxaceae); dicots are (c) *Prunus americana* (hypogeal) and (d) *P. virginiana* (epigeal) (Rosaceae); and monocots are (e) *Asphodelus lusitanicus* (hypogeal) and (f) *A. tenuifolius* (epigeal) (Asphodelaceae). Not drawn to scale. Abbreviations: cot – cotyledon, cs – cotyledonary sheath, eo – eophyll (the expanded blade part of the cotyledon), ep – epicotyl, hp – hypocotyl, sc – scale leaf, pr – primary root or radicle. Redrawn by A. Hoffenberg: (a) from Chick (1903), (b) Rudolf (1974), (c, d) Grisez (1974), and (e, f) from Tillich (2000) with permission from CSIRO.

germination is *hypogeal*. As the seedling grows, the stem above the cotyledonary node is the *epicotyl* and that below, the *hypocotyl*. The hypocotyl is usually distinct in the embryo of epigeal seedlings; in hypogeal seedlings, it is poorly developed and does not elongate during germination (de Vogel, 1980).

Overall, most gymnosperms and dicots are epigeal and most monocots are hypogeal. Hypogeal germination was once considered advanced (Eames, 1961), but both types can be found within the same taxon (Fig. 1.3). Although this brief description suggests that the seedling form is relatively stereotyped, a particular part, such as the cotyledon, can vary considerably in form and function in both dicots (Fig. 1.4) and monocots (Fig. 1.5). Efforts to correlate structure and function provide other levels of classification (e.g. Garwood, 1996). The greatest number of functional types may be found among tropical

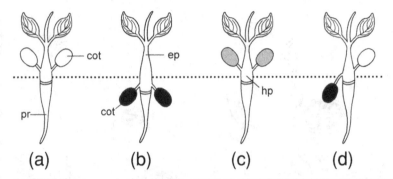

Fig. 1.4 Dicot seedlings showing cotyledon types: (a) photosynthetic, (b) storage, (c) haustorial (see legend Fig. 1.5), and (d) mixed form where only one cotyledon is photosynthetic. The double line indicates the collet, the transition zone between stem and root. Abbreviations: cot – cotyledon, ep – epicotyl, hp – hypocotyl, pr – primary root or radicle. Redrawn by A. Hoffenberg with permission from Bell (1991), copyright Oxford University Press.

woody dicots (Garwood, 1996) and monocots (e.g. palms; Tomlinson & Estler, 1973; Bell, 1991). Focus has been on function and position of cotyledons, but primary roots may vary in site of origin, architecture, and persistence (Tillich, 2000).

1.5 | What seedlings can tell us

Observations from particular environments have relevance across other habitats. For example, Titus and Hoover (1991) observed that in submerged plants, potential challenges to seedling establishment, as well as lack of understanding of the physiology and demography of seed banks and germination, severely limit the predictability of sexual reproductive success in the field. They also note that small seedling size, rapid growth, sparse seed banks, and unfavorable conditions for germination and establishment all contribute to the lack of quantitative data. Garwood (1996) also laments the lack of information about seedlings in reports on germination and other aspects of species biology in tropical environments.

During the past decade, work with *Arabidopsis* (Brassicaceae) mutants has greatly improved understanding of seedling development and physiology (e.g. Leyser & Day, 2003; Achard *et al.*, 2006). Insights have relevance to understanding seedling establishment. For example, *Arabidopsis* studies help explain the basis of etiolation (stem growth in darkness), which raises the cotyledons to the soil surface (Leyser & Day, 2003). Furthermore, understanding seedling requirements may improve the chances for success of restoration projects. In sedge- (Cyperaceae) dominated created wetlands, seedlings may not establish even when high-quality commercial seeds are planted or following transplantation of healthy seedlings (van der Valk *et al.*, 1999). Soil amendments, such as organic matter, are necessary to improve soil moisture and permit establishment (see Chapter 17).

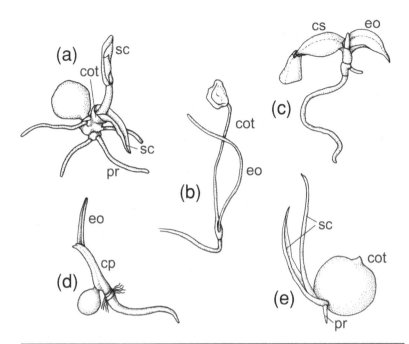

Fig. 1.5 Monocot seedling types distinguished by Tillich (2000): (a) compact cotyledon, characterized by a low sheath and a completely haustorial hyperphyll* that is completely imbedded in the seed and not visible (*Bomarea edulis* Alstroemeriaceae); (b) cotyledon with photosynthetic assimilating, elongated, and upright hyperphyll that raises the seed above the soil surface (*Albuca fastigiata* Hyacinthaceae); (c) cotyledon with wide assimilating sheath (*Pitcairnia corallina* Bromeliaceae); (d) cotyledon with a long coleoptile (*Hypoxis hygrometrica* Hypoxidaceae); and (e) storage cotyledon where, due to absence of endosperm, the hyperphyll has storage function (transient seed coat removed) (*Orontium aquaticum* Araceae). Not drawn to scale.

Abbreviations: cot – cotyledon, cp – coleoptile, cs – cotyledonary sheath, eo – eophyll, sc – cataphyll, pr – primary root or radicle.

*Definitions: *Cataphylls* – first leaves following the cotyledon, sometimes called scale leaves; *coleoptile* – an elongated tubular extension of the sheath above the insertion of the hyperphyll; *eophyll* – first, expanded photosynthetic leaf; *haustorium* – in monocots the leaf blade (or end) of the first leaf, found within the seed, and in gymnosperms and dicots the undifferentiated, colorless suctorial organ that acts as an absorptive organ, transferring nutrients from the endosperm to the growing embryo and developing seedling; *hyperphyll* – part of the cotyledon connecting the haustorium to the sheath, also called the cotyledonary petiole (this may be short in admotive, adjacent germination and long in remote germination); *primary root* – the first root also called the radicle. Redrawn by A. Hoffenberg with permission from Tillich (2000), copyright CSIRO.

Despite numerous studies, literature on seedlings is diffuse and, often, information about seedling ecology is lacking even in species accounts. Also, other stages of the life cycle contribute to seedling success (e.g. Howard & Goldberg, 2001; Chapter 10), including dispersal and maternal investment to seeds. Thus, improved understanding of the roles of seedlings is challenging. Yet, an understanding of plant life cycles underpins the pursuit of knowledge in botanical, ecological, environmental, and agricultural disciplines.

1.6 | The scope of *Seedling Ecology and Evolution*

Seedlings, usually the most transitory of life-history stages, provide opportunities to explore novelties, as well as life cycle continuum features and vulnerabilities and trade-offs that, ultimately, are key to population and community dynamics. The purview of this book is illustrated in Fig. 1.1, with text focusing on seeding diversity; seedling morphology, evolution, and physiology; life history implications; and applications; as well as this introduction and a concluding synthesis.

Part II, Chapters 2–4, considers the breadth of seedling diversity. Chapter 2 focuses on the natural history of seedlings, including morphological and physiological diversity, vivipary, longevity, and dispersal of seedlings. Chapters 3 and 4 explore the boundaries and limitations of seedlings. Specialized seedling strategies in stressful environments, such as shade, litter, cold, heat, salinity, and unstable substrates, are discussed in Chapter 3. Other specialized strategies considered in Chapter 4 include orchids, epiphytes, insectivores, and parasites.

Part III, Chapters 5–9, examines seedling morphology, evolution, and physiology. Chapters 5 and 6 discuss extant plant groups and the fossil record. These chapters establish a foundation for examining seedlings from an embryological perspective and provide a broad scope in which to consider seedling evolution and phylogenetic constraints. Chapter 7 considers the relationships between seedling environment, phytohormones, and phenotypic expression, whereas Chapter 8 examines the strategies – opportunistic and conservative – that seedlings use to attain independence from maternal carbon reserves. Nutritional relationships involving symbioses with fungi and bacteria are discussed in Chapter 9.

Part IV, Chapters 10–13, considers life history implications, focusing on seedlings at several levels. The trade-offs between seed production and seedling survival are considered in the context of the entire plant life cycle in Chapter 10. Chapter 11 discusses seedling recruitment and focuses on populations, including recruitment limitation and genetic structure and selection. Chapter 12 discusses seedlings in a community context and considers internal and external forces that influence regeneration niches, recruitment strategies, and assembly rules. Chapter 13 examines the feasibility of delimiting functional seedling groups, how scale relates to seedling processes, and whether spatial, morphologic, and phylogenetic patterns can be successfully elucidated.

Part V, Chapters 14–17, examines seedlings in the context of ecosystems degraded by anthropogenic activities. The varied strategies of seedlings of invasive species are considered in Chapter 14. In arid lands as shown in Chapter 15, understanding seedling requirements is crucial to understanding patch dynamics, in which human (and natural) disturbances are involved in strong, positive feedback relationships leading to degradation and desertification. The effects of disturbances in forests, which may not be apparent for decades, are related to seed limitation and recruitment; Chapter 16 considers the impact of human disturbances on seedling recruitment, focusing on tropical forests where seedling variation is highest. Worldwide, efforts are being made to restore degraded ecosystems; Chapter 17 explores the factors influencing seedling success in many types of restoration projects. Together, these chapters provide a baseline understanding of regeneration in human-impacted ecosystems.

Part VI, the concluding Chapter 18, provides a synthesis, and explores the trade-offs between phylogeny and recruitment, the significance of safe sites, and the overarching strategies of seedlings. Collectively, these chapters provide insights into the seedling stage and opportunities for fruitful future study.

Part II

Seedling diversity

Chapter 2

Seedling natural history

Mary Allessio Leck and Heather A. Outred

2.1 | Introduction

Consider the following: tidal freshwater marshes along the East Coast of North America in springtime; the deserts near Death Valley, in Africa, and elsewhere following a substantial rainfall; the intermittent wetlands in the arid Australian landscape; and the wheat fields of Europe, North America, and New Zealand. Each landscape is awash with the greens of newly emerged seedlings, each species responding to its particular set of germination cues, each informed by its peculiar evolutionary history. Anyone interested in seed banks and seed germination ecology and physiology, as well as those who garden, are intimately familiar with seedlings. Seedlings are also well known to those who produce seeds for use in agriculture and horticulture and who are concerned with vigor and other seedling attributes (Geneve, 2005; Stephenson & Mari, 2005; Farooq *et al.*, 2006). In this chapter, we explore the diverse and fascinating array of seedlings and seedling natural history. Topics include the seedling stage, morphological and physiological diversity, vivipary, seedling equivalents, seedling longevity and dispersal, and environmental filters and safe sites.

Nomenclature generally follows that of the author and family names (Mabberly, 1997).

2.2 | The seedling stage and fate of seedlings

Contrary to what seems intuitive, the seedling stage is not always easily defined. The success of seedlings is, furthermore, influenced by many environmental factors that determine survival, establishment, and, ultimately, community composition.

The seedling

Seedling is used for a very young individual (Burger, 1972), but problems occur in determining the beginning and end of the stage.

Harper (1977) observed that the ultrastructure of the seed may change within 30 minutes of wetting. Others have suggested the seedling stage to begin with enlargement of the embryo following seed maturation (Tomlinson, 1990), protrusion of the radicle from the seed coat (Wardle, 1984; Kitajima & Fenner, 2000; Fenner & Thompson, 2005), when cotyledons become free of the seed coat (Wardle, 1984), or when cotyledons of epigeal seedlings emerge above ground (Martin & Ogden, 2002).

Plants, however, vary. In *Carex* and *Cyperus* (Cyperaceae), for example, only the cotyledon grows at first; the middle part extends rapidly, pulling the root, which grows later, out of the seed (Lubbock, 1892; Boyd, 1932). In Alismataceae and *Populus* (Salicaceae), the first organ to emerge is the hypocotyl (Lubbock, 1892; Young & Young, 1992). In these, the seedling is initially anchored by rhizoids, which develop from the base of the cotyledon or hypocotyl. Shoot emergence before roots also occurs in aquatics from diverse families (Boyd, 1932; Muenscher, 1944).

Stages of seedling development may differ for epigeal and hypogeal species. For epigeal seedlings, the first stage constitutes growth up to the first leaf; and for hypogeal seedlings, up to the first fully expanded leaf; and the second stage is a plant with almost normal form, young leaves (Burger, 1972).

Before becoming independent, the very young seedling will usually pass through a stage when it is dependent upon stored seed reserves, but like the beginning, the end of the seedling stage also seems difficult to determine. It may coincide with the loss of cotyledons (Philipp, 1992); with the appearance of the first bladed seedling leaf (Tomlinson, 1990) or maturation of the first true foliage leaves, leaving the young plant capable of independent existence (Harper, 1977); or the end of the exponential growth period (Kitajima & Fenner, 2000). The last, however, ignores observations that fluctuating growth spurts often occur (de Vogel, 1980; Chacon & Armesto, 2005). Foresters consider the seedling stage to include young plants to 2.7 m (Whitmore, 1996).

Exhaustion of seed reserves may be a reasonable end of the seedling stage, but in some species the young plant uses reserves stored in the hypocotyl, making delineation between a seedling and a juvenile plant difficult (de Vogel, 1980). The end of dependence on cotyledons and stored reserves appears to correlate with a decline in maximum growth rate (RGR) that follows a bell-shaped curve (Fenner & Thompson, 2005). If widely applicable, the RGR could be used to assess the transition of the seedling to the next stage. However, the availability of seed reserves may exceed the time when seedlings become totally dependent on the environment. For example, *Posidonia australis* (Posidoniaceae) seedlings are capable of independence at 9 months, yet 2-year-old seedlings have 20% of the phosphorus and 14% of the nitrogen derived from the parental source (McComb *et al.*, 1981).

The fate of seedlings

The loss of seedlings in most ecosystems is usually high (e.g. Darwin, 1859; Grubb, 1977; Harper, 1977; Pate & Dixon, 1982; Louda, 1989; Leck & Simpson, 1994; Körner, 2003), their fate determined by many factors that prevent establishment (Chapters 3, 11, 15). Causes of death, whether by biotic or abiotic factors or a combination of both, vary in importance with habitat. In deserts, available moisture, optimal growing season, fire, and grazing determine seedling survival (Pate & Dixon, 1981; Florence, 1981; Gutterman, 1993; Denham & Auld, 2004). In tropical forests, high losses of seedlings may be due to burial under litter and herbivory, washing away by rain, trampling by animals, and starvation due to lack of light or water (e.g. Ng, 1978; Garwood, 1996; Whigham, *et al.*, 1999). In temperate forests, the two major sources of seedling mortality are herbivory and drought (Cook, 1979), but litter and frost heaving may be important, especially for small seedlings (Young & Young, 1992; Myster, 1993; Kostel-Hughes *et al.*, 2005). In alpine tundra, seedlings are susceptible to soil heaving caused by nighttime needle ice and cryogenic processes during winter, as well as drying and heat on bare soil (Körner, 2003).

Wetland habitats provide other challenges. These include anoxic substrates; insufficient aerenchymatous tissue limiting nighttime respiration; unstable substrates, especially where tides and waves occur; light limitations at depths or in shade of established plants; low nutrient availability where competition with neighboring plants occurs or where nutrients are difficult to acquire because of low oxygen or salinity; and susceptibility to pathogens (Titus & Hoover, 1991). To these can be added uprooting, burial or damage to seedlings by litter or flotsam, and, if germination occurs in deep water, lack of resources (energy and time) for survival (Haag, 1983). In brackish and marine habitats (as well as in arid places), salinity offers another level of stress affecting seedling establishment.

Many kinds of organisms can cause death, including differential crab predation in mangrove and coastal terrestrial forests (Lindquist & Carroll, 2004) and rats that reduce establishment of *Pleurophyllum hookeri* (Asteraceae), a subantarctic megaherb (Shaw *et al.*, 2005). Death may also be caused by blight organisms, nematodes, spider mites, grubs, and fungi that cause damping off and root rot (Johnsen & Alexander, 1974). Further, lack of establishment may be due to absence of symbionts required by mycorrhizal and nodulating species (Chapter 9). Where allelopathic compounds that are toxic to mycorrhizal fungi are present, survival of tree seedlings is reduced (Stinson *et al.*, 2006).

Although most seedlings have no special adaptations (de Vogel, 1980), there are differences in vulnerability and ability to repair damage. Young epigeal seedlings are least able to recover because the apical bud is usually lost when cotyledons are eaten. In contrast, seedlings with hypogeal cotyledons have buds that can replace aerial portions, and in certain monocots (e.g. *Cordyline australis* Agavaceae, *Freycinetia banksii* Pandanaceae, *Ripogonum scandens* Smilacaceae), buds in the axils of (scale) leaves permit regeneration after mechanical damage (Tomlinson & Esler, 1973). In alpine tundra, small-seeded species suffer higher mortality (99%) than larger-seeded ones (50%) over 12 months (Körner, 2003).

2.3 | Seedling types

Monocotyledons

Tillich (1995), who considers seedling structure key to detecting phylogenetic relationships, characterizes monocot seedling organs, standardizes terminology, and illustrates phylogenic distributions of selected characteristics. His selected characteristics that call attention to functional attributes of monocot seedlings include: (1) the coleoptile: a highly derived structure, which is a tubular elongation of the cotyledonary sheath, is found in some species (e.g. *Elyna myosuroides* Poaceae), but not others (e.g. *Alisma plantago-aquatica* Alismataceae); (2) photosynthetic capacity of the cotyledon: whether green or not, the upper part of the cotyledon (hyperphyll) is primarily a storage organ and only in *Acorus* (Araceae) is it a truly photosynthetic structure; (3) endospermless seeds: nutrients are stored in the embryo itself, either in the cotyledon or the hypocotyl; (4) conspicuous collar: the collar, the transitional zone between the hypocotyl and the primary root, can be very pronounced and can develop dense trichomes or rhizoids; (5) robust and branched primary root: presumably an ancestral condition, and although generally considered atypical for monocots, occurs variably in a number of families (e.g. in Dracaenaceae, it occurs in *Dracaena* but not *Sansieveria*); (6) velvety root hairs: these are very dense, short root hairs that look like velvet and are most conspicuous under laboratory conditions in moist chambers (Convallariaceae, Luzuriagaceae, Philesiaceae); and (7) seedling without a primary root: a highly derived condition with only four families completely lacking primary roots (Eriocaulaceae, Lemnaceae, Poaceae, Zosteraceae). Roots may have a coleorhiza, or root cover, through which the root grows.

Among the monocots, the Araceae contains the greatest number of seedling types (Fig. 2.1; Tillisch, 1995). The primary root, found in most examples, is lacking in *Pistia* and *Lemna*, indicating a close relationship between these genera. The seedling axis also shows considerable modification, varying from the tuberous corm of *Arisaema* to the elongated internodes of *Pothos*, a liana. Lack of information about seedlings of many subgroups precludes full understanding of relationships and determining probable ancestral types beyond reductionary evolution from a robust, long-lived, branched primary root, ultimately, to total reduction.

In addition to epigeal or hypogeal germination, the cotyledonary axis of monocotyledons may be either extended (remote) or not (adjacent) (Boyd, 1932; Tomlinson, 1990; Bell, 1991). In the latter, the cotyledonary part of the embryo barely extrudes, causing the seedling to

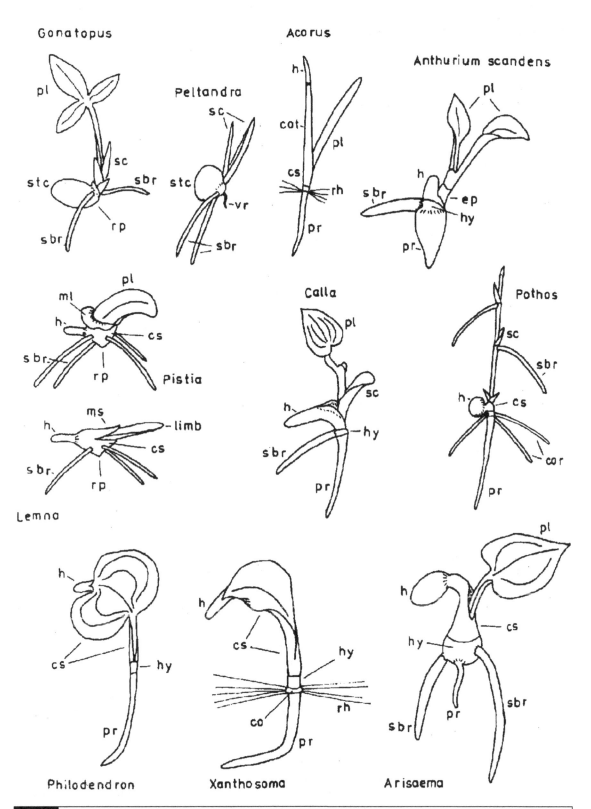

Fig. 2.1 Seedlings as found in the order Arales that have the greatest familial seedling diversity among monocots (*Acorus* Acoraceae, *Lemna* Lemnaceae; others Araceae). Abbreviations: co – collar, cor – collar root, cot – unifacial, assimilating hyperphyll (upper leaf part of cotyledon), cs – cotyledonary sheath, ep – epicotyl, h – haustorium, hy – hypocotyl, in – internode, ml – median sheath lobe, ms – median sheath lobe of *Lemna* cotyledon, pl – primary leaf, pr – primary root, rh – rhizoids, rp – root pole, sbr – shoot-borne root, sc – scale leaf, stc – storage cotyledon, vr – vestigal primary root. Seedlings are not drawn to scale. From Tillich (1995), reproduced with permission, copyright The Board of Trustees of the Royal Botanic Gardens, Kew.

(a) (b)

(c) (d)

(e) (f)

(g) (h)

(i) (j)

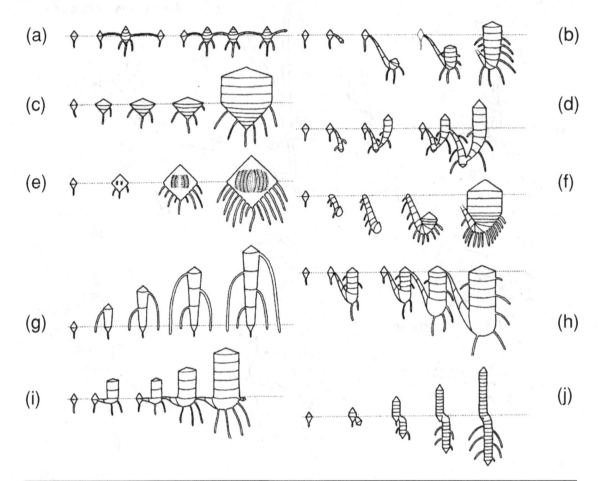

Fig. 2.2 Establishment growth of monocots (see Bell 1991 for details). (a) Stolon production, (b) production of successively larger, short-lived sympodial units, (c) increase in width of successive nodes, internodes short, (d) increase in size of sympodial units that alternate growth direction, (e) increase in width from cambial activity, (f) initial vertical growth, similar to 'c,' (g) prop roots supporting initial vertical growth, internodes long, (h) increase in size and depth of successive long-lived sympodial units, (i) increase in size of successive sympodial units, and (j) production of single downward growing side shoot. From Bell (1991), reproduced with permission, copyright Oxford University Press.

develop next to the seed. In contrast, during remote germination where the plumule develops away from the seed, the extended cotyledon buries the plumule, promoting rooting. In *Lodoicea maldivica* (the massive double coconut, Palmae), the cotyledonary axis, developing over 3–4 years, may extend horizontally for greater than 4 m, assisting dispersal.

In woody monocotyledons, establishment involves various growth behaviors (Fig. 2.2; Tomlinson & Esler, 1973). Because of the lack of secondary growth in arborescent monocotyledons, radicles of fixed diameter would not be able to supply adequate water to a growing aerial shoot; in palms, this problem is circumvented by the production of many adventitious roots (Tomlinson, 1990). Establishment growth also involves increase in diameter of successive internodes, which for the largest palms involves increase from 1 mm to 1 m in adults.

Tomlinson also notes a number of seedling features. First, internode length is typically short and nodes close together. Successive internodes increase in diameter. The transition to the adult phase, characterized by long internodes, is often abrupt. Stilt palms, with consistently long internodes, are an exception. Second, leaves become progressively larger and more elaborate. The leafy crown, as it expands in size, remains at the soil surface. Third, adventitious roots progressively increase in number and diameter. Finally, the vascular bundles of the stem increase in number, increasing the water-carrying capacity of the stem. For additional discussion of monocotyledon seedlings, see Boyd (1932), Eames (1961), Tomlinson (1990), Bell (1991), Tillich (1995, 2000), and Henderson (2006).

Dicotyledons

Examination of a temperate seedling flora (e.g. Muller, 1978; Young & Young, 1992) would suggest that most dicot seedlings are epigeous (96.7% and 94.5%, respectively). Virtually all herbaceous dicot crop and weed seedlings are epigeous (e.g. Kummer, 1951; Reilly, 1978; Uva *et al.*, 1997), with only *Lathyrus*, *Pisum*, and *Vicia* (Leguminosae) being hypogeal. In contrast, an examination of, for example, Burger (1972), de Vogel (1980), and Garwood (1996) who describe tropical tree seedlings, would indicate that both epigeous and hypogeous seedlings are important. Five seedling functional types for tropical forest trees based on exposure, texture, and position of the cotyledons have been identified (Miquel, 1987; Garwood, 1996). These types vary in importance: with planerocotylar*-epigeal-foliaceous ranging from 33 to 56% of tropical forest tree species (eight sites; three continents); followed by planerocotylar-epigeal-reserve, 16–43%; cryptocotylar*-hypogeal-reserve, 7–28%; planerocotylar-hypogeal-reserve, 6–14%; and cryptocotylar-epigeal-reserve, 1–8% (Garwood, 1996; see also Ibarra-Manríquez *et al.*, 2001). Analysis of Malaysian woody plants by de Vogel (1980) showed ≈13% of genera and 46% of families having more than 1 seedling type (of 16); Leguminosae, with both Caesalpinaceae and Papilionaceae; and Myrsinaceae, each having 5 to 6 types. Germination of *Citrus aurantifolia* (Rutaceae) and *Durio zibethinus* (Bombacaceae) may be either hypogeal or epigeal, which in the latter is determined by the orientation of the micropylar end (Enoch, 1980).

In temperate woody plants described by Young & Young (1992), only *Prunus* (Rosaceae) had both epigeal (*P. virginiana*) and hypogeal (*P. americana*) germination types. Seedling type, at least for tropical woody species, is considered a conservative evolutionary trait (Ibarra-Manríquez *et al.*, 2001). Garwood (1996) notes several problems with attempting to relate seedling classification schemes based on morphology to the ecology of (tropical tree) seedlings. Among the points she makes are these: (1) Focus, primarily on the cotyledon, is too narrow because other functions, such as water and nutrient uptake and

Planerocotylar – cotyledons become entirely exposed and are free from the fruit wall and testa; *cryptocotylar* – cotyledons remain enveloped in a persistent fruit wall and/or testa and, if it is present, also in the endosperm; de Vogel, 1980.

anchorage, are also important in establishment. (2) The five types may not be good functional groups. Seedlings with photosynthetic cotyledons may in some species have no stored reserves while in others reserves are stored in endosperm. (3) Focus on only one stage may obscure the functional morphology that is necessary to understand seedling development and establishment. (4) To understand functional morphology, it would be important to focus on how functions change through development and establishment. [See Garwood (1996) for a comprehensive discussion of functional seedling morphology.]

No overall seedling classification scheme with a functional perspective exists. Classification schemes for tropical tree seedlings do not include gymnosperms or monocots, even arborescent kinds that have distinctive establishment strategies (Fig. 2.2) (Tomlinson & Esler, 1973; Tomlinson, 1990; Bell, 1991). Also, there is no inclusion of root architecture, despite the observation that roots are as diverse as shoots (Robinson et al., 2003; but see Tillich, 1995, 2000). Functional types in other habitats, including aquatic and wetland species, may be more obscure. Seedlings of parasitic plants, those with specialized morphologies (e.g. bulbs and vines), and viviparous species, although considered by de Vogel (1980) and Bell (1991), are needed to develop a unified view of functional seedling types. Ibarra-Manríquez et al. (2001) further note, that for improved understanding of functional morphology, it is necessary to determine variation and role of seedling traits, including size, quantity, and quality of maternal reserves; growth rate; allocation of resources among organs; and the relationship between morphology and function of seedling organs.

2.4 | Seedling diversity – morphology

What appear to be inconsequential aspects of seedling morphology may have great adaptive importance in determination of safe sites and, thus, successful establishment (Cook, 1979). The range of morphological features exhibited by seedlings, while not exhaustive, is described in Appendixes 2.1 and 2.2. Examples can be found in which one or more of the typical seedling organs are missing. In some cases, structure mirrors the morphology of the parent, but in others, differences are developmental. Variable attributes reflect evolutionary forces working in diverse habitats. A brief consideration of seedling diversity is given below. For more extensive discussions, see Lubbock (Vol. 1, 1892), Boyd (1932), de Vogel (1980), Tomlinson (1990), Tillich (1995, 2000), Garwood (1996), and Henderson (2006). Kummer (1951) and Uva et al. (1997) also provide descriptions of seedlings.

Seedling organs
Cotyledons
Cotyledons, usually synonymous with seed leaves or seed lobes, are of three types: food-storing, haustorial, and photosynthetic (de Vogel,

1980; Chapter 1). The undifferentiated haustorial cotyledons transfer nutrients from the endosperm to growing regions of the embryo and seedling; they may not be easily distinguished from food-storing ones. Muller (1978) uses cotyledon for the first two, and *seed leaves* for photosynthetic ones. For our purposes, cotyledon will encompass all three (reflecting our temperate biases). Usually cotyledons are of one type, but in certain *Peperomia* (Piperaceae) species, one cotyledon maintains haustorial function and remains within the seed coat, while the other becomes exposed (Hill, 1906, cited in de Vogel, 1980; Bell, 1991) and presumably photosynthetic. Generally, cotyledons occur in constant numbers within a taxon, but numbers can vary (Appendix 2.1).

In most epigeous species, cotyledons are sessile when borne above ground by hypocotyl growth, but petiolate when close to the ground (Lubbock, 1892). In other species, the cotyledons are aerial because of elongated petioles that in *Polygonum bistortoides* and *P. bistorta* (Polygonaceae) form a tube through which the first leaf passes (Allessio, 1967; Muller, 1978).

For certain hypogeous species, cotyledon burial may be accomplished during germination. Extension of the cotyledonary stalk of certain palms pushes the embryo into the ground a considerable distance (Tomlinson, 1990). Except for *Nypa fruticans*, a mangrove species, and *Ravenea musicalis*, the river palm of Madagascar, that are viviparous/cryptoviviparous, burial is necessary to provide support for the trunk; lacking this, support is provided by stilt roots or a broad root-bearing surface below the soil surface (Tomlinson & Esler, 1973; Hallé *et al.*, 1978; Tomlinson, 1990; Beentji, 1993). In temperate and tropical oaks (*Quercus* spp. Fagaceae), seeds are pushed belowground during germination (Ng, 1978).

Typically, hypogenous cotyledons are large, nonphotosynthetic, and lack stomates (Lubbock, 1892; de Vogel, 1980). However, cotyledons of certain hypogeous species (e.g. *Lucuma* sp. Sapotaceae) may turn green when exposed to light at the soil surface. In *Quercus ilex*, if an acorn is deeply buried, the first leaves are scale-like, but if shallow and in light, they become green and foliaceous. In the epigeal peanut (*Arachis hypogaea* Leguminosae), cotyledons, although carried up into the air, never become green (Brown, 1935).

During seedling growth, cotyledons can change (Appendix 2.1; e.g. Lubbock, 1892; Burger, 1972; Muller, 1978; de Vogel, 1980). Opposite cotyledons may become alternate by unequal growth of the stem, sessile cotyledons may become petiolate, shape may change, or one cotyledon may grow while the other does not.

Stems, leaves, and surface features

Seedlings may develop a variety of surface features, including hairs, pores, and secretory structures, as well as odors, colors, and specialized functions (Appendix 2.2). These, as suggested by Fahn (1979) for secretory plant tissues *per se*, appear to be ecological adaptations of two types, mediating either edaphic/climatic conditions or surrounding animal populations. Hydathodes and salt glands are of the first type, permitting loss of water when moisture conditions reduce transpiration or removal of salt for species in saline habitats. High tannin content, odors, and the myrosin cells of Brassicaceae, as well as color variegation (*Byttneria aculeata* Sterculiaceae) (Lee, 2007), provide defense against animals. Special secretory tissues permit the production and accumulation of poisonous substances.

Form may be quite variable (e.g. de Vogel, 1980; Young & Young, 1992). First leaves are usually simpler than those that follow, but this generalization is reversed in species from arid habitats (Lubbock, 1892). Certain gymnosperm seedlings, including *Chamaecyparis lawsoniana* (Cupressaceae), *Cryptomeria* spp. (Taxodiaceae), and *Juniperus* spp. (Cupressaceae), have both juvenile needle-like leaves and appressed adult scale leaves. Cycad seedlings have a shoot apex that produces scale leaves for several years and single foliage leaves at intervals. Over time, the number of foliage leaves increases until crowns of leaves alternate with scale leaves (Chamberlain, 1935). Dicots may have simple leaves followed by compound leaves (e.g. *Aegle marmelos* Rutaceae, *Vitex pubescens* Verbenaceae) and, less frequently, compound leaves followed by simple ones (e.g. *Acacia oraria* Leguminosae). In palms, where leaf form changes following the first green bladed leaf, Tomlinson (1990) distinguishes six transition types. Finally, red-tipped seedlings of *Eusideroxylon zwageri* (Lauraceae), growing from one of the largest seeds in the world (12 cm long × 4 cm wide), exceed 1 m in height before producing leaves (Veevers-Carter, 1991).

Roots

There are three broad groups of root types (Lauenroth & Gill, 2003), conifers and woody dicots, herbaceous dicots with roots organized around a primary (tap) root, and herbaceous monocots with roots not organized around a primary root. However, generalizations are difficult because the tap root of many dicots is augmented by adventitious roots (Kummer, 1951) and some monocots have a primary root (Fig. 2.1; Tillich, 1995, 2000). Root architecture responds to environmental factors, including nutrient and water availability, herbivory, and soil microbes. Root form, in turn, may influence function. The herringbone type (main axis and few laterals), although requiring more resources to construct than those with dichotomous architecture, is more efficient at exploiting resources such as phosphate. In *Picea sitchensis* (Pinaceae), roots growing on the windward site of young seedlings are thicker, longer, and more branched than on the leeward side providing resistance to wind throw (Robinson *et al.*, 2003).

Another feature is root exudates (Inderjit & Weston, 2003). Seedling roots may produce a variety of compounds, including in the case of wheat (*Triticum* sp. Poaceae) phenolic acids and *Sorghum bicolor* (Poaceae) root hair droplets containing a mixture of long-chain hydroquinones. In these crops, the amount of exudate varies with cultivar. Exudates have a variety of roles, including inhibiting seedling growth of other plant species, enhancing inorganic nutrient availability in nutrient-limited habitats, inhibiting nematodes, stimulating nodule and mycorrhizal symbioses, and providing a favorable environment for growth of a beneficial rhizosphere community. They can also stimulate germination and haustorial growth of parasites like *Orobanche ramosa* (Orobanchaceae) and *Striga* species (Scropulariaceae).

We know little about seedling root characteristics (but see Tillich, 1995, 2000). For example, we do not understand the significance of the lack of nodules on *Mora megistoperma* seedlings, although characteristic of many Leguminosae (adults appear to be mycorrhizal) (Janzen, 1983); the lack of root hairs on most hypogeal seedlings (de Vogel, 1980); or the adherence of, for example, *Pilea pumila* (Urticaceae), roots to filter paper (Leck, Outred pers. obs.).

Specialized storage structures

Modified stems, such as bulbs, tubers, and rhizomes, as well as storage roots (Appendix 2.1), are adaptations that provide storage and survival function. In fire-adapted and nutrient-limited communities, lignotubers (woody storage structures) provide protection from fire and aridity (Hallé *et al.*, 1978; Florence, 1981; Boucher, 1983). An interesting example of lignotuber formation occurs in *Quercus oleoides* where the seedling must transfer resources to an underground tuber to win the race against seed consumption with a moth larva, the egg of which was laid just as germination began. The larva begins to eat the inside of the acorn; tuber formation must take place before shoot development can occur (Hallé *et al.*, 1978). In the case of *Ginkgo biloba* (Ginkgoaceae), lignotubers originate from cotyledonary buds that become imbedded in the stem cortex, and, then, following a traumatic event, grow downward to form the lignotuber (Del Tredici, 1997).

Contractile roots or hypocotyls bury bulbs, corms, or other structures (Appendix 2.1) (Pate & Dixon, 1982). Contraction occurs by widening and shortening of cells or by their total collapse (Bell, 1991). In alpine tundra plants, contractile roots gradually pull the shoot apex of young *Lepidium* (Brassicaceae) several centimeters below the surface, an adaptation to withstand the vertical forces associated with freezing when protective snow cover is missing (Körner, 2003). Contractile roots also occur in the grass trees *Kingia australis* (Dasypogonaceae) and *Xanthorrhoea australis* (Xanthorrhoeaceae), where stem apices of seedlings are several cm below the soil surface and production of an aerial stem is delayed many years to protect against fire (Staff & Waterhouse, 1981). The hypocotyl is contractile in *Asclepias tuberosa* (Asclepiadaceae), placing the apical meristem (and tuber formation) belowground (Kummer, 1951). If the hypocotyl takes part in the formation of a storage organ, then both roots and hypocotyls may be contractile [e.g. *Chloraea membranaceae* (Orchidaceae); Skene, 1959].

In certain cases however, burial involves growth rather than contraction. Asymmetrical growth of the present year's corm carries the new *Colchicum autumnale* (Liliaceae) corm deeper into the soil (Boyd, 1932). In *Tulipa* (Liliaceae), the cotyledon base forms a hollow, blunt tube, which penetrates the soil, and as it grows downward, carries the plumule into the ground (Lubbock, 1892; Boyd, 1932). Both the cotyledon and base of the plumule form the first bulb that is buried in the soil. In *Marah* (Cucurbitaceae), burial is accomplished by elongation of the fused cotyledonary petioles that can be 20 cm long and that push the radicle and plumule into the soil (Schlising, 1969). Dropper roots, as occur on seedlings of *Dichopogon strictus* (Anthericaceae), place the storage root well below the soil surface (Pate & Dixon, 1982). In seedlings of *Pinus rigida* and *P. echinata* (Pinaceae), basal crooks develop, forcing dormant basal buds into mineral soil where they are protected from fire (Good *et al.*, 1979). Similarly, the hypocotyl of *Eupatorium perfoliatum* (Asteraceae) bends and reclines, bringing the crown to the soil surface (Kummer, 1951).

Plasticity

Varied seedling responses can occur depending on environment, including inundation, crowding, light, and soil conditions like salinity and compaction. An *Eichhornia crassipes* (Pontederiaceae) seedling germinated in water develops an elongated hypocotyl and a basal ring of rhizoids, but on soil the hypocotyl does not elongate and adventitious roots are produced (Muenscher, 1944). Shoots appear first when seeds of wetland species (e.g. *Peltandra virginica* Araceae, *Phalaris arundinacea* Poaceae) germinate underwater; the leaves of submerged *P. arundinacea* seedlings may have conspicuous oxygen bubbles at their tips (Leck, 1996). The shoot of *Zizania* (Poaceae) seedlings may grow 50–100 cm to reach a lighted position in the water column where photosynthesis can replace stored reserves (Aiken, 1986). In *Mimulus lutea* (Scrophulariaceae), the primary internodes do not develop unless seedlings are crowded, resulting in elongation (Lubbock, 1892). *Dactylis glomerata* and *D. polygama* (Poaceae), in response to reduced light (20–30% full sunlight), double or triple leaf area and allocation to roots, but do not reduce root length, nitrogen uptake, or plant growth (Ryser & Eek, 2000). These examples suggest that phenotypic plasticity appears to maximize resource acquisition and growth for the short term; for shade-tolerant species, higher tissue-mass density and longer leaf-life spans provide long-term adaptations.

Varied soil conditions may contribute to seedling plasticity. *Taxodium distichum* (Taxodiaceae) seedlings respond to increased salinity (0–4 g l^{-1}) by increasing partitioning of biomass to roots, thereby increasing the surface area for water uptake and increasing the possibility of reaching zones lower in salinity (Allen *et al.*, 1997). Dwarf and nondwarf home site differences can cause striking phenotypic differences between reciprocal seedlings transplants of *Pinus rigida* (Pinaceae) (Fang *et al.*, 2006). Additionally, the mechanical resistance of soil, due to surface crusts or compaction, can cause seedlings (e.g. *Lycopersicon esculentum* Solanaceae) to develop thicker hypocotyls, thereby providing greater emergence force (Liptay & Geier, 1983). The greatest compression regime caused the greatest increase in hypocotyl diameter and increased time to emergence. Other types of plastic responses occur. Hypocotyl growth can be significantly reduced by pricking or rubbing cotyledons (e.g. *Bidens pilosus*, *B. dioica* Asteraceae), implicating the rapid transmission of a signal from the cotyledon to the hypocotyl with the underlying mechanisms involving auxin activity and lignification by peroxidases on elongating cells (Desbiez & Boyer, 1981). The *Trapa natans* (Trapaceae) seedling initially has a negatively geotropic hypocotyl and may produce both positively and negatively geotropic roots (Sculthorpe, 1967). In addition, plasticity attributes may change with seedling age. Shade-tolerant woody species, for example, may lose the ability to modify architecture to capture limited light as investment to mechanical structure increases, and shade-tolerant and shade-intolerant species converge to become less shade tolerant with age (Kneeshaw *et al.*, 2006). Behavior of seedling parts may vary. For example, the shoot of *Hedera helix* (Araliaceae), a vine, is photophobic and grows toward dark forms, while the leaves are positively phototropic (Metcalfe, 2005).

Regeneration from cotyledons has been observed in *Gustavia superba* (Lecythidaceae) (Harms *et al.*, 1997) and in *Idiospermum australiense* (Calycanthaceae), in which any of the two to six cotyledons can produce independent seedlings (Edwards *et al.*, 2001). Detached cotyledons of *Bidens laevis* (Asteraceae) can form roots (Leck, pers. obs.).

Polymorphism

Seedlings of a given cohort may exhibit little (e.g. agricultural crops) or much variability in form, depending on genetic and environmental factors. The seedlings of *Geranium sessifolium* (Geraniaceae) may be brown, green, or intermediate (Philipp, 1992). The brown ones have the highest survival rates due to the higher cyanidin-6-glucoside content that protects seedling and juvenile leaves from ultra-violet (UV) light and/or predators (Mooney *et al.*, 1983; Drumm-Herrel & Mohr, 1985). Certain other species produce more than one kind of seed, resulting in marked seedling differences. When large seeds of *Taraxacum hamatiforme* (Asteraceae) germinate, the radicle tends to emerge first, but cotyledons tend to emerge first from small seeds (Mogie *et al.*, 1990). The adaptive value of this is not known. The offspring of amphicarpic species with dimorphic seeds vary in size and other characteristics. For example, *Amphicarpum purshii* (Poaceae) seedlings from aerial seeds were shorter than those from subterranean ones (7.9 ± 0.4 vs. 18.1 ± 0.8 cm), fewer in number (48.4 ± 13.3 vs. 112.4 ± 22.6 m^{-2}), and weighed less (5.1 ± 0.6 vs. 50.8 ± 4.3 mg) (Cheplick, 1982; Cheplick & Quinn, 1988). In *Cardamine chenopodifolia* (Brassicaceae), seedlings from subterranean seeds, which were 6 times heavier than aerial ones, had more rapid root elongation, greater dry weight accumulation, and larger cotyledons and first leaves (Cheplick, 1983). The seedlings of subterranean seeds, because of their faster growth, greater vigor, and earlier reproduction, have selective advantage in temporary unpredictable habitats. For desert species such as *Gymnarrhena micrantha* (Asteraceae) and *Emex spinosa* (Polygonaceae), seedlings from subterranean propagules are more tolerant of water stress (see Gutterman, 1993).

Chasmogamy/cleistogamy and fruit/seed dimorphism are more widespread than amphicarpy and also represent responses to a variable, heterogeneous environment (Cheplick, 1998). In *Impatiens capensis* (Balsaminaceae), cleistogamous flowers, produced in shade, yield fewer and smaller seeds than chasmogamous flowers (Simpson *et al.*, 1985); seedlings from chasmogamous flowers have a greater chance of surviving (Waller, 1984).

In contrast, seed heteromorphism is conspicuous in certain Asteraceae. While important for dispersal, Venable *et al.* (1995) found that seedlings from disk, ray, and intermediate achenes of *Heterosperma pinnatum* did not differ in size, growth, or competitive ability, and embryo biomass at maturity was also similar. In the case of *Polygonum hydropiper* (Polygonaceae), however, polymorphism is related to cross-generation effects; with the seed's architectural position (terminal vs. axial inflorescence) on the parent plant influencing the rate of development and other seedling traits (Lundgren & Sultan, 2005). Responses also varied with plants grown in shade versus full sun.

Other forms of polymorphism occur. Species with polyembryony may produce seedlings of varying form and sizes (e.g. *Citrus aurantifolia* Rutaceae, Enoch, 1980). In *Macfadyena unguis-cati* (Bignoniaceae), a dry forest liana of Costa Rica, two distinct juvenile forms occur (Gentry, 1983). Initially, the seedling is erect with largish opposite, simple leaves; it then develops into a wiry vine with tiny bifoliate leaves having a trifid tendril with hooked tips. The second stage is photophobic and grows toward the closest dark, light-blocking form, usually a tree trunk that it climbs with its cat's claw tendrils. The juvenile may persist for several years.

For dioecious species, the sex of seedlings is not discernable without using histological techniques/DNA markers for species with sex chromosomes (J. Consolloy & J. Quinn, pers. comm.). There are, however, some morphological differences in male and female laboratory-grown *Cannabis sativa* (Cannabaceae) by the time plants are about three weeks old (S. Datwyler, pers. comm.).

Interesting oddities

Seedlings of a number of species defy categorization (Bell, 1991). Those of *Streptocarpus* (Gesneriaceae), which begin apparently as normal dicots, have one cotyledon that enlarges to form a 30- to 90-cm-long and 23- to 50-cm-wide phyllomorph (a leaf-like structure) (Lubbock, 1892; Bell, 1991; Rauh & Basile, 2000; Mantegazza *et al.*, 2007). The plants of Podostemaceae and Tristichaceae, found in fast-flowing streams, superficially resemble algae. Germinating seeds do not produce radicles, and adventitious roots may be produced by the hypocotyl. Roots may elaborate to form a hapteron, a holdfast-like structure, which attaches the plant to the rock surface (Bell, 1991). The young thallus may develop as a lateral outgrowth from the hypocotyl and less often from the cotyledons and plumular leaves (Arber, 1920; Sehgal *et al.*, 1993). Attachment may occur via rhizoids, and a cyanobacterial film on the immobile substrate appears necessary for attachment (Philbrick & Novelo, 2004). The Lemnaceae also do not fit common morphological interpretation (Bell, 1991). Seedlings are exceedingly small (e.g. ≈1.2 mm, *Lemna minor*, Muencher, 1944) and lack differentiation, except in the development of two meristematic areas (usually one in *Wolffia*) that are each in a sunken pocket (but see Tillich, 1995). New fronds on the developing seedling occur in these pockets.

2.5 | Seedling diversity – ecophysiology

Growth patterns

Growth patterns of the seedling, and subsequently of the mature plant, may involve the loss of meristems. In certain temperate tree seedlings, abscission of the terminal bud occurs at the end of the first year followed by substitution of a lateral bud that functions for a year (e.g. *Ulmus effusa* Ulmaceae), but in tropical trees, growth periodicity is not necessarily annual (Hallé *et al.*, 1978). The fate of the meristem influences the subsequent architectural model (Hallé *et al.*, 1978). Moreover, behavior appears related to the amount of food reserve in the seed. Severing the seed from the hypogeous seedling may reduce the length of the orthotropic (vertical) phase; seedlings with little or no reserves quickly become plagiotropic (oblique or horizonal); and in *Sida carpinifolia* (Malvaceae), the orthotropic phase may be extended by growing the seedlings in a very rich medium (Hallé *et al.*, 1978). In some trees, the length of the epicotyledonary axis determines the length of the trunk (e.g. *Aloe* spp. Liliaceae, *Senecio johnstonii* ssp. *johnstonii* Asteraceae). The distribution of carbohydrate appears to trigger lignotuber development in *Ginkgo biloba*, with formation in 50% of horizontal seedlings compared to 0% in vertical ones (Del Tredici, 1997). *Welwitschia mirabilis* (Welwitschiaceae) displays a unique developmental pattern. Following formation of two permanent leaf primordia and so-called scaly bodies (foliar appendages) that persist for 30–40 years, precocious death of the terminal meristem occurs. This is unique among vascular plants because further growth is restricted to the edge and leads to increased girth; the elliptical apex becomes a crater-like depression (Sporne, 1967).

Dormancy/interrupted growth

Once germination has begun, growth of seedlings may not be continuous. Stem growth of woody species may be interrupted with pulses of inactivity (Lubbock, 1892; de Vogel, 1980). How development proceeds, especially the occurrence of resting phases, may be an especially important clue to seedling identification (de Vogel, 1980). Resting stages may be short or long or growth may occur in flushes with several internodes produced during each. In *Embelia viridiflora* (Myrsinaceae), four distinct growth intervals appear in a short period of time during which the primary axis grows the initial 10 cm. Growth may also be interrupted, as seen in *Jacquinia ruscifolia* (Theophrastaceae), and when growth ceases, the bud is protected by small scales. There is a long interval before growth resumes in seedlings of *Mangifera indica*

(Anacardiaceae) following the initial growth phase during which the epicotyl reaches approximately 30 cm and the first leaves have matured (Enoch, 1980). The seedlings of *Mora megistosperma* (Leguminosae) grow 1–2 m in the first 2–4 months and then remain at that height for more than 3 years (Janzen, 1983). In some Burseraceae and Leguminosae, dormancy begins after the unfolding of the primary leaves (van der Pijl, 1982). The tuberous swollen hypocotyl of some *Araucaria* species (Araucariaceae) rests for several months, a feature that enabled the transport of the first specimens of *Araucaria araucana* from Chile to Europe in 1795 (Dallimore & Jackson, 1966).

Root growth may also be interrupted. De Vogel (1980) notes that branching and formation of lateral roots in hypogenous seedlings of tropical species may be delayed and also that elongation of the primary root is delayed in some herbaceous temperate seedlings (e.g. *Downingia pulchella* Campanulaceae). Roots of cacti, with sympodial, determinate growth, cease growing when experiencing water deficits (Dubrovsky, 1997a).

Interrupted germination and, consequently, interrupted seedling growth have been reported for a number of herbaceous species from temperate regions. Some exhibit epicotyl (defined to include apical meristem or bud) and/or radicle dormancy. During germination, the epicotyl and/or radicle require specific dormancy-breaking conditions. See Baskin & Baskin (1998) for a detailed discussion; only two examples will be considered here.

In species with epicotyl dormancy like *Hydrophyllum* species (Hydrophyllaceae), radicle emergence typically begins in autumn and the root system develops slowly, becoming 4–7 cm long depending on species; then in early spring when temperatures are high enough for leaf growth, the cotyledons emerge. In other species, both the epicotyl and radicle may be dormant. This double dormancy was first described for *Trillium grandiflorum* (Liliaceae). Such species require at least two winters and one summer for complete seedling emergence. Moist seeds of *T. grandiflorum* require 3 months at 5–10 °C (1st winter) to break radicle dormancy, followed by 3 months at 20–30 °C (spring and summer) to allow radicle emergence, development of the root system and bud, and finally, 4 months at 4 °C (2nd winter) to break bud (epicotyl) dormancy (Barton, 1944; cited in Baskin & Baskin, 1998).

Young seedlings of a variety of temperate woodland herbs, such as *Arisaema triphyllum* (Araceae), *Lilium superbum* (Liliaceae), and *Mertensia virginiana* (Boraginaceae), become dormant well before the end of the growing season (Phillips, 1985), as do the ephemeral adults. Finally, seedlings of winter annuals may also show interrupted growth. *Chaerophyllum procumbens* var. *shortii* (Apiaceae) of deciduous forests overwinters with cotyledons and one or two leaves (Baskin *et al.*, 2004). Germination in late summer permits establishment of seedlings because deciduous trees lose leaves during autumn, creating canopy gaps.

Likewise, adaptations of wetland seedlings (e.g. Topa & McLeod, 1986; Končalová, 1990; Al-Hamdani & Francko, 1992; de Oliveira Wittmann *et al.*, 2007) permit establishment in various kinds of wetlands, including, for example, such periodically flooded forests as occur along the Rio Negro (Brazil) where *Eschweilera tenuifolia* (Lecythidaceae) and *Macrolobium acaciifolium* (Leguminosae) can tolerate months of complete inundation (S. Mori, pers. comm.).

Seedling environmental chemistry

Like adults, seedlings may produce secondary metabolites, but species vary in seedling palatability (Fenner *et al.*, 1999). Escape from herbivory may involve production of distasteful substances, such as acetophenone by *Stirlingia latifolia* (Proteaceae) (Main, 1981). The high alkaloid and free amino acid content of *Pentaclethra macroloba* (Leguminosae) seeds confers protection, considerably reducing seedling predation and resulting in high seedling densities (Hartshorn, 1983). Similarly, *Mora megistosperma* (Leguminosae) seedlings whose leaves have 23% tannin are not browsed (Janzen, 1983). Seedlings of other mangrove species also contain high concentrations of tannins (Outred, pers. obs.). Finally, young *Phaseolus lunatus* (Leguminosae) plants may emit volatile protective substances in response to herbivores that can elicit defensive responses in neighboring plants (Kost & Heil, 2006). In addition to resistance to herbivores, chemical defenses can protect against other kinds of organisms, and seedlings of *Juniperus* spp. (Cupressaceae) are resistant to damping off and to root-rot organisms, but not cedar apple rust (*Gymnosporangium juniperi – virginianae*) (Johnsen & Alexander, 1974).

Another aspect of protection can be seen in frost resistance. Although seedlings are most vulnerable when cells are rapidly elongating, seedlings of alpine plants can rapidly become frost resistant. Abilities are species specific, dependent on geographic origin and microsite preferences (Sakai & Larcher, 1987). Young plants of species, such as *Embothrium coccineum* (Proteaceae), from habitats with severe radiation frost, harden early and may be more resistant than adults (Alberdi & Rios, 1983). Resistance of *Quercus ilex* seedlings changes with season (winter > summer), age (1 year < juvenile < adult), and tissue (shoot cambium > leaves > root cambium).

2.6 | Vivipary and seedling equivalents

Typically, seedlings are produced after propagule dispersal with the embryo prevented from growing by various dormancy mechanisms. However, under some conditions, germination begins on the parent plant (*vivipary*) and sometimes plantlets are produced in inflorescences without the intervening seed stage (*pseudovivipary*). To further complicate the reproductive picture, various specialized structures (bulbils, turions) or other plant parts produce young plants at any time or serve as hibernacula (Sculthorpe, 1967) that bridge periods unfavorable to growth. This section will explore causes of vivipary, as well as reproductive strategies that serve as seedling equivalents.

Vivipary

Seeds often have a resting stage where desiccation is imposed (e.g. Walbot, 1978), but with vivipary, embryo growth is continuous and germination precocious. Viviparous seedlings are nutritionally dependent on the parent plant, especially during the early stages of growth, and dispersal may be delayed for many months until they have grown to considerable size. Although the viviparous condition is rare in woody plants, in saline tidal habitats of the tropics and subtropics, woody species may have viviparous seedlings that possess some adult characteristics such as salt glands at the time of dispersal. When such advanced forms (e.g. *Rhizophora* spp. Rhizophoraceae) abscise, cotyledons are often left behind in the fruit. In others, the seedling develops within the fruit but does not emerge until after the fruit has fallen (e.g. *Avicennia* Avicenniaceae). Thus, two categories of vivipary may be distinguished: *true vivipary*, where precocious germination within the fruit produces an embryo well advanced toward seedling status, and *cryptovivipary*, where the embryo grows but usually does not emerge from the fruit before dispersal (Appendix 2.3). Regardless of type, radicle elongation and development of roots do not usually take place until after dispersal, and in Rhizophoraceae, root meristems are not differentiated until release from the parent plant (Osborne & Berjak, 1997).

Vivipary occurs only in a very small number of higher plants that are phylogenetically diverse, of varying life forms, and from a wide range of habitats (Appendix 2.3; Farnsworth, 2000). With few exceptions (e.g. *Inga, Medicago* Leguminosae), vivipary occurs only occasionally in most terrestrial plants. It is, however, common in mangrove (mangal) vegetation (Dowling & McDonald, 1982). Of the 54 species designated as mangroves (Waisel, 1972; Hogarth, 1999), 29 are viviparous or cryptoviviparous, and, with the exception of *Pelliciera* (Pellicieraceae), they are all found only in mangrove swamps.

Mangrove swamps have an understory composed almost exclusively of mangrove seedlings. Seedlings grow in a very viscous, muddy, unstable substrate, saturated even between tides, and almost continuously anaerobic due to decomposition of organic material and sluggish water movement; it is toxic with iron sulfides, tannins, and other phytotoxins (Lugo & Snedaker, 1974), and the soil surface between tides has significantly increased salinity. Temperatures are high and seedlings are at risk from herbivores and pathogens. In addition, photosynthesis is reduced by the low light under adult canopies and by silt deposition on leaves (Chapman, 1976). Vivipary is considered an adaptation that protects young seedlings from such habitat conditions, especially from high salinities. With this in view, it seems surprising that vivipary is not more common among mangrove species.

Origins and causes of vivipary

Tomlinson (1986) notes that there is no generally accepted explanation, and Elmqvist and Cox (1996) and Farnsworth (2000) regard vivipary as having independent origins at different times and in a number of lineages. Walbot (1978) has suggested that developmental arrest of the embryo is an imposed condition. Thus, dormancy could be bypassed, allowing viviparous development, provided that physical and environmental constraints were not placed on the embryo. Interestingly, Kidd (1914) suggested that in mangroves autonarcosis (growth cessation) induced by respiratory CO_2 is absent and therefore development continues, resulting in vivipary. However, the pericarp of *Avicennia* (Avicenniaceae) presents a barrier to free gaseous diffusion, allowing CO_2 concentrations, produced by both anaerobic (60%) and aerobic (40%) processes, to reach approximately 6%, and O_2 concentrations to fall to 11% in developing cotyledons, yet viviparous development proceeds (Outred, 1973). The situation may be more complex than sole control by CO_2. Pannier & Rodriguez (1967) identified a ß-inhibitor complex, which decreased continuously during growth of *Rhizophora mangle* seeds, and suggested that decreasing concentrations may allow precocious germination. Furthermore, abscisic acid (ABA) has assumed a central role in understanding the transition to dormancy when embryo growth becomes limited during seed maturation (e.g. Neill *et al.*, 1987; Chapter 7). Reduction in embryo ABA has been linked with viviparous development in seven mangrove genera, except *Avicennia* (Farnsworth & Farrant, 1998); in *Avicennia*, the pericarp contains most of the ABA (Osborne & Berjak, 1997). ABA may control water relations rather than act more directly as a growth inhibitor (Walbot, 1978). It is also possible that ethylene is involved. Fountain and Outred (1990) have shown that it has a fundamental role in controlling water relations and permitting germination of seeds within fruits of *Phaseolus vulgaris* (Leguminosae).

Seedling equivalents

A variety of plant fragments or structures, such as bulbils and turions, are also capable of developing into fully reproductive adult plants. These can be considered *seedling equivalents*. The young plants produced may be anatomically and morphologically similar to seedlings produced from seeds, but unlike most seedlings, have the same genotype as the parent. Production of asexual propagules is advantageous in habitats where the chances of pollen reaching the female organ are remote (e.g. seagrasses, *Zostera* Zosteraceae), or in species with impaired sexual reproduction (*Saxifraga cernua* Saxifragaceae or *Polygonum viviparum*) and are sterile (van der Pijl, 1982; but see Molau, 1992). They are also important in species maintenance, but may be responsible for rapid expansion, allowing species like *Allium ascolonium* (Liliaceae) to become an invasive weed (Chittenden, 1951).

Pseudovivipary, under environmental control, involves abortion of the flowering process and inflorescence proliferation with the formation of propagules such as bulbils or leafy plantlets (Tooke *et al.*, 2005). This phenomenon has been associated with environmental severity (e.g. Elmqvist & Cox, 1996; Baskin & Baskin, 1998; Leck & Schütz, 2005). It occurs, for example, in the High Arctic where 10% of the flora in some areas produces plantlets (Lee & Harmer, 1980); in certain sedges when plants are inundated (Leck & Schütz, 2005); or in *Drosera rotundifolia* (Droseraceae) apparently during drought (Leck, pers. obs.).

The seedling equivalents, as found in pseudoviviparous and other propagules, provide analogous regeneration modes. The various types of propagules may be effectively dispersed. Dispersal may be temporal, as occur in dormant tubers of *Ranunculus ficaria* (Ranunculaceae) and *Sagittaria latifolia* (Alismataceae) (Leck, pers. obs.) or asexual propagules of *Ruppia* (Ruppiaceae), which can survive prolonged dessication (Brock, pers. comm.), or spatial, as seen in *Drosera pygmaea* gemmae (Salmon, 2001), the pseudoviviparous plantlets on reclining scapes of sedges (Leck & Schütz, 2005), and the ballistic bulbils of *Cardamine bulbifera* (Brassicaceae) (van der Pijl, 1982). In *Drosera pygmaea*, gemmae, plantlets that develop from foliar buds are effectively dispersed away from the parent in raindrops; dispersal distance may be up to 2 m, a distance 178 times the diameter of the parent (10 mm).

Although seed plants are the focus of this book, non-seed plants and even algae (Charophytes) have germlings (Brock, pers. comm.). These, too, can be regarded as seedling equivalents in their life cycles.

2.7 | Longevity

Seedling organs
Like other seedling features, longevity of seedling organs is variable (Lubbock, 1892; de Vogel, 1980). Cotyledons may be nonpersistent (e.g. *Atriplex hortensis* Chenopodiaceae, *Matricaria globifera* Asteraceae, *Ptelea trifoliata* Rutaceae), functioning for reserve storage, and having little assimilatory purpose. In *Cucurbita* (Cucurbitaceae), the cotyledons may remain photosynthetic well after the development of the first leaves (Don, 2003). In still other species, cotyledons last for more than 1 year; those of *Mora megistosperma* (Leguminosae), for example, persist for more than 3 years (Janzen, 1983). Usually gymnosperm cotyledons abscise during the first year, but in *Taxus* (Taxaceae), cotyledons persist for 3 years and in *Abies* (Pinaceae) for 4 years (Dallimore & Jackson, 1966). The cotyledons of *Welwitschia mirabilis* seedlings begin to die distally after 10 weeks but slowly enough to remain photosynthetic for 1.5 years (Mabberley, 1997).

Primary roots vary in longevity; persistent ones are found on some perennial herbs (e.g. *Stylidium adnatum* Stylidiaceae) and woody plants (e.g. *Forsythia suspensa* Oleaceae, *Acacia verticillata* Leguminosae, cycads). Roots may also be either annual (e.g. *Hypecoum procumbens* Papaveraceae) or biennial (e.g. *Brassica balearica* Brassicaceae) (Lubbock, 1892).

Seedling banks
Seedling banks are an important component of many forests. They are found, for example, in dry sclerophyll forests of Western Australia, where the occasional seed that escapes ant predation quickly develops

into a lignotuberous seedling that remains suppressed until a disturbance improves light and/or nutrient conditions (Bell *et al.*, 1993).
In the tropics over 32 years, seedlings of *Chrysophyllum* sp. nov.
(Sapotaceae) grew slowly in shade, taking 27 years to double in height
(Connell & Green, 2000). Also, seedlings were long-lived, with 6% of
the initial cohort persisting for the duration of the study, showed little effect of natural enemies, and had high densities six times during
the study. Furthermore, ages of seedlings do not necessarily correspond to height. *Dysoxylum* (Meliaceae) seedlings 2 m high are 15–20
years old (Buddenhagen & Ogden, 2003), conifer seedlings less than
1.3 m may be more than 150 years old (Antos *et al.*, 2005), *Rhopalostylis
sapida* (Palmae) 20 cm high are 10 years old (Enright & Watson,
1992), and New Zealand podocarp seedlings 4.5 m tall are 5 years old
(Ebbett & Ogden, 1998). Following fire, death of a nearby tree, a storm
(Antos *et al.*, 2005), or catastrophic wind events (Whigham *et al.*, 1999;
Baldwin *et al.*, 2001), release of seedlings and saplings is an important
recovery mechanism.

Interestingly, seedlings can influence one another. Conspecific
competition promoted vertical growth in *Shorea leprosula* (Dipterocarpaceae) seedlings, whereas competition with other species, having greater branching, increased branching (Massey *et al.*, 2006). In
addition, such diversity, where neighbors influence growth and differential herbivory due to changes in levels of phenolic compounds,
may have an impact on establishment and, consequently, community
dynamics. The potential importance of seedling banks for nonwoody
species appears not to have been explored despite the persistence of
seedlings of such species. Woodland herbs, for example, may take
several years to reach maturity (Phillips, 1985; Ohara *et al.*, 2006).

2.8 | Dispersal

Water is the most important dispersal vector for seedlings (Ridley,
1930). Dispersal may be related to adaptive characteristics of the
seedlings or to movement of the substrate on which they grow.
Seedling buoyancy has been reported in aquatic and wetland species
representing 19 families and even in nonwetland species representing
seven families (Ridley, 1930; van der Pijl, 1982). While the list is not
exhaustive, it raises two points. First, adaptations for this method of
dispersal arose numerous times in diverse families. Aquatic species,
for example, are believed to have terrestrial ancestors (Cook, 1999).
Also, because nonwetland species also exhibit buoyancy, genes permitting this may be basic to plant survival skills and related (via
development of air cells, aerenchyma) to physiological pathways that
reduce ethanol and other products of anaerobiosis.

Seeds of amphibious plants typically germinate under water and
as soon as seedlings become photosynthetic and develop aerenchyma,
they become buoyant (Cook, 1987). In some cases, however, seedlings
(e.g. *Typha* spp. Typhaceae, *Nymphioides* spp. Menyanthaceae) have no

apparent adaptations for buoyancy. For wetland species whose seeds sink immediately, dispersal by buoyant seedlings, as found in *Baldellia ranunculoides* (Alismataceae), *Lythrum salicaria* (Lythraceae), *Scrophularia aquatica* (Scrophulariaceae), and certain *Juncus* species (Juncaceae) (Sculthorpe, 1967), may substitute for seed dispersal. In *Hottonia palustris* (Primulaceae), an aquatic species, floatability is delayed; germination occurs in the autumn and seedlings rise to the surface in spring (Wernert, 1982). Like seed dispersal, seedling dispersal does not guarantee success. The seedling may not reach a site that permits establishment. Floating seedlings of marine angiosperms are vulnerable to destruction by wave action and currents (Sculthorpe, 1967). However, for seedlings of *Amphibolis antarctica* (Zannichelliaceae), the probability of establishment is enhanced by a comb, composed of several spines, that keeps the seedling upright and anchors it to marine plants, sediment, or debris where it grows roots and rhizomes (Sculthorpe, 1967; Aston, 1977).

Dispersal of seedlings by water may involve substrate movement rather than seedling adaptations. Seedlings may be made buoyant by gasses that cause the surrounding surface mat on which they grow to float or by detachment of sediment and development of floating islands (Cherry & Gough, 2006). Seedlings may float along on wood and other debris on which seeds germinated (Leck, pers. obs.).

Despite the intrinsic or extrinsic floatability of seedlings, the effectiveness of seedling dispersal, except for mangroves (Rabinowitz, 1978a) and tropical várzea trees (de Oliveira Wittmann *et al.*, 2007), appears not to have been explored extensively.

2.9 | Environmental filters and safe sites

Harper (1977) described the *safe site* as a zone that provides the cues required for breaking dormancy, conditions necessary for germination to occur, and resources required during germination (water, oxygen). It is also where hazards, such as predators, pathogens, and competitors, are excluded. Safe sites are discontinuous, resulting in environmental filters that are defined by the heterogeneity of the environment and subtlety of conditions that the seed and seedling experience (see also Grubb, 1977).

Safe sites may be very subtle constructs in space and time, but for a seedling to establish, conditions must be favorable for a sufficient amount of time, keeping in mind that adaptive traits change during ontogeny (Niinemets, 2006). For seedlings in New Zealand forests, the seedling stage can be long and protracted, often decades (Wardle, 1984; Brockie, 1992) and in coastal rain forests of British Columbia, seedlings may exceed 150 years, release coming when light gaps are created (Antos *et al.*, 2005). In a temperate tidal freshwater marsh, where seedling densities can exceed 4500 m^{-2} (Leck & Simpson, 1995), seedling success can be determined in less than a month (Parker &

Leck, 1985), and is related to the ability to tolerate inundation and/or competition, but can vary from year to year (Leck & Simpson, 1995).

Characteristics of safe sites are elaborately multidimensional, reflecting subtle spatial and temporal variation and species traits. A number of examples illustrate this. Across a successional temperate landscape, where soil needle ice formation varied, species segregated; seedlings of *Acer* spp. (Aceraceae) were very sensitive to frost heaving, but those of *Carya tomentosa* (Juglandaceae) were not (Myster, 1993). In mountain grasslands (Switzerland), seedling survival was enhanced by the presence of nearby unpalatable species. *Picea abies* survived best when near *Gentiana lutea* (Gentianaceae) and was less well-guarded by *Circium acaule* (Asteraceae) (Smit *et al.*, 2006). In alpine tundra, seedling success was enhanced by other seedlings or nurse plants (e.g. cushion plants) or by a protective rock (Egerton *et al.*, 2000; Körner, 2003). Similarly, establishment in arid habitats by cacti intolerant of extreme temperatures, such as *Mammillaria gaumeri* (Cactaceae), required the presence of a nurse plant that reduced light intensity and minimized soil temperature fluctuations, improving water storage by the seedling (Cervera *et al.*, 2006). In the case of *Chamaecyparis thyoides* (Cupressaceae), decline in seedling establishment appeared related to decline in *Sphagnum* cover, which in turn was sensitive to anthropogenic modifications, such as elevated nitrogen, and change in hydrology (Ehrenfeld & Schneider, 1981; Allison & Ehrenfeld, 1999). The survival of seedlings of the seagrass, *Halophila stipulacea* (Hydrocharitaceae), was tripled by protecting them from UV radiation and reducing light intensity to 80% of full sunlight, suggesting that absence of this species from the upper intertidal zone could be due to photoinhibition of growth (as well as uprooting by storms) (Malm, 2006). In a coastal California grassland, a primary factor influencing tree seedling establishment was fog (Kennedy & Sousa, 2006). Seedlings of *Pseudotsuga menzeisii* (Pinaceae) were only able to invade arid chaparral vegetation beneath species with which they share symbiotic mycorrizae (Horton *et al.*, 1999). Endophytic fungi may enhance the competitive ability of grass seedlings (Cheplick, 1998). Palm seedling densities were related to crown architecture of canopy palm species, but the relative restriction of seedling recruitment by arborescent palms was limited at individual and landscape scales (Wang & Augspurger, 2006) and growth of tree seedlings was influenced by the architecture of neighboring vegetation (Ladd & Facelli, 2005). In dioecious species such as *Neolitsea sericea* (Lauraceae), although present in higher densities beneath the female parent, seedling survival rate was greater beneath males (Arai & Kamitani, 2005). Seedlings of *Geranium maculatum* (Geraniaceae) from female mothers (from larger seeds) had higher above- and belowground biomass than those from hermaphroditic mothers, although the mothers were indistinguishable in vegetative and reproductive characters (Chang, 2006). In tidal freshwater wetlands dominated by tussock-forming *Carex stricta*, small herbivores heavily graze any plants colonizing intertussock areas, and plants gain an herbivore refuge when growing on top of raised tussocks

(Crain & Bertness, 2005). A final example of subtle environmental effects on seedling recruitment can be seen in a study by Herrera (2000). The probability of an ovule producing an established seedling in the third year was significantly greater for flowers pollinated between 0930 and 1630, compared to those pollinated between dawn–0930 or 1630 and dusk. This was explained by both increased outcrossing and gametophyte competition, related to increased pollinator diversity and visitation.

Throughout the process of establishment, chance is an important component. Recruitment success of a seedling is determined by the occupation of a safe site by a seed. Delivery of the seed to the safe site by dispersal is determined by availability of vector(s). Once there, the response of the seed is based on its genetic core makeup, as determined by its evolutionary history, mitigated by the maternal environment, stage of development, storage conditions (e.g. soil conditions, duration), and degree of dormancy. Suitable conditions, including germination cues and resources for growth, must be sustained for establishment to occur.

The status of the seed in a safe site, accordingly, is determined by intrinsic and extrinsic factors, the interactions of which are complex and yield a particular phenotype. Similarly, the germinated seed experiences a unique set of conditions, resulting in still another level of phenotype. Assuming the ability of the safe site to provide all seedling requirements, the seedling becomes capable of independent existence. Establishment occurs, provided that the safe site, in fact, excluded hazards. The varied seed and seedling phenotypes improve the chances for dispersal to a safe site in a heterogeneous landscape. Within this context, seed–seedling conflicts may be important and the safe site for the seed may not be safe for the seedling (e.g. protection from predation) (Schupp, 1995).

Duration of the seedling stage, like that of the seed (Dalling, 2005), might suggest that sources of mortality play important roles as environmental filters (*sensu* Harper, 1977), influencing both population growth and niche partitioning in communities. Understanding seed fate and ultimately community dynamics requires an integrated life history approach, including close examination of seedlings.

Despite the high selection (death) at the seed and seedling stages, plants typically produce ample numbers of seeds that, by chance, are delivered in time and space to safe sites. The heterogeneity of safe sites, coupled with the heterogeneity of seed and seedling phenotypes, determines establishment success. Thus, seedling variability, reflected in morphological and physiological attributes (Appendixes 2.1 and 2.2), suggests that nature has provided many successful strategies to bridge the gap between seed and postseedling stages and to exploit the safe sites of the varied habitats in which seed plants occur. Even in harsh alpine tundra, sexual reproduction systems are effective, buffered by substantial seed production (Körner, 2003).

Because characteristics of safe sites are not necessarily obvious and are transitory, attempts to describe them by examining the habitats of the mature plant are generally ill-fated, the environment having changed since germination (e.g. Harper, 1977; Cook, 1979; Körner, 2003). For example, in areas of ample rainfall where acidity decreases with soil depth, seeds germinate in a medium more acidic than that in which adults thrive and, thus, sampling the pH of roots of the adult plant may not indicate the preferences of the seedling (Wherry, 1948). The particular soil surface features in alpine tundra caused by disturbance (Chambers, 1995) do not persist. The positive effect of earthworm activity on seedlings (Milcu *et al.*, 2006) could not be established by examining adult plants. Seedlings of wetland species grow best in moist conditions, while adults can tolerate inundation (Bakker *et al.*, 2007). Finally, despite a number of studies and ensuing suggestions, including fire frequency, competition with exotic grasses, and grazing by vertebrates, Allen-Diaz *et al.* (1999) conclude that it is still not clear how California oaks (*Quercus* spp.) successfully regenerated in the past, making future management problematic. Moreover, there may be no association between seedling and adult traits (Shipley *et al.*, 1989).

2.10 | Summary

The considerable morphological and physiological diversity found among seedlings (Appendixes 2.1–2.3; Figs. 2.1–2.2) reflects adaptations for establishment success in heterogeneous habitats. Regeneration alternatives, whether seed bank, seedling, seedling bank, or seedling equivalent, vary in tolerances to stress and to small changes in environment that may have varying relative spatial and temporal advantage (Parker *et al.*, 1989). Interesting oddities may not indicate general principles, but they may stimulate thought about seedlings, their adaptations, ability to acclimate, and how ontogenetic modifications might relate to seedling establishment and, thus, to communities.

2.11 | Acknowledgments

We are grateful to the following: W. Ferren and M. Williams made suggestions that improved this chapter; J. Bryant, R. Deni, C. Leck, and C. McGill assisted in its preparation; and M. Brock (Tasmania, Australia), J. Consolloy (Princeton University), S. Datwyler (Sacramento State University), S. Mori (New York Botanical Gardens), and J. Quinn (Rutgers University) provided personal insights.

Appendix 2.1 | Diversity of seedling features, including cotyledons, roots, stems, leaves, modes of climbing, and origins of organs

Organ	Character	Variation	Examples[a]	Reference
Cotyledon	Range: 0–13	0	*Barringtonia, Bertholletia excelsa, Lecythis zabucajo* Lecythidaceae, *Garcinia* Guttiferae; some Orchidaceae, some Cactaceae	Lubbock, 1892; *de Vogel, 1980
		1	Most monocots; some Cactaceae and *Cyclamen* Primulaceae	Lubbock, 1892
		2	Most dicots; also *Araucariaceae, Cephalotaxaceae, Podocarpaceae, Taxaceae	Dallimore & Jackson, 1966; *Mabberley, 1997
		3	*Acer pseudoplatanus* Aceraceae (often), *Actinostrobus pyramidalis* Cupressaceae, *Coffea arabica* Rubiaceae, *Cryptomeria japonica* Taxodiaceae (70%), *Streptocarpus rexii × S. lutea* Gesneriaceae, *Terminalia megalocarpa* Combretaceae	Lubbock, 1892; *de Vogel, 1980
		4	*Pittosporum crassifolium* Pittosporaceae (2, 3 also common)	Lubbock, 1892
		5	*Pittosporum erioloma* Pittosporaceae, *Utricularia geminiscapa* Lentibulariaceae	Lubbock, 1892; *Muenscher, 1944
		>2 (Range)	Araucariaceae 2–4, Cupressaceae 2–5, Pinaceae 3–18, *Taxodiaceae 2–6; *Pinus strobus* 8–11, *P. pinea* 10–13; **Idiospermum australiense* Calycanthaceae 2–6	Lubbock, 1892; *Dallimore & Jackson, 1966; **Edwards et al., 2001
	Narrow		*Foeniculum vulgare* Apiaceae, *Menispermum canadense* Menispermaceae	Lubbock, 1892
	Petiolate		*Delphinium elatum* Ranunculaceae; elongated petiole: *Microloma* sp. Asclepiadaceae, *Primula denticulata* Primulaceae, *Vitellaria paradoxa* Sapotaceae	Lubbock, 1892; *Bell, 1991
	Sessile		*Delphinium staphysagria* Ranunculaceae, *Laburnum vulgare* Leguminosae	Lubbock, 1892
	Lobed		*Amsinckia intermedia* Boraginaceae, *Ipomoea quamoclit* Convolvulaceae, *Peucedanum sativum* Apiaceae	Lubbock, 1892
	Fused		*Lupinus sulphureus* Leguminosae	Lubbock, 1892
	Phyllotaxy		Usually opposite or whorled; alternate: *Citrus aurantium, Limonia acidissima* Rutaceae	Lubbock, 1892

Dissimilar		*Brassica nigra* Brassicaceae, *Hibiscus trionum*, *H. moscheutos* Malvaceae, **Streptocarpus dunnii* Gesneriaceae (one enlarges to 30–90 cm long, 23–50 cm wide), ***Trapa natans* Trapaceae (large functional cotyledon remains in the seed; the other abortive one, with a shoot bud, is borne upward on the negatively geotropic hypocotyl), *Zilla myagroides* Brassicaceae	Lubbock, 1892; *Leck, pers. obs.; ** also Rauh & Basile, 2000; *** Sculthorpe, 1967
Length		*Bowiea volubilis* Hyacinthaceae (10–23 cm), *Lodoicea maldivica* Palmae (cotyledon stalk 3–4 years old, 1–4 m)	Lubbock, 1892; *in Tomlinson, 1990
Longevity		Not persistent: *Matricaria globifera* (Asteraceae), *Atriplex hortensis* (Chenopodiaceae), *Ptelea trifoliata* (Rutaceae)	Lubbock, 1892
		Persistent > 1 year: *Chimonanthus fragrans* Calycanthaceae, *Crassula quadrifida* Crassulaceae, *Opuntia labouretiana* Cactaceae, *Sideroxylon tomentosum* Sapotaceae	Lubbock, 1892
Axillary buds		Swollen buds: *Forsythia suspensa* Oleaceae, *Lycium afrum* Solanaceae	Lubbock, 1892
		Branches geotropic, forming rhizomes or creeping stems: *Convolvulus sepium* Convolvulaceae, *Linaria cymbalaria* Scrophulariaceae,	Lubbock, 1892
		Growth following removal of apical bud: *Pisum sativum* Leguminosae	Leck, pers. obs.
Tubular/sheathing		*Polygonum bistorta*, **P. bistortoides, P. sphaerostachyum* Polygonaceae	Lubbock, 1892; *Muller, 1978; **Allessio, 1967
Connate		*Lonas inodora* Asteraceae, *Plantago callosa* Plantaginaceae, *Ranunculus hederaceus* Ranunculaceae	Lubbock, 1892
Color	Green	Most with epigeal cotyledons	Good et al., 1979
	Red	*Pyxidanthera barbulata* Diapensiaceae	Lubbock, 1892; *Leck pers. obs.
	Green above, white below	*Fagus grandifolia, F. sylvatica* Fagaceae (large foliaceous cotyledons)	
Roots			
Radicle/primary root – long		Many dicots	
Primary root – short	Many roots form at base	*Downingia pulchella* Campanulaceae	Lubbock, 1892

(cont.)

Appendix 2.1 (cont.)

Organ	Character	Variation	Examples[a]	Reference
	Fibrous roots		Most Poaceae, also *dicots having degenerate tap roots	Lubbock, 1892; *Kummer, 1951
		Long unbranched	Lucuma Sapotaceae	Lubbock, 1892
		Long + laterals	Calotropis gigantea Asclepiadaceae	Lubbock, 1892
		Single root + fibrous	Rheum officinale Polygonaceae, Verbena officinalis Verbenaceae	Lubbock, 1892
	Adventitious roots	From stem nodes	Hodgsonia macrocarpa Cucurbitaceae (primary root does not emerge from seed), *many palms	de Vogel, 1980; *Tomlinson, 1990
	Prop roots	Stem internodes	Stilt palms: Iriartea, Socratea Palmae (in some, contact of the stem with the soil does not persist)	Tomlinson, 1990
	Spiral		Amphibolis antarctica Cymodoceaceae	Aston, 1977
	Radicle	Lacking	Angraecum maculatum Orchidaceae, **Lemna minor Lemnaceae, *Rhyncholacis macrocarpa Podostemaceae (haptera grow from the hypocotyl), Stratiotes aloides Hydrocharitaceae	Lubbock, 1892; *Arber, 1920; **Muenscher, 1944
		Long-lived	Forsythia suspensa Oleaceae, Stylidium adnatum Stylidiaceae	Lubbock, 1892
		Annual	Hypecoum procumbens Fumariaceae	Lubbock, 1892
		Biennial	Brassica balearica Brassicaceae, Melilotus officinalis Leguminosae	Lubbock, 1892
		Non-persistent	Archontophoenix cunninghamiana Palmae	Tomlinson, 1990
		Persistent	Phoenix dactylifera, Washingtonia filifera Palmae	Tomlinson, 1990
	Color	Orange-red	Rubia cordifolia Rubiaceae	Lubbock, 1892
		Yellow, red laterals	Galium saccharatum Rubiaceae	Lubbock, 1892
		Bright red	Sherardia arvensis Rubiaceae	Lubbock, 1892
		Pink or flesh	Amaranthus hypochondriacus Amaranthaceae	Lubbock, 1892
		Brown hairs	Drosera binata Droseraceae	Lubbock, 1892
		Root tips red	*Impatiens capensis Balsaminaceae, Streblus asper Moraceae	Burger, 1972; *Leck, pers. obs.
	Nodules		Legumes: 50% (28 spp. Lubbock, 1892), 42% (32 spp. Burger, 1972); also Myrica californica Myricaceae, Alnus spp. Betulaceae	Lubbock, 1892; Burger, 1972
	Coralloid roots	N-fixing	Macrozamia riedeli (Zamiaceae)	Pate & Dixon, 1982

Section	Category	Subcategory	Example / Notes	Reference
	Adventitious roots		Poaceae, growth of radicle is only temporary; Cyperaceae and others	e.g. Lubbock, 1892; Boyd, 1932; Kummer, 1951
		Cotyledon node	Ranunculus arvensis Ranunculaceae	Lubbock, 1892
		Leaf base	Trollius ledebouri Ranunculaceae	Lubbock, 1892
		Hypocotyl base	Crassula quadrifida Crassulaceae, Impatiens parviflora Balsaminaceae, Limnanthes douglasii Limnanthaceae, Mirabilis dichotoma Nyctaginaceae	Lubbock, 1892
		Stem nodes	Hodgsonia macrocarpa Cucurbitaceae (primary root does not emerge from seed)	de Vogel, 1980
Stems	Undeveloped	Entirely stem	Cuscuta europaea Convolvulaceae	Burger, 1972
		Hypocotyl	Senecio pulcher Asteraceae, Androsace rotundifolia Primulaceae	Lubbock, 1892
	Length		Aleurites moluccana Euphorbiaceae (9–26 cm); certain Rhizophoraceae to 90 cm	Burger, 1972; Joshi et al., 1972; Wells, 1982
	Thickness		Durio zibethinus Bombacaceae (1.25–1.75 cm)	Lubbock, 1892
	Woody		Diospyros embryopteris Ebenaceae, Mimusops balata Sapotaceae; also Alnus, Carpinus, Quercus Fagaceae	Lubbock, 1892
	Hollow		Echinocystis lobata Cucurbitaceae; Proboscidea louisianica Pedaliaceae (enlarging to form peg, hollow belowground)	Kummer, 1951
	Belowground only		Ipomoea leptophylla Convolvulaceae	Kummer, 1951
	Adventitious buds		Euphorbia esula Euphorbiaceae (exogenous buds distributed over most of hypocotyl, endogenous buds (from pericycle) only at proximal end near collet; endogenous buds more successful in growth)	Raju, 1975
	Various		Centaurea dementei Asteraceae (red above, yellow belowground)	Lubbock, 1892
	Colored		266 spp.: ≈53% green or colorless; 27% red or purple; 6.8% brown/gray (generally woody); 4.1% brown; 1.1% yellow/straw; 0.8% flesh colored	Lubbock, 1892
			188 spp.: 66% green; 13% red; < 6% other colors	Burger, 1972

(cont.)

Organ	Character	Variation	Examples[a]	Reference
	Adventitious buds		Linaria spp. (except L. cymbalaria) Scrophulariaceae	Lubbock, 1892
	No stem elongation		Bulbs, biennials, rosette plants	
Leaves	Present at germination		Stratiotes Hydrocharitaceae (up to 10 leaves in embryo, may help anchor seedling); *Phaseolus vulgaris Leguminosae (2 leaves)	Cook, 1987; *Leck, pers. obs.
	Scale		Many hypogeous species, e.g. Quercus Fagaceae, Juniperus Cupressaceae	Schopmeyer, 1974
	Simple		Most species	
	Compound		First leaf of Acacia, Adenanthera, Albizia spp., and most other Leguminosae	Burger, 1972
	Dimorphic		Cephalotus follicularis Cephalotaceae (foliage and pitchers)	Bell, 1991
	Phylotaxy		Opposite, alternate, whorled; may change with age	Various
	Heteroblasty (shape change along a shoot)		Alisma plantago-aquatica Alismataceae, *Arabidopsis thaliana Brassicaceae, Kennedia rubicunda Leguminosae	Bell, 1991; *see Berardini et al., 2001
	Heterophylly		Many aquatic plants, most Nymphaeaceae	Lubbock, 1892
	Specialized function		*Dionaea muscipula, Drosera binata Droseraceae, *Nepenthes khasiana Nepenthaceae (leaves hairy/variously shaped, capable of catching insects)	Lubbock, 1892; *Bell, 1991
	Color	Green	Most species	
		Red, new foliage	*Coleus × hybridus Labiatae, **Dryobalanops cf. lanceolata, **Shorea sp. Dipterocarpaceae, **Endertia spectabilis Leguminosae, Gaura coccinia Onagraceae, **Quercus turbinata Fagaceae	Kummer, 1951; *Reilly, 1978; **de Vogel, 1980
		Lack chlorophyll	Parasites, e.g. Convolvulaceae, Orobanchaceae; Monotropoideae; also albinos	
Climbing	Cirrhi (from axils of petioles)		Passiflora macrocarpa Passifloraceae	Lubbock, 1892

Organ	Origin / Taxa	Reference
Hooked/curved stem tip	*Lathyrus nissolia* Leguminosae	Lubbock, 1892
Leaves	*Lathyrus articulatus* Leguminosae	Lubbock, 1892
Stems	*Humulus japonicus* Cannabaceae, *Ipomoea dissecta* Convolvulaceae	Lubbock, 1892
Tendrils (modified leaflets)	*Cobaea scandens* Polemoniaceae	Lubbock, 1892
Origins of specialized organs		
Bulbs	From cotyledons: *Bowiea volubilis* Hyacinthaceae, *Tulipa* spp. Liliaceae	Lubbock, 1892
Phyllode	From cotyledon: *Streptocarpus* Gesneriaceae	Lubbock, 1892; Rauh & Basile, 2000
Rhizome	From hypocotyl: *Cyclamen persicum* Primulaceae, *Gesnera macrantha* Gesneriaceae, *Polygonum bistorta* Polygonaceae	Lubbock, 1892; *Muller, 1978
Root	From adventitious root: *Asparagus madagascarensis* Liliaceae	Boyd, 1932
Stolons or runners	From axillary buds of leaves: *Mitella breweri* Saxifragaceae, *Potentilla reptans* Rosaceae, *Viola palustris* Violaceae	Lubbock, 1892
Storage organ	From root & hypocotyl: *Rheum officinale* Polygonaceae; from root: *Mirabilis linearis* Nyctaginaceae	Lubbock, 1892; *Kummer, 1951
Thallus	From hypocotyl: *Dicraeia stylosa* Podostemaceae	Sculthorpe, 1967
Tubers	From hypocotyl: **Aponogeton distachyon Aponogetonaceae, *Arisaema triphyllum Araceae, ***Dioscorea batatas Dioscoreaceae, Gesnera macrantha Gesneriaceae, Mirabilis dichotoma, Oxybaphus ovatus Nyctaginaceae	Lubbock, 1892; *Boyd, 1932; **Aston, 1977; ***Mabberley, 1997
	From tips of stolon: *Sagittaria sagittifolia*, *S. latifolia Alismataceae	Sculthorpe, 1967; *Leck, pers. obs.
	From roots: *Araucaria araucana* Araucariaceae, *Platysace cirrosa Apiaceae	Dallimore & Jackson, 1966; *Pate & Dixon, 1981
	From stems: *Nymphaea lotus* Nymphaeaceae, *Podophyllum emodi* Berberidaceae, *Tussilago farfara* Asteraceae	Lubbock, 1892
	From branches axils of aerial cotyledons: *Convolvulus sepium* Convolvulaceae	Lubbock, 1892

[a]Family names are generally from Mabberley (1997). Names are those used by the source. Asterisk(s) before a taxon indicate(s) the appropriate reference.

Appendix 2.2 | Seedling attributes, including surface appendages, odor, latex, secretion, gas exchange, movement, and irritation

Character	Structure	Example	Representative species[a]	Reference
Appendages	Aeroles	Spot with grandular projections	Many Papilionaceae, no Caesalpinioideae or Mimosoideae (of Leguminosae)	Endo & Ohashi, 1998
	Bristles		Stems: Cereus emoryi, Phyllocactus stenopetalus Cactaceae	Lubbock, 1892
	Domatia		Swollen hypocotyls, mainly Rubiaceae (Malaysia and Papua)	Juniper & Jeffree, 1983
	Glands	Open pores	Cotyledon, Senecio vulgaris Asteraceae	Lubbock, 1892
		Sunken black glands	Cotyledon petiole, Embelia ribes Myrsinaceae	Lubbock, 1892
		Submarginate red glands	Leaves: Ardisia crenulata Myrsinaceae	Lubbock, 1892
		Pellucid glands	Leaves: Callistemon rigidus, Rhodomyrtus tomentosa Myrtaceae	Lubbock, 1892
		Short cylindrical glands	Basal sinus of leaves: Dalechampia capensis Euphorbiaceae	Lubbock, 1892
	Hairs	From hypocotyl, anchorage	Alismataceae, Cyperaceae; Anacharis occidentalis Hydrocharitaceae, **Avicennia Avicenniaceae, *Eichhornia crassipes Pontederiaceae, Elatine Elatinaceae, Najas marina Najadaceae, Podostemum subulatus Podostemaceae, Vallisneria Hydrocharitaceae	Muencher, 1944; *Sculthorpe, 1967; **de Vogel, 1980; ***Cook, 1987
		Variably hairy	Cotyledons & hypocotyl glabrous, stem with dense white hairs when young and leaves hairy in older plants: Banksia australis Proteaceae; pubescent when young, hairy when older: Moscharia rosea Asteraceae	Lubbock, 1892
		Glandular hairs	Primula sinensis Primulaceae	Lubbock, 1892
		Hammer-shaped hairs	Malphigiaceae	de Vogel, 1980
	Marginal teeth		Dyckia floribunda Bromeliaceae	Boyd, 1932
	Spines		Stems: Echinocactus viridescens Cactaceae	Lubbock, 1892
	> 1 Appendage	Bulbous short hairs, sessile glands	Phacelia tanacetifolia Hydrophyllaceae	Lubbock, 1892

Rhizoids		Podostemaceae, e.g. *Polypleurum stylosum*, produce a black gummy substance and may have bulbous or forked ends; *certain monocots	Seghal et al., 1993; *Tillich, 1995
Odor	Entire plant	*Agathis dammara* Araucariaceae, *Cananga odorata* Annonaceae, *Chenopodium ambrosioides* Chenopodiaceae, *Eucalyptus globulus, E. tereticornis* Myrtaceae, *Hedeoma pulegioides* Labiatae, *Protium javanicum* Burseraceae	Burger 1972; *Kummer, 1951
	Leaves	*Citrus aurantium* Rutaceae, *Pinus merkusii* Pinaceae	Lubbock, 1892; *Burger, 1972
	Roots	Irritating: *Pittosporum ferrugineum* Pittosporaceae; stink: *Albizia lophantha* Leguminosae, *Toona sinensis* Meliaceae	Burger, 1972
Latex	Stems	*Alstonia spp. Apocynaceae, * Artocarpus spp. Moraceae, *Euphorbia splendens* Euphorbiaceae, *Manilkara kauki*, Sapotaceae, *Palaquium amboinense* Sapotaceae, *Plumeria alba* Apocynaceae, *Streblus asper* Moraceae	Lubbock, 1892; *Burger, 1972
Secretory function	Glands	Secrete white substance: *Crassula quadrifida* Crassulaceae	Lubbock, 1892
		Digestive fluid: *Drosera binata, Drosophyllum lusitanicum* Droseraceae (viscid secretion "may be drawn out some inches")	Lubbock, 1892
		Resin or oil: *Ardisia polycephala* Myrsinaceae	Fahn, 1979
		Salt: several angiosperm families, e.g. Frankeniaceae, Plumbaginaceae, Tamaricaceae; also species of other families and mangroves, e.g. Avicennia Avicenniaceae, *Laguncularia* Combretaceae	Fahn, 1979
	Hairs	*Dionaea muscipula, Drosera capensis* Droseraceae, *Nepenthes khasiana* Nepenthaceae	Bell, 1991
	Ducts	*Pinus halepensis* Pinaceae (resin); also, e.g. Anacardiaceae, Asteraceae, Leguminosae	Fahn, 1979
	Trichomes	*Callitriche stagnalis* Callitrichaceae (mucilage secretion in hypocotyl)	Arber, 1920
	Myrosin cells	Brassicaceae: endogenous secretion of myrosinase to produce mustard oil following cell damage in hypocotyl and cotyledons	Fahn, 1979

(cont.)

Appendix 2.2 (cont.)

Character	Structure	Example	Representative species[a]	Reference
		Nectaries	*Darlingtonia* Sarraceniaceae, *Nepenthes* Nepenthaceae (primary leaves), *Ricinus communis* Euphorbiaceae (cotyledons); carnivorous plants, "alluring" glands are generally nectaries (Fahn, 1979)	In Fahn, 1979;* Lloyd, 1976
		Water gland	*Impatiens fruticosa* Balsaminaceae (apex each cotyledon); underside of floating leaves, e.g. *Callitriche* Callitrichaceae	Lubbock, 1892; *Arber, 1920
		Hydathodes	*Arabidopsis thaliana* Brassicaceae, first leaves with few hydathodes, number increases as seedling grows	See Berardini et al., 2001
		Roots	Cotyledons and stems, *Ginkgo biloba* Ginkgoaceae	Chamberlain, 1935
			7 phenoloic compounds: wheat (*Triticum* Poaceae)	Inderjit & Weston, 2003
		Root hairs	Mixture hydroquinones, *Sorghum bicolor* Poaceae	Inderjit & Weston, 2003
Succulence		Water-storing parenchyma	*Aloe* spp. Liliaceae	Boyd, 1932
Gas exchange		Lenticels	*Alstonia angustiloba* Apocynaceae	Burger, 1972
			On taproots of *Pinus serotina*, *P. taeda* (Pinaceae) under anaerobic conditions	Topa & McLeod, 1986
		Stomates	Aerial portions of seedlings; lacking in hypogeous cotyledons; *roots of *Helianthus annuus* (Asteraceae), *Ceratonia siliqua* Leguminosae (permanently open)	Lubbock, 1892; *Bhaskar, 2003
Movement function	Touch	Hairs	*Aldrovanda vesiculosa*, *Dionaea muscipula* Droseraceae, *Utricularia* spp. Lentibulariaceae	Simons, 1992
	Touch	Motor cells in pulvinus (base of leaflets and leaf)	*Mimosa pudica* Leguminosae	Simons, 1992
	Sleep	Pulvinus cells (flux K+, water)	*Samanea saman* Leguminosae	Simons, 1992
Irritation	Relocation of shoot apex	Contractile roots, hypocotyl	Corm or bulb maintained at particular soil level: *Cyclamen persicum* Primulaceae, *Oxalis hirta* Oxalidaceae	Boyd, 1932; *Bell, 1991; see also Pate & Dixon, 1982
	Stinging	Hairs	*Laportea canadensis*, *Urtica dioica* Urticaceae	Lubbock, 1892; *Kummer, 1951

[a] Family names are generally from Mabberley (1997). Names are those used by the source. Asterisk(s) before a taxon indicate(s) the appropriate reference.

Appendix 2.3 | Viviparous species of mangroves and nonmangroves of various life forms. Indicated are families, height, habitat, and region. In truly viviparous species (V), the seedling penetrates the testa and fruit wall before dispersal; whereas in cryptoviviparous species (CV), it usually does not. The distinction is unclear for many species.

I. MANGROVES

Family	Species	Height	Propagule	Region	Reference
Avicenniaceae	Avicennia spp. (8)	60 cm–30 m	Ovoid capsule 4–2.5 cm	Pantropics – N. New Zealand	Cockayne, 1928; Tomlinson, 1986
Combretaceae	Laguncularia racemosa – CV	Small tree	Lens-shaped 2 × 1 cm	W. Africa, tropical America	Rabinowitz, 1978a
Myrsinaceae	Aegiceras spp. (2)	Small tree	Capsule	Tropics	Joshi et al., 1972
Pellicieraceae	Pelliciera rhizophorae	to 10 m	Large onion-shaped fruit, to 8 cm, ≈82 g	Tropical E. Pacific	Rabinowitz, 1978a, b
Plumbaginaceae	Aegialitis spp. (2)	To 3 m	Elongated capsule	Tropics	Heywood, 1978
Rhizophoraceae	Bruguiera spp. (6) – V	To 18 m	10–26 cm hypocotyl	Tropics	Burger, 1972
	Ceriops spp. (4) – V	To 12 m	To 30 cm hypocotyl	Tropical	Burger, 1972
	Kandelia candel – V	To 8 m	20–47 cm hypocotyl	S. E. Asia	Maxwell, 1995; Sun et al., 1998
	Rhizophora spp. (8) – V	To 18 m	30–90 cm hypocotyl	Tropics	Joshi et al., 1972; Wells, 1982

II. NON-MANGROVES

Family	Species	Height	Habitat	Region	Reference
TREES / SMALL TREES					
Anacardiaceae	Mangifera indica	≈18 m	Forest	Malaysia / Indonesia	Singh & Lai, 1937 in Mayer & Poljakoff-Mayber, 1963
Malvaceae	Montezuma speciosissima	≈9 m	Forest	Mexico	Marrero, 1942; Chittenden, 1951
Caricaceae	Carica papaya	3–6 m	Forest	Tropical America	Guppy, 1906
Connaraceae	Connarus grandis	Small tree	Forest	Tropics	Corner, 1976
Dipterocarpaceae	Dryobalanops camphora	≈30 m	Forest	Borneo, Sumatra	Guppy, 1906

(cont.)

II. NON-MANGROVES

Family	Species	Height	Habitat	Region	Reference
Fagaceae	*Quercus suber*	≈18 m	Woodland	Temperate, e.g. Portugal	Guppy, 1912
Lecythidaceae	*Barringtonia racemosa*	≈9 m	Coastal	Tropical Asia, Africa, Australia	Guppy, 1906; Ng, 1992
Leguminosae	*Inga* spp. (2)	Small tree	Swamp forest	Brazil, Venezuela	van der Pijl, 1982; McCormick, 1995
	Mora oleifera		Coastal	Tropics	Jiménez, 1994
	Pithecellobium racemosum	24–30 m	Forest	Tropical Asia, America, Africa, Australia	Leite & Rankin, 1981
Liliaceae	*Dracaena* sp.	Tree / shrub	Forest	Tropics	Guppy, 1906
Meliaceae	*Carapa guianensis*	≈18 m	Forest	Senegal–Angola	Guppy, 1912
Moraceae	*Artocarpus incisa, A. altilis*	≈15 m	Forest	Malaya, Pacific Islands	Guppy, 1912
	Morus latifolia	3–6 m	Forest	China	Ellis et al., 1985
Myristicaceae	*Myristica hollrungii*	>9 m	Swamp forest	E. Indies, Sri Lanka	von Teichman & van Wyk, 1991
Nyctaginiaceae	*Pisonia longirostris*		Swamp forest	Malaya	Corner, 1976
	P. brunoniana	To 6 m	Coastal forest	N. New Zealand	Salmon, 1991
SHRUBS					
Araceae	*Dieffenbachia longispatha*	1.8–2.4 m	Forest	Tropical America	Croat, 1983
Asteraceae	*Pachystegia insignis* var. *minor*	Stout spreading shrub	Rocky places coastal to montane	New Zealand, endemic genus	Simpson, 1979; Burrows, 1993
Caryophyllaceae	*Scheidea diffusa*		Mesic / wet forest	Hawaiian Islands	Farnsworth, 2000
Cornaceae	*Corokia macrocarpa*	Shrub or tree	Forest margins	New Zealand, endemic genus	Fountain & Outred, 1991
Euphorbiaceae	*Croton* sp.			Warmer regions	Guppy, 1906
Leguminosae	*Medicago lupulina*			S. Europe, Mediterranean	Sidhu & Cavers, 1977

Family	Species	Size/Habit	Habitat	Distribution	References
Rubiaceae	*Coprosma robusta*	To 6 m	Lowland forest	New Zealand	Burrows, 1993; 1995
	Myrmecoidia	Ant-nesting shrub	Forest	S.E. Asia, N. Guinea, N. Australia, Solomons	Guppy, 1906
CLIMBERS / VINES					
Araliaceae	*Hedera helix*	To 30 m	Woodland	Britain, Europe	Guppy, 1912
Celastraceae	*Salaciopsis ingifera*		Coastal forest	Tropical, Sarawak	Corner, 1976
Convolvulaceae	*Ipomoea glaberrima, I. peltata*		Coastal	Tropical	Guppy, 1906
Cucurbitaceae	*Sechium edule* – V	To 3.6 m		W. Indies, tropical America	Chittenden, 1951; Reiche, 1921 in van der Pijl, 1982
Smilacaceae	*Ripogonum scandens*	Stout liane	Lowland forests	New Zealand	Burrows, 1993; 1996
PALMS					
Palmae	*Cocos nucifera*	≈20 m	Coastal	Tropical	Guppy, 1906; 1912; van der Pijl, 1982
	Nypa fruticans	To 10 m	Saline mud	Tropical Far East	Tomlinson, 1971
	Ravenea musicalis – CV		River	Madagascar	Beentji, 1993
BAMBOOS					
Poaceae	*Dinochola* spp. – V		Tropical forest gaps		van der Pijl, 1982
	Melocanna bambusoides – V		Tropical forest gaps		Stopf, 1904; Arber, 1934
	M. baccifera – V				Void, 1962; McClure, 1966
	Melocalamus compactiflorus		Tropical forest gaps		Arber, 1934
HERBS					
Asteraceae	*Abrotenella linearis* var. *apiculata*	Rosulate herb	Montane to subalpine bog and herbfield	New Zealand	Simpson, 1979; Burrows, 1993
Malvaceae	*Hibiscus diversifolius*	Tall perennial		Tropics	Guppy, 1906

(cont.)

Appendix 2.3 (cont.)

II. NON-MANGROVES

Family	Species	Height	Habitat	Region	Reference
Polygonaceae	Fagopyrum esculentum			Temperate N. Hemisphere	Ellis et al., 1985
Scrophulariaceae	Euphrasia disperma	Succulent prostrate herb	Coastal to montane boggy places	New Zealand endemic	Simpson, 1979; D. Lloyd, pers. comm.
BULBOUS PLANTS					
Amaryllidaceae	Crinum capense syn bulbispermum – V; C. longifolium – V		Coastal near streams and marshes	S. Africa	Guppy, 1906, 1912; Skene, 1959 Barton, 1961
	Hymenocallis occidentalis – V		Coastal near riverbanks	S. America, Africa	Whitehead & Brown, 1940; van der Pijl, 1982; Elmqvist & Cox, 1996
	Nerine spp. – V		Steppes	S. Africa	van der Pijl, 1982; Elmqvist & Cox, 1996
Iridaceae	Neomarica gracilis			Brazil	Chittenden, 1951
Liliaceae	Chlorophytum sp.		Grassland	Tropics / subtropics	Chittenden, 1951; van der Pijl, 1982
MISTLETOES					
Loranthaceae	Amyema preisii				Lamont & Perry, 1977; Lamont, 1983
CARNIVOROUS PLANTS					
Lentibulariaceae	Utricularia nelumbifolia		Grows in water-containing leaf rosettes of large Tillandsias	Organ Mts., Brazil	Lloyd, 1976

Family	Species	Life form	Habitat	Distribution	Reference
	U. reniformis		Epiphyte on trees, *Sphagnum* bogs	Brazil	Lloyd, 1976
FRESH WATER					
Butomaceae	*Limnocharis*		Aquatic / marsh	Brazil	Chittenden, 1951; van der Pijl, 1982
Lemnaceae	*Lemna* spp.		Floating		Arber, 1920
MARINE / COASTAL					
Araceae	*Aglaodorum griffithii* – V		Tidal mudflats and freshwater habitats		van der Pijl, 1982
Chenopodiaceae	*Cryptocoryne* spp. – V		Tidal mudflats		van der Pijl, 1982
	Salicornia dolichostachya, S. europaea	Succulent sub-shrub	Salt marshes, coastal rock platforms	Temperate	Ranwell, 1972
Cymodoceaceae	*Amphibolus* (*Cymodocea*) *antarctica* – V		Sublittoral	Tropical / subtropical esp. Indian & Pacific oceans	Sculthorpe, 1967; Isaac, 1969; Ducker & Knox, 1976; Aston, 1977
	C. ciliata		Intertidal pools / sublittoral	E. Africa	Cox & Knox, 1988
	Thalassodendron ciliatum – V		Intertidal pools / sublittoral		Arber, 1920; Aston, 1977; Kuo & Kirkman, 1990
	T. pachyrhizum – V				Cox, 1991
Hydrocharitaceae	*Enhalus acaroides*		Sublittoral	Tropical / subtropical	Sculthorpe, 1967
	Thalassia hemprichii		Sublittoral	Indo-Pacific spp.	Sculthorpe, 1967
Poaceae	*Spartina versicolor*		Estuarine		Arber, 1934

Chapter 3

Specialized strategies I: seedlings in stressful environments

José M. Facelli

3.1 | Introduction

Seedlings are particularly susceptible to harsh conditions. Indeed, the seedling stage is considered to be the most vulnerable stage in the life of the plant (Stebbins, 1971; Fenner, 1987; Fenner & Thompson 2005) because even small reductions in biomass may lead to the death of the plant (Dirzo, 1985; Fenner & Thompson, 2005). Selection has favored strategies that reduce the high risk of the seedling stage primarily in two ways: first, maternal deployment of optimal amount of reserves to ensure maximum likelihood of seedling survival (Smith & Fretwell, 1974; Westoby et al., 1992; Leishman & Westoby, 1994a; Leishman et al., 2000), and second, timing of germination to avoid emergence during periods of high environmental stress as well as during transient favorable conditions too short to ensure postemergence survivorship (Grime, 1979; Baskin & Baskin, 1989, 1998; Fenner & Thompson, 2005). However precise the mechanism to adjust germination to low-risk conditions may be, many environments present inherently high risks for seedlings because the stress is chronic or favorable conditions are intermittent and uncertain (see Table 3.1). Because the ability of the seedling to accumulate or replace biomass decreases as the environment becomes less favorable, seedlings in stressful environments are at higher risk of mortality. Furthermore, when postemergence mortality is highly probable, avoidance of stress *per se* is not a viable strategy and seedlings are selected to tolerate stressful conditions. This can be seen in the seedlings of Australian arid lands that develop belowground storage bulbs, corms, and tubers (Pate & Dixon, 1982), the viviparous seedlings of mangroves (see Chapter 2), or the shade-tolerant seedlings of many rain forest trees.

Adaptations that allow young plants to tolerate stress can be different from those found in adults because the combinations of environmental variables experienced by seedlings and adults of the same population are often quite different. Many late successional forest trees, for example, may grow in an intermediate or high light environment as adults, but seedlings emerge in deep shade conditions. Desert

---

Table 3.1 | Components of stressors for seedlings[a]

A. *Intensity*
    Acute (shock responses possible)
    Chronic
    Gradients
B. *Scale*
    Temporal (e.g. seasonal, drought cycles)
    Spatial (e.g. horizontal, vertical gradients, microhabitat, habitat, landscape, region, biome)
    Individual, population, community, ecosystem
C. *Stage in life cycle*
    Seed, seedling, juvenile, adult
    Seedling stage most vulnerable
    Maternal effects
    Age of seedling organs
D. *Responses*
    Acclimation – phenotypic plasticity (some characters: plastic, reduced growth, reduced biomass)
    Adaptation – population-level genetic changes
    Combination
E. *Physiological action: synergistic, antagonistic, independent*
    Tolerance (e.g. mechanisms to maintain metabolic activity, salt glands in mangroves)
    Avoidance (e.g. reduction in metabolic activities, seedling dormancy)
    Combination
F. *Morphological*
    Aboveground (e.g. change in architecture with competition)
    Belowground (e.g. root growth in response to drought)
    Combination
G. *Other factors influencing stress responses*
    Evolutionary history
    Phylogenetic constraints
    Variable sensitivity (e.g. seasonal frost tolerance)

---

[a]Stressors are aspects of the environment that can induce change in physiology (Chapin *et al.* 1993; Nilsen & Orcutt, 1996).

plants often have deep roots and are able to access more reliable water sources than their relatively shallow-rooted seedlings that experience a far more dynamic soil water environment. Furthermore, small size and physiology of seedlings may preclude the development of some adaptive characters that are viable in larger plants. One such character is modularity, which allows large plants to partition risks and lose biomass without compromising individual survivorship. Mobilization of reserves from one part of the plant to another may enable the plant to survive challenges such as burial by sand or organ destruction by cold temperatures. Seedlings have limited capacity to achieve this or other size-dependent strategies.

As is the case with many concepts in ecology, precisely defining a stressful environment is difficult. Acknowledging this difficulty, Grime (1979) broadly defined stress as the external constraints that limit the rate of dry matter production for all or part of the

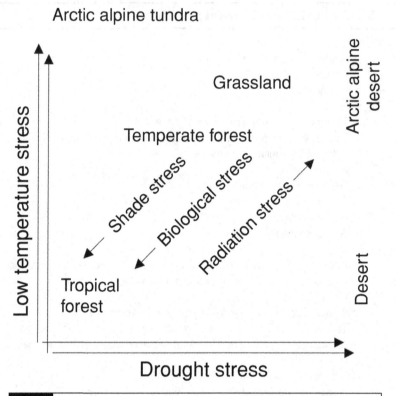

Fig. 3.1 Environmental stressors related to major vegetation types. The major stressors shown are low temperature and drought; other stresses, imbedded in the matrix, are also shown. From Osmond et al. (1987) reproduced with permission, copyright American Institute of Biological Sciences.

vegetation. Clearly, ecosystems differ in frequency and duration of these external constraints and the typical intensity of the events. A logical extension of Grime's stress concept would be to define an environment as extreme if stressful conditions are chronic. This, however, is riddled with pitfalls. While most ecologists would readily agree that the environment of some ecosystems (e.g. sand dunes, deserts, or salt pans) is extreme, it would be difficult for them to reach agreement on when stressful conditions that hinder seedling growth or increase the risk of seedling mortality are strong enough to be considered extreme. Environmental conditions exist as gradients (Fig. 3.1) and it is logically impossible to place limits on a continuous variable. Similarly, it is difficult to determine how much reduction in dry matter accumulation is required for a constraint to be deemed a stress. Furthermore, what is extreme for one species may be a perfectly suitable environment for another. Regardless of these difficulties in establishing end points, stress responses, which result in reduced biomass accumulation (e.g. Grime, 1979; Nilsen & Orcutt, 1996), do occur and should increase in importance as both (e.g. hot–cold, wet–dry) or just one (e.g. salinity, burial) end of the gradient is approached. The precise relationship between intensity and frequency of stressful condition and the degree of adaptations for stress tolerance is yet to be elucidated.

Stressors, environmental factors that can induce physiological change (Nilsen & Orcutt, 1996), vary in intensity, scale, with stage in the life cycle, and in kinds of responses that elicit this change (Table 3.1). The relative importance of various stressors also changes across major vegetation types (Fig. 3.1) and within habitats. In wetlands (not shown), for example, an inundation gradient results in species zonation due to differing tolerances of seedlings to physical stress (Parker & Leck, 1985), while in intermittent wetlands that experience drought for extended periods often exceeding a decade, wetland communities alternate with drawdown communities (e.g. Boulton & Brock, 1999).

The factors that reduce biomass accumulation in seedlings may be abiotic (physical factors include drought, extreme temperature, floods, or wind; chemical factors include soil pH or salinity) or biotic (e.g. competition, allelopathy, herbivory, or pathogens) (Grime, 1979; Nilssen & Orcutt, 1996). Sources of stress related to human activities include air pollution, heavy metals, salinization, sedimentation, or soil erosion triggered by overgrazing or vegetation removal, and, on a global scale, desertification and climate change. The sources of stress, regardless of type, may act independently, but they may also have synergistic or even antagonistic effects.

Chapin et al. (1993) discuss the evolution of stress tolerance and indicate that there seems to be a well-defined suite of adaptations that enable species to tolerate stress. These include characters such as low rates of growth, low photosynthetic and nutrient absorption rates, and large seed size (Grime, 1977; Chapin, 1980; Chapin et al., 1993; Grubb, 1998). Biological parameters are useful as objective measures to determine how extreme an environment is. For example, physiological syndromes that allow plants to maintain a positive carbon balance at very low irradiance or to exclude cations at extremely high salt concentrations could be used to identify adaptations of seedlings from systems where stress is chronic or extreme.

It is not clear whether or how adaptations of adult plants to stress relate to the ability of seedlings to sustain stress. As indicated above, because of their small size, seedlings may be more susceptible, therefore requiring more radical adaptations. Furthermore, some size-dependent stress-tolerance adaptations may not be effective in small seedlings because of allometric constraints. The relative scarcity of large comparative studies and the unsuitable design of many seriously limit our understanding of seedling adaptations. Researchers commonly compare a group of species, or even a single species, from habitats where a particular factor is at stress-inducing levels with those from another location where the stress is reduced or lacking. Abundant examples are found in the shade-tolerance literature where physiological characters of seedlings of a handful of species adapted to establish in shade are compared to species adapted to establish in gaps with high light availability. Frequently, the choice of species is quite arbitrary and the potential for confounding adaptations with phylogenetic trends detracts from the value of these studies. Few studies have been systematically designed to assess adaptations to stressful environments using Phylogenetic Independent Contrasts (PICs). In this approach, species in each group (e.g. salt tolerant or salt intolerant) are matched in a way that ensures that a series of phylogenetic lineages are equally represented in each group (i.e. stress tolerant and stress intolerant) (Burt, 1989). This procedure is essential to separate characters that are intrinsic adaptations to the stressful conditions from those inherent to a particular phylogenetic clade. Indeed, if any given phylogenetic lineage is more heavily represented in one of the groups under study, any difference between groups could be due to preserved ancestral characters, rather than to an actual adaptation evolved under the selective pressure of the stress factor under analysis.

This chapter considers stressful all environments that require special adaptations in the seedling stage to maximize survival, as discussed by Grime (1979), where resources or conditions impose severe limits to growth and establishment. However, it also considers other systems in which survivorship of seedlings is endangered by the high probability of small-scale disturbances on the order of centimeters to a few meters where adaptations *per se* may be lacking. While the possible sources of stress are many, the chapter focuses on how seedlings respond to shade, litter, drought, salinity, cold, and unstable substrates and discusses their interactive effects with biotic interactions, mainly herbivory and pathogenicity. Drought, salinity, and shade are examples of stresses requiring special adaptations while litter and unstable substrates may be considered stresses that do not. The goals of this chapter are to explore the boundaries and limitations, as well as extremes, of several environmental stresses encountered by seedlings and to identify critical gaps in our knowledge.

3.2 | Seedling establishment in dense shade

There is wide variation in the light environments encountered by seedlings after emergence. Some emerge into fairly open conditions (e.g. in early successional sites or deserts), but many emerge under established canopies that intercept various fractions of sunlight. In some environments, the light intensity is close to the photosynthetic compensation point of adult plants (i.e. the light intensity at which respiration equals photosynthesis). At this end of the light gradient, seedlings may be stressed because of reduced carbon assimilation (Chapter 8). Successful establishment requires that the seedling achieve a sustained positive carbon balance. Specifically, photosynthetic gains must be larger than losses due to respiration, herbivory, pathogens, and physical damage. Seed reserves are critical, but when the juvenile plant is subject to prolonged light limitation, its survivorship can be compromised. In addition, biomass destruction reduces the amount of nutrients available for the construction of new resource-acquiring organs. Both nitrogen, in particular, and energy are critical under shaded conditions for maximizing light capture. This section considers adaptations and responses that allow establishment under the extreme light environments created by dense canopies (defined variably in the literature from less than 10% to less than 3% of full sunlight).

Responses of seedlings to extremely low light have received more attention than any other type of stress largely because responses of seedlings and saplings are a key component of successional and gap dynamics in dense forests (Shugart & West, 1980; Bazzaz, 1996). A set of species (gap specialists) are only able to establish and grow when light availability is high. After a disturbance opens a gap, these species grow quickly to close the gap. Seedlings of another set of species can establish under deep shade and then grow slowly until they eventually replace the early successional, shade-intolerant species. Late successional species can gain access to the canopy using different strategies. Some establish in gaps of variable size and persist until the death of gap specialists has created openings, while others act like typical tolerant species (Grime, 1979), establishing under dense shade, growing slowly, and eventually reaching the upper part of the canopy, finally displacing early successional species. However, adaptations that make seedlings of some species tolerant to deep shade are far from clear.

Empirical studies indicate that the degree of differentiation between seedlings of early colonists vs. late successional is not clear-cut, but those of particular successional stages share characters. However, seedling performance seems to be more important than germination responses in establishing a successional pattern. For example, a study of the germination and performance of 26 species from Chilean temperate forests in gaps and deep shade showed that there was no correlation between germination requirements and seedling distribution in a successional sequence, with germination in shade not necessarily corresponding to shade tolerance in seedlings (Figueroa & Lusk, 2001). This suggests that seedling performance rather than germination regulation is the main determinant of seral position of species. Results for tropical forest species on Barro Colorado Island (Panama) also suggest that differential seedling survivorship is paramount. Mortality of seedlings growing in deep shade (2% of full sunlight) was substantially higher for gap species than late successional species (Kitajima, 1994).

Seedlings produced by larger seeds are assumed to be more shade tolerant (Leishman & Westoby, 1994b) because available reserves allow the seedling to survive for a longer period before achieving independence from reserves (Saverimuttu & Westoby, 1996a; Reich et al., 1998; Walters & Reich, 2000). This in itself could maintain the seedling for a limited time and would be advantageous only where gap creation is likely. Grime and Jeffrey (1965) demonstrated a direct negative relationship between rates of mortality of seedlings of trees and grasses grown in deep shade and the weight of their respective seed reserves. In several tropical rain forest studies, shade tolerance seems also related to seed size. Performance of tree seedlings from a tropical rain forest in Queensland (Australia) depended directly on seed reserves (Osunkoya et al., 1994), but for seedlings of trees in tropical west Africa, seed size was only weakly related to seedling responses to light availability (Agyeman et al., 1999) and seed size only marginally predicted ability to survive in shade (Saverimuttu & Westoby, 1996a). In North American hardwood trees, shade tolerance correlated positively with seed size, but this relationship did not exist for gymnosperms, probably reflecting different evolutionary constraints in angiosperms and gymnosperms (Hewitt, 1998). A comparison of seedlings of 23 species, with seed sizes ranging between 0.04 and 22.2 mg, found that the most likely mechanisms through which large seeds gave advantage to seedlings in shaded environments were longer survivorship and ability to grow in height and hence reach higher light availability (Leishman et al., 1994b). Differences were only found at very low light availability (1% of full sunlight). Seed size may not be an adaptation to shade per se, but it may enhance survival from desiccation, animal damage, or burial by litter (Metcalfe & Grubb, 1997). Increased allocation to roots increases depth of the root system, making the seedlings less susceptible to short dry periods.

The physiological adaptations that enable seedlings of some species to persist in extremely low light environments are far from clear. One of the common assumptions is that they have lower compensation and saturation points than sun-adapted seedlings. It is also assumed that they do not perform as well as shade-intolerant seedlings at high light intensities as a trade-off for better photosynthetic carbon balance at low irradiance (Walters & Reich, 1996). Seedlings of nonpioneer species (assumed to be shade adapted) from a west African tropical rain forest grew even at 2% full light but showed little response to increased irradiance (Agyeman et al., 1999). Contrary to expectations, the growth of seedlings of some pioneer species (assumed to be shade intolerant) was depressed at high irradiances (Agyeman et al., 1999). Seedlings of boreal tree species growing in 25% and 5% of full sunlight (representative of large gap and dense canopy understory radiation, respectively) showed no changes in rankings of growth at any light intensity, indicating that there was no trade-off between good performance at high or low light availability (Reich et al., 1998). These studies suggest that factors other than photosynthetic responses of seedlings determine successional patterns and that no single species may have all the characteristics considered typical of shade-tolerant or -intolerant seedlings (Hättenschwiler, 2001).

Seedling performance in various light environments could be affected by the partition of resources between different plant organs and differences in organ morphology. Resource allocation theory predicts that more resources should be allocated to organs that capture the most limiting resources. Accordingly, seedlings growing in shade should allocate more seed reserves to leaves, and the leaves should be broader and thinner (i.e. have a lower specific area) because this leaf morphology optimizes quantum capture at low light intensities (Bazzaz, 1996). Importantly, there should be significant interactions between physiological characters and optimal allocation to different organs. At lower light intensities, shade-intolerant species with high compensation light intensities would be more energy deficient than shade-tolerant species and would, therefore, allocate resources to build broader, thinner leaves. In one study, variation in leaf and root structure had a strong influence on relative growth rate (RGR), leading the authors to suggest that resource allocation may be more important than other physiological traits (Reich *et al.*, 1998). Low irradiance reduced allocation to roots and increased allocation to shoots, but had no effect on the allocation to leaves in all species. Under low light intensities, all species had lower specific root length and increased leaf area per unit of plant mass, consistent with allocation theory (Reich *et al.*, 1998). These different allocation patterns could act to protect plants against other interactive stress factors. Kitajima (1994) observed that seedlings of shade-intolerant species had higher RGR and deeper roots, but lower root:shoot ratios, specific leaf mass, and LAR. No single character could explain higher mortality under low irradiance conditions. The deeper roots, tougher leaves, and higher wood density may, for example, offer protection against herbivores and the lower RGR of seedlings of these species may result from the higher cost of these characters (Kitajima, 1994). Myers and Kitajima (2007) studied the role of nonstructural carbon reserves in seedlings of tropical rain forest trees and found that seedlings with low carbon reserves were far less likely to survive. The storage of carbohydrate in stems and roots seems to enhance survival in deep shade by buffering the seedlings against periods of biotic and abiotic stress.

Given the importance of nitrogen for photosynthetic activity and the fact that N stores in seeds are often low, it is expected that differences in N economy may be involved in seedling adaptation to low light environments. Nitrogen addition, however, appeared to have little or no overall impact on seedling growth of hardwood species with a range of seed sizes from temperate forests of northeastern North America (Walters & Reich, 1996, 2000), but the small-seeded, broad-leaved, shade-intolerant species only had high survivorship and RGR at higher N availability.

The scarcity of studies using PIC and the lack of a coherent protocol (e.g. variation in light intensities used, length of the study, growing conditions) hinder generalization. Future studies should use a range of light intensities, and investigate how N affects the fate of seedlings growing under deep shade and its interaction with seed size.

3.3 | Effects of litter on seedling establishment

After emergence from the soil, many seedlings may have to continue to grow through litter. Litter presents different challenges to seedlings emerging from under deep litter because of its unique physical properties. While moderate amounts of litter have no effect on emergence (or even a positive one), accumulation over 200 g m^{-2} has negative effects on seedling emergence and survivorship (Sydes & Grime, 1981) with the effects becoming stronger as the amount (but not necessarily the thickness) of litter increases (Facelli & Pickett, 1991a). The thickness and physical properties of litter layers are highly variable and may determine which species establish in a given site (reviewed in Facelli & Pickett, 1991a). The type of litter is also important (Xiong & Nilsson, 1999) because of different light extinction coefficients (Facelli & Pickett, 1991b) and different degrees of impedance (Grime, 1979). Even the previous litter history can be important. In systems with snow accumulation, litter may be compressed into a mat that hinders seedling emergence. Hyphal growth that holds together fragments of dead leaves or bark can also increase the physical impedance of litter to seedlings. A meta-analysis of published results (Xiong & Nilsson, 1999) indicated that litter overall has a negative effect on seedling emergence, the exception being systems with slight litter accumulation and where water limits germination (Fowler, 1986; Facelli & Ladd, 1996; Xiong & Nilsson, 1999). While some of the effects of litter are mediated by changes to the environment that directly affect germination, many other mechanisms hinder seedling emergence. For example, reduction in the day–night fluctuations in temperature produced by thick layers of litter inhibits the germination of many seeds whose dormancy is broken by this disturbance-related cue (Grime, 1979; Grime et al., 1981; Fenner & Thompson, 2005) and many ruderal species are inhibited by the presence of litter because of reduced light (Facelli & Pickett, 1991b). Thus, the presence of litter can strongly inhibit establishment of species adapted to disturbances (Grime, 1979; Fenner & Thompson, 2005). In addition, litter may release leachates that may have allelochemical effects on seedling establishment (Bosy & Reader, 1995; Olson & Wallander, 2002).

The fate of seedlings is affected by the placement of the seed at the time of germination. Establishment is more successful when the seeds are at the soil surface because roots in litter can fail to obtain enough water and nutrients to sustain growth. Seedling survivorship of Eucalyptus obliqua (Myrtaceae) was much lower when seeds were placed on top of than below litter (Facelli et al., 1999). However, if litter is composed of broad leaves and pieces of bark, more seeds may be retained on top than if composed of fine materials that allow seeds to move to the ground surface (Peterson & Facelli, 1992; Facelli & Kerrigan, 1996). Size and shape of seeds are important in determining the placement of the seedling above or below the litter layer and may affect the usefulness of large seeds to secure seedling establishment in environments with dense litter accumulation. While large seeds may provide more reserves that allow successful seedling emergence through a dense litter mat, their large size may, however, reduce the chances of the seeds reaching a safe germination site beneath the litter layer before seed predators find them.

Litter can hinder seedling emergence either through physical impedance to growth and/or shading. Seedlings beneath litter expend energy in stem elongation to place their cotyledons or first photosynthetic leaves above the litter layer. This usually reduces the allocation to roots (Peterson & Facelli, 1992) and extends the time before the seedling becomes independent of seed reserves. Under deep layers of litter, seedlings from small seeded plants often exhaust their reserves and die before emerging. Even if seedlings have enough reserves to emerge, they have reduced resources available for allocation to resource-acquiring organs, which may make the seedling more susceptible to negative biotic effects. Seedlings that emerge more slowly are at a competitive disadvantage (Facelli & Facelli, 1993). Insect herbivory (Facelli, 1994) and pathogen-induced mortality (Facelli et al., 1999) also increase. Because of impediments to germination and emergence, in productive woodlands and grasslands, litter may accumulate to the point where it prevents establishment of new individuals until fire, flood, animal scratching, or other disturbance disrupts the litter mat.

Because the shading effects of litter are similar to those of an established canopy (although with a different spectral quality; see Vazquez-Yanes et al., 1990), some adaptations for canopy shade tolerance also contribute to successful emergence through litter. However, while light intensities under litter are lower than under most canopies, the litter layer generally allows emergence in a few hours or a few days, while shaded conditions under canopies can produce chronic carbon starvation in seedlings. Litter, therefore, can have stronger effects through impedance to emergence. Deep litter layers, particularly when matted by compaction after snow packing or by fungal hyphae, can hinder and delay emergence substantially. Little information exists regarding adaptations to the mechanical impedance of litter to seedling emergence, although Grime and collaborators (see review in Grime, 1979) have shown that shoot morphology determines the ability of clonal herbaceous plants to resprout from underneath a thick litter layer. The extent to which seedling shoot morphology contributes to the ability to penetrate thick matted litter layers remains to be investigated.

3.4 | Seedling establishment in dry environments

While most seedlings may face temporarily dry conditions, the risk posed by drought increases with aridity. Xeric environments provide fewer and more unpredictable germination opportunities. Thus, seeds of plants from xeric environments have complex adaptations that prevent germination unless conditions are likely to favor seedling survivorship. These adaptations not only include sensitivity to water availability, but often also seasonal cyclic dormancy. Baskin *et al.* (1993) found that seeds of winter annual plants from the deserts of southwest North America had low germination under controlled conditions when recovered from the field during the summer, but germinated readily when they were retrieved after the hot season ended. The induction of dormancy proved seasonally cyclic, indicating adaptations to avoid germination in the hottest part of the year. Facelli and Chesson (2008) found that time of retrieval and temperature of incubation determined germination fraction in *Carrichtera annua* (Brassicaceae), a winter annual originally from the Middle East and now an invader in South Australia. However, seedlings of summer annual plants from areas with reliable summer rains are often able to emerge at high temperatures (Gutterman, 1990). The germination fraction of arid land annual plants is usually fairly low, buffering the population against total establishment failure (Clauss & Venable, 2000; Facelli *et al.*, 2005). Other species have alternative adaptations to ensure seedling establishment only when water availability is high and likely to continue. In species such as *Erodiophyllum elderi* (Asteraceae) (Emmerson 1999), *Asteriscus pygmaeus* (Asteraceae) (Gutterman & Ginott, 1994), and *Anastatica hierochuntica* (Brassicaceae) (Friedman *et al.*, 1981), seeds are retained in persistent structures with hygroscopic structures. Seedlings only establish when the seeds are released following large and sustained rainfalls. Because only a fraction of the seeds are released with each rainfall event, the mechanism also serves to spread risk of seedling mortality.

Overall, the seedling establishment of desert annuals requires relatively high water availability (Mott, 1972) and most species have adaptations that allow them to time germination to the most favorable and often most predictable part of the year (Elberse & Breman, 1989; Gutterman, 1990, 2002; Briede & McKell, 1992; Baskin *et al.*, 1993; Pake & Venable, 1996; Gutterman & Shem-Tov, 1997a; Clauss & Venable, 2000).

Because of variable seed availability and mechanisms that time germination to extraordinary high rainfall events, the emergence of seedlings of perennial species is usually episodic, creating often well-defined cohorts. In spite of this timing, they will eventually experience dry conditions. Fast root extension is a principal strategy in habitats where the wet front in the soil column recedes at very high rates. Nicotra *et al.* (2002) assessed the patterns of root architecture of several species growing along rainfall gradients, using PICs, and growing the seedlings under standard conditions. Seedlings of species restricted to areas of low rainfall allocated proportionally more resources to the main root axis, but the axis was thinner in diameter. While root systems of species from higher rainfall areas had higher elongation rates because of higher RGR, elongation occurred along more root tips.

Overall, this indicates that plants of arid lands have a genetic component underlying deeper and faster root production by seedlings. In another comparative study, root depth of species from dry habitats tended to increase in drying substrates while the root depth of species from wet environment decreased (Liu *et al.*, 2000). Plasticity of root growth was correlated in another study (Reader *et al.*, 1993) to the ability to sustain shoot growth in drying substrates, suggesting that plasticity in rooting depth contributes to drought tolerance. Because this study compared plants from a single mesic system, conclusions may not be extrapolated to other systems.

Seedling establishment in xeric systems often requires facilitation (see Chapter 15). Canopies of larger plants have a protective effect on seedling emergence and survivorship, as demonstrated initially for the establishment of saguaro (*Carnegiea gigantea* Cactaceae) (Turner *et al.*, 1966). The seedlings of other species of cacti are quite sensitive to elevated soil temperatures and may suffer heat stress and, as a consequence, die (Turner *et al.*, 1966; Valiente-Banuet & Ezcurra, 1991). Many species establish better or exclusively under the canopies of trees and shrubs. As conditions become more severe, shade has an increasingly greater effect on seedling survival (Greenlee & Callaway, 1996). However, several processes may contribute to these patterns, including shade and consequent reduction of extreme temperatures, but also changes in soil properties and hydraulic lift (Franco & Nobel, 1988; Nobel, 1989). Few studies have disentangled the various mechanisms. Because organic matter is much higher, soil under tree canopies can retain more water, which enhances seedling establishment (Facelli & Brock, 2000). Hastwell and Facelli (2003) studied the survivorship and growth of *Enchylaena tomentosa*, a chenopod shrub found almost exclusively under the canopies of trees in the arid lands of southern Australia, using artificial shade structures. Survivorship during the summer increased substantially if the seedlings were protected from full sunlight. Thus, shade assists in seedling establishment independently of the soil changes produced by the presence of trees. Some facilitative effects are not directly mediated by alleviation of water stress, but by interaction with photoinhibition (Werner *et al.*, 1999).

Due to evolutionary traits that control germination and the strong constraints to survivorship, seedling establishment is an important determinant of community structure of both annual and perennial plants in arid lands. For annual plants, timing of rainfall may determine the community structure in any given year, and variation from year to year in rainfall patterns can produce very different assemblages (Facelli *et al.*, 2005). For perennial plants, the paucity of seedling establishment opportunities creates strong age structure in populations. For example, Westbrook (1999) determined that over 160 years (with available climate data) there had been only seven opportunities for the successful establishment of *Myoporum platicarpum* (Myoporaceae) seedlings in arid lands of southern Australia. A similar pattern has been proposed for *Acacia papyrocarpa* (Leguminosae) (Lange & Purdie, 1976). Because of different longevities, changes in the frequency of establishment events, such as those possibly triggered by climate change, could modify the relative abundance of these two species, which have very different ecological roles (Facelli & Brock, 2000).

3.5 | Seedling establishment in saline environments

Salinity affects plant growth in three ways. First, it lowers the soil water potential, making water less available for the plant. Second, salinity produces inorganic imbalances, which result from uptake of toxic ions at external concentrations that overcome membrane selectivity. Third, it produces nutritional imbalances due to competition for transporters at the membrane level between essential nutrients and ions in high concentration in the soil (e.g. Na^+ vs. K^+). The relative importance of changes in water balance and mineral nutrition on seedlings varies depending on species and environmental conditions.

Plants adapt to saline conditions through a combination of mechanisms, including ion dilution (succulence), storage of toxic ions away from metabolically active components, extrusion, and/or increased ion selectivity. Wet and dry saline systems differ in quite substantial ways in the possibilities for adaptations for salt tolerance. In wet saline systems, such as salt marshes or mangroves, plants have access to ample water supply, which makes the use of water-intensive strategies such as salt extrusion more effective. Wet systems, however, add other stresses such as anaerobicity, sedimentation on seedlings, or wave damage. In dry saline systems (saltpans, saline deserts), constraints to plant growth are more extreme and, consequently, the range of viable strategies is far more limited. Salts tend to accumulate in the top few centimeters due to capillary ascent of salty water and subsequent evaporation. These conditions are particularly harsh for seedlings, which may have the totality of the root system in the driest and saltiest part of the soil.

Germination and seedling growth, which depend on absorption of water by the seed, require a gradient in water potential that is lower in the seed than in the surrounding environment. Under saline conditions, the external soil water potential is lowered by the osmotic effect of ions, making water absorption by seeds more difficult. Some plants solve this by accumulating ions in the seed or in the viviparous seedlings of some mangroves. For example, seedlings of *Avicennia marina* (Avicenniaceae) collected from trees growing in tidal flats had an osmotic potential lower than that of seawater, which would allow water movement into the seedling when immersed in seawater-infused substrate (Downton, 1982). K^+, rather than Na^+ (which is more abundant in seawater), was the most abundant inorganic ion. Cl^- concentration was also low, suggesting that organic anions were probably used by the mother plant to balance the ionic composition of the seedling. The resulting ionic composition in the seedling would reduce the problem created by toxicity of some ions in young tissues, and indicates selective accumulation of ions. This active control of ionic composition of the seed requires a substantial maternal investment. When the seedlings became independent, they always maintained a water potential 2 MPa lower than the external solution; however, they used much higher concentrations of Cl^- and Na^+ to generate a low osmotic potential (Downton, 1982). This suggests that seedlings have much less selective ability to prevent Na^+ entry when not under maternal control, which may reflect a reduced energy availability in the seedling. Young seedlings of *Suaeda maritima* (Chenopodiaceae) also maintain a finely regulated osmotic gradient, not only between plant and soil, but between different organs, which ensures continued water uptake and allocation to expanding organs. Leaf osmotic potential declined as the salt concentration increased from 10 to 100% seawater. Furthermore, the seedlings can maintain a gradient with the lowest osmotic potentials in apical leaves, regardless of the external osmotic potential (Clipson *et al.*, 1985).

Salinities that prevent germination do not irreversibly damage seeds (Ungar, 1996), suggesting that in saline environments seedlings are more likely to emerge during periods of relatively low stress. *Typha domingensis* (Typhaceae) can only establish as seedlings during periods of reduced salinity, while adult plants or young rhizome sprouts can continue to grow at much higher salinities (Beare & Zedler, 1987).

Seedlings may actively extrude salts, a mechanism of great importance for adult plants, such as mangroves and several species of *Atriplex* (Chenopodiaceae). Salt exclusion has also been documented in seedlings of the mangroves *Avicennia marina* and *Aegiceras corniculatum* (Myrsinaceae) (Ball & Farquhar, 1984; Ball, 1988). The ability to exclude salts, however, can saturate at relatively low salt concentrations: exclusion increased from 90% at salinities of 50 mol^{-3} to 97% with a tenfold increase in salinity.

Seedlings are probably quite susceptible to salinity because of their small size. Adaptations to tolerate salinity are either energetically expensive or are more effective when plants are larger because they are volume dependent. Quite often, the tolerance to salinity increases as the seedling develops and grows, reaching its maximum in adult plants. For example, salt tolerance increases from germination to advanced sapling stages in *Prosopis flexuosa* (Leguminosae) (Catalan *et al.*, 1994). In mangroves, salinities as low as 50% seawater strongly limit seedling growth, although it can continue at slow rates in 100% seawater (Downton, 1982). These levels of salinity have a much smaller effect on well-established saplings. In viviparous species [e.g. *Rhizophora* spp. (Rhizophoraceae)], germination, which occurs while the seed is still attached to the plant, was suggested to be a period of acclimation to higher chloride concentrations; older, dispersed seedlings grew better than smaller, predispersal stage seedlings in high saline environments (Smith & Snedaker, 1995). In *Atriplex halimus*, salinity levels that strongly reduced germination slightly reduced seedling growth, but stimulated adult growth (Mayer &

Poljakoff-Mayber, 1963). The salt tolerance for growing *Atriplex patula* adults can be one to two orders of magnitude larger than for germination or seedling growth (Ungar, 1996). Rhizome sprouts of *Typha domingensis* can grow under salinity conditions that completely preclude seedling establishment (Beare & Zedler, 1987).

This is not universal, however, because other studies have demonstrated that seedlings of some species have similar or higher tolerance to salinity than adult plants. *Avicennia maritima* was reported to exhibit its greatest tolerance to hypersalinity as seedlings and saplings (McMillan, 1974) while *Atriplex nummularia* (Uchiyama, 1981) and *Kosteletzkya virginica* (Malvaceae) (Poljakoff-Mayber *et al.*, 1994) had similar sensitivity to salinity as seedlings and adults. The mechanism behind the variation in sensitivity of the seedling stage remains obscure and further research on what characteristics of the seeds or adaptations of the seedlings contribute to these differences is warranted.

The effect of salinity on seedling establishment seems to have important effects on determining community structure because differences in tolerance to salinity at the seedling stage, rather than tolerances of the adult plants, can contribute to species distributions along salinity gradients. A comparative study of *Prosopis argentina* and *P. alpataco* suggested that the differences in distribution are due mostly to differences in tolerance to salinity during the seedling and juvenile stages, rather than to their responses of adult plants (Villagra & Cavagnaro, 2005). Similarly, differences in seedling tolerance may contribute to mangrove zonation (Ball, 1988). However, in some cases, the effect of differential establishment in different parts of salinity gradients may be obscured by subsequent clonal growth, a common strategy in wet saline environments. In these species, seedlings establish almost exclusively at times or sites of low salinity, but once established, the plants can continue to expand into sites where seedling establishment rarely occurs. Adult plants that have higher tolerance to salinity can persist long after the benign conditions for establishment have disappeared.

3.6 | Seedling establishment in cold environments

There is wide variability in the sensitivity of plants to cold conditions. While adults of some cold-tolerant plants can be chilled to temperatures below $-30\,°C$ without suffering permanent damage to the tissues, some tropical species are damaged when temperatures drop below $10\,°C$ because of increased viscosity of fatty acids in their membranes (Crawley, 1997a). In plants from temperate environments, below-freezing temperature damage is caused when ice crystals form in cells or intracellular spaces, producing permanent damage to tissues or organs. Frost-tolerant species contain organic solutes that lower the freezing temperature of water in solution (Alberdi &

Corcuera, 1991) and have tissue structures that accommodate ice crystals with minimum damage to the cells. Freezing also disrupts the continuity of the water column in xylem (cavitation), potentially causing dessication death of plant tissues or organs that become disconnected from the water flow from the roots. The ability to restore the continuity of the water column in the xylem (decavitation) is an important adaptation to cold. Faster decavitation of conifer xylem is one of the factors contributing to their dominance in colder environments (Alberdi & Corcuera, 1991). Accumulated resources that allow for faster regrowth in spring or after cold damage are also important adaptations to cold tolerance (Crawley, 1997a).

Cold conditions can affect seedlings by reducing growth rate, producing frost damage, and inducing photoinhibition. Under some conditions, seedlings may be damaged by soil frost heaving. Many plants growing in environments subject to potentially damaging cold conditions for seedlings have evolved strategies to prevent germination in autumn or early winter. Indeed, cold storage (stratification) is a key dormancy-breaking mechanism for species of cold environments. Response to cold storage and high minimum germination temperatures ensures emergence only after the risk of frost has declined (Fenner & Thompson, 2005), increasing the chances of seedling survivorship.

As with other stressful conditions, the seedling stage is often the most susceptible to cold. Seedling biomass is confined completely within a few centimeters of the air–soil interface, a space that experiences the greatest extremes of temperature and is much colder at night than areas above or below it. Furthermore, seedling stems lack the insulation that bark provides to adult plants. While cold-induced photoinhibition effects on seedlings have been widely studied, other physiological effects of cold temperatures are still poorly understood (Crawley, 1997a). Tolerance to cold conditions seems to increase with age, but the mechanisms involved are not well understood. While adults acclimatize to low temperatures, it is not clear how quickly or efficiently this happens in seedlings. Cold damage occurs when sudden chills affect seedlings. In some temperate species, seedling mortality during the first winter can be as high as 80% (Osumi & Sakurai, 2002). Carbon limitation, low temperature injury, and lower photosynthetic capacity may all contribute to poor seedling performance and mortality (Osumi & Sakurai, 2002; Piper *et al.*, 2006).

While frost may damage or kill seedlings directly, seedlings exposed to cold also become susceptible to cold-induced photoinhibition, which can subsequently reduce growth or even endanger survivorship if the seedling is challenged by other stresses or negative biotic factors. Photoinhibition occurs when stressed plants are exposed to high irradiance and the mechanisms that protect the photosynthetic apparatus against excess energy are overcome (Lambers *et al.*, 1998). The excess energy can cause temporary or permanent damage to the photosynthetic apparatus (Osmond, 1981). Because of their small photosynthetic capacity and lack of substantial reserves, seedlings are particularly susceptible to this effect (Blennow & Lindkvist, 2000). Ball *et al.* (1991) found that seedlings of *Eucalyptus pauciflora*, growing close to its altitudinal limit, suffered severe reduction in quantum yield due to exposure to high irradiance following low temperatures. Seedlings growing in the shade were not affected although they were subjected to the same low winter temperatures. The effects of photoinhibition were strong enough to reduce seedling survivorship in exposed sites. Subsequent experimental studies demonstrated a causal link between shade and protection against photoinhibition (Ball *et al.*, 1997; Egerton *et al.*, 2000). Similarly, at the limits of their range, conifer seedlings are favored in shaded environments that protect seedlings from cold-induced photoinhibition (e.g. Alberdi & Rios, 1983), making facilitation an important factor in their establishment in extreme cold environments.

The differential susceptibility of seedlings to cold conditions contributes to species altitudinal and probably latitudinal distribution. Often seedlings are unable to establish under conditions that adults can tolerate. Damage to seedlings of the herbaceous perennial *Digitalis purpurea* (Scrophulariaceae) by cold was shown to be the main determinant of its altitudinal distribution in Germany (Bruelheide, 2002). Overall, there is ample evidence that direct and indirect effects of cold determine seedling establishment, and, subsequently, the position of the tree line (Ball *et al.*, 1991; Germino *et al.*, 2002).

While little information exists on adaptive reproductive strategies of species in extremely cold systems, large seed size may contribute to seedling success. Osumi and Sakurai (2002) attributed the high mortality of *Betula maximowicziana* (Betulaceae) seedlings to small seed size, while experimental work on seedlings of *Quercus rubra* (Fagaceae) demonstrated that growth and survival after exposure to freezing temperatures increased with seed size (Aizen & Woodcock, 1996).

In extreme cold environments, seedling establishment can be a rare occurrence. Cooper *et al.* (2004) reported that only a small fraction of species found at a field site in the Arctic territory of Svalbard (Norway) could establish as seedlings, suggesting that establishment of the others occurred during periods or years of more benign temperatures, and most populations persisted in the site mostly through longevity and clonality. A study of the genetic structure of *Carex curvula* (Cyperaceae) in the European alpine zone found that clones were approximately 2000 years old (Steinger *et al.*, 1996). Similar ages were estimated for clones of *Juniperus sabina* (Cupressaceae) in Mongolia, which showed no sexual reproduction (Wesche *et al.*, 2005), leading the authors to speculate that seedling establishment had only occurred in the distant past. Because harsh cold conditions limit plant growth, this increases the time it takes for plants to reach the reproductive stage and to accumulate resources for seed production. Pre-dispersal abortion of seeds seems to increase toward the cold limits of a species range (Garcia *et al.*, 2000). The combination of constraints on seed production with the low success of seedling establishment may have contributed to the evolution of clonality as the main mechanism for persistence of many species in Arctic and alpine environments. Sexual reproduction can be limited to rare favorable conditions.

3.7 | Physical damage to seedlings

Seedlings often establish in systems where physical damage can be an important source of mortality. Small-scale disturbances a few centimeters to a meter and of low intensity may produce only a slight reduction in biomass and may not have any obvious direct impact on the whole plant community. However, they can have profound effects on the survivorship of seedlings and hence can have, in the long term, profound effects on community structure. In environments as contrasting as sand dunes and snowdrift areas, seedlings can be buried or exposed, while in tropical rain forests a large number of seedlings can be killed by falling limbs and epiphytes (Clark & Clark, 1989). In cold systems, heaving of freezing soil can uproot seedlings, sometimes producing substantial mortality and damage (Goulet, 1995).

In open systems, such as young sand dunes, with loose particles on the surface, wind and water erosion can affect seedlings through repeated root denudation or shoot burial, sometimes alternating, thereby strongly affecting survivorship. Exposed tissues are subject to the abrasive effect of blown sand. Many plants have evolved adaptations to withstand these conditions and some, such as *Ammophila arenaria* (Poaceae), even seem to perform better when subjected to moderate sand accumulation (Eldred & Maun, 1982).

Once dispersed, seeds can be buried at variable depths. Maun (1985) found that depth of burial of the same cohort by the end of the winter following dispersal varied between 1 and 29 cm. Depth of burial controls germination and burial beyond a threshold induces enforced dormancy, the degree of which can increase with depth (Pemadasa & Lovell, 1975; Zhang & Maun, 1994). Depth of burial has, however, a stronger effect on emergence than on germination (Maun, 1998), and seedlings of various sand dune plants have different abilities to emerge from deep burial (Maun, 1998). The main determinant of seedling ability to emerge from depths below 5 cm is seed size or endosperm reserves (Weller, 1985; Zhang & Maun, 1990). This effect of seed size on seedling emergence occurs both between and within species (Zhang & Maun, 1993).

After emergence, seedlings can still be subject to substantial substrate movement. Partial exposure of roots often kills the seedling because the length of exposed root cannot obtain water, and becomes desiccated. Any secondary roots connected to the shoot lose function. Thus, exposure of even a small fraction of the root system may lead to seedling death. On the other hand, a gradual accumulation of sand on shoots has little effect on survivorship, even if some photosynthetic area is covered, because seedling shoot growth and elongation of sand dune plants usually can match the rate of sand accretion

(Maun *et al.*, 1996). When accumulation happens at higher rates or in one large event, the fate of the seedling depends on the degree of burial (Maun, 1998). While partially buried seedlings tend to survive, albeit growing more slowly, fully buried seedlings can survive only if they have enough reserves to reemerge or if wind exposes the shoot before all the reserves are consumed (Harris & Davy, 1987). The ability of seedlings to survive in darkness can be a good predictor of their ability to survive burial, although some species have differential ability to grow when buried irrespective of their shade tolerance.

A moderate level of sand accumulation seems to promote the growth of many sand-adapted species (reviewed in Maun, 1998). While these effects have been mostly studied for mature plants, there is strong indication that they also act at the seedling stage. Several factors may contribute to this response, including reduced root temperatures, accumulation of nutrient-rich organic and inorganic material, increased soil volume for root exploration, sudden placement of the roots in a deeper layer better able to retain water, escape from pathogens, or better growing environment for mycorrhizal fungi.

Soil heaving in cold, wet environments also creates an unstable substrate for seedlings. Seedlings can be heaved out of the soil by the formation and growth of ice crystals near the soil surface. This phenomenon requires freezing air and soil surface temperatures, high soil water content, no snow cover, and little or no decomposing organic matter (Miller, 1972). The consequences for seedling survivorship in forestry and agricultural situations have been widely studied (reviewed in Goulet, 1995). Seedling damage can range from reduced growth, due to root tearing and partial exposure of roots, to death (Legard, 1979). Root characters such as extent of the root system, depth of branching, mechanical resistance to tension, and ability to regrow after damage seem all to contribute to tolerance of frost heaving (Goulet, 1995). For example, *Medicago* (Leguminosae) seedlings with long, unbranched taproots are more likely to be heaved out from the soil than seedlings with more lateral roots (Shimada *et al.*, 1982). While the importance of frost heaving has been largely documented in agricultural and forestry studies, it can have important effects on the establishment of seedlings in systems where ground cover is sparse. Using artificial seedlings as models, McCarthy and Facelli (1990) found that a large proportion of the models placed in an early successional field, but not those in a late successional site or in an old growth forest, were frost heaved during one winter. While the technique is likely to overestimate the incidence of heaving damage to real seedlings, results indicate that frost heaving can be extensive, and therefore, sizable numbers of seedlings may be affected. Furthermore, visual observations confirmed the importance of frost heaving for actual seedlings.

In some systems such as tropical rain forests, falling limbs and epiphytes can cause severe damage or kill significant numbers of seedlings. Because it can be difficult to tease apart mortality produced by biotic effects, such as herbivory or pathogens, from that produced by physical damage, Clark and Clark (1989) used artificial plastic seedlings to assess physical damage and found that falling limbs and epiphytes damaged 20% of the models at the end of one year. In a subsequent study using actual seedlings, the authors confirmed that damage by falling debris accounted for at least 20% of the mortality of stems less than or equal to 1 cm in diameter (Clark & Clark, 1991). The authors concluded that physical damage constitutes an important ecological factor in tropical rain forests, and speculated that differences in susceptibility to falling debris can affect floristic composition. This is dramatically illustrated by *Socratea exorrhiza* (Palmae), the "walking" palm from the Peruvian Amazon. When logs or limbs fall on seedlings or saplings of this species, they can "crawl" from under fallen logs by elongating their stems (Bodley & Benson, 1980). The new trunk is elevated by the developing stilt roots, and the old stem and roots rot away. Bodley and Benson (1980) reported that almost half of all juveniles in a half-hectare plot had "walked" between 10 and 200 cm from the spot they had established originally. Because this behavior seems to involve specific adaptations, it can be assumed that the damage of falling debris may be important enough to have triggered evolutionary changes, but this requires more study.

3.8 | Interactive effects of stress agents and biological interactions

All seedlings are subject to potentially lethal or damaging effects of biotic interactions such as herbivory, parasitism, and competition. While chronic stress may reduce the effect of competition in some systems because it prevents the overlap of resource acquisition zones (Grime, 1979), seedlings subject to abiotic stresses may be more susceptible to the negative effects of herbivory and parasitism. Competitive effects during or after a stressful period in otherwise productive systems may also further reduce seedling growth and increase mortality.

Pathogens and herbivores may have stronger effects in stressful environments because they can destroy biomass that is difficult to replace under growth-limited conditions. The high concentration of structural secondary compounds in plants from stressful environments indicates a strong selective pressure of herbivores (Chapin et al., 1993) and probably pathogens. Seedlings of trees seem to be more susceptible to herbivory when growing in shade (Dirzo, 1984; McGraw et al., 1990), not only because of limited ability to accumulate biomass (McGraw et al., 1990), but also because shade may reduce the ability of the seedling to produce chemical defences (Baraza et al., 2004). Seed size and concentration of secondary compounds were identified as important determinants of herbivory impact on young seedlings (Dirzo, 1984). Herbivory has been also shown to reduce survivorship of seedlings that are exposed after suffering from sand burial with the likelihood of death for a given degree of defoliation increased with the time of burial. Shade may also predispose seedlings to the damaging effects of pathogens. Damping off affects more seedlings of smaller seeded trees when growing under the shade of litter (Grime & Jeffrey, 1965), although increased humidity around the base of the stem may also contribute to mortality (Facelli *et al.*, 1999). Similarly, Augspurger (1984a) found that seedlings of shade-tolerant species from Barro Colorado Island (Panama), when growing in deep shade, had a lower proportion of seedlings dying from disease than shade-intolerant ones.

A consequence of these interactive effects between physical stress and negative biotic interactions is that the evolution of specialized strategies to cope with specific stressors may not ensure successful establishment. Instead, a general adaptation, such as large seed size or a syndrome of adaptations that, in general, protect seedlings against the compounded effects of combined stress agents, biomass destruction events, and limited ability to replace biomass and resources are important (Grime, 1979; Chapin *et al.*, 1993).

3.9 | Overview: adaptations of seedlings to stressful environments

The seedling stage is subject to very high mortality, especially in stressful environments. Clearly, timing of emergence to periods of minimal stress, the ability to survive under severely limiting growth conditions, and the ability to grow quickly within environmental constraints are critical. Various morphological and physiological adaptations have evolved to tolerate extreme and chronic environmental constraints like shade, drought, cold, and salinity. However, the evidence from a variety of systems where stress is prevalent suggests that a common feature for successful seedling establishment relies on large seed size representing abundant seed reserves available to sustain the seedling (Kidson & Westoby, 2000).

It has been proposed that a general mechanism for large seeds to increase seedling survival is based on the slow use of a larger amount of reserves, thus leaving uncommitted resources that can be mobilized during periods of negative carbon balance, either because of stressful conditions, loss of biomass due to herbivory, or combinations thereof (Kidson & Westoby, 2000). However, not all the available evidence supports this model (Green & Juniper, 2004a). As a general process, uncommitted reserves can be applied to the establishment of seedlings under any combination of conditions that curtail the ability to reach and sustain a positive carbon balance. Dry mass accumulation is seldom limited by a single process (Moles & Westoby, 2004a). Seedlings growing in extremely stressful environments have a narrow margin for any reduction in resource acquisition and/or biomass loss. Even small initial differences in seedling size may have consequences for survivorship far beyond the time when the seedling becomes independent from maternal resources. Because of the exponential nature of seedling growth, small initial differences in size may subsequently result in differences in seedling size sufficient to produce changes in survivorship. Smaller seedlings resulting from smaller seeds are less likely to survive further stress, defoliation, or physical damage. There is evidence for the potential importance of having reserves to survive extreme shade and partial defoliation (Dirzo, 1984; Moles *et al.*, 2003). A better understanding of the physiological adaptations involved and the evolutionary trade-offs required is needed.

A common feature of some stressful systems is the paucity of seedling recruitment because establishment may be constrained by reproductive limitations. Restrictive growth conditions for adults limit the production of biomass, which when added to the other energetic demands, particularly allocation of resources to protect expensive organs, curtail resource availability for sexual reproduction. In addition, the evolution of large seeds, which seems to be essential for establishment under these conditions, limits the numbers of seeds produced (see Moles *et al.*, 2003, for a discussion of the benefits of trade-offs between seed numbers and seed sizes; Chapter 10). Finally, in stressful environments, seedling establishment is risky and opportunities for successful recruitment through seedlings may be limited. Clonal spread may contribute more to population persistence than sexual reproduction and is common in a number of stressful environments. For example, even though *Ammophila breviligulata* produces large seed crops, successful seedling establishment is episodic and the population grows fundamentally through rhizome extension (Maun, 1989). Indeed, part of its range in the most unstable foredune is unsuitable for seedling establishment and its presence there depends on clonal growth from plants established as seedlings meters away under less stressful conditions. Similarly, *Iva imbricata* (Asteraceae) is able to germinate, but unable to survive in the foredune, where populations are maintained exclusively by clonal growth (Colosi & McCormick, 1978).

The fact that seedling establishment is not a frequent event in environments with chronic and extreme stressful conditions actually makes seedling establishment even more important because aspects of population structure and dynamics can be determined during extraordinary episodes. Events that produce differences in the relative contribution of seed genotype, genetically determined differences in germination fractions, or differential survivorship can have profound effects on the genetic makeup of a population that may persist for long periods through clonal reproduction.

The special role of seedlings in establishing the genetic makeup of the population, as well as other aspects of population dynamics, could have important consequences for the evolutionary rates of species in stressful environments. While slow growth and limited sexual reproduction promote intrinsic genetic population inertia, adaptation to stressful conditions could preadapt these species to a suite of stresses

(Chapin *et al.*, 1993). The documented age of some plants in extreme environments indicates that they established when the climate was different from the present one (Steinger *et al.*, 1996; Wesche *et al.*, 2005). In the context of global climate change, the potential of these organisms to survive may be based on their evolved tolerance to a range of conditions.

Overall, it is clear that specialized strategies that allow seedlings to cope with specific stressors may not be enough to secure successful establishment (e.g. Climent *et al.*, 2006); rather, they must be combined with more general adaptation (particularly large seed size) or a syndrome of adaptations (e.g. resulting in slow growth, highly protected organs, and parsimonious allocation of resources) that in general protect seedlings against the compounded effects of combined stress agents, biomass destruction, limited ability to replace biomass and resources, and infrequent reproduction (Grime, 1979; Chapin *et al.*, 1993). Although some work has already addressed the role of the carbon economy of seedlings in stressful environments, little is known regarding the importance of mineral nutrition, which could be also important in determining the success of seedlings subject to chronic stress. Given the prevalence of large seeds in stressful and extreme environments and that the impending global environmental changes are likely to increase the frequency and intensity of stress-inducing events, there is an urgent need to increase our understanding of the mechanisms involved in seedling tolerance to stressful conditions, and attendant evolutionary pathways and inherent tradeoffs. These features, and their interaction with stressful conditions, may be main determinants of future vegetation structure.

3.10 | Acknowledgments

The comments of two anonymous reviewers, and of J. Prider, R. L. Simpson, V. T. Parker, and, especially, M. A. Leck helped improve this chapter.

Chapter 4

Specialized seedling strategies II: orchids, bromeliads, carnivorous plants, and parasites

Dennis F. Whigham, Melissa K. McCormick, and John P. O'Neill

4.1 Introduction

This chapter focuses on phylogenetically diverse groups of plants that do not have typical life history strategies as seedlings, juveniles, and mature individuals. Plants that live on other plants are classified as epiphytes, and they include both vascular and nonvascular species (Benzing, 1990). Tropical orchids and bromeliads comprise the vast majority of epiphytic flowering plants. However, orchids are global in their distribution and many are terrestrial (Dixon *et al.*, 2003). Carnivorous plants are also globally distributed, occurring in many types of ecosystems (Lloyd, 1976), as are parasitic plants (Press & Graves, 1995). The ability of seedlings to establish in habitats with extreme limiting resources is one of the few factors that links the diverse plants covered in this chapter. Epiphytes, for example, must initially become established on structures (i.e. branches) where resources are scarce. Once established, epiphytes may have to deal with combinations of stresses, including aridity, few available nutrients, and either high or low light conditions. Carnivorous plant species and many terrestrial orchids occur in habitats where nutrients or light are limiting. Many parasitic plants (e.g. mistletoes) also occur in resource-limited environments.

One of our approaches to organizing this chapter is to determine if seedlings differ in their physical or ecological characteristics in a manner similar to mature plants. The literature has few examples of investigations focusing specifically on seedlings. As an example, in the seminal book on Bromeliaceae, Benzing (2000) did not consider seedlings as a separate heading. Stems, roots, vascular cells, foliage, and trichomes were all considered separately, but information on seedlings was limited.

We organize the relatively sparse information on seedlings from these diverse groups of plants into four sections, each of which covers plants that are linked phylogenetically or functionally.

- The first section focuses on terrestrial and epiphytic orchids. All species in this family have a life history stage, the protocorm, which separates them from most other plants. We focus on the morphology and ecology of the protocorm life history stage, and also consider the ecology of seedlings that develop from protocorms. Emphasis is given to the importance of orchid–fungus interactions.
- The second section reviews epiphytic Bromeliaceae, focusing on seedling morphology and issues related to seedling ecology (e.g. seed dispersal, seedling establishment and survival, resource acquisition, mycorrhizae). Benzing (2000) provides a more comprehensive compendium on all aspects of bromeliads.
- The third section focuses on carnivorous plants, a polyphyletic group of plants functionally related by their ability to trap and consume prey as an important element of resource acquisition. Here we also cover most of the same topics presented in the section on bromeliads.
- Parasitic plants are the focus of the fourth section. This section proved the most difficult to organize as little information on seedling ecology is available. Historically, most research on parasitic plants has focused on interactions between parasites and their host plants. The format of this section differs in that it has been divided into four subsections: mistletoes, Orobanchaceae, *Cuscuta*, and mycoheterotrophic plants. In each subsection, we focus on seedling morphology and aspects of seedling ecology. The chapter concludes with a brief summary and commentary on future research needs.

4.2 | Epiphytic and terrestrial orchids

The family Orchidaceae, with an estimated 20 000–35 000 species, is arguably the largest and most widespread among the monocotyledons (Cribb *et al.*, 2003) and DNA studies suggest it may be one of the oldest plant families (Chase, 2005). As diverse as the Orchidaceae is, there are likely hundreds or thousands of unidentified species left to be described (Pridgeon, 2003). Although the greatest concentration of orchid species occurs in the tropics, the family is cosmopolitan with orchids occurring on all continents except Antarctica (Stoutamire, 1974; Dressler, 1981). In tropical regions, most orchids are epiphytic and include species that are evergreen (e.g. *Epidendrum*), seasonally deciduous (e.g. *Catasetum*), and leafless (e.g. *Polyradicion*), where the entire plant may be reduced to a congested system of photosynthetic roots (Benzing *et al.*, 1983). In temperate regions, orchids are restricted to the terrestrial environment but their leaf phenologies are similarly diverse. At our forested site in Maryland (USA), it is not uncommon to find evergreen (*Goodyera*), summergreen (*Galearis*), wintergreen (*Tipularia*), and leafless (*Corallorhiza*) species growing within a few meters of each other.

Despite the great diversity, several characteristics, mostly related to reproductive structures, are diagnostic for the Orchidaceae

(Dressler, 1983; Arditti, 1992), and include the abundant production of dust-like seeds that contain a simple embryo with little if any endosperm (Leroux *et al.*, 1997; Arditti & Ghani, 2000). Most plants with such limited resources begin to photosynthesize immediately after germination (Fenner, 1985). In contrast, Orchidaceae has evolved a different strategy, initially developing a protocorm, a separate life history stage that is present after germination, and eventually gives rise to the seedling (Rasmussen, 1995). Functionally, the protocorm is heterotrophic, and growth is dependent on an external source of carbohydrates, obtained *in situ* from a symbiotic fungus (Richardson *et al.*, 1992; Leake, 1994; Rasmussen, 1995; Peterson *et al.*, 1998), or from simple sugars in laboratory culture (Harrison & Arditti, 1978).

From seed to protocorm

As in all angiosperms, germination of orchid seeds is initiated by water entering the seed (Fenner, 1985). The seed coat becomes permeable to water by natural weathering processes or by chemical treatment in the laboratory and the testa splits (Yoder *et al.*, 2000). Some species can germinate without an appropriate fungus, but germination occurs faster with a fungus. Other species require a fungus (e.g. Rasmussen, 1995; McKendrick *et al.*, 2002; Whigham *et al.*, 2006). The embryo, which is morphologically undifferentiated in all but a few species, begins to swell and develops into a protocorm, which is equivalent to the radicle and hypocotyl of seedlings of other plants (Rasmussen, 1995). The protocorm forms rhizoids, which are epidermal hairs, and fungal infection occurs through either the rhizoids or the suspensor on the embryo (Rasmussen, 1995).

Protocorm to seedling

Most of our knowledge about orchid protocorms comes from *in vitro* studies, which demonstrate that most protocorms share the same basic structure (Fig. 4.1) comprising a basal part that becomes infected with an appropriate mycorrhizal fungus and an apical part that contains the meristematic cells from which a seedling develops (Fig. 4.2). Protocorms are rarely observed in nature, but when encountered, they are most often adjacent to established plants (Curtis, 1943; Willems, 1982) or have established in specialized habitats such as decomposing wood (Rasmussen & Whigham, 1998a).

Protocorms of tropical epiphytic and temperate terrestrial orchids differ somewhat in shape. Protocorms of temperate terrestrial species (Fig. 4.3a) are often initially cone-shaped with a tapered end and flattened area at the apex. During growth, however, they become much more variable in shape. Older protocorms of some species resemble juvenile plants in shape and vary from sausage-shaped (e.g. Fig. 4.4a – *Goodyera pubescens*) to coralloid and branched (e.g. Fig 4.4b – *Corallorhiza odontorhiza*), but *Tipularia discolor* has bone-shaped protocorms (e.g. Fig. 4.4c) that appear very different from juvenile plants. In contrast, protocorm morphology within tropical epiphytic species varies little with the protocorm being a somewhat flattened sphere (Fig. 4.3b).

Fig. 4.1 Developing seedling of *Dactylorhiza lapponica* with the leaf bud and initial root at the apex of the protocorm. Rhizoids (epidermal hairs) are visible on the lower portion of the protocorm. Photo by J. O'Neill.

Fig. 4.2 Stages of orchid seedling development. (a) Unimbibed seed, (b) imbibed seed with cracked testa, (c) germination and production of rhizoids, (d) enlargement of protocorm and initiation of leaf primordium, (e) leaf expansion, and (f) seedling with leaf and initiation of root. Adapted from Batty *et al.* (2001b), reproduced with permission, copyright CSIRO 2001, CSIRO Publishing, Melbourne, Australia.

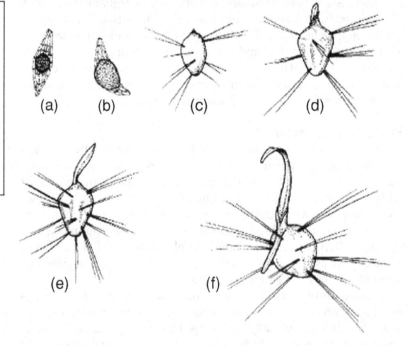

Regardless of protocorm shape, meristematic cells at the apical end give rise to the first primordial leaf. Subsequently, the first root emerges from the same part of the protocorm (Fig. 4.1 and Fig. 4.4c). As the seedling continues to develop, protocorm tissues persist for some time before decaying and, as a result, one often finds seedlings with protocorm tissue still attached (Fig. 4.2f and Fig. 4.4c). Achlorophyllous orchids (e.g. Fig. 4.4b – *Corallorhiza*) differ from this general

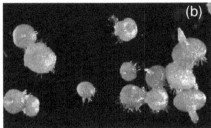

Fig. 4.3 Initial protocorm development of (a) *Dactylorhiza lapponica*, a temperate terrestrial and (b) *Encyclia tampensis*, a tropical epiphyte. Photos by J. O'Neill.

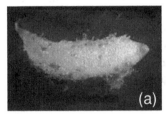
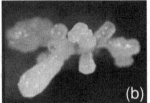
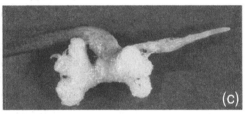

Fig. 4.4 Advanced protocorm morphology of three temperate orchids, (a) *Goodyera pubescens*, (b) *Corallorhiza odontorhiza*, and (c) *Tipularia discolor*. Photos by M. McCormick & J. O'Neill.

pattern of protocorm development by having several meristems on branched structures, each of which produces scale-like leaves at the tips.

During the protocorm stage and at least until initial leaves form, orchids are entirely dependent on fungi for the resources required for growth and development (Harvis & Hadley, 1967; Hadley & Williamson, 1971; Beyrle *et al.*, 1991, 1995). Hyphae of mycorrhizal fungi enter the protocorm, primarily through rhizoids, and grow into orchid cells where they form tight coils of hyphae called pelotons. The orchid then digests pelotons through a process that involves lysis of fungal cell walls. The obligate nature of the protocorm–fungus interaction has been demonstrated in both laboratory and field-based studies. Protocorms grown *in vivo* without an appropriate fungal partner fail to develop to the seedling stage (Anderson, 1991; Rasmussen, 1995; Batty *et al.*, 2006). Field experiments have also demonstrated the obligate nature of the mycorrhizal interaction as protocorms uninfected by fungi cease development following germination (e.g. Rasmussen, 1994; McKendrick *et al.*, 2000a, 2002). In all cases that have been examined, the fungi associated with naturally occurring protocorms represent a very specific subset of fungi in their environment (e.g. Rasmussen, 1995; McKendrick *et al.*, 2000a, 2002; Batty *et al.*, 2001a; McCormick *et al.*, 2004). However, this specificity may be less marked under laboratory conditions (e.g. Hadley, 1970; Masuhara & Katsuya, 1994). There is evidence that fungi associated with roots of adult orchids are not always the same fungi as those associated with protocorms (Rasmussen, 1995; Peterson *et al.*, 1998; McCormick *et al.*, 2004), suggesting that some orchids switch fungal symbionts during the transition from the protocorm to the seedling stage. With very few species adequately studied, the generality and ecological significance of fungal switching between these two life history stages is not clear. McCormick *et al.* (2004) found that the fungi associated with protocorms of *Tipularia discolor* belonged to a clade of fungi basal to the Sebacinaceae and differed from the fungi associated with adults. In a few instances, fungi associated with *Tipularia discolor* protocorms have also been found in adults (R. Burnett & M. McCormick, per. comm.), suggesting that this may not be a switch, but rather an accumulation of additional fungi. In contrast, both protocorms and adults of *Goodyera pubescens* and *Liparis liliifolia* have the same fungal associations.

There is no universal pattern that describes the longevity of the protocorm stage (Rasmussen, 1995). In tropical and subtropical orchids, seedlings may appear quickly following protocorm development. Field studies of temperate species using seeds sown in seed packets have found that the protocorm stage varies, ranging from 6 months for *Goodyera pubescens* (McCormick et al., 2006) to a year or more in species of *Corallorhiza* (McKendrick et al., 2002). Other field studies also describe the variable length of the protocorm stage. Willems (1982) found that it took up to 3 years for *Orchis simia* seedlings to develop while, at the other extreme, protocorms of *Liparis loeselii* mature after 3–4 months (Mrkvicka, 1990).

Seedling establishment and survival

Most seedlings, juveniles, and adult orchids examined *in vivo* continue to form mycorrhizae, although the fungal relationship may no longer be obligatory. With the exception of achlorophyllous orchids, which are fully mycoheterotrophic, orchid seedlings are similar to those of other mycorrhizal monocots and growth *in vitro* is similar to that encountered in the field (Curtis, 1943; Stoutamire, 1974). Leaves of seedlings are morphologically similar to leaves of juveniles and adults, and the roots are typically fleshy with no or few secondary roots. Roots of epiphytic and terrestrial seedlings function as storage organs and roots of many epiphytic species enable the plants to remain attached to the substrate on which they occur.

Few data exist on orchid seedling survival (Batty et al., 2001a), but seedlings are rarely observed in long-term population studies (Willems, 2002) and appear to have low survival rates similar to seedlings of most other plants. Curtis (1943) found that only 0.5% of the seedlings of *Cypripedium reginae*, a North American terrestrial species, survived for 4 years with highest mortality occurring between 1- and 2-year-old seedlings. Less than 1.0% of the seeds of a temperate Australian terrestrial species (*Caladenia arenicola*) produced a seedling that was capable of surviving the dry season (Batty et al., 2001a). Mortality is also high for tropical species. Zotz (1998) estimated that out of two million seeds of the epiphytic *Dimerandra emarginata*, only one was likely to become a seedling. Additionally, among established seedlings, only 50% survived the first dry season. Ackerman et al. (1996) also found low survival rates of *Tolumnia variegata* with 32.8% of the seedlings surviving for 1 year and 14.5% for 2 years.

Seedling dormancy

In vivo studies of orchids indicate that one or more periods of dormancy are required for continued development of protocorms and seedlings (Rasmussen, 1995). Field studies of orchid populations indicate that individuals become dormant for part of the growing season (i.e. aboveground parts senesce). In temperate zones, dormancy of individual plants is triggered by periods of environmental stress such as decreasing temperatures or a combination of high temperatures associated with a prolonged dry season (Dixon, 1991). In recent years, studies have investigated another type of adult whole-plant dormancy in terrestrial orchids (Shefferson *et al.*, 2001, 2005; Tali, 2002; Kéry *et al.*, 2005). This type of dormancy is defined as a condition in which the rootstock of a perennial herbaceous plant fails to produce annual shoots during the growing season (Shefferson *et al.*, 2005). Several orchid species have been shown to exhibit this type of dormancy (see citation in Rasmussen, 1995; Shefferson *et al.*, 2005), but it also occurs in other plant families. We found no references to protocorm or seedling dormancy of this type, but it seems probable that both life history stages are capable of surviving for more than one growing season in a dormant state because, similar to larger individuals, they are mycorrhizal.

4.3 | Bromeliads

Seedling morphology

We are unaware of any systematic study of the morphology of bromeliad seedlings but descriptions found in Benzing (1980, 1990, 2000) suggest that most seedlings emerge from seeds with typical seedling structures. Seedlings of bromeliads in the Tillandsioideae, however, appear to be atypical in their morphology, especially atmospheric species that have no roots or reduced root systems (Benzing, 2000). Figure 4.5 shows examples of two bromeliad seedlings. *Canistrum lindenii* and *Pitcairnia flammea* have seedlings that are similar to many plants producing seed. The first structure to emerge from the seed is the hypocotyl, followed by the radicle and shoot axis (epicotyl). Primary and secondary roots eventually replace the radicle. In contrast, seedlings of *Vriesea scalaris* appear to be typical of atmospheric epiphytes in the Tillandsioideae in that they do not produce roots for weeks or months and the resultant roots are specialized for anchorage.

Anchorage may be the primary function of roots in the majority of epiphytes, especially bromeliads, and many species have sclerified roots that are durable and strong (Benzing & Renfrow, 1974; Benzing, 2000). Studies by Brighigna *et al.* (1990) and Fiordi *et al.* (2001) further support the generalization that roots of most epiphytic bromeliads are used primarily for anchorage. Physiological studies of nutrient and water uptake (e.g. Benzing & Renfrow, 1974) provide further evidence that roots are primarily involved in anchorage.

Compared to roots, the rates of nitrate reductase induction was higher in leaves of *Vriesea hieroglyphica* that were given nitrogen in the form of calcium nitrate (Nievola & Mercier, 1996). Laube and Zotz (2003) also provided evidence for water and nutrient uptake by leaves. They measured the growth response of *Vriesea sanguinolenta* to variations in water, nutrient, and light supply, including small individuals that had grown past the seedling stage (i.e. they had started to produce leaves that retained water). Because roots were removed when first collected and no regrowth occurred, all experimental responses resulted from uptake through leaves. The smallest plants had the greatest potential to adjust relative growth rates in response to treatment variables, providing evidence that seedlings responded to variation in resource availability, primarily by uptake of water and nutrients through leaves and not roots.

Seedling establishment and survival

For many bromeliads, aridity and low levels of nutrient availability represent the primary stresses at all life history stages (Hernández *et al.*, 1999; Benzing, 2000). Moisture stress is less important in montane tropical environments and temperate rain forests, which have abundant moisture (Madison, 1977). Aridity would be expected to influence seed germination, as most bromeliads have short-lived seeds that are nondormant and germination depends on the presence of adequate moisture. Little is known, however, about seed germination among bromeliads (Baskin & Baskin, 1998), and Benzing (2000) noted that "seed viability and germination rank among the least-studied aspects of bromeliad reproduction." One would expect that germination rates would generally be low and highly variable in space and time because most seeds would fall on unsuitable arboreal habitats (Benzing, 1978). Additionally, short- and long-term variation in microclimatic conditions (e.g. availability of moisture) would control patterns of germination. Germination of *Tillandsia paucifolia* seeds, for example, was always less than 4.0% on several potential host trees (Benzing, 1990) and seedling survival was low (Benzing, 1981).

Many factors influence patterns of seedling survival and growth, not the least of which is the stability and surface of the substrate where the seed lodges. Bromeliad seeds, while typically smaller than most seed-bearing plants (Benzing, 2000), have a variety of structures, such as hooked coma hairs, that enable them to remain in the tree canopy (Benzing, 1990).

Seedling survival is likely highly variable in space and time. For example, Benzing (1981) found that recruitment of *Tillandsia paucifolia* occurred most frequently on trees on which maternal plants occurred. Several factors may account for the higher recruitment on trees that already support mature plants. Most seeds have leptokurotic dispersal patterns (Madison, 1979 as cited in Benzing, 1990; Baskin & Baskin, 1998), assuring that the majority of seeds will disperse within the host-tree canopy. This dispersal pattern increases the probability that seeds will be dispersed to a suitable microhabitat and results in a clumped distribution (Hietz & Hietz, 1995). In addition, host trees that support higher numbers of epiphytes grow in locations with micro-climatic conditions that favor germination, seedling establishment, and growth. Zimmerman and Olmsted (1992) found that *Tillandsia dasylirifolia* was most abundant on trees with branches that overhang water in seasonally flooded forests of the Yucatan Peninsula (Mexico). Genetic variation within host-tree species (e.g. bark roughness) could also be important.

One would expect that seedling and juvenile establishment and survival would be low because both life history stages are vulnerable to predation as well as physical stresses associated with life in the forest canopy. Dessication is likely an especially important factor in survivorship of small individuals because of their low surface:volume ratios. We only found survivorship information for seedlings and juveniles of two bromeliads and there was no consistent pattern. Benzing (1990) reported low survival rates of seedling cohorts of *Tillandsia paucifolia*. In contrast, Schmidt and Zotz (2002) found that mortality rates averaged 16.5% for all life stages of *Vriesea sanguinolenta*, a bromeliad occurring in the outer canopy of trees in Panama. In fact, Schmidt and Zotz found that mortality was highest in medium-sized individuals, not seedlings and juveniles.

Some bromeliad seedlings store water in cotyledons, hypocotyls, and the first roots (Benzing, 1990), but most deal with aridity through adaptations that absorb water from the atmosphere or protect them from desiccation. Seedlings of some tank-forming bromeliads do not form tanks for 1–5 years, but a dense layer of trichomes covers juvenile foliage. Juveniles of *Vriesea geniculata* had at least twice the density of trichomes on leaves and the percent of the leaf surface covered by trichome shields was 30–40 times greater on juveniles (Benzing, 2000). As a result, they respond physiologically more like atmospheric bromeliads (Benzing, 1980), which are able to photosynthesize at a higher rate than adults and use water more efficiently under conditions of drought stress (Adams & Martin, 1986). Laube and Zotz (2003) found similar results for *Vriesea sanguinolenta*. In fact, seedlings of some bromeliads may be so well-adapted to aridity that they die when exposed to too much moisture. *Tillandsia circinnata* exhibits this adaptation and mortality occurs when plants are not able to dry out (Benzing, 1978).

Seedlings and mycorrhizae

Similar to orchids, other epiphytes benefit from mycorrhizal interactions through increased resource acquisition. However, because many bromeliads have reduced root systems or roots that are not the primary organ involved in water and nutrient uptake, interactions with mycorrhizal fungi may not occur or be limited in extent. There are no reports of mycorrhizal colonization of seedlings in non-orchid epiphytes, although there is also no record that seedlings have been examined. Lesica and Antibus (1990) examined 50 taxa in 14 families of epiphytes on fallen trees and tree branches in two Costa Rican forests. Most adult orchids and species of Ericaceae and Gesneriaceae had colonized roots but only one genus (*Pitcairnia*) in the Bromeliaceae had colonized roots. The standard deviation for the measured metric (% colonization) was similar to the mean, indicating that colonization was highly variable. No colonization or low levels of colonization have been reported for bromeliads in Mexico and Malaysia (Allen *et al.*, 1993; Nadarajah & Nawawi, 1993). In contrast, Rabatin *et al.* (1993) found that roots in 16 of 19 bromeliads in a Venezuelan cloud forest had arbuscular mycorrhizae fungi, with infection occurring at higher levels in more mesic sites. Rains *et al.* (2003) also found high levels of infection in epiphytic Araceae, Clusiaceae, Ericaceae, and Piperaceae in a lower montane cloud forest in Costa Rica. None of these studies, however, reported on infection in seedlings. Thus, the relationship between bromeliad seedlings and mycorrhizal fungi requires study.

4.4 | Carnivorous plants

Rice (2006) reported that there are about 650 described species that can be categorized as carnivorous. They occur in all regions of the world from the Arctic to the tropics, and grow over a wide range of habitats from tree canopies to wetlands, in nutrient-poor to nutrient-rich habitats. The species are distributed, however, among a relatively small number of families (12) and genera (19). Carnivorous plants are polyphyletic, having multiple origins but also demonstrating a high degree of convergent evolution (Albert *et al.*, 1992; Ellison & Gotelli, 2001). Well-known examples are species of *Darlingtonia* (Sarraceniaceae), *Sarracenia* (Sarraceniaceae), and *Nepenthes* (Nepenthaceae) that trap prey passively in pitcher traps (Juniper *et al.*, 1989). Carnivorous plants that use glandular secretions to capture prey are species of *Drosera* (Droseraceae) (e.g. Crowder *et al.*, 1990) and *Pinguicula* (Lentibulariaceae) (e.g. Lüttge, 1983). The monotypic genera *Dionaea* (Droseraceae) and *Aldrovanda* (Droseraceae) have the most active prey-capturing mechanisms with leaves that function as snap-traps. The genus *Utricularia* (Lentibulariaceae) uses a trapdoor mechanism (suctioning bladders) to capture small prey (Schnell, 1976; Forterre *et al.*, 2005).

Seedling morphology

Few references, mostly in older publications, described the seedling morphology of carnivorous plants. A general pattern emerges among a wide range of species in this functional group of plants. The first leaves of seedlings of most species are already capable of trapping prey. Small seedlings of *Dionaea muscipula* (Smith, 1931; Schnell, 1976), *Sarracenia purpurea* (Shreve, 1906), *Cephalotus follicularis* (Cephalotaceae) (Boon, 2006), and species of *Nepenthes* (Geddes, 1893) and *Drosera* (Crowder *et al.*, 1990) have insect-trapping leaves. Seedlings of *Utricularia* spp. quickly form prey-trapping bladders (Kondo *et al.*, 1978). The early formation of functional leaves in the seedling stage indicates the importance of prey-capturing as an ecologically adaptive strategy among carnivorous plants.

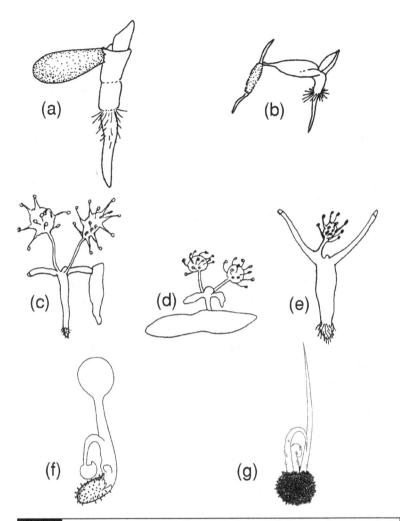

Fig. 4.5 Seedlings of bromeliads (a) *Canistrum lindenii* and (b) *Pitcairnia flammea*, and of carnivorous plants (c) *Drosera anglica*, (d) *D. rotundifolia*, (e) *D. intermedia*, (f) *Utricularia striatula*, and (g) *U. radiata*. From Benzing (2000) (a & b); Crowder *et al.* (1990) (c–e); and Kondo *et al.* (1978) (f & g). Reproduced with permission (f & g) from New York Botanical Garden, Bronx, NY.

The second general characteristic of carnivorous plant seedlings is that most have weakly developed root systems or no roots at all (Emerson, 1921; Baskin & Baskin, 1998). Seedlings of *Cephalotus follicularis* produce a single, small tap root (Boon, 2006) and *Dionaea muscipula* seedlings have a tap root from which arise numerous root hairs (Smith, 1931). *Drosera* seedlings also have poorly developed root systems (Fig. 4.5), and the tap root is replaced by relatively few adventitious roots with few root hairs and persists for less than a year (Crowder *et al.*, 1990). Seedlings and mature plants of *Utricularia* spp. do not produce roots (Fig. 4.5; Shannon, 1953; Adamec, 2000) and roots of *Aldrovanda vesiculosa* stop growing when they are less than 1 mm long (Kondo *et al.*, 1978).

Seedling establishment and survival

Most carnivorous species for which seedling information is available occur in low nutrient and relatively open wetland habitats, typically bogs, where moisture is adequate and competition is low. In addition, little information is available on patterns of seedling distribution in relation to patterns of seed dispersal. Ellison and Parker (2002) found that most seedlings of *Sarracenia purpurea* occurred near parent plants, but suggested that long-distance dispersal of seeds likely also occurred, accounting for the broad dispersal of *Sarracenia purpurea* following the end of the last ice age. Similar rare long-distance dispersal, perhaps by animals or water, must have occurred in other carnivorous genera species (e.g. *Drosera*, *Utricularia*) that are broadly distributed across much of the boreal zone. We found no quantitative dispersal data for seeds of other species of carnivorous plants. The review of *Drosera* for the *Biological Flora of the British Isles* (Crowder et al., 1990), for example, contained no information on seed dispersal and only suggests that wind, the feet of birds, and water were likely dispersal agents. Schnell (1976) suggested that seeds of *Sarracenia* species are water dispersed because of their relatively large size, but provided no supporting evidence. Longer-distance seed dispersal may occur in species of *Nepenthes* that are wind dispersed (Schatz, 1996).

Seedling survival is probably low in most habitats, but similar to data on patterns of seed dispersal, there have been few studies of seedling survival. Ellison and Parker (2002) found that only 5% of the observed seedlings of *Sarracenia purpurea* survived an entire growing season. Only 5–7% of the *Dionaea muscipula* seed sown by Luken (2005a) survived as seedlings compared to between 72 and 85% survival for transplanted adults.

Seedlings of *Drosera intermedia* had the highest rates of survival in cut-over bogs with more available nitrogen (N) and low phosphorous (P) (Sansen & Koedam, 1996). Seedlings of *Drosera capillaris* and *Dionaea muscipula* were most abundant in hollows in a bog system, a habitat that is consistently wet and largely devoid of litter and competing vegetation (Luken, 2005a). In a separate study, Luken (2005b) found that removal of vegetation by mowing in a low-nutrient Carolina Bay (North Carolina, USA) resulted in greater survival of already established *Dionaea muscipula* plants and that establishment of *Dionaea*, *Drosera capillaris*, and *Utricularia subulata* seedlings increased. Seedling densities of species of *Drosera* and *Sarracenia alata* have also been shown to increase significantly following burning of pine-savanna wetland ecosystems in southeastern USA (Barker & Williamson, 1988; Brewer, 1999a,b) although fire exhibited little effect on seedling survival of *Sarracenia alata* (Brewer, 2001).

Nutrient acquisition

Carnivorous plants obtain varying degrees of nitrogen from captured prey (see Table 1 in Ellison & Gotelli, 2001). Ellison and Gotelli concluded that the amount of N obtained from prey increases with increasing complexity of the carnivorous structures. Species with sticky leaves obtain less N from prey compared to species with pitchers. Carnivorous plants also obtain a significant amount of nutrients other than N from captured prey (e.g. Jaffe *et al.*, 1992; Schulze *et al.*, 2001; Adamec, 2002). Additionally, root uptake provides significant resources (Karlsson *et al.*, 1987). In a few instances, the process of capturing prey resulted in increased root uptake of nutrients. Hanslin and Karlsson (1996) found that root uptake of N increased following prey capture by *Pinguicula alpina, P. villosa, P. vulgaris*, and *Drosera rotundifolia* in a subarctic ecosystem. Adamec (2002) found similar results working with three species of *Drosera* and suggested that "the main physiological effect of leaf nutrient absorption from prey is a stimulation of root nutrient uptake."

While most studies of nutrient uptake in carnivorous plants have been based on mature plants (e.g. Adamec 1997, 2002; Schulze *et al.*, 1997; Thorén & Karlsson, 1998; Ellison & Gotelli, 2001), two studies examined nutrient acquisition in smaller individuals, including seedlings. Schulze *et al.* (2001) found that seedling survival of *Dionaea muscipula* increased after plants had reached the six-leaf stage, a size that facilitates the capture of prey. Seedlings and small plants had less N in leaves but their $\delta^{15}N$ signature suggested that insect capture continued to be an important source of N. Adamec (2002) applied mineral nutrients to leaves of seedlings of three species of *Drosera* and found that they were able to assimilate nutrients at rates that were similar to those obtained from prey. Except for P, the roots were responsible for most of the nutrient uptake in fertilized plants, primarily through increased root biomass rather than increased root uptake.

Seedlings and mycorrhizae

Carnivorous plants appear to be nonmycorrhizal, although few species have been examined, and we found no studies that focused on seedlings. *Dionaea muscipula* (Roberts & Oosting, 1958), *Pinguicula* spp. (Harley & Harley, 1987), and *Drosera* spp. (Crowder *et al.*, 1990) have all been reported to be nonmycorrhizal. Efficient uptake of nutrients from captured prey and from root systems that assimilate nutrients more efficiently following prey capture in leaves may have precluded the benefits of mycorrhizal interactions that are common in flowering plants.

4.5 | Parasitic plants

There are 3 000–4 000 species of direct plant parasites, making up somewhat less than 1% of all angiosperms (Musselman & Press, 1995; Nickrent, 2003; Press & Phoenix, 2005). Parasitic species occur in most plant communities worldwide and include members of 22 plant families, including trees, shrubs, vines, and herbs (Musselman & Press, 1995; Nickrent *et al.*, 1998; Phoenix & Press, 2005). Roughly 60% of

(a)

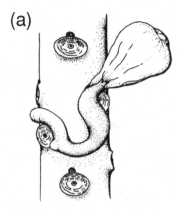

(b)

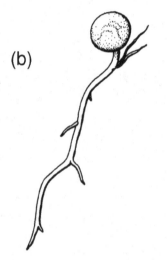

(c)

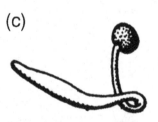

Fig. 4.6 Seedlings of parasites. (a) *Phorodendron densum* (Viscaceae) seedling with radicular lobes, (b) young seedling of *Lathraea clandestina* (Scrophulariaceae) showing branching of the emerging radicle, and (c) young seedling of *Cuscuta gronovii* showing coiling growth pattern during early development. From Kuijt, 1969, reproduced with permission.

parasitic plants are root parasites and 40% are shoot parasites. A few are important crop pests, but many others can have dramatic impacts on the composition of plant communities (Riches & Parker, 1995; Pywell *et al.*, 2004; Phoenix & Press, 2005; Press & Phoenix, 2005). Approximately 20% of parasitic plants lack chlorophyll and are holoparasitic, with the rest being hemiparasitic (Musselman & Press, 1995). In addition to these direct plant parasites, which are united by their ability to form haustoria, there are also more than 400 species of mycoheterotrophic plants in 11 families that parasitize other plants through specialized mycorrhizal connections (Leake, 1994).

The seedling stage is clearly critical for parasitic plants because that is the stage at which a host connection must be established (Musselman & Press, 1995). The morphology and ecology of seedlings have only been studied in a few groups, primarily the mistletoes, members of Orobanchaceae, and *Cuscuta* (Convolvulaceae). These have focused on a few economically important crop pests within each group. Of the mycoheterotrophic plants, we will focus on the most extensively studied group, the Monotropoideae. Obviously, there are many groups not addressed, but we hope that those presented give an indication of the range of seedling stages present. For additional information on some of these groups, we refer the reader to *Biology of Parasitic Flowering Plants* by Kuijt (1969) and *Parasitic Plants* edited by Press and Graves (1995).

Mistletoes

The mistletoes are a phylogenetically diverse group of plants, primarily belonging to the families Loranthaceae and Viscaceae, with a few members of the Misodendraceae, Eremolepidaceae, and Santalaceae (Restrepo *et al.*, 2002) represented. The best studied are the Loranthaceae and Viscaceae (Reid *et al.*, 1995). Primitive mistletoes (*Atkinsonia*, *Nuytsia*, *Gaiadendron*) are terrestrial root parasites or facultatively epiphytic shoot parasites, whose early development is typical of nonparasitic flowering plants. The radicle, with a typical radicular cap, grows out of the seed to penetrate the soil, soon after which the hypocotyl elongates and lifts the seed from the ground. The cotyledons dissociate themselves from the endosperm, the shoot begins to elongate, and the root begins to branch. After a root forms, it may form a haustorium (discussed in detail below), but always as a lateral organ (Kuijt, 1969).

Some primitive mistletoes persist independently for up to a year before contacting a host (Kuijt, 1982). In contrast, more advanced and typical mistletoes have only a brief free-living phase. Often only a few days occur between germination and infection after which the seedling is dependent on the host. Growth during this free-living period involves primarily the radicle, which emerges from the seed and grows in search of a host (Fig. 4.6a). Radicle growth precedes cotyledon emergence (if they emerge), and is physiologically dependent on mobilization of the endosperm, except *Aetanthus* (Loranthaceae), *Psittacanthus* (Loranthaceae), and *Lepidoceras*

(Eremolepidaceae), where the cotyledons function as storage organs (Kuijt, 1982). In these typical mistletoes, the radicle lacks a root cap and develops into the primary haustorium. Radicle growth depends on temperature and moisture, with growth being higher in moderate temperatures and high humidity (Reid et al., 1995). In many species, radicle growth is negatively phototropic and geotropic, allowing it to grow toward the host from the underside of a branch on which they are often deposited (Kuijt, 1969). Radicle growth may be very limited in some species, especially tropical ones. In other species, radicles can be extremely long. When the tip of the radicle contacts a solid surface, it flattens against the surface and begins to develop into a haustorium (Dobbins & Kuijt, 1974a).

Seedling morphology

The haustorium is a defining character in direct plant parasites and its formation is critical for seedling survival. With the exception of the primitive genera, mistletoes that are independent of a host cannot persist for long. Haustoria in the mistletoes come in a range of types but all are similar in some respects (Kuijt, 1969). The tip of the radicle appears to be thigmotropically sensitive. Once the radicle contacts a host branch, its tip thickens to form a club-shaped or hemispherical disk (Kuijt, 1969). The radicle tip is made up of several layers of tissue, including an epidermis and a hypodermis, each one cell layer thick, that surrounds an inner core of parenchyma (Dobbins & Kuijt, 1974a). At the point of contact with the host, the epidermal cells sometimes disintegrate with surrounding epidermal cells forming finger-like projections reaching toward the host surface and becoming an interwoven network. Both the radicle tip and developing haustorium contain chloroplasts and appear green (Dobbins & Kuijt, 1974a).

Host penetration is carried out in a structurally complex contact zone. The network of epidermal cell projections develops very dense cytoplasm in thin-walled tips that terminate at the host surface. Although this association is not yet fully described, the tips may then lyse with their cellular contents aiding in adhesion to the host (Dobbins & Kuijt, 1974a). A defining characteristic of mistletoe haustoria is the formation of the gland. This lens-shaped mass of dense cells forms in the center of the haustorium near the contact zone and is filled with fibrillar material that is secreted onto the host surface (Dobbins & Kuijt, 1974b). Tissue from the haustorium is thought to penetrate host tissue using a combination of enzymatic digestion and mechanical pressure (Thoday, 1951). After formation of the gland, a layer of dense parenchyma cells in the center of the haustorium contiguous with the gland region begins to differentiate into xylem vessels (Dobbins & Kuijt, 1974b). If initial host penetration is unsuccessful, a second contact zone forms with the expanding cells displacing the previous contact zone laterally (Dobbins & Kuijt, 1974b).

Once the haustorium reaches the host vascular cambium, direct xylem-to-xylem contact is quickly established and the mistletoe xylem is subsequently embedded in the host wood. Further mistletoe growth occurs by an intercalary meristem that allows correlation between host and mistletoe growth (Kuijt, 1969). The point at which the mistletoe begins to draw on host plant resources defines the end of the seedling stage in mistletoes (Kuijt, 1982).

Cotyledon emergence, if it occurs, happens somewhat after initiation of haustorium formation. In most advanced mistletoes, the epicotyl and cotyledons remain within the seed until a host connection is established (Kuijt, 1969). Cotyledon morphology varies markedly among mistletoes. Cotyledons are free and leaf-like in the primitive genus *Gaiadendron*. More advanced genera have varying combinations of cotyledon fusion (gamocotyly) and cotyledons that do not emerge from the seed (cryptocotyly). In some cases, cotyledons remain within the seed as a fused absorptive organ and never become photosynthetic (Kuijt, 1969). Small-flowered Loranthaceae generally have two large, green, spreading cotyledons that appear to function as storage organs (Kuijt, 1982). The large-flowered groups of Loranthaceae are somewhat more variable. In this group, seedlings of *Aetanthus* and *Psittacanthus* have large, fleshy cotyledons that appear to have replaced the endosperm of the seed as seeds are large but almost lacking in endosperm. Seedlings in the genus *Psittacanthus* have a variable number of cotyledons, ranging from 2 to as many as 14 that vary in shape and size (Kuijt, 1982). Cotyledons in the genus *Tristerix* (Loranthaceae) are cryptocotylar and range from being fused only at the tips to complete fusion. The fused portions are often shed with the endosperm when the shoot emerges from the seed.

Seedling establishment and survival

Much of mistletoe seedling establishment depends on host suitability. Mistletoe seeds are dispersed by birds and deposition on appropriate hosts depends on which trees the birds encounter (Aukema, 2003). Viscin, which coats the seeds, adheres them to branches until they germinate and form haustorial connections. Not all trees are suitable hosts (Roxburgh & Nicolson, 2005). Host preference in mistletoes can range from specialists with only one appropriate host species, to generalists that can use most of the tree species in a forest (Barlow, 1981; Reid et al., 1995; Norton & Carpenter, 1998). In general, temperate mistletoes are more host-specific than tropical mistletoes, perhaps because specificity is disadvantageous in very diverse tropical forests (Norton & Carpenter, 1998). Among mistletoes with more specific host requirements, host range tends to include several closely related hosts. Similarly, mistletoes that are closely related also tend to infect closely related hosts. The compatibility of a host is determined by a combination of host attachment and attempts at host penetration (López De Buen & Ornelas, 2002). Hosts can resist mistletoe colonization by having bark that is impenetrable or exfoliating, forming wound periderm that prevents penetration, or by resisting xylem growth once penetration has occurred (Yan, 1993).

For most mistletoes, the uptake of host-derived water and nutrients occurs through open tracheids and vessels that offer little flow resistance. In many cases however, specialized parenchyma cells form the primary link between host and parasite xylem. Many holoparasitic species even link to host phloem elements (Riopel & Timko, 1995). Mistletoes are often divided into heterotrophic and autotrophic groups: dwarf mistletoes, which contain little chlorophyll and tap into both host xylem and phloem, and other mistletoes (all Loranthaceae and most Viscaceae), which have substantial chlorophyll and generally access only host xylem. Still, even photosynthetic mistletoes are at least partially heterotrophic (Room, 1971; Raven, 1983; Marshall & Ehleringer, 1990; Pate *et al.*, 1991; Schulze *et al.*, 1991; Pate, 1995; Press, 1995).

Adult parasite transpiration is generally much greater than host transpiration and 5–68% of parasite carbon, and essentially all nitrogen and water (except in root parasitic species) is derived from the host. Less is known about transpiration in seedlings. It is clear that, except for primitive species, mistletoe seedlings do not survive long without connecting to a host. It is reasonable, therefore, to suppose that once seed reserves are exhausted, the great majority of seedling water, nitrogen, and, especially in species without foliar cotyledons, carbon is derived from host sources at least until the first true leaves are produced. For holoparasitic species, most, if not all, seedling nutrients are likely host derived. Parasite concentrations of most mineral nutrients are substantially higher than their hosts. Although it is not certain, this may be a reflection of greater parasite transpiration rates, which allows concentration of nutrients. It is likely that seedlings also transpire at the high rates seen in adult mistletoes (Press, 1995).

Once established, the growth rates of mistletoes are host dependent, growing very quickly on one host and slowly on another (Reid *et al.*, 1995). Seedlings are recruited episodically in some species, while in others, tropical species in particular, recruitment is continuous (Reid *et al.*, 1995).

Orobanchaceae

Unlike the mistletoes, which reach their greatest diversity in the tropics, members of the Orobanchaceae are primarily temperate (Musselman & Press, 1995). Members of this family include some of the most obvious root parasitic plants in temperate forests. With a few exceptions, *Striga* among them, these plants are largely absent in the tropics (Kuijt, 1969). While many species were previously classified as Scrophulariaceae (e.g. Musselman & Press, 1995), all parasitic and hemiparasitic species have been reclassified as Orobanchaceae (Olmstead *et al.*, 2001; Bennett & Mathews, 2006). However, in much referenced literature, the species are still separated into two families. We consider them together here under the Orobanchaceae, which includes holoparasites and both obligate and facultative hemiparasites.

Seedling morphology

Germination of most species is epigeous, with the hypocotyl growing down into the soil (Kuijt, 1969; Westbury, 2004). In many

Orobanchaceae, host-root exudates stimulate seed germination and also attract the emerging radicle. As in advanced mistletoes, the seedling radicle is generally capless (Kuijt, 1969). Less than 24 hours after contact with a host plant, it swells, produces a ring of haustorial hairs, and quickly forms many branches (Kuijt, 1969; Fig. 4.6b). Haustoria formation in *Striga asiatica* seedlings begins with a redirection of cellular expansion at the root meristem from longitudinal to radial, starting roughly 8 hours subsequent to exposure to the host-derived chemical 2,6-dimethoxy-*p*-benzoquinone (DMBQ; Smith *et al.*, 1990). This cell expansion resulted in a swollen radicle tip followed by the formation of prominent haustorial hairs (Musselman & Dickison, 1975 and references therein; Riopel & Timko, 1995; Tomilov *et al.*, 2005). These haustorial hairs had surface papillae of hemicellulose, which bond the parasite to the host tissue (Riopel & Timko, 1995). A similar sequence of events occurs in *Triphysaria versicolor* (Tomilov *et al.*, 2005).

Haustoria formation differs somewhat between holo- and hemi-parasitic plants. In hemiparasites, which make up the majority of the family, haustorium development occurs laterally, just behind the radicle meristem. Normal growth resumes after haustorium development and additional haustoria may be formed (Tomilov *et al.*, 2005). In holoparasites, a primary haustorium forms at the meristematic tip and halts further growth (Okonkwo & Nwoke, 1978). Additional lateral haustoria, similar to those of hemiparasites, may form after the primary haustorium has established a connection with the host (Tomilov *et al.*, 2005).

Differentiation of vascular tissues within the haustorium is variable and influenced by host roots. Following the penetration of host tissues by either release or induction of cellulases, pectinases (Reddy *et al.*, 1980, 1981), and/or acid phosphatases (Toth & Kuijt, 1977), elongated haustorial tip cells divide longitudinally, pushing aside host tissues (Stephens, 1912; Maiti *et al.*, 1984). Once the host endodermis is penetrated, haustorial cells enter the host vessel lumen through pits in the side wall or dissolution of the vessel wall with hydrolytic enzymes. Cells in the haustorial tip differentiate into xylem elements, and quickly establish a connection between host and parasite xylem (Riopel & Timko, 1995; Bouwmeester *et al.*, 2003). Musselman and Dickison (1975) found that phloem elements were generally lacking in the haustorium, though specialized parenchyma cells may connect host and parasite phloem, as in *Striga*.

Seedling establishment and survival

After haustorium penetration, the holoparasite seedling is entirely dependent on the host and can trigger root proliferation by altering host hormone balances (Musselman & Press, 1995). Most holoparasites are cryptocotylar and cotyledons never emerge from the seed. In many hemiparasitic species, the cotyledons emerge from the seed shortly after germination as in typical nonparasitic plants. Some species, such as *Tozzia*, live entirely belowground for several years as parasites, only producing a green shoot when flowering (Kuijt, 1969).

Host specificity varies dramatically in the Orobanchaceae, from species that associate with hundreds of hosts to those that associate with only a single species, such as *Epifagus virginiana* (Musselman & Press, 1995). Most species exhibit some level of host preference (Musselman & Press, 1995). Seeds are thought to be primarily dispersed by a combination of wind and water, which washes seeds into soil crevices, but ants have also been shown to aid in dispersal (Kuijt, 1969).

Cuscuta

In addition to the families mentioned above, which consist mostly or entirely of parasitic species, there are also several small parasitic families, including the holoparasitic Lennoaceae (e.g. *Lennoa* and *Pholisma*), and hemiparasitic Krameriaceae (e.g. *Krameria*). Other parasitic species occur in largely autotrophic families. These include *Cuscuta* in the Convolvulaceae and *Cassytha* in the Lauraceae. Dodders of the genus *Cuscuta* are the best known of these groups, with more than 150 species (Kuijt, 1969). Little or nothing appears to be known about the seedling stages of *Cassytha*, Lennoaceae, and Krameriaceae. Members of the genus *Cuscuta* occur on all continents except Antarctica, and are important agricultural pests in many areas (Runyon *et al.*, 2006).

Seedling morphology

Largely or completely lacking chlorophyll, dodders are entirely parasitic. Seedlings are rootless and must forage for host plant tissues. The shoot that emerges from the seed exhibits directed growth toward volatile compounds produced by potential host plants (Runyon *et al.*, 2006; Fig. 4.6c). Soon after initial contact with a host, thigmotropic responses produce a narrow coil that bears multiple haustoria. After establishment of initial haustoria, the radicle dies and subsequent growth is in the form of wide coils that can contact the host in another location or contact a new host (Kuijt, 1969). Haustoria in dodder are structurally similar to those in the mistletoes and Orobanchaceae, but dodder haustoria also include a bridge of parenchymatous cells that link host and parasite phloem elements (Dörr, 1990). Embryos of dodders lack cotyledons but have two laterally arranged scale leaves.

Seedling establishment and survival

Some dodder species have minute quantities of chlorophyll and are capable of fixing limited amounts of CO_2, but most depend on host-derived carbon in addition to water, nitrogen, and other nutrients (Pate, 1995; Press, 1995). Adult dodder haustoria can divert substantial quantities of host carbon, resulting in crop yield reductions of 23–57% or more (Riches & Parker, 1995). As with many other parasitic plants, it is clear that the seedling stage, during which dodder must locate a host and establish haustorial connections, is the critical point in its life history. Recognizing this, most agricultural control methods focus on the seedling stage (Runyon *et al.*, 2006). The seedling stage also likely determines population dynamics in natural environments.

Dodder can parasitize a wide range of plants, but is capable of exploiting a narrower range (Musselman & Press, 1995) and shows distinct host preferences in its foraging behavior (Runyon *et al.*, 2006).

Mycoheterotrophic plants

Mycoheterotrophism has apparently evolved independently many times in the plant kingdom. These plants are often considered epiparasites because they exploit a fungus involved in an ectomycorrhizal relationship with woody plants. Mycoheterotrophism is especially common in the Orchidaceae and several families within the Ericales (Leake, 1994; Bidartondo, 2005). Mycoheterotrophism here is defined as applying only to nonphotosynthetic plants, not the many plants that supplement their photosynthetic carbon with carbon from mycorrhizal fungi or those that are only initially mycoheterotrophic. Some Orchidaceae and Gentianaceae and all Monotropoideae and Triuridaceae are mycoheterotrophic (Bidartondo, 2005). As orchids are covered in section 4.2, we will limit the discussion here to the most extensively studied group of mycoheterotrophic plants, the monotropes (Ericaceae).

Seedling morphology

After germination triggered by molecules released by specific fungi, monotrope seedlings are morphologically quite different from non-parasitic plants (Bidartondo, 2005; Bidartondo & Bruns, 2005). They do not form haustoria as the other parasitic groups mentioned here, but rather resemble seedlings of mycoheterotrophic orchids, perhaps due to their minute seeds with undifferentiated embryos that lack cotyledons or nutrient resources.

Prior to fungal infection, but likely stimulated by fungal chemical signals, the embryo/hypocotyl emerges from the cracked seed coat (Bidartondo & Bruns, 2005). Once an embryo is successfully colonized by an appropriate mycorrhizal fungus, it quickly begins to branch, coming to resemble the dense aggregation of mycorrhizal roots which can adventitiously produce inflorescences. Colonization by an appropriate mycorrhizal fungus takes place immediately after germination such that, unlike direct plant parasites, connection to a host is present throughout the seedling stage and cannot be said to define the end of this stage. Although a few individuals of *Monotropa uniflora* (Monotropoideae) start to flower after 2 years, it is clear that most plants remain entirely belowground for several years before producing an inflorescence that is devoid of chlorophyll and has only minute scale leaves.

Seedling establishment and survival

Monotrope seedlings obtain all nutrients from their mycorrhizal fungi and all monotropoid mycorrhizal fungi form ectomycorrhizae with trees. Thus, the monotropes are effectively epiparasites on the tree hosts of this three-way interaction. Mycorrhizal associations involving a mycoheterotrophic plant are also very specific. Each mycoheterotroph studied only associated with a narrow range of fungi.

Each monotrope lineage required fungi from a single clade of obligate ectomycorrhizal fungi (Bidartondo & Bruns, 2005). The transfer of nutrients between fungus and monotrope may occur by lysing hyphal ingrowths and absorbing their nutrients, in a process similar to hyphal digestion in orchids (section 4.2). However, in monotropes where there is little intracellular fungal colonization, it seems likely that more transfer of nutrients occurs by a disruption of plant and fungal cell membranes or by source-sink dynamics across a permeable membrane (Bidartondo, 2005).

It is clear, given the large numbers of dust seeds produced by mycoheterotrophs compared to the small number of emerging inflorescences, that the life history stages that experience the greatest mortality are those that occur underground (Rasmussen & Whigham, 1993; Bidartondo & Bruns, 2005). For monotropes, which emerge only to flower, the underground phase includes the seed, seedling, and a vegetative stage of unknown duration. Bidartondo and Bruns (2005) found that the fungal specificity in monotropes begins at the time of germination and seedlings associated exclusively with the same fungi that formed successful mycorrhizae with adult plants. This suggests that germination and the seedling stage are critically important to monotrope population dynamics.

4.6 | Summary and future needs

This chapter focuses on phylogenetically diverse groups of organisms that demonstrate a high degree of convergent evolution in several life history characteristics, apparently associated with their occurrence in habitats with severe environmental constraints. Most species in the groups of plants considered have small seeds (Baskin & Baskin, 1998) and many have simple or undifferentiated embryos. Following germination, embryos take divergent paths as they grow and develop into seedlings. Seedlings of carnivorous species assume adult structures and are capable of capturing prey (Fig. 4.5c–g). Embryos of orchids and mycoheterotrophic parasitic plants develop into a life history stage that is very different from juveniles and adults, and is completely dependent on mycorrhizae for nutrient acquisition. Seedlings of most epiphytic bromeliads and parasitic plants are similar in that they do not have mycorrhizal associations, but differ in how seedlings become established. Most bromeliads allocate resources to aboveground tissues through which most resources are acquired; roots are often highly reduced and primarily provide anchorage. In contrast, seedlings of parasitic species allocate most of their resources to haustoria, an adaptation that is important because seedlings must quickly become morphologically and physiologically attached to their host plant.

There are, of course, exceptions to these general descriptions but it is important to note that knowledge about seedlings of any of the groups described in this chapter is minimal. Seedlings of species in each of these groups acquire resources through nontraditional pathways. Seedlings of parasitic plants obtain resources directly from their hosts. Protocorms and seedlings of mycoheterotrophic plants including achlorophyllous orchids that interact with ectomycorrhizal fungi also obtain all of their resources from host plants, but rather than being connected directly to the host, they use host resources that have first been obtained by fungi. Seedlings of carnivorous plants obtain resources from prey and in the process of trapping prey, nutrient uptake through roots is enhanced. Seedlings of epiphytic bromeliads allocate most of their resources to nonroot structures and a variety of leaf modifications allow resources to be captured in tanks and other modified leaves.

As very few species in any of the groups of plants covered in this chapter have had extensive studies performed on seedlings, further research should focus on seedling morphology, physiology, and ecology. Particular attention, however, should be given to quantitative studies of seedling establishment and survival. This information must be obtained if we are to fully understand the ecology of these groups, and so that we may apply that knowledge to conservation and restoration issues. Perhaps the best example of this need is seen in the orchids where there is a paucity of information on the fungal requirements of protocorms, including the identification of fungal species that are essential for their growth and survival. There is even less information on the distribution and substrate requirements for nonectomycorrhizal fungi that protocorms and seedlings of many species, especially epiphytes, require.

The main conclusion is that we still know very little about the seedling ecology of these interesting groups of plants. An unlimited number of themes and challenges await our attention as we seek to integrate knowledge about the seedling ecology into our understanding of population ecology.

4.7 | Acknowledgments

Materials for this review were obtained through the Smithsonian Institution and funded by a grant to MM and DW from the National Science Foundation. Some information on orchid protocorms and seedlings came from our laboratory, which greatly benefited from the work of several undergraduate interns (Janie Becker, Robert Burnett, Brendon Hodkinson, Kelly O'Malley, Dan Sloan, Sarah Werner), of which RB and BH were supported by an REU grant from the National Science Foundation. DW especially thanks Katsu Kondo, Aaron Ellison, and Lubomir Adamec for discussions and for sharing information about carnivorous plants.

Part III

Seedling morphology, evolution, and physiology

Chapter 5

Embryo morphology and seedling evolution

Karl J. Niklas

5.1 | Introduction

The goal of this chapter is to review seedling morphology and evolution within a broad phylogenetic perspective. Its contents, therefore, are derivative of numerous and widely scattered publications.

Traditionally, *seedling* refers to the juvenile seed plant sporophyte after its emergence from the seed coat, which immediately evokes the concept of the seed itself, that is, an indehiscent, integumented megasporangium (Bierhorst, 1971; Gifford & Foster, 1988). Within this limited phylogenetic framework, a review of the morphology and evolution of the seedling is necessarily restricted to the seed plant lineages (i.e. spermatophytes) represented in contemporary floras by *Ginkgo biloba*, cycads, gnetophytes, conifers, and angiosperms.

However, the thesis advocated in this chapter is that a much broader phylogenetic perspective is required to understand seedling morphology and evolution fully because many of the features that characterize the seedling *sensu stricto* evolved well before the appearance of the first seed plants. Perhaps the most important of these features is the physical retention and physiological nurturing of the developing sporophyte within gametophytic tissues. This feature is characteristic of all land plants (i.e. embryophytes) by virtue of their (1) *diplobiontic* life cycle, in which a multicellular *diploid sporophyte* alternates with a multicellular *haploid gametophyte* to complete the sexual reproductive life cycle, and (2) their archegoniate condition, in which the developing diploid embryo is retained, protected, and nurtured within an archegonium (or its presumed vestigial remnants, e.g. synergids). Across all extant vascular plants (i.e. tracheophytes), the juvenile sporophyte eventually emerges from its enveloping megagametophyte and assumes an independent life style. Likewise, across all nonvascular embryophytes (i.e. bryophytes), the sporophyte reaches reproductive maturity and eventually dies while attached to its gametophyte. These features are absent from all known representatives of the unequivocal sister group to the embryophytes (the charophycean algae; Graham & Wilcox, 2000) and thus must trace their evolutionary origins to the time the embryophytes made their debut during early Paleozoic times (Stewart & Rothwell, 1993; Taylor & Taylor, 1993).

Data from the fossil record additionally indicate that each major tracheophyte lineage evolved the heterosporous condition (i.e. the production of both unisexual egg and sperm-producing gametophytes), which was subsequently modified to achieve a seed-like condition independently in some lineages. The most notable example of this convergent evolution among the seedless tracheophytes (i.e. pteridophytes) occurred when lycophyte megasporangia became surrounded by their leaf laminae and retained a single functional megagametophyte within which the sporophyte embryo developed (e.g. *Lepidocarpon*; see Phillips, 1979). Although fossil embryos are rare throughout the Phanerozoic, they can provide invaluable insights into the convergent evolution of the seed and seedling morphology.

A broad phyletic (and temporal) perspective is also helpful to any attempt to draw functional analogies among the different parts of land plant embryos and juvenile sporophytes. Early plant embryologists employed much the same terminology to describe seemingly similar structures produced during early sporophyte development in very different plant lineages (e.g. foot and suspensors). Even today, this terminology can imply (but does not assert) homologies among embryo body parts. This conflation in terminology unfortunately resulted in often curious and unproductive debates. It continues to obscure attempts to determine homologies among embryo body parts, particularly in the context of the evolution of the vascular plant shoot system (Gifford, 1983; Albert, 1999; Kato & Akiyama, 2005). Homologies may be resolved ultimately only by means of genomic and proteomic techniques, but a comparative approach remains useful to determine functional analogies even if homologies remain problematic.

For these reasons, this chapter reviews sporophyte embryogeny and morphology across the entire spectrum of extant embryophytes. A review of embryophyte phylogeny is presented first to establish a phyletic framework with which to map the distribution of evolutionarily significant transformations of embryo and seedling characteristics. Among the transformations that will be treated is the reversal of embryo-polarity when bryophytes are contrasted with seed plants, the presumed transfer of function from the foot of bryophytes and some pteridophytes to the suspensors and cotyledons of seed plants, and the evolution of well-defined seed dormancy especially in the angiosperms (Fig. 5.1). This briefly introduces some of the more important characteristics that will be mapped onto a recent cladistic hypothesis for embryophyte phylogeny. The salient details of sporophyte embryogeny and "seedling" structure are then reviewed, using extant taxa drawn from each of the major land plant lineages as examples. The final section presents a discussion of some of the more prominent morphological and developmental patterns emerging from a broad comparative phyletic approach. Passing reference to

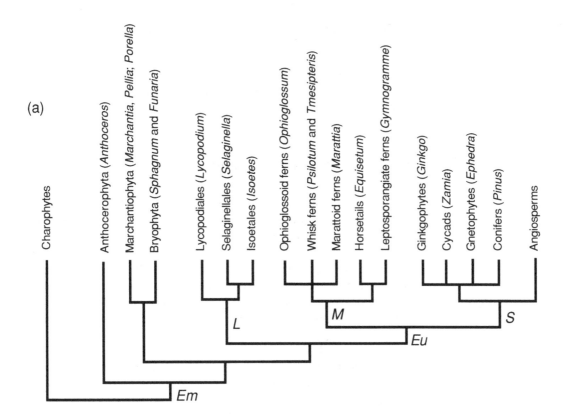

(a)

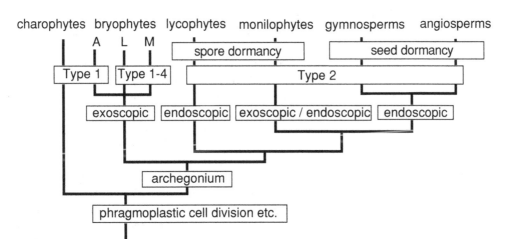

(b)

Fig. 5.1 (a) Phylogenetic hypothesis for the relationships among embryophyte lineages and their sister-group, the charophycean algae. Representative genera from each lineage treated in this review are shown in parentheses. Phylogenies redacted from Goffinet (2000), Graham & Wilcox (2000), Pryer et al. (2001, 2004), and Shaw & Renzaglia (2004). Em = embryophytes, L = lycophytes, Eu = euphyllophytes, M = monilophytes; S = spermatophytes. (b) Phylogenetic distribution of exo- and endoscopic embryogeny, gametophyte–sporophyte junction types, and seed dormancy. Phylogenetic hypothesis redacted from Fig. 5.1a and unresolved for bryophytes (A = hornworts, L = liverworts, M = mosses), monilophytes, and gymnosperms. Gametophyte–sporophyte junction types based on occurrence of transfer cells: Type-1 (on gametophyte side), Type-2 (both sides), Type-3 (sporophyte side), Type-4 (absent on both sides). See text for details.

the fossil record is made throughout these sections when appropriate. However, no attempt is made to review the fossil record in detail.

5.2 | Embryophyte phylogeny

Comparative molecular and morphological studies consistently demonstrate that the embryophytes are monophyletic and shared a last common ancestor with the extant charophycean algae (see Goffinet, 2000; Graham & Wilcox, 2000; Palmer *et al.*, 2004). However, the charophytes lack the diplobiontic life cycle in which the zygote and developing sporophyte embryo are retained within and provided nutrients by the gametophyte. It is nevertheless clear that the gametophyte generation of these plants provides protection and nutrients to developing eggs and zygotes (Graham & Wilcox, 2000).

The embryophytes were traditionally divided into the nonvascular (bryophyte) and vascular (tracheophyte) land plant lineages. However, the monophyletic concept of the bryophytes (the mosses, liverworts, and hornworts) has little recent support from analyses of 5S rRNA nucleotide sequences (Hori *et al.*, 1985), data from sperm cells (Garbary *et al.*, 1993; Renzaglia *et al.*, 2000), and even reevaluations of classical morphological data (Bremer, 1985). The current thinking is that the mosses, liverworts, and hornworts are a paraphyletic assemblage of plants. The basal-most lineage among the three bryophytic groups remains problematic (see Goffinet, 2000; Shaw & Renzaglia, 2004), although the hornworts are more consistently identified as the basal-most embryophyte group of the nonvascular plant lineages, based on analyses of nuclear and chloroplast nucleotide sequences (Mischler *et al.*, 1994; Nishiyama & Kato, 1999) and classical morphological and anatomical details (Renzaglia *et al.*, 2000). The tracheophytes are currently considered monophyletic. However, molecular studies (recently reviewed by Pryer *et al.*, 2004) support the paleobotanical perspective that the extant vascular plant lineages represent the result of a deep phylogenetic dichotomy, which occurred by early Devonian times (approximately 400 million years ago; see Taylor & Taylor, 1993; Niklas, 1997), which separated the lycophytes from all other vascular plant lineages (Fig. 5.1a). The extant lycophytes include the homosporous Lycopodiales (club mosses) and the heterosporous Selaginellales (spikemosses) and Isoetales (quillworts), which produce leaves with an intercalary meristem (microphylls) and have stems with protosteles with exarch xylem maturation. The lineage also includes a vast number of extinct life-forms, such as the arborescent lepidodendrids of Carboniferous and Permian age. The fossil record supports the view that the lycophytes are the descendents of the zosterophyllophytes, an extinct group of late Silurian and Devonian plants.

The sister group to the lycophytes is the euphyllophytes (Fig. 5.1a). All extant species in this clade are characterized by leaves that develop from marginal and apical meristems (euphylls or megaphylls)

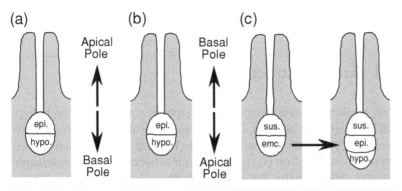

Fig. 5.2 Main types of polarity establishment during embryogeny of nonvascular and vascular plants. Shaded area denotes outline of archegonium neck (at top) and venter (at bottom). Directions of apical–basal growth indicated by arrows. (a) Exoscopic type, (b) endoscopic type lacking suspensor, and (c) endoscopic type with suspensor. epi. = epibasal cell, hypo. = hypobasal cell, sus. = suspensor, and emc. = embryonic cell. See text for details.

and a 30-kilobase inversion in the large single-copy region of their plastid genome (Raubeson & Jansen, 1992). Some cladistic analyses support the notion that the euphyllophytes are composed of two monophyletic clades, the monilophytes and the spermatophytes (Pryer *et al.*, 2001; Pryer *et al.*, 2004) (Fig. 5.2a). The former include all modern-day horsetails, eu- and leptosporangiate ferns, and the genera *Psilotum* and *Tmesipteris* (Pryer *et al.*, 2001, 2004; Renzaglia *et al.*, 2000). The monilophytes share a distinctive vasculature in which the protoxylem is confined to the lobes of the xylem strand (Stein, 1993). The spermatophytes include all modern-day seed plants, which are additionally characterized by having stems with eusteles and a bifacial vascular cambium (unless secondarily lost, e.g. monocots). The fossil record indicates that a plexus of late Silurian-Devonian trimerophytes gave rise to all modern-day euphyllophyte lineages, including the eu- and leptosporangiate ferns, horsetails, and seed plants.

However, the monophyletic status of the monilophytes is problematic when viewed in terms of the deep-time divergence of the horsetails from those that gave rise to the various fern lineages, which was nearly as ancient as the early Paleozoic lycophyte-euphyllophyte dichotomy. The inclusion of fossil taxa in formulating cladistic hypotheses can profoundly change the topology of cladograms exclusively formulated with reference only to extant taxa. For example, in their evaluation of competing phyletic hypotheses for tracheophytes (Pryer *et al.*, 2004), Rothwell and Nixon (2006) obtained very different topologies by including fossil taxa, suggesting that the ferns *sensu lato* are polyphyletic. The phyletic relationship of the horsetails and the monophyletic status of the monilophytes, therefore, remain problematic.

For convenience, however, the cladistic relationships proposed by Pryer *et al.* (2001, 2004) are accepted as a working hypothesis with which to map the phyletic distributions of various embryo, seed, and seedling characters (Fig. 5.1).

5.3 | Embryo structure

This section describes some of the embryological and morphological features that bear on the interpretation of early sporophyte or seedling development across the entire range of extant embryophytes. Emphasis is placed on (1) the distinction between exo- and endoscopic embryo development, because the former characterizes all known bryophytes, whereas the latter describes all spermatophyte embryos, and (2) terminology that unwittingly implies homologies (where none may exist) critical to the interpretation of embryophyte embryo evolution and form–function relationships.

Zygote division and embryo polarity

The first division of the embryophyte zygote is transverse to the longitudinal axis of the archegonium, with the exception of comparatively few bryophyte and tracheophyte species (e.g. the hornwort *Anthoceros* spp. and the leptosporangiate fern *Gymnogramme*). Among those species with embryos lacking suspensors (see next subsection), the cell at the bottom of the venter is traditionally called the hypobasal cell, whereas the cell nearest the neck is called the epibasal cell (Fig. 5.2a). Subsequent development of these two cells establishes one of two types of polarity with respect to the orientation of the axis of the developing embryo: (1) *exoscopic polarity*, in which the apical pole of the embryo develops from the epibasal cell, and (2) *endoscopic polarity*, in which the apical pole of the embryo develops from the hypobasal cell (Fig. 5.2a–b).

All extant bryophytes have exoscopic polarity; most vascular plants have endoscopic polarity. The principal exceptions among tracheophytes are the whisk ferns (*Psilotum*, *Tmesipteris*), horsetails (*Equisetum*), the leptosporangiate fern *Actinostachys*, and some ophioglossoid ferns, all of which have exoscopic embryogeny. Considerable variation also exists in the establishment of polarity among tracheophyte species assigned to the same genus (e.g. *Botrychium*). Neither developmental type occurs in species whose zygotes initially divide longitudinally or obliquely (e.g. *Equisetum*), in spheroidal embryos that lack axial polarity by virtue of their geometry (e.g. *Riccia*), nor in zygotes that undergo free nuclear divisions (e.g. *Zamia*).

The apical pole of all bryophyte embryos gives rise to cell lineages that contribute to the formation of the sporangium; the apical pole of all vascular plant embryos gives rise to the shoot either by means of a single apical meristematic cell or a multicellular apical meristem. With few exceptions, the basal pole of the bryophyte embryo gives rise to the foot. The basal pole of the tracheophyte embryo contributes to the cell lineages that give rise to the first root. However, Bower (1935) concluded that the roots of pteridophytes develop laterally with respect to the longitudinal axis of the embryo. Embryos with this type of root orientation were called homorhizic by Goebel (1928) as opposed to the allorhizic embryos of seed plants in which the shoot and root poles develop on opposite ends of the embryo axis. It is worth noting that the fossil record and current cladistic hypotheses about embryophyte phylogeny indicate that roots evolved independently in the lycophytes and monilophytes because these two groups shared a last common ancestor that lacked roots (and leaves). It is not clear whether this convergence is the result of a shared genomic repertoire.

The foot

As noted, the basal pole of the vast majority of bryophyte embryos gives rise to a multicellular structure called the foot. Bryophyte embryos lacking this structure are believed to have developmentally lost the capacity to form one. Some tracheophyte embryos also develop a structure called the foot (e.g. *Lycopodium*, *Equisetum*, and *Botrychium*), which typically occupies the same position as that of the bryophyte foot relative to the apical–basal embryo axis. It is not clear whether the structures produced by bryophytes and pteridophytes are developmentally homologous. However, across all embryophytes, the foot appears to function as a haustorial structure that provides a gametophyte–sporophyte metabolic junction (Ligrone *et al.*, 1993; Frey *et al.*, 1996). In the case of bryophytes, this functional role persists throughout the lifetime of the sporophyte; among tracheophytes, the foot, if present, functions only during the early stages of embryo development.

The suspensor

The basal pole of some bryophytes and many tracheophytes also gives rise to structures called suspensors. It is not clear whether those of bryophytes are homologous with those of tracheophytes, although they occupy the same position with respect to the basal pole as do the suspensors of tracheophyte embryos. Likewise, the functions of suspensors may differ among the different embryophyte lineages (see below).

The suspensors of bryophytes are unicellular and lie beneath the cells that ultimately give rise to the foot (e.g. *Pellia* and *Porella* spp.; see Fig. 5.3b,c). Among vascular plants with suspensors, the zygote divides transversely to give rise to two cells (as it typically does among tracheophyte embryos that lack suspensors). However, the uppermost cell in the venter is the parent cell of the future suspensor, which may enlarge and subsequently divide to become multicellular; the lowermost cell facing the venter of the archegonium is called the embryonic cell, which subsequently divides to produce the apical and basal poles from the hypo- and epibasal cells, respectively (Fig. 5.2c).

The growth of the suspensor is believed to orient and maintain the embryo in contact with its enveloping gametophyte from which it obtains nutrition (Gifford & Foster, 1988). This hypothesis is compatible with the growth of spermatophyte suspensors, which elongate

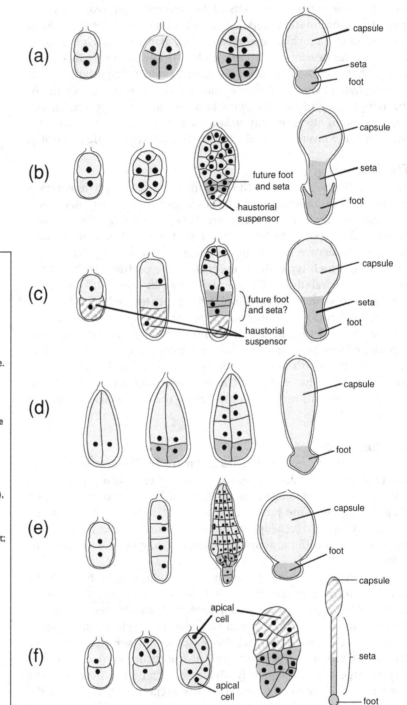

Fig. 5.3 Exoscopic embryo development of representative bryophytes beginning with first division of zygote (on left) to mature sporophyte (far right). Orientation of archegonial neck (top) and venter (bottom) is indicated by thin line in each stage. Outlined areas with black dots denote cells; outlined areas without dots denote multicellular portions of the embryo or mature sporophyte. Differently shaded areas are used to indicate shared cell lineages. (a) *Marchantia polymorpha* (thalloid liverwort; Marchantiophyta, Marchantiaceae), (b) *Pellia epiphylla* (leafy liverwort; Marchantiophyta, Pelliaceae), (c) *Porella bolanderi* (leafy liverwort; Marchantiophyta, Jungermanniaceae), (d) *Anthoceros erectus* (hornwort; Anthocerophyta, Anthocerotaceae), (e) *Sphagnum subsecundum* (moss; Bryophyta, Sphagnaceae), and (f) *Funaria hygrometrica* (moss; Bryophyta, Funariaceae). Redrawn from Parihar (1962) and Smith (1955) and references therein.

and push embryonic cells deeper into the megagametophyte, particularly among conifer species. However, there is little evidence in support of a single function for this structure across the embryophytes, nor does there appear to be any polarity in the evolution of the suspensor from the presumed ancestral bryophytic condition to the presumed derived pteridophytic or seed plant condition. Moreover, the suspensors of many pteridophyte species are small, fail to elongate, or have little or no observable affect on the orientation or contact of the embryo with its gametophyte (e.g. *Helminthostachys* spp.; see Fig. 5.5b); many endoscopic tracheophyte embryos lacking a suspensor grow and develop as vigorously as those having one (e.g. *Isoetes lithophila*, *Botrychium virginianum*, *Equisetum arvense*); and the suspensors of exoscopic bryophyte embryos may actually elevate the foot of the embryo out of the archegonium (e.g. *Pellia epiphylla*).

Cotyledons

Older studies of pteridophyte embryos often referred to the first embryonic leaves as cotyledons (e.g. Bower, 1935; see Smith, 1955). Other workers opine that the term should be reserved for the first foliar organs of spermatophyte embryos (e.g. Wagner, 1952; Gifford & Foster, 1988). There are two primary but interconnected reasons for this: (1) a homology across the first leaves of all embryophyte embryos has not been proven, and (2) the cotyledons of many seed plants function as haustorial organs, which absorb nutrients from the megagametophyte or endosperm (in the case of gymnosperms and angiosperms, respectively), whereas the first embryonic leaves of pteridophytes tend to function as foliar (photosynthetic) organs.

Regardless of the terminology employed, the size and number of embryo leaves vary widely across tracheophytes. Among lycophytes, embryonic leaves range from very small or rudimentary (e.g. *Lycopodium*) to visibly large (e.g. *Isoetes*). The first leaves of most eu- and leptosporangiate ferns are large; many are photosynthetic while still enveloped by gametophytic tissue. As far as can be determined, none of the embryonic leaves of seedless vascular plants function haustorially and all are emergent as the juvenile sporophyte grows out of the confines of its parental gametophyte. In contrast, the cotyledons of seed plants often function to absorb or store nutrients provided by the megagametophyte. Thus, the two cotyledons of cycad seedlings remain permanently within the seed; the tips of *Ginkgo* cotyledons remain within the nutritive tissues of the megagametophyte; and the cotyledon of graminoid monocots (the scutellum) never emerges from the endosperm against which it is pressed. The number of cotyledons varies dramatically among seed plants. Although the number varies widely across *Pinus* species, most conifers have two cotyledons due to the suppression or lateral fusion of cotyledonary initials early in embryogenesis (e.g. species within the Cupressaceae and Podocarpaceae, respectively). Likewise, the monocotyledonous condition of some dicot species appears to result from the abortion of a second cotyledonary set of initials (e.g. Ranunculaceae and Portula-caceae spp.).

5.4 | Embryophyte embryogenesis

The following subsections review the early embryology and growth of the sporophytes of bryophytes, seedless vascular plants, and spermatophytes. An exploration of the adaptive and evolutionary significance, if any, of the features discussed here is presented at the end of this section.

Bryophytes

The embryology of most liverwort, hornwort, and moss species is characterized by exoscopic development; cells resulting from early zygotic cell divisions that ultimately give rise to the capsule trace their lineage to an epibasal cell. This brief account of bryophyte embryogenesis is illustrated by a few representative species, shown in Fig. 5.3. The embryology for the genus *Marchantia* is well documented (Fig. 5.3a). After fertilization, the egg enlarges until it fills the venter cavity. The first division of the zygote is transverse to this longitudinal axis to form an upper epibasal and a lower hypobasal cell. Subsequent development varies among *Marchantia* species. Based on analyses of cell lineages, the epibasal tier gives rise to the cells that will eventually form the capsule and the hypobasal tier gives rise to the foot and seta (Durand, 1908). In *Marchantia domingensis*, the derivative cells of the epibasal tier produce the capsule and the upper part of the seta.

In a broad sense, while different in detail, the embryology of other bryophytes is roughly similar. In *Pellia epiphylla* (Fig. 5.3b), the hypobasal cell forms a suspensor that is reported to function eventually as a haustorium; all the cells of the sporophyte are derived from the epibasal cell. The foot of *Pellia epiphylla* is distinct in possessing a collar whose edge projects toward the neck of the archegonium. As with *Pellia epiphylla*, the hypobasal cell of *Porella bolanderi* does not divide and appears to function as a haustorium (Fig. 5.3c). The demarcation between cell lineages of the seta and the sporangium is problematic and the embryo segments that give rise to the capsule cannot be precisely identified. The embryology of the hornworts is illustrated by the genus *Anthoceros* (Fig. 5.3d). In *A. erectus*, the basal tier of cells divide to give rise to the foot; the other tiers of cells give rise to the capsule and the seta (Mehra & Handoo, 1953). The embryology of the mosses is illustrated by *Sphagnum* and *Funaria*, two taxa that have been extensively studied (Fig. 5.3e,f). In *Sphagnum subsecundum*, the first division of the zygote is transverse (Bryan, 1920). Additional transverse divisions give rise to a filamentous embryo. In *Funaria hygrometrica*, the zygote divides to form an epi- and a hypobasal cell, each of which subsequently divides longitudinally and then obliquely to produce two apical cells directed in opposing directions (Fig. 5.3f).

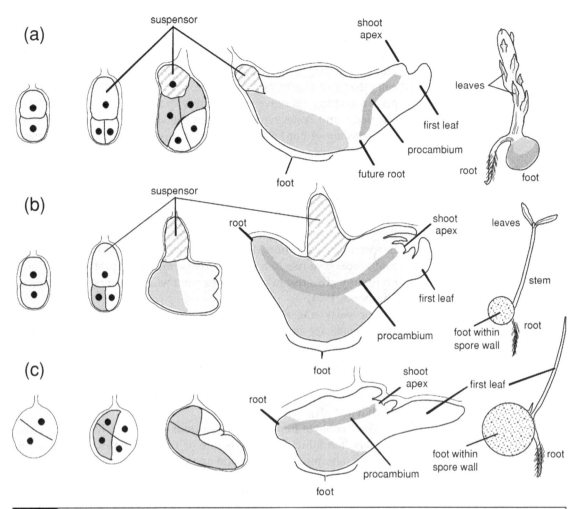

Fig. 5.4 Endoscopic embryo development of representative lycophytes, beginning with first division of zygote (on left) to young sporophyte (far right). Orientation of archegonial neck (top) and venter (bottom) indicated by thin line in each stage. Outlined areas with black dots denote cells; outlined areas without dots denote multicellular portions of the embryo or young sporophyte. Differently shaded areas used to indicate shared cell lineages. (a) *Lycopodium clavatum* (homosporous; Lycopodiales, Lycopodiaceae), (b) *Selaginella martensii* (heterosporous; Selaginellales, Selaginallaceae), and (c) *Isoetes lithophila* (heterosporous; Isoetales, Isoetaceae). (a,b) redrawn from Bruchmann (1909, 1912) and Smith (1955), and (c) Bierhorst (1971).

Thus for this species, the establishment of two apical growing points occurs early and cell lineages are almost equally divided in terms of the apical cells to which they trace their origin.

Although the fossil record of Mesozoic and Tertiary Age bryophytes is reasonably robust, we currently know nothing about the embryology of any fossil nonvascular embryophyte taxon.

Lycophytes

As in the case of extant bryophytes, considerable variation exists in the sporophyte embryology of extant lycophytes. However, all species thus far examined have endoscopic embryo development, wherein the foot comes to lie closest to the venter and the shoot and root apices are located near the neck of the archegonium (or displaced at right angles to it) (Fig. 5.4). Each of the embryos examined has procambial cells that give rise to vascular tissue, and bipolar growth from shoot and root apices that give rise to leaves and stem or roots. The foot of *Selaginella* and *Isoetes* remains within the spore wall. The embryology of *Lycopodium* is not known with great certainty because the gametophytes of many species are subterranean, significant developmental events are widely separated chronologically, and appear to be variable even within individual species (Bierhorst, 1971). The embryology of the subterranean gametophytes of *Lycopodium clavatum* is the best known of all species (Fig. 5.4a) (Bruchmann, 1898). The early stages in the embryology of species with surface-living gametophytes are similar with the exception that the four cells that form the shoot and root apices give rise to a bulbous tissue mass (the protocorm), which eventually gives rise to a stem apex from which the first root emerges (Bierhorst, 1971).

The early stages in the embryology of *Selaginella martensii* and *S. kraussiana* are similar to those of *Lycopodium clavatum* (Bruchmann, 1909, 1912; Smith, 1955; Bierhorst, 1971). The shoot apical meristem develops between the first pair of leaf primordia and the root apex develops between the bases of the first roots. Like those of *Selaginella* species, the embryos of *Isoetes* develop endosporically (Smith, 1955; Bierhorst, 1971).

The lycophyte lineage, which extends back to early Devonian and possibly late Silurian times, is one of the oldest vascular plant lineages to have morphological and anatomical fossil records (Stewart & Rothwell, 1993; Taylor & Taylor, 1993). Currently, we know nothing about Devonian lycophyte embryos, but reports of Carboniferous embryos, although rare, suggest that the early stages of embryogenesis including bipolar growth may not have differed substantially from those of *Selaginella* and *Isoetes* (Stubblefield & Rothwell, 1981). Phillips *et al.* (1975; Phillips, 1979) published the first definitive evidence for *Lepidocarpon* embryos in various stages of development. The vascularized embryos of this arborescent lycopod possessed a foot and branched xylary strand suggesting that bipolar growth of the embryo was likewise initiated early in embryogenesis at the root and shoot apices. Secondary growth, which is unknown among extant lycopods, also appears to be precocious, because *Lepidocarpon* embryos developed secondary tissues before they exited the confines of their megaspore walls (Phillips, 1979).

Monilophytes

Although the monilophytes, a sister group to the lycophytes (Fig. 5.1a), are hypothesized to be monophyletic (Pryer *et al.*, 2001, 2004), from an embryological perspective, the lineages grouped under this rubric are exceptionally diverse and include species, even in the same genus, with exoscopic and endoscopic embryogeny.

The embryology of the horsetail *Equisetum*, which is the only surviving genus of a once taxonomically and ecologically diverse lineage, illustrates exoscopic development (*E. arvense*; Fig. 5.5a) (see Sadebeck, 1902; Laroche, 1968). Whether exoscopic development occurs in all 15 or so extant *Equisetum* species is unknown. Growth and elongation of the embryonic stem and root are rapid (Smith, 1955). The root grows directly through the gametophyte and into the soil underneath. The stem emerges through the neck of the archegonium and ceases growth after developing 10 to 15 internodes. A secondary branch is produced at the base of the first, reportedly originating from primary root tissues. This second branch then produces roots at its base, a process that is reiterated to establish the rhizomatous growth that is characteristic for the majority of species.

Among the eusporangiate ophioglossoid ferns, examples of both endoscopic and exoscopic development are known (Fig. 5.5b). Indeed, this group of plants is outstanding in its extensive variation, even within the same genus. For example, *Botrychium virginianum* embryos (shown in Fig 5.5b) develop exoscopically while those of *B. dissectum* develop endoscopically (and possesses a two-celled suspensor). The gametophytes of *Ophioglossum* and *Botrychium* are subterranean and the development of their sporophytes can require 1 or more years to complete.

The embryology of the leptosporangiate ferns (Fig. 5.5c) is less varied compared to that of the ophioglossoid ferns. For these ferns, early embryos consist of four cell quadrants. The foot and root portions develop from the inner and outer posterior quadrants, respectively; the shoot apical meristem and first leaf develop from the inner and outer anterior quadrants, respectively. Detailed accounts of subsequent embryology are available (Hofmeister, 1862; Campbell, 1928; Vladesco, 1935). However, Bierhorst (1971) opines that subsequent embryo development has not been sufficiently documented or reinvestigated.

Fossil monilophyte gametophytes have been reported throughout Phanerozoic strata (for a comprehensive summary, see Taylor & Taylor, 1993). Unfortunately, however, I am unaware of any reports of monilophyte embryos that provide developmental details of early events sufficient to draw meaningful comparisons with related extant taxa.

Spermatophytes – gymnosperms

Considerable diversity has been reported for gymnosperm embryogeny. This is illustrated here by four taxa (Fig. 5.6): the cycad *Zamia*, the only surviving ginkgophyte *Ginkgo biloba*, the gnetophyte *Ephedra*, and the conifer *Pinus*, whose embryology is well known.

The early embryology of cycads involves free nuclear divisions, which are synchronized through eight successive divisions resulting in 256 nuclei (Fig. 5.6a). Subsequent nuclear divisions are not synchronized. Cell walls ultimately form and separate nuclei starting at the base of the embryo closest to the venter. In *Zamia*, the free nuclear condition is maintained near the neck of the archegonium

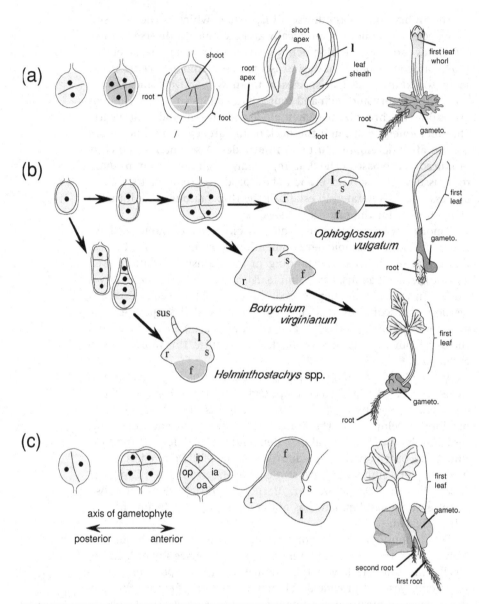

Fig. 5.5 Stages in the embryo development of representative monilophytes, beginning with first division of zygote (on left) to the young sporophyte (far right). Orientation of archegonial neck and venter is indicated by thin line in each stage. Outlined areas with black dots denote cells; outlined areas without dots denote multicellular portions of the embryo or young sporophyte. Differently shaded areas are used to indicate shared cell lineages. (a) Exoscopic development of *Equisetum arvense* embryo (horsetail; Equisetaceae). (b) Exoscopic and endoscopic embryo development of three eusporangiate ophioglossoid ferns (*Ophioglossum vulgatum* and *Botrychium virginianum*, and *Helminthostachys* spp., respectively; Ophioglossales). L = first embryonic leaf, r = embryonic root, s = site of future shoot apex; f = foot; gameto. = gametophyte. (c) Endoscopic development of the leptosporangiate fern *Gymnogramme sulphurea* (Filicales). Note reversed orientation of archegonial neck with respect to (a–b) to indicate ventral location of archegonia. Cell quadrats, located relative to anterior–posterior axes of the gametophyte and inner–outer axes of the archegonium: ip = inner posterior cell, ia = inner anterior cell, op = outer posterior cell, oa = outer anterior cell. Redrawn from Sadebeck (1878) and Laroche (1968) (a), Bierhorst (1971) (b), and various sources (c).

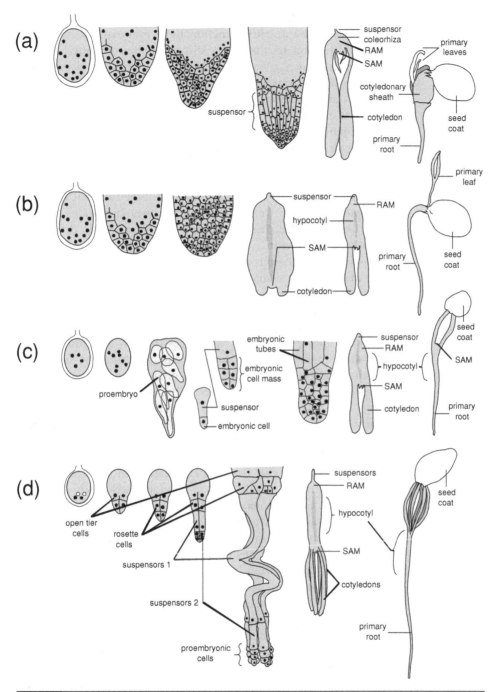

Fig. 5.6 Embryo development of representative gymnosperms, beginning with first division of zygote (on left) to the seedling (far right). Orientation of archegonial neck (top) and venter (bottom) is indicated by thin line in each stage. Outlined areas with black dots denote cells; outlined areas without dots denote multicellular portions of the embryo or seedling. Differently shaded areas are used to indicate shared cell lineages. (a) *Zamia pumila* (cycad; Cycadophyta, Zamiaceae). (b) *Ginkgo biloba* (Ginkgophyta, Ginkgoaceae). (c) *Ephedra trifurca* (Gnetophyta, Gnetaceae). (d) *Pinus* spp. (Pinales, Pinaceae). SAM = shoot apical meristem, RAM = root apical meristem. Redrawn from Bryan (1952) and Chamberlain (1935) (a), author's observations (b), Khan (1943) and Lehmann-Baerts (1967) (c), and a variety of sources (d).

(Bryan, 1952); the free cellular and highly vacuolated region of the embryo may serve a nutritive function, but it eventually disintegrates. Active cell division results in the differentiation of a meristematic zone of cells, which gives rise to the main embryonic organs, and a more posterior zone of elongating cells, which forms a massive multicellular suspensor. In addition, a distinctive cap of superficial cells forms around the tip of the *Zamia* embryo that is pushed into its enveloping megagametophyte.

Although cleavage of the apex can produce separate embryonic cell clusters, only one of the isogenic embryos typically survives (Chamberlain, 1919, 1935). Except for *Ceratozamia*, whose embryos possess a single cotyledon, most cycad embryos have two that flank the shoot apical meristem. Embryos may form several bud scales and first foliage leaf primordia. The radicle develops from cells in direct contact with the suspensor.

Among cycads, there is no fixed period for seed dormancy; germination may occur immediately after seeds fall to the ground. However, cycads typically have underdeveloped embryos and thus manifest *morphological dormancy*; three *Zamia* spp. manifest what has been called *morphophysiological dormancy* (Baskin & Baskin, 1998). During germination, the micropylar end of the seed is the first to rupture as the coleorhiza elongates and the root tip emerges from the seed coat. The haustorial cotyledons remain within the seed coat and absorb nutrients from the megagametophyte. The cotyledonary sheath and primary leaf, however, emerge soon after the seed coat is ruptured.

Early *Ginkgo* embryogeny is similar to that of cycads by being characterized by a period of rapid free nuclear divisions (Fig. 5.6b). However, after eight nuclear divisions, cell walls form from the outside toward the center to give rise to a completely cellularized embryo, which, unlike that of cycads, fails to develop a well-defined suspensor. As in the case of *Zamia*, the basal end of the *Ginkgo* embryo assumes the status of a meristem pressed against the venter from which the shoot apical meristem and cotyledons develop. The radicle develops from cells immediately behind this meristematic region (Ball, 1956a,b). Germination is similar to that of cycads. The primary shoot and root emerge after the micropylar end of the seed is ruptured, and the tips of the cotyledons, which are haustorial, remain in contact with the megagametophyte. Fertilization and early embryo development are reported to occur after immature ovules drop to the ground. Dormancy is characterized as a form of morphophysiological dormancy (Baskin & Baskin, 1998).

The embryology of the three genera in the Gnetales is as diverse as the morphology and anatomy of adult plants (Gifford & Foster, 1988). The embryology of *Ephedra* is reviewed here to the exclusion of that of *Gnetum* and *Welwitschia* because it is best known (see Khan, 1943; Lehmann-Baerts, 1967).

As in the case of cycads and *Ginkgo*, a series of free nuclear divisions soon follows the fertilization of the egg (Fig. 5.6c). Each of the resulting diploid nuclei is subsequently enveloped by a densely staining sheath of cytoplasm that later forms a cellulosic wall. This process can be viewed as a highly precocious form of simple polyembryony because each of the resulting proembryonic cells can develop independently to form an embryo. Each elongates and then divides transversely to form a suspensor cell facing the neck of the archegonium and an embryonic cell that is pushed toward the venter where it eventually ruptures the egg cell wall and enters the megagametophyte. The embryonic cells subsequently divide to form an embryonic cell mass.

Further development results in a shoot apical meristem flanked by two cotyledons and a root apical meristem beneath the suspensor cells. Typically, only one of the many embryos survives in each ovule; the others are resorbed. The report of *Ephedra trifurca* seeds germinating while still attached to mature plants (Land, 1907) has not been verified, but seeds appear to have physiological dormancy (Baskin & Baskin, 1998). As in the case of cycads and *Ginkgo*, the cotyledons remain within the seed coat in contact with the megagametophyte.

Like that of most conifers (and all cycads and *Ginkgo biloba*), the *Pinus* zygote undergoes free nuclear divisions, resulting in four free nuclei located toward the venter (Fig. 5.6d). The second set of nuclear divisions is synchronized and accompanied by wall formation to produce four nuclei sharing the same cytoplasm but partially separated by wall in-growths (the open tier 'cells') and a quartet of nuclei completely separated by cell walls adjacent to the venter. The cells of this closed tier divide in sequence. The quartet of cells closest to the open tier, called the rosette cells, may undergo a limited number of cell divisions. The tier of cells closest to the rosette cells elongate and divide to form one or two sets of suspensor cells that push terminal clusters of proembryonic cells deeper into the megagametophyte. The occasional separation of these embryonic cells results in simple polyembryony, as in the case of cycads. Typically, only one *Pinus* embryo develops within a seed. It consists of a whorl of cotyledons, varying in number with species and surrounding the shoot apical meristem surmounting a short hypocotyl with a root meristem at its base. The entire *Pinus* embryo emerges from the seed coat, which may adhere to the tips of cotyledons.

This embryology is highly conserved across pine species. However, it is not representative of the majority of other conifers whose embryologies are more reminiscent of intermediate stages of *Ephedra* embryo development (Doyle, 1963). For most conifers, the first two free nuclear divisions of the zygote nucleus are followed by other divisions to produce a cell with 16 to 32 nuclei. These nuclei migrate to well-defined positions before cell wall formation is initiated to produce a group of primary embryonic cells and a primary upper cell tier (Fig. 5.7). All of these cells divide, doubling the number of embryonal tube cells that function as suspensors by pushing embryonic cells deeper into the megagametophyte. As in the case of *Pinus*, simple polyembryony may occur (e.g. *Abies* and *Picea*).

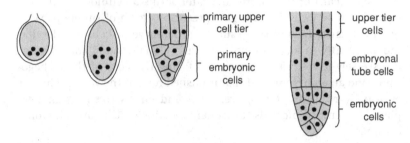

Fig. 5.7 Stages in early embryogeny of conifers (compare with *Pinus* Fig. 5.6). Orientation of archegonial neck (at top) and venter (at bottom) indicated by thin line in first two stages. Redrawn from Doyle (1963).

The oldest known gymnosperm seeds are those of the late Devonian fossil *Elkinsia polymorpha* (Rothwell *et al.*, 1989). Unfortunately, *E. polymorpha* embryos have not been found, nor are there any reports of other Devonian seed plant embryos. However, the fossil record indicates that massive suspensors, simply polyembryony, and embryos with multiple cotyledons are comparatively ancient attributes. Mapes *et al.* (1989) describe seeds from the uppermost Carboniferous containing embryos with six cotyledons surrounding an epicotyl; Smoot and Taylor (1986) describe *Plectilospermum* seeds containing multiple embryos (polyembryony) with suspensors.

Spermatophytes – angiosperms

The embryological details of diverse flowering plants species have been extensively studied and synoptically summarized (Schnarf, 1929; Johansen, 1950; Maheshwari, 1950). Although evolutionarily highly conservative in many respects, the embryology of monocots and dicots evinces remarkable variation and diversity as, for example, the prolific free nuclear divisions and resulting polyembryony attending the embryogenesis of *Paeonia* (Yakovlev & Yoffe, 1957) and the development of the so-called monocotyledonous dicots (e.g. *Claytonia virginica* Portulacaceae; see Haccius, 1954; Haccius & Lakshmanan, 1966).

The extensive literature treating angiosperm embryogenesis indicates that the endoscopic embryo typically begins with an unequal transverse division of the zygote (Fig. 5.8a), although exceptions such as *Juglans nigra* (Juglandaceae) and species of the Piperales involving oblique or longitudinal zygotic divisions are well documented (see Johansen, 1950; Maheshwari, 1950). The smaller and larger of the two cells are called the terminal and basal cell, respectively. In the case of the dicot *Capsella bursa-pastoris* (Brassicaceae), the terminal and basal cells divide longitudinally and transversely to yield a tetrad of cells. A filamentous suspensor develops from the basal cell and its derivative cell, which are located closest to the micropylar end of the ovule; a spherical cluster of proembryonic cells develops from successive divisions of the two terminal cells. The suspensor cell closest to the embryonic cell cluster is an important locus of differentiation for the

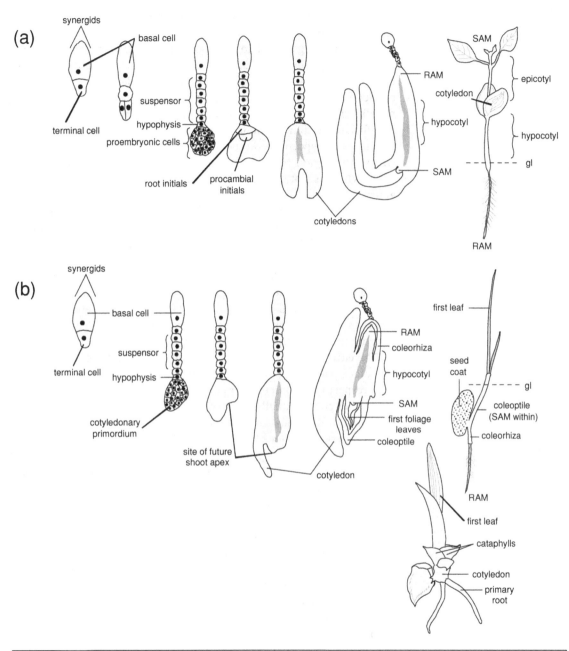

Fig. 5.8 Embryogeny of angiosperm embryos (oriented with respect to synergids), beginning with the division of the zgote (on left) and to the seedlings (far right). Outlined areas with black dots denote cells; areas without dots denote multicellular portions of the embryo or mature sporophyte. (a) Development of *Capsella bursa-pastoris* (Brassicaceae) and a stereotypical dicot seedling with epigeal germination. (b) Development of *Ottelia alismoides* (Hydrocharitaceae) and a stereotypical monocot seedling with hypogeal germination. SAM = shoot apical meristem, RAM = root apical meristem, gl = ground level. Developmental stages redrawn from Souèges (1919) (a) and Haccius (1952) (b).

root apical meristem. Its derivatives are the initials for the root cortex, protoderm, and root cap (Souèges, 1919). Likewise, the embryonic cluster cells lying closest to the root cortex initials become procambial initials (Souèges, 1919). Two ridges of cells, which develop at the free end of the now heart-shaped embryo, demarcate the pair of cotyledonary primordia. Undifferentiated cells between the two primordia give rise to the future shoot apical meristem. Cell divisions and differentiation in the basal portion of the heart-shaped embryo produce the hypocotyl; subsequent cell divisions and differentiation within the region of the shoot apical meristem give rise to the epicotyl, which is not well developed until seed germination. The root apical meristem develops at the base of the hypocotyl.

With the exception of some dicot embryos with one or three cotyledons (see Chapter 2), the vast majority has two lateral cotyledons between which lies a shoot apical meristem. In contrast, the monocot embryo typically possesses one lateral cotyledon (and an opposing structure, the epiblast, that has been interpreted by some to be a reduced second cotyledon (however, see Tillich, 2000). Its shoot apical meristem, which appears to be lateral, is considered to be terminal in origin for those taxa that have been studied carefully (e.g. Haccius, 1952; Haccius & Lakshmanan, 1966).

The early stages of monocot embryogeny parallel those of the typical dicot (Fig. 5.8b). The first zygotic division, which is transverse with respect to the micropylar chalazal axis of the megagametophyte, establishes two unequally sized cells, the basal and terminal cell; subsequent cell divisions give rise typically to a filamentous suspensor and a spherical cluster of embryonic cells.

Nevertheless, the appearance of the mature monocot embryo, as well as the typical mode of seedling establishment, differs from those of the dicot. Perhaps the most striking specializations of monocot embryology are the coleoptile and coleorhiza, which respectively surround the shoot and root apices of some monocot embryos. Another is the appearance of the scutellum, a specialized discoid haustorial cotyledon (Fig. 5.8b). Features, such as the coleoptile, appear to be functionally adaptive. For example, the hypogeal germination of *Triticum aestivum* (Poaceae) seedlings results in the concertina-like mechanical collapse of leaves when the intercalary growth of the hypodermic-like coleoptile fails to penetrate the soil crust (Niklas & Paolillo, 1990).

5.5 | Phylogenetic patterns

The variation in embryogeny reviewed in this chapter indicates that the details of sporophyte development have been altered evolutionarily in diverse and often dramatic ways without apparent lethal results, which supports the widely held view that plant development is capable of more nonlethal plasticity than animal development. The sole, apparently nonnegotiable, feature is to keep the embryo in contact with its megagametophyte until the sporophyte achieves physiological independence in the case of tracheophytes or reproductive maturity (attended or followed by death) in the case of bryophytes. Yet, even this feature has been achieved in one of several evidently equally successful ways during the course of land plant evolution.

Nevertheless, a few phylogenetic patterns emerge when embryological features are mapped onto current cladistic hypotheses about the relationships among the various embryophyte lineages (Fig. 5.1b). Among the most striking of these are (1) the reversal of embryo polarity when bryophytes are contrasted with seed plants, (2) the presumed transfer of function from the foot of bryophytes and some pteridophytes to the suspensors and cotyledons of seed plants, and (3) the evolution of well-defined dormancy in spermatophytes.

Before these features are reviewed individually, it is useful to draw a distinction between an *evolutionary trend* and a *phylogenetic pattern*. As used here, a *trend* implies that ancestral to derived character transformations manifest some sort of directionality during the course of evolution, whereas a *pattern* indicates that the distribution of one or more characters within a clade can be interpreted in ecological or functional terms. The distinction between the two terms is important because many of the features discussed below are poorly preserved in the fossil record, which makes speculations about their ancestral character states highly problematic.

Embryo polarity

It is generally agreed that the most ancient embryophytes had a diplobiontic life cycle, archegoniate gametophytes, and lacked vascular tissues (Stewart & Rothwell, 1993; Taylor & Taylor, 1993; Niklas, 1997). This view is consistent with current phyletic hypotheses that identify, albeit not invariably, the bryophytes as a paraphyletic assemblage (Goffinet, 2000; Shaw & Renzaglia, 2004). The basal status of the bryophytes suggests that these three lineages are the survivors of numerous evolutionary experiments spawned by an ancient plexus of plants, which was in some important respects perhaps not too different from modern-day charophycean algae. If true, then exoscopic embryogeny is the ancestral condition of the first land plants and the endoscopic embryogeny of the seed plant lineages is a derived condition. This speculation is consistent with the embryological heterogeneity observed across modern-day pteridophyte lineages, which include endoscopic and exoscopic species. The retention of exoscopic development in seemingly very different tracheophyte lineages (e.g. the horsetails and ophioglossoid ferns) suggests that endoscopic embryogeny evolved independently.

Importantly, it is clear that embryo polarity is not established by the plane of zygotic cell division. This is clearly illustrated by comparing the embryogeny of *Anthoceros* and *Funaria* (see Fig. 5.3d,f). The former begins with a longitudinal division. The latter begins with a transverse division. Yet, the foot and sporangium occupy the same positions with respect to the axis of the archegonium and the maturing embryo. Accordingly, polarity is established by the cytoplasmic attributes of the zygote. Unfortunately, little is known about the mechanisms responsible for this establishment. In *Arabidopsis*, Long *et al.* (2006) suggest that polarity is a two-stage phenomenon with axis formation occurring during the first cell divisions (and relying on axisymmetric polar auxin transport; also see Friml *et al.* 2003) and subsequent axis fixation requiring a chromatin-mediated transcriptional repression system for axis stabilization. This scenario is conceptually similar to the mechanism proposed for polarity establishment in the brown alga *Fucus*, where axis formation and fixation are temporally distinct events. Whether it is a generalized phenomenon remains unclear.

Speculations regarding the functional or adaptive value, if any, of either type of embryogeny must be tempered with the recognition that both types occur in the same genus (i.e., *Botrychium*) and are equally successful. However, the exoscopic development of bryophyte embryos is eminently compatible with the emergent growth of the short-lived sporophyte generation and its aerial release of spores. By the same token, endoscopic development may arguably be seen as a precondition for the evolutionary acquisition of the seed habit in which sporophyte development initially proceeds within the confines of the megagametophyte (see Fig. 5.4). This speculation resonates with the convergence on the seed habit by some ancient lycophytes and the presence of endoscopic embryos in all modern-day lycophytes.

Placental junctions and the foot

Comparative studies of the gametophyte–sporophyte junctions of bryophytes and pteridophytes, which are typically evident around the foot region of embryos, have revealed structurally significant differences among all the major groups of bryophytes and those pteridophyte species thus far examined (Ligrone *et al.*, 1993; Frey *et al.*, 2001; Carafa *et al.*, 2003; Duckett & Ligrone, 2003). At the grossest level of comparison, this junction is lost after embryo maturation among tracheophyte species, whereas, among bryophytes, it is permanent, prominent, and highly differentiated structurally.

One the most important diagnostic features of the junction is the distribution of transfer cells, which has resulted in a *Type* classification system (Ligrone *et al.*, 1993; Carafa *et al.*, 2003) (Fig. 5.1b). The presence of the Type-1 junction, where transfer cells occur only on the gametophytic side, in the hornworts (and many other bryophytes) and of a similar kind of junction at the gametophyte–zygote interface of the charophycean alga *Coleochaete* (Graham & Wilcox, 2000) suggests that this distribution of transfer cells is ancient. Likewise, the phyletic distribution of the Type-2 junction, with transfer cells on both sides, in some mosses and liverworts as well as the lycophytes, monilophytes, and spermatophytes suggests that this junction type is a shared derived feature of all other embryophyte lineages (Fig. 5.1b).

If true, then junctions Type-3 (transfer cells only on the sporophyte side) and Type-4 (no transfer cells) are derivatives, reflecting the loss of transfer cells on one of the two junction sides.

The classification of junctions into only four types undoubtedly belies considerable diversity within each type and cautions against far-reaching speculations about evolutionary trends. Nevertheless, in all cases examined, the junctions of bryophytes and tracheophytes lack evidence of symplastic transport by means of plasmodesmata; metabolic trafficking thus appears to involve apoplastic diffusion followed by active membrane transport.

Suspensors

The preceding indicates that a variety of developmental modifications have evolved that establish and maintain to varying degrees a close physiological interface between the two multicellular generations in the diplobiontic life cycle of land plants. In this context, it is worth noting that the endoduplication of DNA and resulting polyploidy occurring in the suspensor cells of most if not all seed plants have been interpreted to provide gene products required for the growth of the embryo (Raghavan, 1986). In addition to orienting and pushing the embryo into the megagametophyte or endosperm, the suspensor cells of seed plants also appear to function as transfusion tissue that absorbs nutrients from surrounding tissues. This feature parallels the establishment of the gametophyte–sporophyte junction from the foot and suspensors of bryophytes and the suspensor derivatives in the developing embryos of *Lycopodium cernuum* and other pteridophytes (Duckett & Ligrone, 1992).

The haustorial function of angiosperm suspensors, which vary in size from filaments to biseriate columns of short cells to massive elongate structures six cells wide, is nowhere more evident than in those species in which suspensors penetrate the nucellus and integuments before seed germination (Schnarf, 1929). As in the case of bryophytes and pteridophytes, there is no evidence of symplastic transport across the gametophyte–suspensor cell interface; wall in-growths, if present, may amplify the surface areas of cell membranes, thereby expediting active transport at membrane surfaces.

Dormancy and the seed habit

Unlike bryophytes, pteridophytes, and most extant gymnosperms, the embryos of angiosperms typically undergo a period of *innate dormancy*, here defined as the biologically programmed suppression of germination under conditions that are otherwise favorable to growth by virtue of some property of the embryo or its associated tissue, a condition that is prefigured among pteridophytes by the production of spores capable of delayed germination (Fig. 5.1b).

The various mechanisms responsible for seed dormancy, which are treated extensively in the plant physiology literature, include (1) incomplete development of the embryo, (2) control by a biochemical or physical trigger, (3) the removal of an inhibitor, (4) physical restriction of water or air, and (5) genetic controls (Harper, 1977). Not all angiosperm species produce dormant seeds, and examples of dormancy can be found among many extant gymnosperms. The latter are by no means rare and include species that retain seeds in cones until scorched by fire (e.g. *Pinus contorta*), release ovules or seeds with immature embryos (e.g. *Ginkgo biloba*), produce seeds that require low temperatures or specific wavelengths of light, and species that form mechanically restrictive seed coats (e.g. *P. koraiensis*) or that block air from reaching embryos (e.g. *Juniperus virginiana*) (Schopmeyer, 1974; Young & Young, 1992; Baskin & Baskin, 1998).

The prevalence of seed dormancy among extant species and the apparent absence of it among the most ancient seed plant fossils beg the question, what were the selective advantages of the seed habit before seed dormancy acquired its present-day broad taxonomic distribution? This question is answered traditionally by drawing attention to the benefits of the indehiscent megasporangium, which provides mechanical protection to the megagametophyte and thus the embryo(s) that develop within it. This stage in the evolution of the seed habit was achieved by the lepidodendrid lycophytes (Phillips, 1979) for which endoscopic rather than exoscopic embryogeny might have provided an additional selective advantage. The subsequent or concurrent evolutionary reduction and eventual elimination of the megaspore wall would have provided an opportunity for the exchange of water and metabolites among the mature sporophyte, its retained megagametophyte(s), and the embryonic sporophyte(s) within, that is, a sporophyte–gametophyte–sporophyte junction. Finally, the evolution of a megasporangium with integuments (i.e. the ovule *sensu stricto*), which has been the subject of intense speculation, would have conferred additional protection to the megagametophyte as well as an opportunity to modify integuments for seed dispersal.

Although this adaptationist hypothesis suggests that the evolution of the seed habit was a linear procession from one stage to the next, it is equally reasonable to argue that all or most of these posited character transformations occurred in a comparatively brief span of evolutionary time. The explosion of morphological variation in ovule structure during late Devonian and early Carboniferous times suggests a rapid episode in early spermatophyte evolution and subsequent winnowing by natural selection (Andrews, 1963; Niklas, 1983, 1997).

Cotyledons and seedlings

As noted, early plant embryologists referred to the first embryonic leaves of pteridophyte embryos as cotyledons. This practice is currently eschewed and the term is reserved exclusively for embryonic structures of spermatophytes. This distinction is predicated more on functional differences than on the basis of problematic homologies among the photosynthetic organs of embryophytes, which have evolved independently at least three times (in bryophytes, lycophytes, and euphyllophytes). The embryonic leaves of pteridophytes almost invariably function photosynthetically; the cotyledons of spermato-phytes function as absorptive (haustorial) organs, storage organs, and/or as photosynthetic structures depending on species.

Among extant gymnosperms, cotyledons function as photosyn-thetic and/or absorptive organs. However, conifer cotyledons become green even in the dark and are reported to be absorptive, whereas those of dicots require exposure to light to develop functional chloro-plasts. Cotyledons of monocots can function primarily as photosyn-thetic, protective, storage, or absorptive organs (Tillich, 2000, and ref-erences therein). The presumed ancestral condition is a *compact cotyle-don* "characterized by a low sheath (hypophyll) with or without small or inconspicuous appendages and a completely haustorial hyperphyll which is embedded in the seed" (e.g. species in the Alstromeriaceae and Dracaenaceae); the most derived condition occurs in species with seeds lacking endosperm, where the cotyledonary hyperphyll serves as a storage organ (e.g. species in the Alismatidae and Araceae) (Tillich, 2000). Tillich opines that the ancestral monocot condition also includes a vigorously growing, freely branched, and persistent primary root. The stereotypical (and presumably highly derived) grass embryo (with a coleoptile, coleorhiza, and scutellum architecture) and seedling (with hypogeal germination and ephemeral juvenile root), therefore, belies a far more complex and diverse repertoire among the monocots.

Attempts to map the potentially manifold functions of cotyle-dons onto phylogenetic hypotheses for dicots are more problematic. For example, based on a long-term study of the seedlings of 1251 dicot species in 740 genera and 157 families, Ji and Ye (2003) propose the following ancient-to-derived trend for cotyledon function: absorp-tive, absorptive and photosynthetic, photosynthetic, photosynthetic and storage, and storage. Yet, within many dicot families, closely related species can possess cotyledons that function very differently in part because of the mechanical constraints and physiological require-ments correlated with hypo- or epigeal germination (e.g. *Pisum* and *Phaseolus*).

The storage function of some cotyledons, and the biological impo-sitions of epi- versus hypogeal germination bear on attempts to define the seedling concept in the context of dicots. For example, in their study *Pisum sativum* and *Helianthus annuus* (a hypogeal and epigeal species with nonphotosynthetic and photosynthetic cotyle-dons, respectively), Hanley *et al.* (2004) report that relative growth dur-ing the germination of *Pisum sativum* reaches its maximum rate as the cotyledonary reserves of nitrogen and phosphorus become exhausted, which coincides with the stage at which the removal of cotyledons had no noticeable affect on subsequent plant growth. Although the relative growth rates of *Helianthus annuus* also reach their maximum at the stage when the removal of cotyledons has no effect, this stage in plant establishment was not correlated with declines in some essential nutrients stored in these structures, presumably because *Helianthus annuus* cotyledons import nutrients during photosynthesis that have been absorbed by roots. Hanley *et al.* (2004) conclude that the attainment of maximum relative growth rates, which is a distinctly identifiable event, provides a reliable marker that can be used to define the dicot seedling stage (Chapter 8). Whether a similar definition holds true for the seedlings of monocots or parasitic species remains unclear.

5.6 | Concluding remarks

This review highlights a number of morphological and embryological patterns that emerge when some well-studied character states are mapped onto current phyletic hypotheses about the evolutionary relationships among extant embryophyte lineages. Care has been taken to make a distinction between the terms *pattern* and *trend*. The former is used to describe the distribution of characters on the terminal branches of cladograms (Fig. 5.1b). As such, it does not imply ancestral-to-derived character transformations, as does the word *trend*.

This distinction is important because (1) in many cases, the fossil record does not provide unambiguous evidence for the ancestral condition of many important embryological or morphological features, (2) phylogenetic hypotheses are often subject to major revision as new information becomes available, and (3) seemingly dramatic developmental transformations may involve relatively minor and reversible changes. The latter is well illustrated by a comparative study of seedling germination across wild and domesticated cassava (*Manihot* Euphorbiaceae) species by Pugol *et al.* (2005), who report that hypogeal germination and the retention of cotyledons within the seed testa characterize wild subspecies (e.g. *M. esculenta* spp. *flabellifolia*), whereas the seedlings of the domesticated subspecies (*M. esculenta* ssp. *esculenta*) have epigeal germination and produce foliaceous cotyledons. Pugol *et al.* (2005) speculate that these seedling transformations are the result of human selection for high seedling growth rates, which favored epigeal germination and photosynthetic cotyledons. For these reasons, the validation of evolutionary trends is notoriously difficult and often the subject of excessive speculation.

The resolution of many substantive issues, such as the establishment of homologies among the parts of embryophyte embryos, has been elusive despite intensive and detailed embryological investigations spanning well over one century. In part, these efforts have been hampered by the search for broad-sweeping generalizations where none may exist, particularly in light of the manifold variations in embryophyte embryology that achieve non-lethal results. The current molecular revolution in our understanding of plant development, coupled with more refined and reliable phylogenetic hypotheses, may resolve many of these issues. Until that time, a conservative approach to interpreting the evolution of the seed and seedling is warranted.

The patterns that emerge from this review are nevertheless suggestive of evolutionary trends. As such, they provide fertile ground for study. Among these, two were emphasized because they appear to attend major evolutionary innovations across the long history of the embryophytes: (1) the bryophyte-to-spermatophyte transformation from an exoscopic to an endoscopic embryogeny, (2) the apparent transfers of function from the foot-suspensor gametophyte–sporophyte junction to the suspensor–cotyledon junction of spermatophytes, (3) the evolution of apparently more well-defined seed dormancy types across gymnosperm and angiosperm species, and (4) an ancient-to-derived transition in cotyledonary function from absorptive to storage for both monocots and dicots.

Chapter 6

Regeneration ecology of early angiosperm seeds and seedlings: integrating inferences from extant basal lineages and fossils

Taylor S. Feild

6.1 | Introduction

Flowering plants – angiosperms – are presently the most ecologically significant lineage on the green-plant tree of life. With the exception of high latitude and some upland regions, angiosperms dominate both the species number and biomass of the world's major biomes. Furthermore, the chemistry, productivity, and structure of angiosperms are the foundation of ecological webs of biotic interactions that generate and sustain terrestrial biodiversity.

Yet, despite their entrenched and manifold roles in today's ecosystems, angiosperms represent the youngest major group of terrestrial plants (Wing & Boucher, 1998; Boyce, 2005). Compared to other major lineages of plant evolution that appear in the fossil record during the mid- to late Paleozoic, angiosperms are relative newcomers with the first undisputed fossils appearing nearly 100 Ma later during the Early Cretaceous. This pattern raises the question: How did angiosperms achieve such tremendous diversity and nearly singular ecological dominance over a relatively brief geological interval ($\approx$100–60 million years ago; Wing & Boucher, 1998; Lupia et al., 1999; Nagalingum et al., 2002)? Answering this question is difficult because the deep time phases of angiosperm evolution, preeminently their early biology and causes of initial success, remain uncertain (Wing & Boucher, 1998). Ultimately, the early angiosperm enigma clouds the resolution of the biological and environmental mechanisms sparking the rise of the modern day angiosperm epoch and diverse biotas that co-evolved with them.

However, one important evolutionary pattern appears clear. The fossil record suggests that angiosperms did not vault to dominance by simply expanding into ecological niches emptied out by a catastrophic mass extinction event such as a bolide impact. Rather, the Early Cretaceous debut of the angiosperms and their early diversification appears to have been embedded in a complex community context (Wing & Boucher, 1998; Lupia *et al.*, 1999). Pre-angiosperm Mesozoic communities were codominated by a diverse mid- to late-Paleozoic-originating assemblage of gymnosperm (cycads, conifers, ginkgoes, and many other nonextant lineages) and free-sporing (lycopod and fern) clades (Knoll, 1986; Retallack & Dilcher, 1988; Bond, 1989; Lupia *et al.*, 1999; Nagalingum *et al.*, 2002; Boyce, 2005). From the late Early Cretaceous onward (Knoll, 1986; Lupia *et al.*,1999; Boyce, 2005), the deep phylodiversity of plant communities pinholed as most older gymnosperm and fern clades diminished in abundance or diversity, or became extinct. Angiosperm communities appearing in the Cretaceous, however, were distinctly different from pre-angiosperm ones. These communities exhibited much greater species diversity, and they contained radically new growth forms and ecophysiological performances (i.e. ruderal herbs). Thus, ecology – including biotic and physical forces as well as interactions between them – was intimately associated with the ascent of angiosperms. But how did this ecology permit the initial roothold of early angiosperms in intact Mesozoic communities and drive their subsequent radiation (Feild *et al.*, 2004; Feild & Arens, 2005, 2007)?

Considering the possible functional contexts in which to couch the ecological radiation of the angiosperms, the ecological dynamics associated with seeds and seedlings stand out as particularly important. For example, the regeneration phases of plant life history, encompassing seed germination, seedling emergence, and seedling establishment, are critical in determining the fate of individual plants that in turn shapes community dynamics and species co-existence (Augspurger, 1984b; Garwood, 1996; Kitajima, 1996a; Lusk & Kelly, 2003; Poorter, 2007). Indeed, some have speculated that the invasion of angiosperms into the regeneration niches of conifers as well as ferns and other gymnosperms was the ecological mechanism responsible for the decline of these groups during the Cretaceous (Bond, 1989; but see Brodribb *et al.*, 2005). Yet, until recently, little has been known about the ecology of early angiosperms and the habitat context under which such regenerative replacement may have occurred.

The aim of this chapter is to review and synthesize the wealth of new phylogenetic, paleobotanical, and organismic discoveries that are providing fresh perspectives on the regeneration niches of early angiosperms. I begin by briefly considering previous opinions on the morphology, functional biology, and ecology of early angiosperm seeds and seedlings. Next, I discuss how the identification of extant "basal" angiosperm lineages by molecular phylogenies and understanding the ecology of their seeds and seedlings has recast the imagery of how early angiosperms functioned in their environments and the directions of their early ecological evolution. I examine throughout the chapter how this emergent picture relates to the Early Cretaceous fossil record of early angiosperms. I finish with some thoughts on possible directions of future research on the ecology and evolution of angiosperm regeneration biology.

6.2 | Previous views of ancestral angiosperm ecology and seed/seedling morphology

The woody Ranales or woody magnoliid hypothesis

One of the earliest views on the ancestral ecology of angiosperms is the woody Ranales or woody magnoliid hypothesis (Arber & Parkin, 1907; Bews, 1927). This hypothesis was associated with a view of the ancestral angiosperm flower as a large, numerous-parted (strobiloid) structure. Such flowers are found only in living Magnoliales (e.g. Annonaceae, Magnoliaceae, Myristicaceae; Fig. 6.1) and Winteraceae (Arber & Parkin, 1907). Taking the ecology of these groups at face value, the first flowering plants were proposed to be stately, canopy-reaching trees or perhaps large shrubs that grew and matured slowly in wet tropical lowland rain forests or cloud forests (Bews, 1927). Although no proponents of the woody magnoliid hypothesis explicitly discussed the regeneration ecology of ancient magnoliids (Table 6.1), extending the analogy of living Magnoliales and Winteraceae to early angiosperms predicts that early angiosperms possessed large seed sizes, cotyledons functioning in storage, and monopodial (single growth axis) seedling architecture (Blanc, 1986; Ibarra-Manriquez et al., 2001; Feild et al., 2004; Moles et al., 2005b; Feild & Arens, 2005, 2007). This suite of functional traits is most closely related to regeneration in wet, stable (nonerosional and possibly infrequent stem damage), light-limited understory habitats choked with heavy leaf litter (Garwood, 1996; Grubb & Metcalfe, 1996; see Chapters 8, 10).

Analogizing early angiosperms to extant Myristicaceae, Corner (1976) conjectured that the first angiosperm seeds were moderately large (i.e. 10–15 mm long) and possessed complex testa anatomy. Corner suggested that seed size and structural complexity decreased in tandem as angiosperms diversified. Also relating specifically to seed morphology, Martin (1946) proposed that Magnolia-like seeds housing small to minute, underdeveloped embryos embedded in copious endosperm were primitive among angiosperms.

Disturbance–adaptation: angiosperms as ancestrally weedy

The 1960s marked a dramatic shift in favored opinions on early angiosperm ecology (Stebbins, 1965). Instead of viewing early angiosperms as marginally competitive slow-growing trees, several workers converged on the idea that angiosperms were initially highly competitive, and adapted to unstable, disturbed habitats. In these zones, angiosperms were proposed to have originated with high competitive ability and able to grow and reproduce much faster, cheaper, and therefore "better" than co-occurring Cretaceous gymnosperms and ferns (Stebbins, 1965, 1974; Doyle & Hickey, 1976; Hickey & Doyle, 1977; Doyle & Donoghue, 1986; Bond, 1989; Taylor & Hickey, 1992, 1996; see Feild & Arens, 2005 for a review). Under the conceptual umbrella of early disturbance adaptation, two different ecological

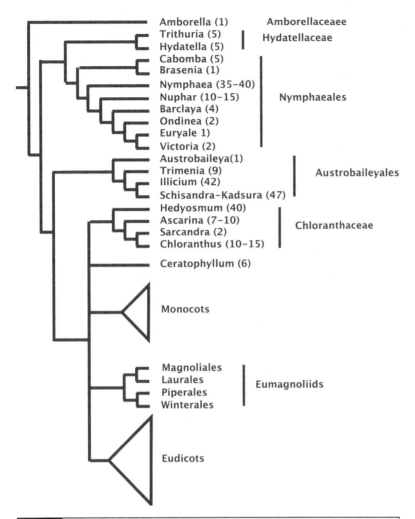

Amborella (1) — Amborellaceaee
Trithuria (5) — Hydatellaceae
Hydatella (5)
Cabomba (5)
Brasenia (1)
Nymphaea (35–40)
Nuphar (10–15) — Nymphaeales
Barclaya (4)
Ondinea (2)
Euryale 1)
Victoria (2)
Austrobaileya(1)
Trimenia (9) — Austrobaileyales
Illicium (42)
Schisandra–Kadsura (47)
Hedyosmum (40)
Ascarina (7–10) — Chloranthaceae
Sarcandra (2)
Chloranthus (10–15)
Ceratophyllum (6)
Monocots
Magnoliales
Laurales — Eumagnoliids
Piperales
Winterales
Eudicots

Fig. 6.1 A consensus tree of the phylogenetic systematics of extant basal angiosperm lineages identified by recent molecular and morphological studies. For details on taxon-sampling, indices of branch support and phylogenetic analysis, as well as molecular and morphological characters, please see the following references: Mathews & Donoghue, 1999; Doyle & Endress, 2000; Zanis et al., 2002; Saarela et al., 2007. The numbers beside each clade denotes the number of species.

images were sketched out. Some suggested that early angiosperms were drought-tolerant, sun-loving shrubs adapted to upland, xeric conditions and unstable substrates (Stebbins, 1965, 1974). Another related view of early angiosperms was that they occurred as lowland tropical herbs adapted to sunny, riparian zones (Taylor & Hickey, 1996; Table 6.1). The impetus for positing early disturbance adaptation was initially conceptual. Stebbins (1965, 1974) argued that adaptation to seasonal drought and/or adaptation to rapidly varying resource availabilities would favor the origin of reproductive and vegetative hallmarks of angiospermy (i.e. condensation of reproductive axes, reduction of gametophyte size, origins of net-veined leaves and xylem

Table 6.1 | Summary of hypotheses concerning the ancestral ecology of seeds and seedlings of early angiosperms. Sources for ecological details under each hypothesis are noted in the footnotes. A "?" denotes that no specific predictions were stated by the authors regarding a particular character, but may be predicted using contemporary ecological theory.

Functional trait	Woody magnoliid[a]	Disturbed xeric shrub[b]	Ruderal "paleoherb"[c]	"Damp, dark, and disturbed" shrub[d]
Seed size/morphology	?, Large, (5–10 mm long); complex, elaborate testal structure; minute embryo	Very small to small	Very small to small	Very small to small
Habitat of seedling establishment	Closed forest floor	Open, disturbed, variable water supply (talus slopes, stream banks)	Open, large-sized disturbed, riparian zones and floodplains	Small-scale disturbance, shade in the forest understory
Seedling ecophysiology	Shade tolerant	Shade intolerant, disturbance tolerant	Shade intolerant	Shade tolerant; high light intolerant
Germination mode	Hypogeal?	?, Epigeal	Epigeal	Epigeal
Seedling cotyledon function	?, Storage/haustorial	?, Photosynthetic	Photosynthetic	Photosynthetic
Juvenile architecture	Monopodial	?, Sympodial	Sympodial	Sympodial
Adult growth habit	Large canopy to tree	Shrub	Herb, rhizomatous	Shrub or small tree
Modern analogs	Woody Magnoliales and Winteraceae	Extinct, replaced by derived angiosperms	Paleoherb clades	Terrestrial extant basal angiosperm lineages

[a] Bews (1927); Martin (1946); Corner (1976).
[b] Stebbins (1965, 1974).
[c] Taylor & Hickey (1992, 1996).
[d] Feild et al. (2004; this chapter).

vessels). In addition to explaining the origin of the morphological gulf between angiosperms and all other living seed plants, these sorts of habitats were predicted to generally foster intense species diversification through decreased generation time (Stebbins, 1974; Taylor & Hickey, 1996; Wing & Boucher, 1998). With the discovery of the oldest vegetative remains of angiosperms in sediments interpreted as unstable, stream-side deposition zones, the view of early angiosperms as disturbance-adapted received near universal acceptance (Doyle & Hickey, 1976; Hickey & Doyle, 1977; Taylor & Hickey, 1992, 1996; Wing & Boucher, 1998).

In regeneration ecology, all disturbance hypotheses predicted that early angiosperms were highly competitive in producing abundant amounts of small, morphologically simple seeds (Table 6.1). Also, some suggested that angiosperm seedlings were ancestrally multiaxial (sympodial) with rhizomatous herbaceous growth (Table 6.1). This complex of characters was suggested to favor colonization of exposed and unstable substrates as well as habitats subject to frequent removal of plant parts (Taylor & Hickey, 1996). Nonetheless, up until the late 1990s, volatile phylogenetic topologies undermined confidence in hypotheses on whether early angiosperms were disturbance adapted.

6.3 | The phylogenetic revolution: inferences on early angiosperm regeneration ecology from extant basal angiosperms

Beginning in 1999, a spectrum of molecular studies dramatically changed previous opinions regarding which living lineages formed the most basal clades in angiosperm phylogeny (Fig. 6.1). These molecular phylogenies identified *Amborella* (one species), Nymphaeales ($\approx$65 species; the "water lilies" (including Cabombaceae and Nymphaeaceae), and Austrobaileyales ($\approx$250 species; including Austrobaileyaceae, Trimeniaceae, Illiciaceae, and Schisandraceae) as successive sister groups to the core angiosperms: the clade representing the vast majority of angiosperms – eudicots (180 000 species), monocots (60 000 species), and eumagnoliids (8 200 species; Fig. 6.1; Mathews & Donoghue, 1999; Zanis *et al.*, 2002). Recent phylogenetic results indicate that Chloranthaceae (70 species) are another near basal lineage (Fig. 6.1; Doyle & Endress, 2000), but their exact phylogenetic position requires much more work. A recent astonishing discovery is that the Hydatellaceae ($\approx$10 species), an aquatic family, are not derived monocots but are instead the sister group to water lilies (Saarela *et al.*, 2007).

While additional phylogenetic breakthroughs are possible, the quantum leap enabled by new phylogenies is that a robust sequence of living transitional lineages has been identified between extinct stem group angiosperms, distantly related extant gymnosperms (Burleigh & Mathews, 2007), and the vast majority of extant and highly diverse major angiosperm clades. Because relationships of extinct and extant gymnosperms (Doyle, 2006) and extinct angiosperms (i.e. *Archaefructus*, Sun *et al.*, 2002; Terada *et al.*, 2005) to extant angiosperms remain so uncertain, many workers have used the recent revision of extant basal angiosperm relationships to infer many putative aspects of Early Cretaceous angiosperm biology. Here, the form and function of basal lineages are surveyed and specific traits mapped onto hypotheses of angiosperm phylogeny to expose in detail the ancestral states and early transformations of organismal biology underlying angiosperm diversification (but see discussions by Feild & Arens, 2005, 2007 for caveats that extant basal angiosperms equal *early* angiosperms). The new phylogenetic edifice has inspired a major re-evaluation of the earliest phases of the angiosperm radiation from diverse perspectives, including rates of diversification, reproductive developmental biology, and ecology.

In the next section, I consider our current understanding of the functional biology of seeds and seedlings of early angiosperms by examining patterns of character evolution among recently identified basal angiosperm taxa. Following this review, I present new phylogenetic character reconstructions in order to examine the distribution of seed and seedling ancestral characters as well as their functional significance. Where possible, I discuss the relative support of each extant phylogenetic pattern by the fossil record and phylogenetic data.

6.4 | Functional biology of basal angiosperm seeds

Size

Seed size varies widely among extant angiosperms. In terms of mass, seed sizes span nearly twelve orders of magnitude (Harper *et al.*, 1970; Moles *et al.*, 2005a,b, 2007). This variation underlies many ecological functions, including interspecific differences in environmental conditions of seedling establishment, life history traits (growth form, life span, dispersal biology), and seed bank ecology (Chapter 10; Augspurger, 1984b; Moles *et al.*, 2005a,b, 2007).

Early in their evolution, angiosperms appear to have explored a limited range in seed size morphospace. Phylogenetic character reconstruction analyses of seed size data from extant basal angiosperms indicated that the common ancestor of extant flowering plants possessed moderately small seeds, with a volume of approximately 3 mm^3 (Feild *et al.*, 2004). This conclusion is robust despite the recent addition of the tiny-seeded, moss-like aquatics Hydatellaceae to the basal grade (Fig. 6.2; Saarela *et al.*, 2007). Overall, seed volumes in extant basal angiosperms range from 0.019 mm^3 in the aquatic herb *Hydatella filamentosa* to 73 mm^3 in *Austrobaileya scandens*, a tropical cloud forest liana (Fig. 6.2; T. S. Feild, unpubl. obs., 2007). These results provide additional support for the conclusion of Moles and colleagues (2005a), based on a much larger analysis of seed plants, that angiosperms are ancestrally small-seeded.

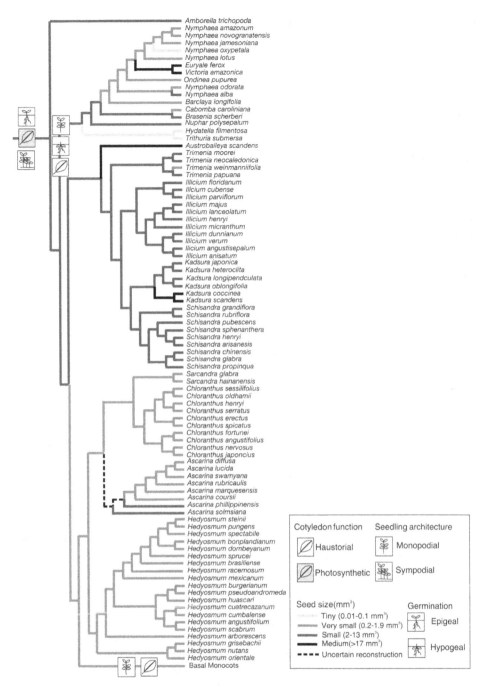

Fig. 6.2 Phylogenetic mapping of morphological and functional characters of seeds and seedlings among extant basal angiosperms. The most parsimonious distribution of ancestral character state reconstructions on internal nodes were reconstructed as described earlier (see Feild et al., 2004 for details). The character matrix for basal angiosperm taxa is available from the author by request. Also, note that seed size categories were determined by analysis breaks in the frequency distribution values across all taxa measured. The inclusion of additional data may change these ranges. The tree was assembled by grafting multiple species-level phylogenies of basal taxa onto a molecular backbone tree determined from comparisons of nuclear, mitochondrial, and chloroplastic markers (see Doyle & Endress, 2000; Mathews & Donoghue, 1999; Zanis et al., 2002; Saarela et al., 2007). Species-level phylogenies include: Nymphaeales (Lohne et al., 2007), Trimenia (S. Graham, S. Mathews, & T. S. Feild unpubl. data, 2007), Illicium (Morris et al., 2007), Schisandra-Kadsura (Denk & Oh, 2006), and Chloranthaceae (Eklund et al., 2004). For clarity, the phylogenetic positions of eudicots and eumagnoliids are removed. Basal monocots are represented by Acorus.

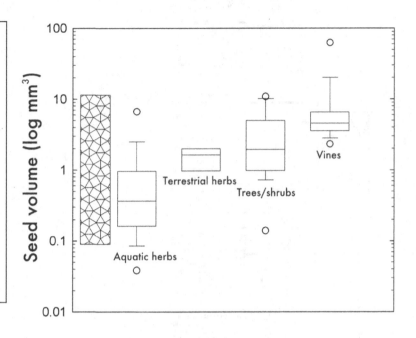

Fig. 6.3 Box plots of seed volume (mm³) variation among growth forms of extant basal angiosperms in comparison to the range of seed volumes (left bar) measured for Early Cretaceous angiosperms. Data were collected from 95 extant species of basal angiosperms (T. S. Feild, manuscript in preparation), as well as mined from the literature (Schneider, 1978; Schneider & Ford, 1978; Wiersma, 1987; Saunders, 1998, 2000; Oh et al., 2003; Feild et al., 2004; Denk & Oh, 2006). Fossil sizes of Early Cretaceous seeds are from the literature (see Wing & Boucher, 1998; Eriksson et al., 2000a,b).

Recent work indicates that phylogenetic divergences in seed size are most closely associated with evolutionary shifts in growth form (Moles et al., 2005a, 2007). Variation in climate, site productivity and function, and dispersal mode appear to play much lesser correlative roles with seed size evolution. Moles and colleagues (2005b) found that larger plants (i.e. trees) had consistently larger seeds as compared to herbs. One explanation for this pattern is that larger plants require a longer juvenile stage to reach maturity. To have a better chance of surviving a longer and precarious juvenile phase, larger seed sizes may have been selected because they produce larger sized seedlings that can be more resistant to hazards (herbivory, trampling, limb falls, disease; Moles et al., 2005a). A preliminary analysis of variation in seed size with life form among basal angiosperms reveals a rough pattern of seed size associations with growth form (Fig. 6.3). For example, seed sizes of aquatic basal angiosperm herbs are smaller than terrestrial woody taxa. The seed sizes of terrestrial herbs (represented only by most members of *Chloranthus*), however, overlap with the range of trees and shrubs (Fig. 6.3). Seeds of vines tended to be larger than trees, shrubs, and herbs (both aquatic and terrestrial) among basal taxa.

Interestingly, the range of seed sizes among basal angiosperms is similar to ranges reported from several Early Cretaceous floras (Fig. 6.3; Tiffney, 1986, 2004; Wing & Boucher, 1998; Eriksson et al., 2000a,b; Feild et al., 2004). From a well-studied Barremian-Aptian site (mid-Early Cretaceous; 112 to 132 Ma) in Portugal, mean seed volume was 0.78 mm³, with values ranging from 0.02 to 6.86 mm³ (Eriksson et al., 2000b). This average value nests well with the lower end of the range of seed volumes found among extant basal angiosperms. Seed volumes, like those of extant basal taxa, continue to characterize Cretaceous angiosperms up to the Mesozoic–Cenozoic boundary, even though these seed floras certainly contained a diversity of eudicots, monocots, and eumagnoliids (Eriksson *et al.*, 2000a,b). At the Mesozoic–Cenozoic boundary, seed size diverges rapidly toward smaller and much larger sizes (Tiffney, 1986, 2004; Wing & Boucher, 1998; Eriksson *et al.*, 2000a,b). Finally, the range of anatomies of the oldest known Early Cretaceous seeds most closely resembles that of several extant basal angiosperms (Frumin & Friis, 1999; Oh *et al.*, 2003; Denk & Oh, 2006). This suggests that relatives of extant basal taxa occurred during the Early Cretaceous (Frumin & Friis, 1999; Eriksson *et al.*, 2000b).

The intriguing pattern of Cretaceous stasis and Paleocene disparity in seed size has inspired two different hypotheses on the ecological mechanisms underlying it. Because many extant eudicot angiosperms with small-sized seeds are stereotypically sun- and disturbance-adapted herbs (Eriksson *et al.*, 2000a,b; however see below), it has been suggested that early angiosperms adopted such habits and remained constrained to these zones until near the close of the Cretaceous (Wing & Boucher, 1998; Eriksson *et al.*, 2000a,b). The appearance of large-sized seeds then marks the first development of angiosperm canopy trees that formed angiosperm-dominated closed forest habitats.

There are several complications with this hypothesis. First, the broad overlap of seed volumes, within the ranges of volumes exhibited by fossil Early Cretaceous angiosperm seeds, in extant basal taxa with diverse growth habits undermines attempts to argue that early angiosperms were a specific type of growth form or ecology type based solely on seed size (Taylor & Hickey, 1996; Wing & Boucher, 1998; Eriksson *et al.*, 2000a,b; Feild *et al.*, 2004) when basal angiosperms are present. Extant basal angiosperms of a similar-sized seed are almost equally likely to be aquatic or terrestrial, herbaceous or woody, as well as sun- or shade-tolerant (Fig. 6.3; Feild *et al.*, 2004). Also, seed size variation among extant basal taxa indicates that canopy tree growth forms are possible for small-seeded plants. For example, several large chloranth trees, such as *Hedyosmum peruvianum* and *H. cuatrezacanum*, form large canopy trees (up to 20 m tall) in mid-montane cloud forests, yet seed volumes are 2.1 and 2.4 mm^3, respectively (T. S. Feild, unpubl. data, 2007 Peru). Given the probable antiquity of *Hedyosmum* based on the fossil record, presence of canopy angiosperm trees in the mid-Early Cretaceous cannot be ruled out. Also, if closed forest shade is the trigger for development of large seeds, then why did angiosperms not develop them in closed forest, gymnosperm-dominated forest communities that existed prior to the Late Cretaceous? Although dinosaurs were likely a major disturbance in Early Cretaceous communities and could have prevented forest development of dry lowland and floodplains (Feild & Arens, 2005), wet, tropical montane communities could have readily supported closed habitats dominated by conifers, as well as a diversity of gymnosperms and ferns in the understory. Indeed, the ecology of early angiosperms retrodicted from extant basal taxa suggests that early angiosperms likely got their start in wet, montane regions and in the understory of closed forests (see below; Feild *et al.*,

2004; Feild & Arens, 2005, 2007). Instead, shade adaptation in early angiosperms seems to have followed a small-seed route (see below).

The other hypothesis is that a series of Late Cretaceous faunal (i.e. birds, mammals) radiations sparked novel dispersal co-evolutionary interfaces that selected for large seed size divergence (Tiffney, 1986, 2004). However, the discovery of Early Cretaceous bird fossils as well as molecular clock dating of crown groups to the mid-Early Cretaceous for these key dispersing groups suggests that ecological constraints on dispersal were unlikely (see discussion by Eriksson et al., 2000a). Indeed, seeds of extant basal angiosperms are dispersed by a wide range of faunal vectors as well as abiotically, yet seed sizes tend to be small (Roberts & Haynes, 1983; Todzia, 1988; Saunders, 1998, 2000; Feild & Arens, 2007). It may be significant that some of the largest divergences in seed size, such as in *Austrobaileya scandens*, *Kadsura coccinea*, and *K. scandens*, are linked to megafaunal dispersal (Saunders, 1998, 2000; J. Williams, unpubl. obs. of germinating seedlings of *A. scandens* in cassowary dung, 2006). Taken together, seed evolution, at least as seen through the lens of extant basal angiosperms, indicates that both hypotheses are incomplete in explaining the tempo of seed volume variation during the Cretaceous.

A final wrinkle on the issue of seed size evolution involves importance of extinct groups on the inference of evolutionary processes. In contrast to the remarkable phylogenetic progress on the question of "Which groups of extant flowering plants are at the base of the phylogenetic tree?," progress on understanding the question of "Which gymnosperms are the closest relatives of angiosperms?" remains uncertain as ever. The main issue that makes the gymnosperm-relative question so hard is that as angiosperms branched out in the Cretaceous, a great dying of diverse and candidate gymnosperm relatives to angiosperms ensued (Knoll, 1986; Bond, 1989; Wing & Boucher, 1998; Boyce, 2005). Consequently, the surviving remnants of gymnosperm diversity are a poor, heterogeneous sample of former phylogenetic diversity, and none appear to be closely related to extant angiosperms (Doyle, 2006; Burleigh & Mathews, 2007). Consequently, inferences of ancestral states and patterns of seed/seedling character evolution below the node of the common ancestor to living flowering plants remain unresolved because extant out-groups are relatively unrelated to angiosperms. This pattern is particularly important to bear in mind for attempting to infer adaptation in seed size evolution across seed plants.

Based on extant taxa alone and under the monophyletic gymnosperm hypothesis, a very large (59-fold) divergence in seed mass between the common ancestor of extant angiosperms (small-seeded) and extant gymnosperms was reconstructed (Moles et al., 2005b). If living Gnetales are placed as sister to angiosperms (a phylogenetic position that seems not to be the case; Burleigh & Mathews, 2007), the divergence is still very large (16-fold; Moles et al., 2005b). This pattern suggests that a massive reduction in seed size occurred somewhere along the angiosperm stem lineage. However, several extinct

Cretaceous gymnosperms had remarkably small seeds ($\approx$5 mm^3; see Tiffney, 1986). Also most of these groups, such as *Pentoxylon*, *Caytonia*, and Bennettitales, can be, under various morphological topologies, placed along the line leading to extant angiosperms. Thus, it is currently unclear whether downsizing of seed volume reflects much deeper processes operating among gymnosperms or if it is a unique adaptation of the angiosperms. There is also the additional problem that the habitats and ecologies of extinct gymnosperms with small seeds remain unclear. Without a robust phylogenetic hypothesis incorporating fossil taxa, it will be difficult to unravel how ecology versus developmental constraint bears on small seed size stasis in angiosperms throughout much of the Cretaceous.

Morphology and dormancy

Forbis and colleagues (2002) reconstructed the ancestral morphology of angiosperm seeds from extant basal angiosperms. They found that ancestral angiosperm seeds were albuminous, filled almost entirely by endosperm. Ancestral seeds also possessed a poorly differentiated embryo at the time of dispersal. The reconstructed angiosperm ancestor had a small embryo relative to the size of its seed (E:S = 0.16). However, it is important to point out that the Nymphaeales and Hydatellaceae have radically different systems of embryo-nourishing tissues compared to the reconstructed seed of the extant angiosperm common ancestor. In these taxa, the endosperm development is limited to a band forming a fleshy sac enclosing the embryo. The rest of the seed volume is occupied by perisperm (storage tissues of nucellar origin). At present, the fossil record is mute on the internal embryo structure of the earliest angiosperm seeds. While fossilized embryos are highly improbable, their discovery is not impossible in permineralized fossils (Stockey & Rothwell, 2003). Discovery of such rare fossils will have to provide the ultimate information necessary to test these hypotheses based on living plants.

The significance of low E:S allocation lies in its developmental effects on the regulation of dormancy (Baskin & Baskin, 1998; Forbis *et al.*, 2002; Finch-Savage & Leubner-Metzger, 2006). Because embryos are underdeveloped at the time of dispersal, further differentiation and maturation of internal tissues within the seed is required before germination can commence. Thus, developmental delay imposes an interval of morphological dormancy on germination (Baskin & Baskin, 1998). Limited data on the germination times in basal angiosperms are consistent with the occurrence of some sort of dormancy process. For example, germination of *Amborella*, *Austrobaileya*, *Chloranthus*, and *Sarcandra* seeds requires 3 to 12 months (T. S. Feild, unpubl. data, 2000, 2001, 2007). It is also probable that physiological dormancy (Baskin & Baskin, 1998) occurs in many basal angiosperms. Anecdotal reports suggest that in cold-adapted temperate taxa, low temperature pretreatments are necessary for germination (i.e. *Ascarina lucida*, *Nuphar*, *Nymphaea*; Smits *et al.*, 1990; Barat-Segretain, 1996; Martin & Odgen, 2005). Regardless of the exact mechanisms, the occurrence of prolonged dormancy in basal taxa may enable development of a soil seed bank, provided that seeds can avoid predation and desiccation in the field. Thus far, the only convincing evidence for development of a seed bank in basal angiosperms comes from *Ascarina lucida* in temperate New Zealand (Martin & Ogden, 2002). Seeds of this species may be able to remain in the soil for several years.

6.5 | Functional biology of basal angiosperm seedlings

Throughout the following discussion, I follow Garwood's (1996; p. 60) definitions of a *seedling* to mean: *an early developmental stage that contains at least some still functioning structures produced from the initial seed reserves* and *initial morphology to indicate the form of the seedling at the time the first entirely photosynthetic organs have fully expanded.*

Architecture
Seedlings of terrestrial basal angiosperms are architecturally strange when compared to the vast majority of temperate and tropical tree seedlings (Hara, 1985; Blanc, 1986; Feild et al., 2004; Feild & Arens, 2005). These plants exhibit several growth behaviors that revolve around a common theme of flexible juvenile growth patterning. For example, *Amborella trichopoda* seedlings pass through an initial creeping establishment phase, referred to as "pseudorhizomatous" growth (Feild et al., 2004). This growth mode is characterized by the early development of a decumbent habit. Once the main hypocotyl with cotyledons and first-formed leaves expands, one or more auxiliary shoots are produced from basal buds along the decumbent root, resulting in a sympodial seedling habit. This architecture is also widespread in *Trimenia*, *Illicium* (Schisandraceae), and Chloranthaceae, and is reconstructed as present in the common ancestor of extant angiosperms (Fig. 6.2; Blanc, 1986; Feild et al., 2004; Feild & Arens, 2005, 2007). Some other basal angiosperm trees, for example *Trimenia papuana* and *Illicium parviflorum*, have seedlings that are best described as viney. In seedlings of these species, the allocation to lateral spread is more pronounced, and the transition zone between root and shoot becomes highly twisted as seedlings navigate through debris on rotting logs or meander along the surface of the forest floor. It can take several years of lateral spread before the apical growth is channeled into vertical growth (T. S. Feild, unpubl. obs., 2001–2003). In contrast to aboveground architecture, precious little information is currently available on the patterns of root system architecture in basal angiosperms. In Chloranthaceae, a main prop root with minimal branching often forms, which orients the seedling upright (T. S. Feild unpubl. obs. on *Hedyosmum translucidum*, *H. maximum*, and *H. goudotianum* in Peru, 2007). Root architectural patterns in Chloranthaceae bear a curious superficial resemblance to the early root system development in seedlings of many basal monocots clades (Blanc, 1986; Tillich, 2000).

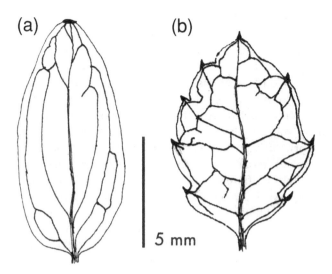

(a) (b)

5 mm

Fig. 6.4 An example of cotyledon (a) and first-formed leaf (b) morphological divergence in *Amborella trichopoda*. (a) In the cotyledon, leaf venation is extremely rudimentary and variable. A single hydathode, with epithem tissue, is developed at the apex of the cotyledon. (b) In the first formed leaf, venation is increasingly ramified and hydathodes are developed along the leaf edge. Drawing by the author.

The seedling architectures of basal angiosperm aquatics provide stark contrasts to those of terrestrial clades. In all water lilies examined so far, the cotyledons remain in the seed coat, and the epicotyl and first leaf grow upward as a long thread-like organ (Conard, 1905; Haines & Lye, 1975; Schneider, 1978; Tillich, 1990). Thus, the initial morphology consists of a single (monopodial) shoot axis. Rhizomatous spread begins later and is characteristic of plants that have outgrown the seedling phase (Conard, 1905; Nieuwland, 1916). Establishment growth and initial morphology of Hydatellaceae remains largely unstudied. Initial growth in one species of *Trithuria* appears to be monopodial (Cooke, 1983).

Cotyledon form and function

Cotyledons are the most conspicuous feature of the initial morphology of angiosperm seedlings (Kitajima, 1992a, 1996a; Garwood, 1996; Tillich, 2000). Cotyledons are also important determinants of the functional performance of angiosperm seedlings by influencing disturbance tolerance, carbon gain, and/or the mobilization of seed reserves during seedling establishment (Chapter 8; Kitajima, 1992a, 1996a; Garwood, 1996; Zanne *et al.*, 2005; Baraloto & Forget, 2007).

Of the basal terrestrial species known, all exhibit epigeal germination with the development of two cotyledons (Fig. 6.2; Doyle & Endress, 2000). As the cotyledons emerge from the seed coat, they expand into thin, small leaf-like green organs. For example, the cotyledons of *Amborella*, several Chloranthaceae (*Chloranthus erectus*, *Hedyosmum translucidum*, *H. goudotianum*, *H. maximum*, *Sarcandra glabra*), *Trimenia moorei*, and *Schisandra chinensis* are thin and ovate shaped, varying from 7 to 10 mm long and 4 to 7 mm wide (Fig. 6.4). Cotyledons also develop sparse, simple venation patterns with a hydathode at the apex and are very different from first-formed leaves (Fig. 6.4). Hydathodes in basal angiosperms are involved in the passive release of guttation sap, driven by root pressure under nontranspiring conditions (Feild *et al.*, 2005). Compared to *Amborella*, Chloranthaceae, and *Trimenia*, cotyledons of *Illicium floridanum* and *Kadsura coccinea* are larger, up to 18 mm long and 25 mm wide, and thicker and more fleshy in construction. Anecdotal observations also suggest that cotyledons can remain attached and green for a year or longer (T. S. Feild, unpubl. obs., 2006). While direct observations of cotyledon photosynthesis and respiration are lacking, the structural character of terrestrial basal angiosperm cotyledons – thin construction, green color, and long lifespan – suggests that these organs function in net carbon gain rather than primarily mediating the transport of seed resources.

The cotyledons of basal aquatics, such as Nymphaeales and Hydatellaceae, exhibit a different pattern of form and function. Germination in these plants is hypogeal (Conard, 1905; Nieuwland, 1916; Cooke, 1983; Doyle & Endress, 2000). Also, the cotyledons remain sandwiched inside the germinating seed and are membraneous and achlorophyllous (Conard, 1905; Nieuwland, 1916; Haines & Lye, 1975; Schneider, 1978; Schneider & Ford, 1978; Tillich, 1990). This indicates that cotyledons of basal aquatics function in a haustorial role, with cotyledon respiration functioning in the metabolism of seed reserves (i.e. endosperm and perisperm) for growth (Garwood, 1996). The common ancestor of monocots may have also been aquatic (Feild & Arens, 2005, 2007). Similar to Nymphaeales and Hydatellaceae, the cotyledon of *Acorus* is initially haustorial, but germination is epigeal (Buell, 1935).

While much more information needs to be gathered on how cotyledons function, the distribution of this character state among extant basal angiosperms refutes previous suggestions (Ji & Ye, 2003) that cotyledons of the earliest angiosperms functioned as haustorial organs (Fig. 6.2). Instead, early angiosperms are reconstructed as having epigeal germination with photosynthetic cotyledons that were subsequently modified convergently in basal aquatic lines for absorptive functions.

Habitats

The phylogenetic distribution of seedling habitats among extant basal taxa indicates that their common ancestor recruited in a specific habitat, namely disturbed microsites in the understory of wet (generally >2000 mm rainfall a year) primary forests (Kwit et al., 1998; Feild et al., 2004; Duchok et al., 2005; Feild & Arens, 2005, 2007). Modern examples of these damp, dark, and disturbed (i.e. 3-D) habitats are diverse, including exposed litter-free and well-drained soils of ravines near understory-meandering streams, tip-up soil mounds near the roots of understory/subcanopy fallen trees, in between boulders, as well as on steep micro-slopes generated by soil washouts and by mega fauna movements (Roberts & Haynes, 1983; Kwit et al., 1998; Saunders, 1998, 2000; Feild et al., 2004; Feild & Arens, 2005, 2007). Although basal taxa clearly show a preference for high rainfall sites and wet pockets in drier forests, seedlings are particularly vulnerable to soil anoxia. Based on germination fatalities in the greenhouse, flooding rapidly induces root death in *Schisandra* (*S. glabra*, *S. chinensis*), *Chloranthus japonicus*, as well as *Illicium floridanum* seedlings (T. S. Feild, unpubl.

obs., 2003–2007). Many species of Chloranthaceae, for example *Ascarina lucida* and *Hedyosmum goudotianum*, as well as *Trimenia papuana* also recruit seedlings on microsites elevated above the shady understory forest floor. Examples of these sites include moss mat crevices on rotting logs and tree fern trunks. These microsites are also prone to instability because logs eventually rot and/or roll and tree ferns do not live as long as the basal angiosperm species on them can. In contrast, large-scale disturbances, such as landslips, forest gaps, fire, and logging, that result in large increases in light availability and vapor pressure deficit eliminate many basal angiosperms from the landscape.

In 3-D zones, traits that permit recruitment on disturbed sites and allow seedlings to persist in the face of root and shoot damage are essential for survival. One recruitment trait is small seed size. Although dogma holds that large seeds are necessary for successful establishment in dark (<5% full sunlight) habitats because abundant seed reserves allow seedlings to weather a depauperate shady life, the seeds of extant basal angiosperms are paradoxically small (Feild *et al.*, 2004). In a variety of eudicot tropical angiosperm taxa, however, recent work demonstrates that small seeds (volumes less than ca. 10 mm^3) are advantageous in shady, disturbed habitats, specifically steep understory slopes, nurse logs, exposed mineral soil washouts, and stream banks – exactly the habitats where most extant basal angiosperms occur, as opposed to undisturbed forest floor microsites with a thick litter layer (Metcalfe & Grubb, 1997; Metcalfe *et al.*, 1998; Lusk & Kelly, 2003; Bellingham & Richardson, 2006). Despite a lower resource pool, small seeds are geometrically advantageous in the colonization of microsites. Small seeds lodge more readily in small crevices and are less prone to fall out of such sites as compared to larger seeds (Metcalfe *et al.*, 1998; Lusk & Kelly, 2003). While these processes favor the probability that small seeds can land in understory microsites, how do such small seeds establish successfully under such strong light limitation? While no data have been collected on seed allocation energetics of basal angiosperms, efficient packaging of energy-rich molecules (i.e. oils) or specialized provisioning of the endosperm with high N may help stretch the budget of small packaging during the initial achlorophyllous heterotrophic germination phase (Kitajima, 1996a). In addition, it is likely that when the cotyledons emerge, their photosynthesis is very efficient because adult foliage has been demonstrated to be strongly shade-adapted (Feild *et al.*, 2001, 2003, 2004; Feild & Arens, 2005, 2007). The generally low amounts of litter in 3-D sites means that greening-up time is not likely delayed for very long because the epicotyls do not have to penetrate much litter before reaching light.

Once seeds germinate, pseudorhizomatous growth may enhance seedling establishment in disturbed sites (Feild *et al.*, 2004). The chief cost of establishment on unstable substrates is that reoccurring disturbance can kill (Bellingham & Richardson, 2006). However, multi-axis seedling architecture may assuage disturbance damage by lowering the probability that a single trauma will kill a seedling. Returning to the issue of maintaining positive carbon balance under dim light, the early development of lateral spread along the forest floor is fascinating because it comes at the expense of height. Although greater height means more light in the strongly light-limited habitats of many basal taxa with small seed size, such a seedling allocation pattern is particularly striking (Feild *et al.*, 2001, 2003, 2004; Feild & Arens, 2005, 2007). However, even in the adult phase of many basal taxa, reaching the canopy or even the sub-canopy seems unimportant. Many woody basal taxa can set fruit in the understory (e.g. *Amborella*, *Austrobaileya*, *Chloranthus* species) and some (*Illicium floridanum*) lose the ability to flower if exposed to too much light (T. S. Feild, unpubl. obs., 2000). All of these observations seem to point to an ontogenetic strategy of carbon-efficient growth in basal angiosperms that begins with seedling establishment (Feild & Arens, 2007). Moreover, flowering can occur in plants that are relatively small (<40 cm; for example *Amborella*, *Ascarina*, *Hedyosmum*, *Illicium floridanum*) and probably of relatively young age. This suggests that ephemeral sites in the understory can be effectively exploited for sexual reproduction.

The aquatic seedling establishment conditions of water lilies and Hydatellaceae are considerably different than terrestrial basal taxa (Nieuwland, 1916). Anoxia, ever wet conditions, bicarbonate as the primary CO_2 source, high buoyancy, and diffusion constraints on transport are the norm underwater. On the other hand, many Nymphaeales, like their close terrestrial relatives, combine generally small seed size (Fig. 6.3) with germination in habitats that are very dark. Seeds of *Brasenia*, *Nuphar*, and *Nymphaea* can germinate in 6 to 15 cm of muck with 30 cm of water (or more) above (Nieuwland, 1916; Smits *et al.*, 1990; Barrat-Segretain, 1996). Once the seed germinates, a long thread-like epicotyl (up to 20 cm long) grows upward through the underwater muck to find light. Other Nymphaeales, such as *Barclaya longifolia*, may germinate in aquatic soils beneath clear, shallower water ($\approx$5 cm deep), but they do so in the tropical rain forest understory! There are limits to shade establishment in water lilies, however. Cloudy water or erosion disturbance that results in deep burial of seeds in the soil (>15 cm) will eliminate seedling recruitment in the field (Smits *et al.*, 1990; Barrat-Segretain, 1996).

6.6 | Outlook and recommendations for future research

Diverse disciplines are searching for answers on how early angiosperms functioned in an ecological context in order to reconstruct the mechanisms responsible for the extraordinary rise of angiosperm dominance during the Cretaceous. Seed and seedling habits are among the most important pieces needed to solve this early angiosperm enigma because these life history stages impact so many ecological processes. In this review, I have tried to illustrate the ways that phylogenetic, extant organismal, and paleobiological approaches can be intertwined to shed new light on the ancestral regeneration ecology of angiosperms. The picture of early angiosperm regeneration ecology that emerges from this synthesis is a curious hybrid of previous hypotheses (Table 6.1), but with several subtle and important differences.

Under this hypothesis, disturbance-adaptation is axiomatic for early angiosperm evolution, but the types of disturbance events, particularly the microclimatic context, differ from earlier predictions. The mix of functional/structural characteristics of seeds and seedlings of basal angiosperm suggests that early angiosperms were effective at establishing in the damp, dark understory. However, the scope of their regeneration ability appears to be quite limited. Early angiosperms appear to have been adapted to specific "Goldilocks" microsites – zones with not too much disturbance, light, or water. Thus, instead of a key to immediate massive diversification, adaptation of seedling function to the damp, dark, and disturbed lifestyle may have been a way for angiosperms to intercalate themselves initially into well-established Mesozoic communities. Perhaps exploitation of the 3-D niche in a new way opened up a new opportunity to make an ecological beachhead, although the zone appears to have afforded little immediate room for species and ecophysiological diversification (Feild et al., 2004; Feild & Arens, 2005, 2007). Consistent with this pattern, angiosperm diversification in the fossil record does appear to lag for up to 20 million years following the initial appearance of chloranthoid fossil pollen (Lupia et al., 1999). Consequently, competitive interactions between angiosperms and many gymnosperms during seedling regeneration probably remained relatively diffuse in the Early Cretaceous, but for how long is unclear. Indeed, forest understory shade appears to be a great equalizer of competition among extant tropical conifers (podocarps), ferns, and derived angiosperms (Brodribb et al., 2005).

However, much ecological, paleobotanical, and anatomical work remains to be done to flesh out and ultimately test the hypotheses developed here on the regeneration ecology of early angiosperms based on living plants. We need to answer the question: Do seedlings of extant basal taxa equal early angiosperm seedlings? Without this information we cannot gauge how accurate any reading of early angiosperm ecology from extant basal lineages may be (Feild & Arens, 2007). Only by integrating our understandings of community, environmental, and phylogenetic perspectives during the Mesozoic will we develop a clear view of the ecological mechanisms guiding the rise of the angiosperms. Below, I outline some of the next research steps that include extant and extinct organismal approaches. Hopefully, progress along these paths, and many others, will greatly advance the development of an integrative picture of early angiosperms and their regeneration niches.

(1) Population dynamics, structure, genetics, and conservation. Future empirical studies on the demography, population structure (including

genetics), and regeneration processes of natural populations of basal angiosperms are needed for several reasons. First, we do not have a grasp on how important seedling recruitment dynamics are to the population dynamics of basal taxa. For example, anecdotal observations suggest that many basal taxa appear to be characterized by episodic seed set, and these taxa may rely heavily on clonal reproduction for vegetative spread (Stone, 1968; Feild *et al.*, 2004). Indeed, many natural populations of basal taxa, particular members of *Trimenia*, Schisandraceae, and *Illicium*, can be tiny and composed of a handful of plants that may be extremely long-lived. In many cases, it is not clear whether this pattern is the result of recent land-use change and disturbance that have removed the opportunities for 3-D regeneration or if local pollinators have become extinct or if the observed population structures characterized these taxa for a long time. Other basal taxa exhibit population structures at the other end of the spectrum, with high population sizes and frequent seedling recruitment (particularly *Amborella* and the chloranth genera *Ascarina* and *Hedyosmum*). To disentangle the web of pattern and process, we need direct observations on seed germination and seedling establishment in the field as well as the processes determining mortality and how ecophysiological variation influences these. We also need indirect studies that assess the spatial variation in molecular diversity (e.g. RFLP analyses) in established populations to parse out the relative roles that sexual and asexual reproduction play in population structure, as well as the timing or frequency of population bottlenecks. Ultimately, knowledge on the dynamics governing population genetic structure will expose important new understanding of the evolutionary and ecological processes underlying speciation in basal angiosperms. Exposing the processes of speciation in basal taxa is essential for understanding how labile ecological variation in these taxa may be to population dynamics, which bears on the question of whether extant basal angiosperms equal early angiosperms.

Beyond academic interest, understanding seedling regeneration and population dynamics are important for conserving these species. While basal taxa, with their ugly flowers (c.f. Verdcourt, 1986) and shaggy growth habits, are unlikely to become emblematic flagship species for national parks and conservation programs anytime soon, important drugs against bird flu (Tamaflu®, derived from star anise *Illicium verum*) and other diseases are being developed from these taxa and phytochemical diversity across the basal angiosperm grade holds considerable promise for additional discoveries (Saunders, 1998, 2000). Beyond the developed world, folk medicine and local economies of several developing world villages depend on the harvesting of basal angiosperm products (e.g. bark and seeds of *Illicium griffithii* in India, Duchok *et al.*, 2005). Presently, exploitation of basal taxa is far from sustainable, and research into the population resilience in the face of over-harvesting is urgently needed.

(2) The ecology of small-seededness under deep shade. One of the general ecological issues exposed by extant basal angiosperms is that more

functional understanding is needed of the biology of small seeds in shady habitats (see Metcalfe & Grubb, 1997; Metcalfe *et al.*, 1998). The 3-D zones favor a curious blend of *K*- and *r*-selected strategies in plants, such as the juxtaposition of small seeds and low photosynthetic productivity. While 3-D microsites are a common feature of many temperate and tropical forests (Metcalfe & Grubb, 1997; Metcalfe *et al.*, 1998; Lusk & Kelly, 2003; Bellingham & Richardson, 2006), our ecological understanding of how these are exploited remain unclear. A likely important mechanistic key to understanding how small seeds are viable under deep shade is seed packaging. In this context, basal angiosperms offer a diverse assortment of seed morphological types, which are underlain by the unparalleled diversity of developmental programs that give rise to endosperm in comparison to eudicots and monocots (Williams & Friedman, 2002; Friedman, 2006). In addition, basal taxa vary profoundly in the extent that endosperm versus perisperm storage (found in Nymphaeales and probably Hydatellaceae) tissues are used to nourish germinating seedlings. Is this diversity of provisioning the mark of an inefficient developmental system yet to be canalized (Friedman, 2006) or are these diverse patterns instead specific adaptations to the unique demands of 3-D habitats? It is curious that *Piper*, another lineage that forms perisperm in the seed, also has radiated profusely in 3-D zones. Microcalorimeteric measurements of perisperm versus endosperm energetic content along the impressive morphological gradient of seed size (Fig. 6.2) and embryo:seed-size allometry (Forbis *et al.*, 2002) offered by basal angiosperms would be an informative approach to access the importance of storage tissue energetics to seed germination in relation to light environment.

(3) Fossil studies. Finally, new research on the ecology of Early Cretaceous fossil seeds is needed. A particularly interesting avenue to explore would involve a more thorough and phylogenetically structured analysis of patterns of seed-size evolution among seed plants with particular attention paid to seed diversity among extinct gymnosperm groups placed as putative angiosperm out-groups. Attention to fossil seeds has waned in recent years because seeds tend to provide weak calibration time points to scale molecular branch lengths in phylogenetic studies. However, understanding the morphology of fossils is important for accessing paleoecological function which bears on hypotheses emerging from living plants.

6.7 | Acknowledgments

I thank Eithne Feild for her enduring support of this research program. I also thank Sarah Mathews and Joe Williams for helpful discussion related to the functional biology of early angiosperm seeds. This research was supported by a research grant from NSF (IOB-0714156) to the author.

Chapter 7

Physiological and morphological changes during early seedling growth: roles of phytohormones

Elizabeth J. Farnsworth

7.1 Introduction: phytohormones, molecular biology, and the "real world" of early seedling ecology

The seedling stage is arguably the busiest phase in a plant's lifetime. A miniscule individual, possessing only a rudimentary cotyledon and radicle, must quickly make its way in the world before a host of environmental challenges and larger competitors bully it out of existence. The maternal reserves from the seed must be mobilized to provide the fuel for producing the first pair of photosynthetically competent leaves and proliferating roots that will assimilate nutrients and firmly secure the seedling to its new home in the soil. These critical activities are initiated within minutes of germination, requiring a flurry of rapid signal transduction, protein synthesis, and tissue differentiation. To grasp how the seedling accomplishes all this demands an examination of physiological mechanisms that govern life after germination.

Studies at the molecular level are making great headway in understanding the complex and fascinating physiological events that mark the early infancy of plants. Phytohormones (a broad term that encompasses a range of organic and polypeptide compounds) are increasingly recognized as the primary integrators of environmental signals that enable the nascent seedling to evaluate and adjust to existing conditions. Biologists working across the spectrum of organizational levels – from cells to trees, from genes to populations – are now in an exciting position to elucidate how phytohormones modulate phenotype, and seedlings from a variety of taxa are natural model systems for study and synthesis (Table 7.1). The time is ripe to extend studies from laboratory species to plants growing in field settings and to explore how these molecular messengers shape plant ecology and evolution.

Table 7.1 | A sampling of families and species for which actions of at least one phytohormone have been studied and for which genetic maps or expressed sequence tags are available; the latter species are denoted by an asterisk.

Plant Family[a]	Species[b]
Aceraceae	*Acer saccharum*
Actinidiaceae	*Actinidia deliciosa*
Anacardiaceae	*Mangifera indica*
Apiaceae	*Bupleurum falcatum, Coriandrum sativum, Daucus carota*
Apocynaceae	*Catharanthus roseus, Rauvolfia serpentina*
Araceae	*Amorphophallus albus, Xanthosoma sagittifolium, Zantedeschia aethiopica**
Arecaceae	*Nypa fruticans, Phoenix reclinata*
Asphodelaceae	*Aloe arborescens*
Asteraceae	*Artemisia annua, Helianthus annuus*, Lactuca sativa*, Zinnia elegans*
Betulaceae	*Betula pendula**
Boraginaceae	*Lithospermum erythrorhizon*
Brassicaceae	*Arabidopsis thaliana*, Brassica campestris, Brassica napus*, Brassica oleracea, Brassica rapa*, Raphanus sativus*
Cactaceae	*Rhipsalidopsis sp., Schlumbergera sp.*
Campanulaceae	*Campanula rotundifolia, Physoplexus comosa*
Caricaceae	*Carica papaya*
Caryophyllaceae	*Dianthus caryophyllus, Phlox paniculata, Phlox setacea, Stellaria longipes*
Chenopodiaceae	*Beta vulgaris*, Suaeda monoica*
Convolvulaceae	*Cuscuta sp., Ipomoea batatus, Pharbitis nil*
Cornaceae	*Camptotheca acuminata*
Cucurbitaceae	*Cucumis melo, Cucumis sativus*, Cucurbita pepo, Cucurbita texana, Momordica charantia*
Dipterocarpaceae	*Hopea odorata*
Euphorbiaceae	*Manihot esculenta*, Ricinus communis**
Fagaceae	*Fagus sylvatica, Quercus robur**
Geraniaceae	*Pelargonium peltatum*
Hippocastanaceae	*Aesculus hippocastanum*
Hydrocharitaceae	*Vallisneria americana*
Iridaceae	*Iris ensata, Iris hexagona*
Lamiaceae	*Lavandula stoechas, Salvia miltiorrhiza, Stachys sieboldii*
Lauraceae	*Persea americana*
Leguminosae	*Arachis hypogaea, Canavalia ensiformis, Chamaecytisus proliferus, Cicer arietinum*, Dolichos lablab, Glycine max*, Lens culinaris, Lotus japonicus*, Lupinus albus*, Medicago sativa*, Medicago truncatula*, Mimosa pudica, Phaseolus lunatus, Phaseolus vulgaris*, Pisum sativum, Robinia pseudoacacia, Trifolium repens, Vicia faba, Vigna radiata*
Lemnaceae	*Lemna minor, Spirodela sp.*
Liliaceae	*Allium cepa*, Allium porrum, Chlorophytum arundinaceum, Hyacinthus orientalis*
Linaceae	*Linum usitatissimum**
Malpighiaceae	*Galphimia glauca*
Malvaceae	*Gossypium hirsutum**
Musaceae	*Musa acuminata**
Myrsinaceae	*Aegiceras corniculatum, Ardisia escallonioides*
Myrtaceae	*Eucalyptus globulus, Eucalyptus saligna*

(cont.)

Table 7.1 (cont.)	
Plant Family[a]	Species[b]
Oleaceae	*Olea europaea*
Orchidaceae	*Dendrobium moschatum*
Orobanchaceae	*Orobanche cumana, Orobanche ramosa*
Pedaliaceae	*Harpagophytum procumbens, Harpagophytum zeyheri*
Pinaceae	*Abies alba, Larix x eurolepis, Pinus pinea, Pinus radiata*, Pinus taeda*, Pseudotsuga menziesii*
Plumbaginaceae	*Aegialitis annulata, Limonium peregrinum*
Poaceae	*Andropogon virginicus, Avena sativa*, Bromus inermis, Hordeum vulgare*, Oryza sativa*, Pennisetum glaucum, Phragmites australis, Secale cereale, Sorghum bicolor*, Triticum aestivum*, Zea mays**
Polygonaceae	*Rumex acetosa, Rumex palustris*
Primulaceae	*Primula glaucescens*
Proteaceae	*Grevillea sp., Leucospermum glabrum*
Rhizophoraceae	*Bruguiera spp., Cassipourea elliptica, Ceriops spp., Rhizophora spp.*
Rosaceae	*Malus x domestica, Prunus persica, Pyrus communis, Rubus fruticosus**
Rubiaceae	*Cephaelis ipecacuanha, Oldenlandia affinis*
Rutaceae	*Citrus aurantifolia*, Citrus reticulata*, Citrus sinensis*, Zanthoxylum stenophyllum*
Salicaceae	*Populus deltoides*, Populus nigra**
Scrophulariaceae	*Antirrhinum majus, Triphysaria versicolor*
Simmondsiaceae	*Simmondsia chinensis*
Solanaceae	*Atropa belladonna, Capsicum annuum, Nicotiana attenuata, Nicotiana tabacum, Nierembergia caerulea, Petunia hybrida, Solanum khasianum, Solanum lycopersicum*, Solanum tuberosum**
Sterculiaceae	*Theobroma cacao**
Typhaceae	*Typha latifolia*
Ulmaceae	*Ulmus glabra*
Valerianaceae	*Valeriana glechomifolia*
Vitaceae	*Vitis vinifera**
Welwitschiaceae	*Welwitschia mirabilis*

[a]Nomenclature and family affiliation follow Mabberley (1997).

[b]Species were compiled from a sample of 915 papers published between 2000 and 2006; data are from a Web search conducted on the keyword "phytohormone" in Science Citation Index (www.isiknowledge.com), and the National Center for Biotechnology Information (http://www.ncbi.nlm.nih.gov/).

This chapter considers the multifaceted roles that phytohormones play in helping a seedling sense and adapt to four major environmental cues: light, temperature, water, and nutrients. Obviously, these factors covary and should not be considered in isolation because diverse environmental signals trigger similar phytohormones and transduction pathways. Thus, recurring themes and links between pathways will emerge in the discussion that follows. Although Chapter 2 points out that there are a dizzying diversity of seedling types and shapes, commonalities are emerging among taxa that suggest that many physiological mechanisms have been evolutionarily conserved through time. By better comprehending the actions of phytohormones, expanding beyond the classic *Arabidopsis* model to encompass many taxa, and integrating physiological mechanisms with optimality theory, ecologists are poised to answer fundamental questions such as these: how does a seedling assume the formidable responsibilities of an autotroph and manage other ontogenetic transitions? What intrinsic factors constrain plasticity over the short term and adaptation over the evolutionary term? How do suites of correlated traits arise and evolve?

Phytohormones are low-mass molecules that perform regulatory functions and serve as signal messengers, usually beyond the tissue in which they are produced (see Davies, 1995, for reviews and molecular structures). The general mode of phytohormone action begins with an environmental stimulus that causes upregulation (initiation or acceleration) of phytohormone synthesis or induces a localized change in tissue sensitivity to a particular phytohormone. Phytohormones are thought to induce a cascade of phenotypic effects by binding with specific receptors situated in organelle membranes and/or by binding with carrier molecules that transport the phytohormone into the cytosol (Fig. 7.1). The receptors for auxin (Napier *et al.*, 2002) and ethylene (Chen *et al.*, 2005) have been studied in detail; other receptors, such as those for abscisic acid (ABA), remain to be fully characterized (Shen *et al.*, 2006). These binding reactions cause changes in transcription of the proteins that mediate cellular responses to the stimulus (e.g. the scaffold RACK1 protein; Chen *et al.*, 2006), that serve a protective role in maintaining cellular homeostasis (e.g. chaperone proteins; Sangster & Queitsch, 2005), or whose breakdown serves to relieve inhibition of metabolic processes (e.g. DELLA transcriptional regulators; Alvey & Harberd, 2005; Achard *et al.*, 2006). Frequently, ion second messengers are linked to phytohormone levels. For example, rapid effluxes of hydrogen ions (H^+) alter plasma membrane polarity and permeability, whereas cyclical oscillations in the concentration of calcium ions (Ca^{2+}) in the cytosol can serve as signals in their own right that set in motion a cascade of other physiological activities (Sanders *et al.*, 2002; Xiong *et al.*, 2002; Fig. 7.1). When the environmental stimulus passes, phytohormones quickly break down and their concentrations decline, but their downstream effects on plant morphology and on plant conditioning to future stimuli may be long-lasting.

Traditionally, plant physiologists have recognized five classes of phytohormones: cytokinins (zeatins), abscisic acid (ABA), gibberellins (GA), auxin (IAA), and ethylene (reviewed by Davies, 1995; Farnsworth, 2000). Each governs a suite of *ad hoc* responses and developmental modifications in the growing seedling and there is much cross-talk between their individual signaling pathways. In very general terms, cytokinins, gibberellins, and auxin tend to stimulate cellular activities, including expansion and division, whereas ABA and ethylene typically serve more conservative functions in slowing metabolism and girding cells during times of stress. In combination, phytohormones can act additively, synergistically, or antagonistically (Table 7.2).

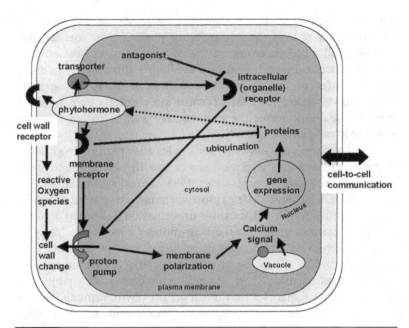

Fig. 7.1 Diagram illustrating some common general processes in phytohormone signal transduction in a generic plant cell. A phytohormone can bind to a receptor in the plasma membrane or cell wall, or it can be actively transported via a transporter into the cytoplasm. Here, it may bind with an intracellular organelle receptor; antagonists such as other phytohormones can compete for these binding sites. This binding may result in the activation of a proton pump, which hyperpolarizes the membrane and alters permeability, stimulates release of calcium, and upregulates gene expression and protein production. Enzymes may catalyze synthesis of phytohormones. Binding may also cause the ubiquination* of proteins or deactivation of other receptors. Phytohormones can also alter cell wall structure by promoting acidification (involving the proton pump), promoting production of radicals that cleave polysaccharides and loosen the wall, or influencing gene expression. (*Ubiquination is the first step in the targeted destruction of proteins, a critical mechanism by which protein products are broken down and recycled in the cell. Ubiquitin, a small protein, highly conserved among species, attaches to a lysine residue in the target protein when catalyzed by target-specific activating enzymes. This binding marks the protein for destruction by proteasomes.)

In recent decades, additional substances of importance to plant signaling have been identified, enriching our understanding of several complex plant behaviors. Salicylic acid (SA) induces systemic acquired resistance (Grant & Lamb, 2006), enabling plants with a prior history of attack from pathogens to be less susceptible to future onslaughts. Jasmonic acid (JA) upregulates production of defense proteins involved in inducible responses to wounding and herbivory (Creelman & Mullet, 1997). There is evidence that these two hormones overlap in aspects of their signaling pathways (Hatcher et al., 2004). Moreover, they are both influenced in turn by ethylene (Devoto & Turner, 2005), and SA can inhibit the action of JA in inducing resistance to mechanical injury (Cipollini et al., 2004).

Polypeptide hormones also help coordinate wounding responses and cellular differentiation in the meristem (Ryan et al., 2002). The

Table 7.2 | Summary of some ecologically significant phenotypic effects and interactions of selected common phytohormones (*sensu lato*) in seedlings.

Phytohormone	Actions in seedlings
Abscisic acid (ABA)	Promotes production of chaperone proteins; closes stomata; prepares seedlings for dormancy; influences heterophylly; sugar sensing; antagonistic with brassinosteroids and auxin
Auxin (IAA)	Cell wall loosening resulting in expansion growth; formation of leaf and root primordia; phototropic bending; lateral root formation; photoperiodic and thermoperiodic stem growth; promotes nodulation; synergistic with brassinosteroids
Brassinosteroid	Enhances generalized resistance to environmental stress and pathogens; synergistic with auxin; antagonistic with ABA
Cytokinin (CK)	Chloroplast maturation; enhances mitosis; promotes lateral bud formation; promotes root production (with auxin); nitrate sensing; enhances biosynthesis of ethylene
Ethylene	Slows stem growth; maintains apical hook; leaf epinasty; aerenchyma formation; regulates nodulation
Gibberellic acid (GA)	Stem elongation during etiolation; circadian rhythms and flowering cues
Jasmonic acid (JA)	Upregulates defensive proteins during wounding; adjustment to osmotic stress; synergistic and antagonistic with ethylene and SA, depending on tissue
Polyamines	Cell division; protein synthesis; possible synergy with auxin in root formation; concentrations increase in injured tissues
Polypeptide hormones (systemin, PSK, CLAVATA, ENOD40, etc.)	Coordinate wounding response (with JA); cellular differentiation in shoot meristem; cell proliferation; adventitious root formation; early phases of nodule formation
Salicylic acid (SA)	Systemic acquired resistance; competitor with JA

polypeptide systemin triggers increased production of JA, which then translocates through the phloem to communicate the wounding signal to the whole plant (Matsubayashi & Sakagami, 2006). Whereas systemins have been well characterized from members of the Solanaceae including potato (*Solanum tuberosum*), tomato (*Solanum lycopersicum*), and bell pepper (*Capsicum annuum*), efforts to isolate it (or any receptors that are sensitive to systemin) from other plant species have been largely inconclusive to date. In contrast, the phytosulfokine (PSK) polypeptide is more widely distributed in the plant kingdom. It promotes cell proliferation and differentiation, two processes of fundamental importance to fast-growing seedlings. Similarly, the CLAVATA3 polypeptide helps to determine cell fate in the developing root and shoot meristems of *Arabidopsis* seedlings (Clark, 1997). Other polypeptides participate in early phases of nodule formation and relay signals involved in recognizing self-incompatible pollen (Ryan *et al.*, 2002). These hormones have been widely documented as crucial in animals, but their presence and critical roles in plant development are only now coming to light.

Brassinosteroids constitute another class of phytohormone with important influences on early plant growth. Found in plants, animals, and fungi, they are likely very ancient in their evolutionary origins. However, the proteins they upregulate in plants are quite variable among plant families, indicating that their signaling functions have diversified during the process of plant speciation (Vert et al., 2005). From an ecological standpoint, brassinosteroids are particularly interesting because they enable plants to resist environmental stress. In seedlings exposed to drought, elevated salinity, extreme temperatures, and pathogen attack, brassinosteroids appear to confer protection by regulating membrane permeability and upregulating production of protective molecules, called chaperones, which stabilize the folding structure of other proteins (Krishna, 2003). Brassinosteroids permit cell expansion, vascular differentiation, and seedling growth even under these adverse conditions, and young tissues are particularly sensitive to applications of these steroids (Clouse & Sasse, 1998). In this way, they differ fundamentally from other stress hormones like JA, SA, and ABA that act more defensively to conserve cell structure and suppress growth and metabolism. Indeed, ABA is noted as an antagonist of brassinosteroid action (Krishna, 2003). Auxin, on the other hand, works synergistically with brassinosteroids, and the two may share several components of their gene-signaling pathways (Vert et al., 2005).

Polyamines, such as putrescine and spermidine, share certain characteristics with phytohormones. They are small molecules that promote a range of cellular activities including cell division and protein synthesis in growing seedling shoot and root meristems (Bais & Ravishankar, 2002). Conversely, amine oxidases, which break down polyamines and release hydrogen peroxide, appear to help lignify and reinforce cell walls and thus to enhance cellular defenses to pathogens. In so doing, they also depress growth of seedling stems in high light (Cona et al., 2006). Polyamine levels tend to correlate positively with auxin concentrations (whereas auxin counteracts polyamine oxidases) and they may interact synergistically with this phytohormone to induce root formation (Bais & Ravishankar, 2002). Unlike conventional phytohormones, however, polyamines must attain much higher cellular concentrations to exert noticeable phenotypic effects and there is no evidence as yet that they are transported throughout the living plant. Thus, their roles as growth regulators sensu stricto remain to be explained.

7.2 | Seedling responses to light

Phytohormones and phytochrome modulate early growth cues

Within seconds of bursting the dark confines of the seed coat, the seedling either receives at least a faint signal of light or, if not, it must elongate quickly in near darkness to reach the soil surface and detect light. There is extensive evidence that multiple hormones control responses of seedlings to both dark and light, and that they are active during different periods of time.

In dicots, an apical hook forms as cells on the abaxial side of the stem elongate faster than those on the adaxial side. As the hooked portion of the stem pushes its way up through soil or leaf litter, it protects the tender apical meristem from abrasion. Both the quality and quantity of incoming light exert profound effects on the formation and longevity of the apical hook. Seedlings growing in full dark or exposed to a pulse of far-red light following growth in low fluxes of red or blue light are slow to unfold the apical hook and show other signs characteristic of etiolation, including exaggerated vertical growth of the hypocotyl, lack of chlorophyll, an abundance of the pigment phytochrome, and slow root growth. In contrast, seedlings grown in higher irradiance rapidly de-etiolate, straightening and becoming positively phototropic. Gibberellins, which are highly active during seed germination, continue to direct hypocotyl elongation as the seedling etiolates to escape dark conditions (García-Martinez & Gil, 2002). DELLA proteins, which normally act as growth suppressors, are destabilized during the GA response phase, permitting growth to occur (Vandenbussche et al., 2005). The gaseous phytohormone ethylene causes an exaggeration of the apical hook, but only during a small developmental window (e.g. 60–72 hours after germination in Arabidopsis [Vandenbussche & Van Der Straeten, 2004]). Ethylene also reduces hypocotyl growth and thickens the stem, producing a stunted, stocky seedling. Together, these phenotypes comprise the so-called triple response to ethylene (Vandenbussche et al., 2005) and may enhance the structural strength of the plant. Cytokinins contribute to an accumulation of ethylene in the hypocotyl by stabilizing the primary enzyme (ACC synthase) involved in ethylene biosynthesis, thus contributing to the triple response. Conversely, auxin opposes ethylene action, suppressing formation of the apical hook and encouraging hypocotyl elongation through longitudinal cell expansion (Vandenbussche & Van Der Straeten, 2004). Brassinosteroids also hasten hypocotyl elongation (Vandenbussche et al., 2005).

Once the apical hook straightens, however, the seedling must prepare itself for vertical growth that is optimized for the existing light environment. Sunlight passing through a plant canopy is rich in far-red wavelengths and comparatively poor in red and blue wavelengths. A red light stimulus causes an increase in the concentration of active P_{fr} (the far-red absorbing form of phytochrome) relative to P_r (the red-absorbing form of phytochrome) in the cytoplasm. When P_r is photoconverted to P_{fr}, a portion of the P_{fr} pool migrates into the nucleus where it regulates gene transcription. There are five forms of phytochrome (PHYA-PHYE). Phytochrome B (PHYB), alone or in concert with PHYD and PHYE, coordinates the process of de-etiolation when the seedling is growing in abundant light. These photoreceptors can sensitively detect the presence of neighbors that subtly influence the red:far-red ratio of incident light (Vandenbussche *et al.*, 2005). Vines, which rely on the ability to locate neighbors for climbing and thus move skototropically toward shade, may have evolved novel photoreceptive mechanisms for responding to phytochrome signals, but few studies have addressed this intriguing question to date (Orr *et al.*, 1996). PHYA, which is highly sensitive to transient high irradiances of far-red light, tends to predominate in shady conditions and may enable the seedling to efficiently capitalize on sunflecks. There is now considerable evidence from field-grown plants that populations of seedlings growing in sun and shade show adaptive divergence for heritable genetic variation in phytochrome-mediated shade-avoidance responses (Schmitt *et al.*, 2003; von Wettberg & Schmitt, 2005). Likewise, plant populations differ in the plasticity of light-response traits depending on the predominant light environment experienced by seedlings (Weinig, 2000).

Phytochrome photoreceptors influence the locus and timing of hormone action in seedlings as they surmount leaf litter or a canopy. As seedlings reach the light, phytochrome may regulate the inactivation of GA and/or shut down its biosynthesis, thus slowing stem elongation (García-Martinez & Gil, 2002). Auxin encourages phototropic movement of the shoot toward light and negative movement away from gravity. The redistribution of auxin away from roots and into the shoot promotes stem elongation and a reduction in the root:shoot ratio under a high far-red to red ratio (Vandenbussche *et al.*, 2005). Interestingly, hypocotyl elongation occurs on a circadian rhythm, with seedlings growing faster during dusk, when diminished light levels may elicit growth patterns typical of shade, and arresting growth at the first perception of dawn. Levels of auxin in stems also cycle, paralleling these growth spurts in time (Jouve *et al.*, 1999).

Blue, short wavelengths are at the other end of the light spectrum. Plants are highly sensitive to blue light, especially after they have emerged out of dark conditions. Two classes of photoreceptors that are ubiquitous in eukaryotes, phototropins and cryptochromes, sense blue light. Phototropins promote positive bending toward blue light, stomatal opening, and chloroplast migration to optimize light capture, whereas cryptochromes mediate de-etiolation and aspects of photoperiodism, such as the rhythmic growth in hypocotyls mentioned above (Lin, 2002). Cryptochromes are required to maintain homeostatic levels of auxin, further evidence of the role of auxin in regulating circadian clocks (Nozue & Maloof, 2006). Early establishment of a circadian rhythm is crucial to the seedling. Plants that correctly anticipate light–dark cycles perform significantly better in growth, photosynthesis, and competition than nonentrained plants (Dodd *et al.*, 2005).

Once light is sensed by the seedling, important developmental transitions begin in earnest, namely, the differentiation and production of the first true leaves and roots. KNOX homeobox genes both regulate phytohormone biosynthesis and are in turn regulated by phytohormone levels in a feedback. KNOX genes are positive regulators of cytokinins and negative regulators of GAs in the shoot apical meristem, and it is a combination of high concentrations of cytokinins relative to GAs that maintains the undifferentiated state. When KNOX genes are downregulated in the peripheral zone around the shoot apical meristem, the affected tissues become leaf primordia and sinks for auxin. These leaf primordia deplete auxin from adjacent tissues such that new leaf primordia can only form at a minimum distance from existing primordia where sufficient auxin accumulates. The spatial distribution of auxin around these sinks in the peripheral zone may explain the development of a regular phyllotaxis along the stem (Kepinski, 2006). Auxin may also modulate the ratio of GAs and ABA in developing leaves. This ratio appears to influence the morphological type of leaf that is formed (Young et al., 1995; Lumba & McCourt, 2005) and may help explain heterophylly in certain species. Asymmetrical distribution of auxin within developing new leaves may also influence morphological differences between the abaxial and adaxial sides of the leaf.

Auxin also tightly regulates lateral root initiation, first by briefly upregulating expression of NAC genes in the root tip and lateral root primordium and then by directing the breakdown of NAC transcriptional products by both ubiquination and microRNAs (Kepinski, 2006). Studies with mutants have also demonstrated that increased auxin concentrations and/or sensitivity to auxin in developing roots are also correlated with the strength of gravitropic bending and growth (Firn et al., 2000; Teale et al., 2005).

Coping with changing light environments as the seedling grows

Plants – especially long-lived woody species – can experience highly variable light environments over their lifetimes as new canopy gaps open or as the canopy is reached (Chapter 16). As Schmitt et al. (2003) demonstrate, a genetic capacity for plastic responsiveness to light is adaptive in seedlings growing in changeable environments. Other work suggests that seedlings are more flexible in a variety of responses than saplings or mature trees, as indicated by large morphological and physiological differences among sun-grown and shade-grown populations (Farnsworth & Ellison, 1996). Because both photoreceptors and phytohormones are intimately involved in shaping light responses, it would be of interest to chart the changes in the roles of these factors as plants age. Several genes involved in the phase transition from juvenile-type to adult leaves and overall plant morphology have recently been identified and their interplay with phytohormones and environmental signals are ripe for exploration (Chuck & Hake, 2005; Wu & Poethig, 2006).

7.3 | Seedling responses to temperature

The quantity and type of light that reaches seedlings profoundly influence the air and soil temperatures they experience. The kinetics of all biologically relevant chemical reactions depend on ambient temperatures that lie between a minimum threshold and a maximum above which proteins denature. Often, germination gets underway as the embryo senses that external temperatures are fluctuating within a narrow range that will be conducive to seedling growth (Baskin & Baskin, 2001). Some seedlings can enter a temporary state of arrested development, controlled by ABA, immediately following germination, if conditions are not conducive to vegetative growth (Finkelstein et al., 2002; Lopez-Molina et al., 2001). Seedlings inhabiting extremely cold or hot environments must protect their cellular machinery while performing basic metabolic actions. Seedlings that germinate shortly before the onset of winter must prepare to enter a period of partial quiescence and simultaneously integrate cold and warm signals that prepare them for renewed activity and reproduction in the spring.

Similar to its role in transducing light signals, auxin translates temperature cues to signals that regulate cell expansion and hypocotyl elongation (Heggie & Halliday, 2005). Rising temperatures are correlated with increasing levels of active auxin in the shoot and act synergistically with an increasing proportion of far-red light to promote stem elongation in the presence of neighbors (Weinig, 2000). Much as seedlings exhibit a photoperiodic rhythm of growth, their growth also cycles thermoperiodically. Elongation in several species increases positively with the difference between daytime and nighttime temperatures. This periodicity in growth may be controlled both by auxin and by levels of GAs (Heggie & Halliday, 2005).

Seedlings exposed to predictable seasonal climates must prepare to protect vulnerable tissues from dramatic changes in temperature. Many species that grow in wintry latitudes or in areas affected by high summer temperatures undergo a period of induced dormancy. This phase of arrested development requires slowed or suspended metabolic processes – including photosynthesis, water uptake, and growth – and requires that cell membranes be stabilized against fluctuating osmotic conditions (Kozlowski & Pallardy, 2002). Preparation for dormancy frequently begins well before the onset of extreme temperatures, when the plant accumulates starch and phospholipid reserves in leaf bases and translocates them underground to storage organs (Volaire & Norton, 2006). Phospholipids are also cleaved to produce inositol 1,4,5-trisphosphate (IP_3), a second messenger that triggers Ca^{2+} release in the cytosol (Xiong et al., 2002; Wang, 2004). On exposure to heat or cold, the plant increases production of protective dehydrin proteins that protect cell membranes against a host of stressors, including tissue desiccation (Close, 1997). The phytohormone ABA accelerates release of IP_3 and upregulates dehydrin production, and temperature cues stimulate the activity of genes responsible for steps in the ABA biosynthetic pathway (Knight et al., 2004). Interestingly, precursors for ABA synthesis include the protective pigments zeaxanthin and violaxanthin. These carotenoids help to protect fragile photosystems from excess light energy and may enable certain evergreen species to withstand the dual stressors of cold and high light (Adams et al., 2004).

ABA similarly stimulates production of heat shock proteins (Finkelstein et al., 2002). These chaperones are ubiquitous among living organisms and throughout plant cells (Sangster & Queitsch, 2005). Mutants unable to produce certain heat shock proteins show myriad growth abnormalities, as well as reduced tolerance of thermal extremes (Sangster & Queitsch, 2005), indicating that these chaperones are crucial to normal growth of leaves, stems, and roots. Although some are constitutively produced, many are highly inducible, especially under high-temperature conditions that can disrupt protein folding. Heat shock proteins also function as sensors of the levels of reactive oxygen species, such as hydrogen peroxide, that tend to increase during many kinds of stressful conditions, from high temperatures to pathogen attacks (Miller & Mittler, 2006). Ecological studies have documented genetic correlations between frost resistance and resistance to herbivory in wild radish populations subjected to natural selection (Agrawal et al., 2004). One could speculate that these correlated traits are linked by the pleiotropic roles of heat shock proteins.

7.4 | Seedling responses to water

The minute most seedlings emerge from the protection of the seed coat to face the world on their own, they are vulnerable to water stress. Lest it succumb to drought, a seedling must quickly send out roots to forage for water in the soil while it unfurls leaves, opens stomata, and initiates photosynthesis and transpiration that will draw water up through its growing xylem. As noted earlier, auxin facilitates the early production of roots by the seedling (Kepinski, 2006). Auxin moves directionally from sites of synthesis in the stem to the elongation zone of the root, where it promotes cell expansion and hastens cell division to lengthen the primary root and encourage growth of lateral roots (Teale et al., 2005). Auxin flow becomes canalized in a narrow band of cells in the growing root, thus influencing the differentiation of cambial strands into vascular strands (Fukuda, 2004). Auxin also promotes gravitropism and hydrotropism, the means by which roots descend into the soil and find water and nutrients (Firn et al., 2000; Eapen et al., 2005).

Once again, cross-talk between auxin and other phytohormones figures in this phase of early root growth. High ratios of auxin to cytokinins in meristems favor root production over shoot production. ABA also influences root morphology by antagonizing auxin and suppressing the production of root hairs (De Smet et al., 2006). At the same time, ABA may promote the efficiency of water uptake by developing roots as it increases membrane permeability to water. ABA also conveys drying signals to the rest of the plant, sensitively controlling the closing of stomata under desiccating conditions (Finkelstein et al., 2002).

Much of our understanding of how phytohormones regulate the adjustments of young seedlings to normal conditions of soil and air moisture comes from studies of plants subjected to drought or flooding stress, during which time ABA is a central player (Chapter 3). ABA is upregulated during the onset of dormancy in plants growing in seasonally arid environments, although its responses to drying appear to operate independently of its role in temperature-sensing (Volaire & Norton, 2006). Many of the protective measures that seedlings take in high and low temperatures mirror their responses to drought or salinity stress, particularly the buildup of compatible solutes that maintain osmotic balance in cells, proteins that stabilize membranes, reactive oxygen species and their scavengers, and ABA (Kozlowski & Pallardy, 2002; Xiong et al., 2002). Recent investigations of seedling responses to salinity indicate that ABA also enhances the activity of DELLA proteins that suppress plant growth and extend the duration of the vegetative/juvenile phase of plant life history (Achard et al., 2006).

Jasmonic acid also increases in seedlings exposed to drought and increased salinity (Creelman & Mullet, 1997; Devoto & Turner, 2005). Seedlings growing in saline conditions may have elevated JA contents that enable them to react rapidly to herbivory because JA also mediates wounding responses. In fact, seedlings of the glycophyte *Iris hexagona* exposed to elevated salinity show heightened resistance to herbivores (Mopper et al., 2004) and *Rhizophora mangle* seedlings growing in hypersaline habitats experience low rates of herbivory compared to populations growing in less saline conditions (Farnsworth & Ellison, 1991). The precise effects of JA on plant water relations are largely unknown, but it likely acts indirectly to impose a general stress reaction by influencing the synthesis of other phytohormones such as ethylene (Devoto & Turner, 2005; Stepanova & Alonso, 2005).

More is known about the role of ethylene in mediating plant responses to water stress, particularly flooding, and some comprehensive ecological studies have addressed floodplain species (Pierik et al., 2006; Voesenek et al., 2006). When seedlings are submerged, they are subjected to multiple stresses: attenuated light spectra that are suboptimal for photosynthesis, reduced partial pressure of carbon dioxide around leaves, lack of oxygen availability to roots, and buildup of bacterial byproducts of anoxia, such as sulfur dioxide, in the root zone. Ethylene gas rapidly increases in concentration around the plant within minutes after submergence. Roots and sometimes stems of flooded plants often show increases in the amount of aerenchyma tissue, which facilitates the passive diffusion of oxygen through intercellular air spaces. Although obligate wetland species tend to produce aerenchyma continually, many plants can flexibly produce aerenchyma when flooded. So-called lysigenous aerenchyma forms when sets of cells collapse to form large gaseous spaces, with ethylene initiating this programmed cell death (Evans, 2004).

Plants exposed to frequent, shallow flooding also tend to increase shoot length and reorient leaves to more vertical positions in an effort to outgrow deepening water. Depressed levels of oxygen around leaves sensitizes petioles to the presence of ethylene, amplifying its effects. Ethylene enhances stem and petiole elongation directly and by increasing the ratio of GAs to ABA in leaves; together, GAs and ethylene upregulate the production of expansin proteins in vegetative tissues, which effectively loosen cell walls and promote cell expansion (Voesenek *et al.*, 2006). Both ethylene and ABA are involved in signaling the cellular status of sugars, which function simultaneously as membrane stabilizers during osmotic stress and as nutritional energy sources for respiration (Gazzarrini & McCourt, 2001).

7.5 | Seedling responses to nutrients

A germinating seedling initially relies on remaining seed reserves and storage cotyledons to supply its early nutritional needs. However, it must soon attain the capacity to absorb nutrients from the soil via mass flow (transpiration-regulated uptake to the roots) and via active proliferation of roots in nutrient-rich sectors of the soil. The phytohormones that guide root morphogenesis, particularly auxin, cytokinins, and ethylene, facilitate acquisition of nutrients by encouraging cell division at the primary root meristem, formation of lateral roots, and expansion of root surface area through production of root hairs (López-Bucio *et al.*, 2003). High uniform soil concentrations of phosphorus or nitrate tend to inhibit lateral root production by many species of plants, whereas a paucity of nutrients or a localized spike in availability of nutrients in part of the rhizosphere causes plants to proliferate lateral, foraging roots. Both auxin concentrations and expression of phosphate transport proteins are elevated in root tips of seedlings grown in low-phosphate conditions and auxin increase is correlated with the production of phosphate-absorbing proteoid roots (cluster roots) (Skenel, 2001; López-Bucio *et al.*, 2003). Likewise, phosphate-starved root tips are more sensitive to auxin application than well-fed root tips. Sulfate- and iron-limited seedlings also increase production of auxin (Skenel, 2001). There is extensive evidence that auxin and ethylene act interdependently to regulate root growth, for example, in response to iron availability (López-Bucio *et al.*, 2003; Stepanova & Alonso, 2005).

The picture for nitrogen sensing and uptake is murkier, but recent work suggests mechanisms by which its availability and limitation are communicated throughout the plant. Nitrate itself upregulates the production of cytokinins in roots. Nitrate and ammonium trigger production of two different adenosine phosphate-isopentenyltransferases (cytokinin precursors), respectively: IP3 in the phloem, which contributes directly to cytokinin production, and IP5 in the root primordium (Sakakibara *et al.*, 2006). Cytokinins then negatively feed back on the production of nitrate transporters, thus scaling back nitrate acquisition activities when nitrate concentrations are high in the soil. Cytokinins may also serve to communicate whether nitrate status is sufficient to enable uptake of less common nutrients such as sulfate (Sakakibara *et al.*, 2006). Cytokinins are translocated from the site of synthesis at the root phloem to the xylem and thence to the rest of the plant, where they stimulate stomatal opening and thus enhance transpirational uptake of nitrogen and other nutrients through mass flow.

Sugars as signals

Nitrogen metabolism is tightly linked with carbon metabolism in plants. Indeed, the internal ratio of carbon:nitrogen may be the primary, physiologically relevant parameter that seedlings monitor in adjusting growth to resource availability (Coruzzi & Zhou, 2001). Internal carbon-sensing mechanisms couple with nitrogen-sensing to permit a seedling to increase nitrogen uptake when supplies of photosynthetic assimilates are abundant and conversely to shut down nutrient uptake when carbon resources are running low. At high concentrations of exogenous glucose or sucrose, ABA levels rise in young seedlings and cause a slowdown in growth (Gibson, 2002), a response that occurs during only a short time and may be symptomatic of osmotic stress (Yuan & Wysocka-Diller, 2006). Ethylene, on the other hand, can antagonize or reverse this growth inhibition by sugars (Gazzarrini & McCourt, 2001). In addition to their interaction with the ABA/ethylene cross-talk, sugars may also feed into the auxin/cytokinin regulatory antagonism, influencing the relative production of roots and shoots in the growing seedling (Rolland *et al.*, 2002; Rashotte *et al.*, 2005). Studies of the interacting effects of sugars and nitrogen on phytohormone levels and morphogenesis are in their infancy and to date mainly involve rather artificial methods such as augmenting levels of exogenous sugars in growth media (Gibson, 2005). Once more nonlethal nutritional mutants can be developed and *in vivo* concentrations of sugars can be more sensitively manipulated, much can be discovered about how seedlings maintain nutritional homeostasis.

Seedlings and their nutritional symbionts: mycorrhizae, nodulation, and parasitism

Many plant species need to establish early symbioses with a bacterial or fungal partner that facilitates nutrient uptake, while others initiate parasitic relationships with host plants from which they draw nutrients directly (Chapters 4, 9). Phytohormones are integrally involved in translating the traffic of signals between seedlings and their partners and in producing specialized structures like nodules and haustoria that seal the partnership itself. Incidentally, symbionts are not the only organisms that communicate hormone-like substances to plants. Earthworms have been shown to influence root growth by exuding phytohormone mimics (El Harti *et al.*, 2001) and as-yet unidentified substances that promote systemic resistance to nematodes (Blouin *et al.*, 2005) and stimulate nodulation (Doube *et al.*, 1994).

Nodulating legumes produce flavonoids that activate rhizobial bacteria to produce so-called Nod factors (lipochitooligosaccharides) that in turn elicit a variety of responses on the part of the plant, including swelling and branching of root hairs and increased cell division in the root cortex (Esseling & Emons, 2004). Ethylene generally inhibits nodulation by negatively impacting early steps in Nod factor signaling, except under certain waterlogged conditions in which ethylene both influences the type of nodule formed and the mode of bacterial entry into the root (D'Haeze *et al.*, 2003).

Auxin increases in concentration in the first dividing cells that form primordial nodules, as Nod factors inhibit auxin transport toward the tip of the root (Mulder *et al.*, 2005). Accumulating auxin may stimulate cell division to form the nodule primordium, and is likely involved in regulating the development of vessels in legume nodules. Cytokinins are similarly implicated in this nodule-forming step and can actually substitute for Nod factors in initiating nodule organogenesis (Cooper & Long, 1994). Sucrose can also alleviate the suppression of nodulation by an overabundance of nitrate (Gibson, 2005), perhaps by stimulating activity of cytokinins.

Phytohormones also mediate communication between ectomycorrhizae and plants and may do so by mechanisms that share many steps with those observed in nodulating plants (Albrecht *et al.*, 1999). These overlaps in signaling pathways are perhaps not surprising, given that the legume–rhizobial symbiosis likely evolved from the more ancient arbuscular mycorrhizal symbiosis some 65 million years ago (Kistner & Parniske, 2002). Mutant mycorrhizae that overproduce auxin and deliver large amounts of IAA to plants induce production of unusually large numbers of ectomycorrhizal roots (Gay *et al.*, 1994). In contrast, several mycorrhizae also have been identified that convey an IAA antagonist, indole alkaloid hypaphorine, to roots, which inhibits root hair elongation and appears to fine-tune the action of IAA at the site of infection (Jambois *et al.*, 2005). Hypaphorine may also affect the activity of endogenous IAA elsewhere in the plant, but the nature of its action is currently unknown.

Localized auxin and ethylene upregulation also figures in the establishment of parasitic haustoria between seedlings and host plants (Tomilov *et al.*, 2005). Phenolic exudates from the host plant roots elicit haustorial formation by the parasite, which entails a cessation of regular root tip elongation and proliferation of adhesive hairs that bind the parasite to its victim. Soon after exposure to haustoria-inducing factors, auxin and ethylene build up at the root tip, working together to initiate haustorial hair growth and radial expansion of hair cells (Tomilov *et al.*, 2005). Once the haustorial connection is made, auxin may be exported from the parasite to the host root where it induces cell elongation (Loffler *et al.*, 1999; Birschwilks *et al.*, 2006). Nothing appears to be known about the fascinating physiological mechanisms behind the subsequent loss of the photosynthetic apparatus in parasites once infection has occurred. It is plausible that ethylene also plays a role in such deprogramming. Parasitism has evolved independently in several disparate families (more than 4000 plant species), yet the early cues and phytohormonal mechanisms appear to have been evolutionarily conserved. Here, then, is an intriguing opportunity to study convergence among phytohormonal pathways in multiple taxa (Yoder, 2001).

7.6 | Insights and common themes

What have we learned during this brief exploration of seedling physiology? This review highlights the fact that life as a seedling involves the complex interplay of all the major phytohormones documented to date. This chapter has focused on light, temperature, water, and nutrients, and has not considered environmental cues, such as gravity, mechanical stress (Telewski, 2006), or fire, that make or break seedlings, or the action of other versatile signaling molecules like nitrogen oxides (Lamattina et al., 2003). Nor have we considered the contributions of maternal phytohormones to early seedling growth and their role in conferring so-called maternal effects (Mousseau & Fox, 1998), a rich field for future study.

What is a student of seedlings to make of all this? Despite the complexity of the picture, a few common themes do emerge, which could inform future integrative work on seedlings from an ecological standpoint. What follows are necessarily a smattering of broader concepts and interesting ideas for further study, reflecting my own interests as an ecophysiologist, but this is by no means an exhaustive list of topics.

Synergy and antagonism

Many phytohormones act in concert to fine-tune responses quickly to changing conditions. We have observed, for example, that auxin and cytokinins work together in determining cell fate in shoot and root meristems and both are sensitive to nutrient cues. An ecologist might now ask: How do they influence the well-known patterns of tradeoffs in root:shoot allocation by seedlings under varying nutrient, water, and carbon availabilities?

One phytohormone, many phenotypes

Single phytohormones are implicated in both the transitory appearance of plastic phenotypes and the evolutionary fixation of correlated traits. The limits on response imposed by one phytohormone can influence the range of phenotypes that can be shaped by others. We have observed, for example, that JA may help seedlings cope with osmotic challenges while simultaneously conferring some protection against herbivores. An ecologist might now ask: Do certain seemingly neutral or even maladaptive traits persist in populations because they are inextricably linked by the action of a single phytohormone (Gould & Lewontin, 1979; Farnsworth, 2000)?

Costs of constitutive versus inducible expression

Most phytohormones are present in consistently low levels in all seedling tissues, but all can respond if a sudden need arises. We have observed, for example, that ABA upregulates the production of chaperone proteins in response to extreme heat or cold, but it may be energetically costly or maladaptive to produce large amounts of these proteins continuously. An ecologist might now ask: What are the costs and benefits of phytohormone-mediated tolerance (constitutive expression to maintain fitness in the long term) versus resistance (inducible expression to reduce damage) under conditions of constant versus transient stress (Agrawal *et al.*, 2004; Jackson *et al.*, 2004)?

Developmental windows and tissue specificity

Phytohormones produce different phenotypic effects in different tissues, and tissue sensitivity to particular phytohormones may only arise during brief stages in the development of a seedling. We have observed, for example, that ethylene controls the formation of the seedling apical hook within hours of germination, but later can enhance stem elongation in older seedlings subjected to flooding. An ecologist might now ask: How do phytohormones shape ontogenetic shifts in seedlings and older plants as they respond to herbivory (Boege & Marquis, 2005), alter leaf display under changing light environments (Farnsworth & Ellison, 1996), or produce heterophyllous leaves (Pigliucci, 1998; Chuck & Hake, 2005)?

Evolutionary conservatism among signaling pathways

Many of the proteins involved in phytohormone signaling are evolutionarily ancient. Likewise, several hormones such as brassinosteroids are shared among animals, plants, fungi, and other organisms. We have observed, for example, that, together with KNOX genes, auxin is a key player in directing meristematic growth and tissue determination. An ecologist might now ask: Can we trace convergent changes in phytohormone action within and among lineages that will help us understand both evolutionary innovations and constraints in suites of stress-related traits (Chapin *et al.*, 1993; Farnsworth & Farrant, 1998) or the repeated evolution of the woody growth habit (Groover, 2005)?

Variation in phytohormone effects among species

As the range of species in which phytohormones are studied expands, it is becoming clear that not only do evolutionarily conserved commonalities exist, but also great interspecific diversity in the timing, mechanism, and effect of phytohormone action. Just as interspecific differences in the external phenotype can yield insights into the ways in which organisms adapt to environmental conditions, diversification in phytohormonal physiology reveals much about species' divergence and the evolutionary process. We have observed, for example, that sensitivity to GA and ethylene differ among species that hail from contrasting flooded and dry habitats (Voesenek *et al.*, 2004). An ecologist might now ask: How do interspecific differences in phytohormones action translate to niche partitioning, such as the ways in which co-occurring species deploy roots at different soil depths or initiate germination in response to different cues?

New tools for understanding the "eco-evo-devo" of seedlings

Phytohormones regulate germination and the processes of extending the tiny organism upward, displaying the first set of leaves to capture light, and deploying roots to find nutrients or symbionts. The brief window of seedling stage (for most plant species that do not exhibit dormancy or developmental arrests) is amenable to experimentation with the new physiological and genetic tools at an ecologist's disposal. However, it is critical to apply these tools to a greater variety of species under a range of field conditions.

Much of our current understanding of the phenotypically relevant functions of phytohormones comes from elegant studies of genetic mutants of *Arabidopsis thaliana*, *Zea mays*, and *Solanum lycopersicum* that underproduce or overproduce particular phytohormones or that are hypersensitive or insensitive to the signals that phytohormones transduce (McCourt, 1999). Mutants like these can be identified in wild populations and cultivated; likewise, targeted knockout mutants could be created in a wider array of species. Applications of exogenous phytohormones or their inhibitors also can cause many of the same symptoms as genetic mutations. Moreover, seedlings can be grown under a variety of ecological conditions that stimulate endogenous production of phytohormones, providing further evidence of the specific roles of these substances in shaping plant responses.

Molecular biologists are rapidly developing new tools for teasing apart phytohormone response pathways. The identification of plant homeobox proteins such as KNOX, which regulates a cascade of gene expression in the shoot apical meristem, has led to new discoveries about how phytohormones can influence indeterminate growth and tissue differentiation in both corn (*Zea mays*) and *Arabidopsis* (Kepinski, 2006). We now understand that many signal transduction pathways involve similar steps that alter or destroy protein structure through the action of protein kinases and ubiquitin activases (Vandenbussche & Van Der Straeten, 2004). Novel technologies for visualizing localized transcriptional activity (e.g. reporter genes that fluoresce in specific plant tissues when activated) allow researchers to pinpoint the location of phytohormone activity. Microarrays, which can screen thousands of gene products at a time, can be used to characterize and compare gene expression patterns under different levels of phytohormones (Gibson, 2002; Nemhauser et al., 2006). Finally, the advent of bioinformatics facilitates online comparison of expressed gene products and sequence tags with an ever-expanding variety of databases, allowing multispecies comparisons that illuminate evolutionary relationships and encourage creative conceptual links to be drawn between plant and animal systems (e.g. similarities between juvenile hormone, retinoic acid, and ABA; McCourt et al., 2005).

The majority of phytohormone studies to date have focused on a few model species, such as *Arabidopsis thaliana*, as well as important crop species, including *Pisum sativum*, *Oryza sativa*, *Zea mays*,

Populus spp., *Antirrhinum majus*, and *Pinus taeda*, whose genetic profiles are well-characterized and for which several experimental breeding lines have been established (Table 7.1). Nevertheless, the diversity of species for which genetic information is available, especially searchable expressed sequence tags (ESTs), is increasing dramatically and currently spans at least 129 species in 82 genera and 37 families (National Center for Biotechnology Information; http://www.ncbi.nlm.nih.gov/; see also Table 7.1). The availability of broad data and new model systems creates opportunities for ecologists to apply genetic techniques to the understanding of phytohormones in seedlings of many more species in field settings (Chao *et al.*, 2005). We could begin, for example, with field studies of congeneric species that are closely related to existing laboratory models (e.g. Mitchell-Olds, 2001).

To date, the field of molecular ecology is largely unexplored and published studies of phytohormones in wild plant species are few (Mopper & Agrawal, 2004). Many researchers who have measured phytohormones in an ecological context have sought to characterize the factors involved in responses to herbivory, particularly SA and JA (Mopper *et al.*, 2004; Cipollini, 2004), and in responses to flooding or osmotic challenges (Voesenek & Blom, 1996; Farnsworth, 2004). Comparative studies of mangrove species have described convergent evolutionary changes in ABA dynamics in developing embryos that may explain the occurrence of precocious germination (vivipary and cryptovivipary, see Chapter 2) in multiple species (Farnsworth & Farrant, 1998). Other researchers have studied wild populations of *Arabidopsis* spp. to understand better the evolution of plant defenses (Mitchell-Olds, 2001) and latitudinal clines in flowering time (Stinchcombe *et al.*, 2004), focusing on gene expression but not phytohormones *per se*.

Given this nascent enthusiasm, it is perhaps surprising that plant ecologists have not begun to explore phytohormones in earnest, particularly in the context of understanding correlated traits. However, it is important to acknowledge that such studies cannot be performed overnight. It is challenging to isolate pure phytohormones from plant tissues, given the multitude of secondary compounds that can complicate extraction. Even physiologists and geneticists themselves are beginning to acknowledge that signal transduction pathways are complex and that cross-talk between phytohormones complicates the traditional simple picture of functionality (Nemhauser *et al.*, 2006). Regarding ABA, McCourt *et al.* (2005: 147) recently quipped that "the signalling pathway looks more like the Tokyo subway system, in which competing trains run in different directions and there are many routes to the same place. Even if you know the destination, you feel unsure, because the names of the trains and stations are written in a language that [many] of us do not understand." Nevertheless, the potential for collaboration among ecologists, evolutionary biologists, geneticists, and physiologists is promising.

Traditional disciplinary divides must be crossed. Physiologists can help ecologists hone in on mechanism, while ecologists can help physiologists apply their laboratory findings to relevant field conditions. Tools for analyzing phytohormones and gene products are increasingly easy to apply and market demands will undoubtedly lead to the creation of still more user-friendly protocols. Despite the very real challenges, it is time for integration across many levels of biological organization.

7.7 | Summary

Phytohormones are small molecules that integrate environmental cues, relay chemical signals throughout the seedling, influence gene expression, and coordinate transient and long-term plant responses. Physiologists have long recognized the critical functions of five classes of growth regulators, auxin, cytokinins, gibberellins, abscisic acid, and ethylene, but more substances have recently been identified that regulate important seedling activities, including salicylic acid, jasmonic acid, brassinosteroids, polypeptides, and polyamines. Fluxes in secondary signals such as calcium ions and sugars amplify the effects of phytohormones on gene expression. There is considerable crosstalk among phytohormone signaling pathways that result in synergy, antagonism, and additive effects among phytohormones, suggesting a significant level of evolutionary conservatism among plants and living organisms generally, even though a myriad of diverse phenotypes exists. Phytohormones shape seedling strategies such as:

- Formation and opening of the apical hook following germination
- Etiolated growth in shade, far-red-enriched light, and the presence of neighbors
- Development of leaf and root primordia from meristem
- Entry into dormancy under conditions of heat or cold
- Elongation of foraging roots and proliferation of root hairs
- Slowing of growth and metabolism in response to flooding or drought
- Plastic production of emergent versus submerged leaf types
- Programmed cell death leading to aerenchymatous tissues
- Growth of proteoid roots under low-phosphorus conditions
- Establishment of symbioses with bacteria, mycorrhizae, and host plants

Ecologists can make use of many newly available tools and develop cross-disciplinary collaborations with geneticists and physiologists to explore the roles of phytohormones in guiding and constraining seedling establishment.

7.8 | Acknowledgments

Research and writing were begun during a Bullard Research Fellowship at Harvard University; field studies were funded in part by grants from the National Science Foundation (IBN-9623313, DGE-9714522, and DGE-0123490) and the Deland Fund of Harvard University. My thanks to A. M. Ellison and two anonymous reviewers for their very helpful comments on this chapter.

Chapter 8

Seedling ecophysiology: strategies toward achievement of positive net carbon balance

Kaoru Kitajima and Jonathan A. Myers

8.1 | Introduction

Many plant ecophysiological issues, such as trade-offs associated with resource acquisition strategies, are shared between seedlings and larger plants. Yet, seedlings face several unique challenges in their struggle to achieve positive net carbon balance necessary for growth and survival. First, seedlings go through dynamic physiological changes from complete dependency on seed reserves to dependency on light and external nutrients. Second, because seedlings are small, modest changes in carbon allocation patterns will have large consequences on whole plant carbon balance and survival. Third, seedlings experience intense mortality from a wide range of abiotic and biotic factors, including strong asymmetric competition from larger neighbors, herbivory, disease, and disturbance (Moles & Westoby; 2004c; Fenner & Thompson, 2005; Kitajima, 2007; Chapter 10). Finally, phylogeny exerts a particularly strong influence on seed size (Moles *et al.*, 2005b) and morphological traits (Saverimuttu & Westoby, 1996b; Ibarra-Manríquez *et al.*, 2001; Zanne *et al.*, 2005) of young seedlings. These unique aspects undoubtedly influence the evolution of resource allocation strategies in relation to the regeneration niches of species (Grubb, 1977).

In this chapter, we discuss both theoretical and technical issues important in evaluating carbon allocation strategies of seedlings. We place a large emphasis on inter- and intraspecific trade-offs due to constraints in resource allocation. Presumably, such trade-offs reflect specializations in ecological space defined by spatial and temporal heterogeneity, and ultimately restrict the range of environments in which seedlings establish (Chapters 3, 10). We explore key concepts pertaining to carbon balance strategies. Many of our examples are drawn from comparative studies in species-rich tropical forests with high morphological and physiological diversity. Yet, these concepts should be relevant in many plant communities. The topics roughly follow the ontogeny of a seedling, starting with issues related to seed reserve utilization, followed by sections on interspecific variation in carbon balance strategies and phenotypic plasticity. Our goal is not to provide a comprehensive review of all relevant literature. Rather, we hope to stimulate interest in conceptual issues relevant for carbon balance strategies of seedlings within and across species and communities.

8.2 | Seed reserve utilization

Developing embryos and seedlings depend completely on resources stored in endosperm or reserve-type cotyledons for their energy and nutrient demands until sufficient development of photosynthetic organs and roots. How long seedlings depend on seed reserves is influenced by many factors, including developmental rates, cotyledon functional morphology, and environment (Kitajima, 1996a), as well as the resource in question (e.g. energy vs. nitrogen and other mineral nutrients). Concentrations of various resources stored in seeds are not completely in balance with demands by young seedlings. Fenner (1986) evaluated the relative durations of dependency for mineral elements stored in seeds by comparing the final sizes of seedlings achieved without supply of individual mineral elements. This method, applied to various temperate legume and nonlegume species, revealed that seedlings show a tendency to shift to external dependency on nitrogen before other mineral elements (Fenner, 1986; Fenner & Lee, 1989; Hanley & Fenner, 1997). Although these approaches help us understand the relative order in which seed reserves become exhausted, they cannot be used to quantify the time course of gradual changes from heterotrophic dependency on seed reserves to autotrophic energy gain.

Gradual decreases in seed reserve dependency for energy as well as for mineral elements can be analyzed by comparing growth curves under contrasting resource levels (Fig. 8.1; Kitajima, 2002). While seedlings are completely dependent on seed reserves, seedling growth rates are independent of light and soil nutrients. Once seedlings start acquiring carbon by photosynthesis and mineral nutrients by roots, their growth is influenced by the external availability of these resources. In three tropical woody species in the Bignoniaceae with contrasting cotyledon functional morphologies, dependency on light developed before dependency on soil nitrogen. In addition, the species with photosynthetic cotyledons (Fig. 8.1b) started to utilize light as the energy source earlier than the other two with reserve-type cotyledons (Fig. 8.1d,f). This was expected because thin photosynthetic cotyledons have high net photosynthetic capacity similar to true leaves on a unit-mass basis (Kitajima, 1992b). Active photosynthesis requires high tissue nitrogen concentration for construction of various enzymes and membrane-bound proteins. Yet, duration of dependency on seed nitrogen was not related to cotyledon type

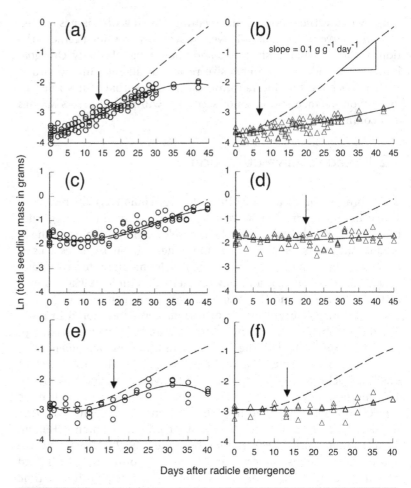

Fig. 8.1 Seedling biomass growth (= natural log of total biomass plotted against time) during the transition from dependency on seed reserves to autotrophy using external resources for three tropical woody species in Bignoniaceae: *Tabebuia rosea* (a, b), *Callichlamys latifolia* (c, d), and *Pitheoctenium crucigerum* (e, f). In each panel, reference growth of seedlings, individually potted in washed sand under optimal light (27% of total daily photon flux density, PFD) and soil nitrogen supply (complete nutrient solution), is shown by broken line (third-order polynomial fit, individual data points not shown for clarity). The instantaneous slope along the fitted curve is the relative growth rate (RGR); RGR for the linear growth phase under optimal growth supply is indicated in panel (b). The left panels (a, c, e) show the effects of eliminating nitrogen from nutrient solution, whereas the right panels (b, d, f) show the effects of shading (1% PFD). Arrows indicate the end of the complete seed reserve dependency for nitrogen and energy, i.e. the age after which the 95% confidence intervals of the growth curves under single-resource limitation no longer overlap with the reference growth curves. Adapted from Kitajima (1992a; see 2002).

among the three species, but it appears to be influenced by seed nitrogen concentration (Fig. 8.1c for the species with high seed nitrogen concentration, compared to the other two species shown in Fig. 8.1a,e; see Kitajima, 2002). This study shows that a seedling's dependency on seed reserves for a particular resource is influenced not only by seed size, but also by its concentration in seed, cotyledon functional morphology, and growth constraints by other resources.

Elevated concentrations of mineral elements may have energetic advantages in environments in which the cost of soil nutrient acquisition is high. In infertile soils in Australia, some species in Proteaceae often have very high nitrogen and phosphorus concentrations in seeds (Pate *et al.*, 1985). The prolonged duration of seed reserve dependency for these elements may be advantageous for seedling establishment in infertile sandy soils, especially following fires (Stock *et al.*, 1990; Milberg & Lamont, 1997). In Amazonian caatinga, tall trees have smaller seeds with higher concentrations of phosphorus and magnesium than tall trees on fertile soil in other lowland rain forest sites (Grubb & Coomes, 1997). High seed nitrogen concentration is also found among shade-tolerant tropical woody species with relatively large seeds (Kitajima, 2002), possibly because it may help initial development of deep tap roots or minimize the energy expended on soil nutrient uptake for an extended period. In contrast, a study of 12 congeneric grass species in New Zealand found no differences in mineral nutrient concentrations in seeds of species from fertile and infertile soils (Lee & Fenner, 1989). How elevated mineral nutrient reserves contribute to seedling establishment in contrasting habitats is a research area that deserves further attention.

Seeds of different species exhibit widely different concentrations of starch and lipids, the two forms of energy storage. Higher lipid content results in higher energy content per gram of seed mass, but do elevated lipid concentrations in seeds confer energetic advantages for seedling establishment (Levin, 1974)? Although energy per gram of seed tissue increases with higher lipid concentrations, it probably does not mean prolonged energetic support for seedlings, because energy retrieval from beta-oxidation of lipids in plants is not as efficient as from starch (Chapin, 1989). A 100% increase in energy content per gram of seed tissue by increasing lipid concentrations may result in only a 40% increase in terms of initial seedling mass (Kitajima, 1992a). Hence, storing starch costs less to the parent than storing lipids for a unit amount of energy endowed to the seedling. Furthermore, isotopic analysis suggests that energy transfer from lipids in seeds to developing seedlings may be negligible (Kennedy *et al.*, 2004). Then, why are lipid-rich seeds common among many plant species even though lipids are not used as energy storage in vegetative tissues in general? Perhaps, the dispersal advantage associated with low seed weight outweighs the disadvantage of added energy costs to the parent (Lokesha *et al.*, 1992).

8.3 | Ontogenetic trajectories of seedling carbon balance

From an energy balance perspective, early seedling growth is a process of transforming the total energy pool in seeds (embryo plus endosperm) into seedlings with gradual development of photosynthetic autotrophy. Thus, there is a strong allometric relationship between seed mass and seedling mass within and across species (e.g. Green & Juniper, 2004b; Paz et al., 2005). Departure from a perfect allometry is caused by species difference in net carbon balance and its determinants including cotyledon functional morphology.

Net carbon balance per unit seedling mass can be quantified as relative growth rate (RGR), the instantaneous slope of the natural log of total seedling mass plotted against time (Fig. 8.1; see Poorter & Garnier, 2007, for technical details). RGR continuously changes throughout early seedling development: initially negative or zero, then becoming positive following development of photosynthetic autotrophy. In a constant environment with adequate resource supply, RGR is expected to reach a maximum for some period (log-linear phase in growth curves, Fig. 8.1), after which RGR typically starts to decline. This ontogenetic trajectory of RGR against time after germination can be generalized as a bell-shaped curve (Hanley et al., 2004; Poorter & Garnier, 2007). Unlike RGR, net assimilation rate (NAR), which quantifies net carbon balance per unit leaf area, is ecologically meaningful only after seedlings become fully autotrophic.

Prolonged dependency on seed reserves, especially in species with large storage cotyledons, requires special considerations in the calculation and interpretation of RGR. Regardless of cotyledon types, cotyledon mass must be included as a part of the total seedling mass for the purpose of quantifying net carbon gain. Otherwise, early RGR will be arbitrarily inflated in species with nonphotosynthetic cotyledons. During early development, significant proportions of carbohydrates are transferred from seeds for storage in stems and roots (Kabeya & Sakai, 2003; Myers & Kitajima, 2007). But, large cotyledons that remain attached in some temperate and tropical woody species stop transferring energy reserves to seedlings after the first set of leaves become fully functional (Ichie et al., 2001; Kennedy et al., 2004; Myers & Kitajima, 2007). For purposes other than quantification of net carbon balance, other measures of growth (e.g. RGR of biomass excluding storage carbohydrates, height or total leaf area, and absolute growth rate) may be more ecologically meaningful variables.

Even under constant availability of light and nutrients, an eventual decline of seedling RGR with increases in size is expected for two reasons (Givnish, 1988): (1) a decrease of the ratio of photosynthetic mass relative to nonphotosynthetic mass due to greater allocation to stems for stability of taller stems, and (2) greater degrees of leaf overlap and self-shading in larger plants. For example, Walters et al. (1993a) found a decrease in allocation to leaf mass to be the main reason for a decrease in RGR with age and size of temperate tree seedlings. In contrast, Lusk (2004) showed that shade-tolerant evergreen seedlings continue to accumulate leaf area with size and age. For seedlings of an evergreen conifer in Chilean rain forests, Lusk et al. (2006) found that increasing leaf area results in greater self-shading and lower light use efficiency, but that light interception efficiency per unit mass remains similar due to increase in leaf area ratio (LAR, the ratio of leaf area to total plant mass). This result suggests that self-shading increases as new leaves are added, whereas the need to increase stem allocation for greater mechanical stability may not develop until much later in environments in which seedlings grow slowly. In summary, allometric analyses combined with functional growth analyses can help illuminate trade-offs associated with increases in plant size and ontogenetic shifts in seedling carbon balance.

8.4 | Species differences in inherent relative growth rate (RGR)

In fully autotrophic seedlings, RGR is a function of species traits (morphology, physiology, and allocation patterns), which may be compromised by suboptimal resource supplies, and biotic and abiotic stresses. Inherent maximum RGR, measured as seedling RGR during the log-linear growth phase under optimal conditions, differs greatly among species, from values less than 0.02 g g^{-1} day^{-1} found in shade-tolerant tree seedlings to values much more than 0.1 g g^{-1} day^{-1} found for herbaceous plants and woody pioneers (Poorter & Remkes, 1990; Kitajima, 1996b; Shipley & Almeida-Cortez, 2003). It is well known that seedling RGR is negatively correlated with seed size across diverse taxa and habitats (see Shipley & Peters, 1990 for a review), for example, pasture grasses and legumes (Fenner & Lee, 1989); species from a variety of climatic conditions in Australia (Jurado & Westoby, 1992; Saverimuttu & Westoby, 1996b; Swanborough & Westoby, 1996); woody plants in temperate (Cornelissen et al., 1996; Walters & Reich, 2000) and tropical climates (Kitajima, 1994; Huante et al., 1995; Valladares et al., 2000; Poorter & Rose, 2005); and Mediterranean annuals (Maranon & Grubb, 1993). In some cases, negative correlations between RGR and seed size is also found within species (e.g. Meyer & Carlson, 2001; Paz et al., 2005).

Although the physiological and genetic bases for the trade-off between RGR and seed size within species are unclear, there are several functional explanations to explain this tradeoff among species.

(1) Large-seeded species tend to have reserve-type cotyledons (Hladik & Miquel, 1990; Kitajima, 1996a; Wright et al., 2000; Zanne et al., 2005) and delayed development of photosynthetic area relative to biomass.

(2) Seedlings of large-seeded species tend to develop more slowly, in terms of time between radicle emergence and shoot extension, time until the first leaf expansion, and time between successive leaf development (Kitajima, 1992a).

(3) Large-seeded species may store greater amounts of carbohydrate in reserve-type cotyledons, stems, and roots (the large-seed-slower-deployment or reserve effect; Saverimuttu & Westoby

1996b; Kidson & Westoby, 2000; Green & Juniper, 2004b; Myers & Kitajima, 2007).

(4) Large-seeded species tend to allocate more to roots and stems (Kitajima, 1994; Walters & Reich, 2000; Paz, 2003) both in terms of nonstructural carbohydrates and structural mass (Canham et al., 1999).

(5) Large-seeded species tend to have leaves with low specific leaf area (SLA, the ratio of leaf area to leaf mass) and leaf area ratio (LAR), the two traits that are known to be positively correlated with interspecific differences in RGR under standardized conditions (e.g. Poorter & Remkes, 1990; Kitajima, 1994; Reich et al., 1998; Wright et al., 2000; Shipley & Almeida-Cortez, 2003; Poorter & Garnier, 2007).

Of the explanations above, those due to seed reserve allocation and cotyledon functional morphology are likely to disappear with time, while differences in biomass partitioning (e.g. SLA, LAR, and root allocation), photosynthetic physiology (e.g. NAR), and developmental rates should become increasingly strong predictors of RGR (Poorter & Garnier, 2007). The relative importance of these morphological and physiological traits as determinants of species differences in RGR differ among environments (e.g. Poorter, 1999) and continuously change with size (Walters et al., 1993a; Lusk & Del Pozo, 2002; Delagrange et al., 2004; Baraloto et al., 2005; Claveau et al., 2005). Yet, life history correlations of seed mass with these functional traits (Reich et al., 2003; Wright et al., 2007) may continue to reinforce the negative correlation between RGR and seed mass, especially in resource-rich environments in which photosynthetic gain per unit leaf area is greater. One interesting functional correlation in seedlings is the positive relationship between cotyledon SLA and true-leaf SLA (Kitajima, 1992a); seedlings with thin photosynthetic cotyledons continue to grow faster even after true leaves become their main source of autotrophic carbon gain. In a meta-analysis of comparative studies of rain forest tree species, Poorter and Rose (2005) found that the negative interspecific correlation between seed mass and RGR is stronger under higher light availability in which growth is faster and seedling size quickly becomes decoupled from seed mass, remaining detectable for 1–2 years. Seed mass is known to be negatively associated even with diameter growth rates of adult trees (in four out of the five neotropical forests, as well as across 219 species in the pooled dataset; Poorter et al., in press).

8.5 | Opportunistic versus conservative strategies

Trade-offs between growth rates and survival have been demonstrated for seedlings of various life forms in diverse communities, including, for example, temperate trees (Grime & Jeffrey, 1965; Reich et al., 1998), tropical trees (Kitajima, 1994; Dalling &

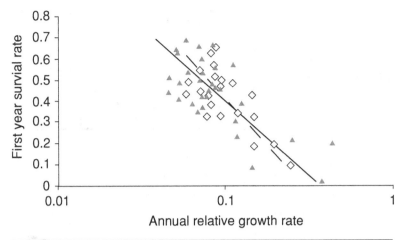

Fig. 8.2 The trade-off between first year survival and relative height growth rates (log-scale) for seedlings of neotropical trees (solid triangle, solid line, n = 31 species) and lianas (open diamond, dashed line; n = 22 species) on Barro Colorado Island (Panama). From Gilbert *et al.* (2006), with permission from the Ecological Society of America.

Hubbell, 2002), and rosette-forming perennial herbs (Metcalf *et al.*, 2006). Growth-survival trade-offs can vary continuously among coexisting species, as shown for trees and liana species in a neotropical forest (Fig. 8.2) (Gilbert *et al.*, 2006) and for temperate forest trees (Pacala *et al.*, 1996). Such trade-offs are considered to arise from continuous variation in growth, defense, and storage strategies from the very opportunistic (i.e. fast growth, low survival) to the very conservative (i.e. slow growth, high survival) ends of the trade-off continuum. However, the functional traits that underlie growth-survival trade-offs are still debated, mainly because researchers differ in their views about carbon balance strategies that enhance survival (e.g. Sack & Grubb, 2001, 2003; Kitajima & Bolker, 2003; Lusk, 2004).

Maintenance of positive net carbon balance is a prerequisite for survival and growth once seedlings survive beyond the initial period of heavy seed reserve dependency. However, maintenance of positive net carbon balance does not necessarily mean that species are selected to maximize short-term carbon gain rates. Indeed, long-term maintenance of positive net carbon gain may be achieved by either of the two contrasting strategies, opportunistic or conservative, as long as plants are alive with sufficient reserves to recover from traumatic tissue losses, represented by the sharp declines in Fig. 8.3. Opportunistic species emphasize growth over defense in their allocation patterns. As a result, they achieve fast growth rates in the absence of physical and biotic disturbance. However, because of low defense, they suffer severe tissue loss and high mortality when they experience physical damage or attack by natural enemies. In contrast, conservative species that emphasize defense over growth experience less frequent and only moderate tissue loss. One underlying assumption of this scenario is

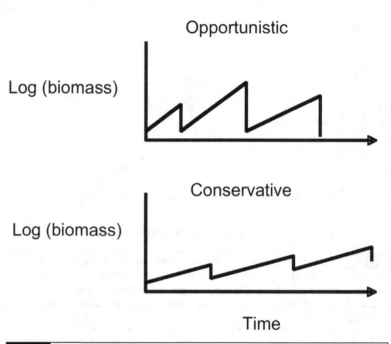

Schematic diagram to contrast opportunistic vs. conservative strategies for maintenance of positive carbon balance, in which loss of tissue to natural enemies or periodic environmental stresses impose negative carbon balance, shown as sudden drops in the trajectories from time to time.

that defense, in particular, physical defense achieved through thick cell walls and tough tissues, is incompatible with fast growth due to allocation constraints.

The opportunistic strategy is represented by a suite of functional traits optimized to achieve fast RGR over the short term (Fig. 8.4). This strategy, analogous to r-strategy in life history theory (Pianka, 1970), is common among many successional species that quickly colonize gaps in vegetation. However, these species may experience large setbacks in biomass from tissue loss to herbivory and disturbance (represented by large declines in Fig. 8.3). As long as they grow in a high resource environment and as long as the tissue loss is not too great, they may recover quickly due to their fast inherent growth rates. However, in low resource environments, such as forest understories and infertile soils, realized carbon income may be too slow for recovery before reserves in remaining tissues are exhausted. This explains the failure of such opportunistic species to persist in low-resource environments, even though their short-term RGR may be greater than conservative species when they are compared under the same low resource environments (Kitajima, 1994; Fine et al., 2004). Indeed, in such low resource environments, species with conservative strategies (K-strategy) dominate despite their slow RGR. As a broad generalization, opportunistic and conservative species exhibit opposite suites of seed and seedling traits (Fig. 8.4). These contrasting trait syndromes

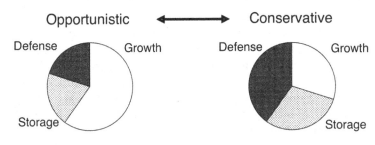

Many small seeds | Few large seeds
Photosynthetic cotyledons | Reserve cotyledons
Thin leaf (= high SLA) | Thick leaf (= low SLA)
High LAR | Low LAR
Short leaf lifespan | Long leaf lifespan
Low tissue density | High tissue density
Low fiber contents | High fiber contents
Low tissue toughness | High tissue toughness
Small carbohydrate storage | Large carbohydrate storage

Fig. 8.4 Overall contrasts in allocation patterns associated with opportunistic vs. conservative carbon allocation strategies. Also shown are individual traits of seedlings typically associated with these strategies. (SLA = specific leaf area, LAR = leaf area ratio.)

parallel those between typical early vs. late successional species (Bazzaz, 1982) and ruderal vs. competitive species (*sensu* Grime's life history classification; Grime, 1979; Westoby, 1998).

Which of these two strategies is most likely to achieve long-term maintenance of positive net carbon gain depends on the interplay of resource competition with natural enemies. The role of natural enemies in mediating growth–survival trade-offs among high and low resource specialists was elegantly shown by Fine *et al.* (2004) in the Peruvian Amazon, where fertile alluvial soil and nutrient-poor white sand support contrasting tree communities. In a factorial experiment, they demonstrated that seedlings of alluvial soil specialists grew faster than white sand specialists in both soil types when they were protected from herbivores. However, when they were grown in white sand without protection from herbivory, alluvial soil specialists suffered greater herbivory and achieved less net leaf area growth than white sand specialists. Other studies also found that species that dominate rich soils grow fast in both rich and poor soils compared to those from poor soils when they are protected from herbivores (Huante *et al.*, 1995; Lusk *et al.*, 1997; Schreeg *et al.*, 2005). Species from infertile soil tend to have lower SLA and greater leaf life span than species from rich soils (Wright *et al.*, 2002). Low SLA species characteristically have thick leaves with high tissue density and thick cuticles that confer protection against herbivores (Wright & Cannon, 2001). These results suggest that avoidance of tissue loss is selected in resource-limited environments, in which replacement of lost tissue is constrained by limited resources (Coley *et al.*, 1985).

Increased allocation to defense is expected to enhance survival, even though this strategy may result in decreased growth rates. Physical and chemical defenses are important for survival in a wide range of seedlings, from temperate forb seedlings that experience mollusk herbivory (Hanley *et al.*, 1995) to tropical tree seedlings attacked by soil-borne pathogens (Augspurger, 1984a,b). Both physical and chemical defenses may incur carbon costs, and there may be a trade-off between the two (Hanley & Lamont, 2002). Physical defense may be of particular importance for seedlings damaged by fallen litter and disturbance by vertebrates (Clark & Clark, 1989). Yet, comparative studies of seedling physical defense are surprisingly rare. The importance of physical defense for seedling survival was demonstrated for eight neotropical tree species (Alvarez-Clare & Kitajima, 2007); the greater the fracture toughness of seedling stems, the greater the first-year survival in the understory (Fig. 8.5a). This study also found that fracture toughness was highly correlated with tissue density (mass per unit volume) as well as fiber contents (expressed per unit volume or mass). These results, supported by data from an additional 70 species at the same study site (K. Kitajima, unpublished data) and studies in temperate forests (Cornelissen *et al.*, 1996), suggest that seedlings that survive well in the shaded understory are physically well defended with high tissue density in both leaves and stems, even very early in ontogeny. Similar survival advantages associated with high tissue density and stem mechanical strength have also been reported for saplings (Muller-Landau, 2004; Poorter & Bongers, 2006; van Gelder *et al.*, 2006).

Despite patterns of ontogenetic shifts across broad taxa, tissue density of stems and leaves are correlated between young and older seedlings (Alvarez-Clare & Kitajima, 2007), as well as between seedlings and saplings (K. Kitajima & L. Poorter, unpublished data). Mechanical strength obviously prevents stem breakage and herbivory, but it also is important for crown development to achieve efficient light interception, especially in older juveniles to minimize self-shading (Sterck *et al.*, 2006). High tissue density is also important in protecting seedlings against soil-borne pathogens (Augspurger, 1984a). Even well-defended conservative species, however, may experience occasional negative carbon balance due to stress and disturbances (Fig. 8.3).

8.6 | Carbohydrate reserves

Carbohydrate reserves not only allow seedlings to survive through periods of negative carbon balance imposed by abiotic and biotic stresses, but also facilitate rapid recovery from tissue losses. In this way, carbohydrate reserves contribute to the maintenance of positive carbon balance in the long term even though allocation to carbohydrate storage may reduce short-term growth rates. For temperate

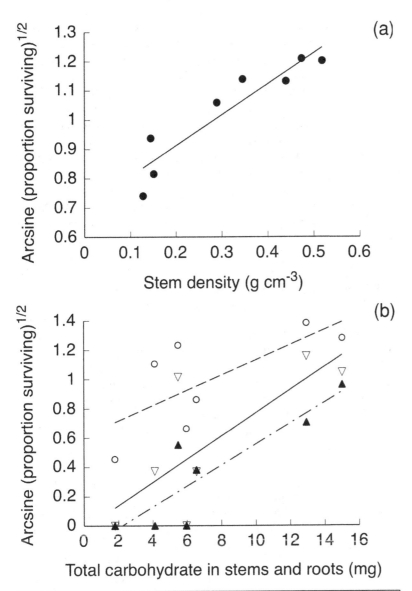

Fig. 8.5 Survival advantage of physical defense and carbohydrate storage in seedling stems and roots of neotropical tree species. (a) Stem density (dry mass per volume) of 6-month-old seedlings is positively correlated with 0–6 month survival in the natural forest understory (r = 0.93, P = 0.008). Arcsine square root transformation was applied to proportion of seedlings surviving. (b) Seedlings were grown in the forest understory (0.8% PFD) and under three treatments (open circle = control, open triangle = 90% light reduction for 2 months, closed triangle = complete defoliation). The total of nonstructural carbohydrate in the stems and roots (average per plant) at the beginning of treatment was positively correlated with survival (r = 0.71, 0.75, and 0.87, and P = 0.07, 0.04, and 0.008, respectively). Adapted (a) from Alvarez-Clare & Kitajima (2007), which also demonstrates high correlation of tissue density with biomechanical strength, and (b) from Myers & Kitajima (2007), reproduced with permission from Blackwell Publishing.

deciduous and evergreen trees, carbohydrate reserves are important for survival of seedlings subjected to stresses of simulated herbivory or winter (McPherson & Williams, 1998; Canham et al., 1999). In tropical savannas, survival and recovery of seedlings subjected to fire depends heavily on carbohydrate reserves (Miyanishi & Kellman, 1986; Hoffmann et al., 2004). It is less obvious why having large carbohydrate reserves may be advantageous for survival in shade, in which carbohydrate allocation is costly relative to potential photosynthetic income. However, Kobe (1997) argued that opportunity costs (i.e. fitness costs of growth reduction) of carbohydrate allocation is smaller in shade than in treefall gaps. Indeed, saplings of shade-tolerant species have higher stem concentrations of carbohydrate reserves compared to saplings of light-demanding species both in temperate (two deciduous and two evergreen species; Kobe, 1997) and tropical dry forests (85 species, Poorter & Kitajima, 2007).

Seedlings under canopy shade in the humid tropics experience occasional periods of negative carbon balance due to variation in cloudiness, seasonal drought, physical disturbance, and attack by natural enemies. In a comparison of seven neotropical tree species, Myers and Kitajima (2007) found that seedlings of more shade-tolerant species had a greater total amount (but not concentration) of sugar and starch in stems and roots. Furthermore, survival of species that received experimental defoliation and heavy shading was enhanced by large pool sizes of nonstructural carbohydrates (Fig. 8.5b). The majority of the carbohydrate reserves in stems and roots must have been transferred from seed reserves rather than produced by photosynthesis, as the experiment was initiated only 2 weeks after full expansion of the first leaves in deep shade (0.8% of light available above the forest). Interestingly, seedlings apparently did not depend on residual carbohydrates in cotyledons, but only those in stems and roots. This finding contrasts with the results of Kabeya and Sakai (2003), in which clipping of shoots during the initial leaf flush resulted in a substantial draw of carbohydrates from cotyledons. Lusk and Piper (2007) also found a positive effect of carbohydrate concentration in leaves and roots, but not in stems, for survival of seedlings naturally recruited in the shaded understory of a temperate rain forest. Their study also suggested a potential role of carbohydrate reserves in long-lived leaves of shade-tolerant species, which may become more important with increases in biomass allocation to leaves in these species during ontogeny. The shifting relative importance of different storage organs for growth and survival is an understudied aspect of seedling ecophysiology.

Total carbohydrate reserve size is ultimately constrained in young seedlings because of their small size. This may constrain small seedlings from adopting opportunistic allocation strategies in low resource environments. At the sapling stage (e.g. juveniles taller than 1.3 m), however, either opportunistic or conservative strategies can lead to high survival. Some shade-tolerant saplings have leaves with high SLA that are more efficient in photosynthetic light utilization

(King, 2003), whereas high survival of other shade-tolerant saplings can be attributed to well-defended low SLA leaves (Poorter & Bongers, 2005), or carbohydrate reserves (Kobe, 1997; DeLucia *et al.*, 1998). However, for seedlings in both temperate and tropical forests, shade tolerance is associated with low SLA (Kitajima, 1994; Reich *et al.*, 1998; Poorter, 1999). *Alseis blackiana* (Rubiaceae), a tropical tree species with small seeds and high SLA, presents an exceptional case. As seedlings, it is a pioneer species unable to establish and survive in shade, but as saplings, it is a shade-tolerant species that can recover from repeated trauma of shading and stem breakage (Dalling *et al.*, 2001). Carbohydrate storage and its dynamics in changing light environments may offer a better understanding of the functional basis of shade tolerance in this and many other species.

Carbohydrate accumulation is influenced not only by current environmental conditions that affect photosynthetic rates, but also by the history of environments experienced by individual seedlings. Seedlings grown in high light have a higher concentration and total amount of carbohydrate compared to seedlings grown in shade (Johnson *et al.*, 1997; Kitajima *et al.*, 2006; Myers & Kitajima, in prep.). Lusk *et al.* (2007) found that larger seedlings of light-demanding species have higher carbohydrate concentration in roots than shade-tolerant species. This pattern may not only reflect species differences, but also the higher light levels that surviving individuals of light-demanding species must have experienced on average compared to shade-tolerant species. Carbohydrate dynamics of seedlings experiencing changing light environments is an ecologically important topic that has only recently received attention (Veneklaas & den Ouden, 2005; Myers & Kitajima, in prep.).

8.7 | Phenotypic plasticity

Individual seedlings adjust their morphological and physiological traits in response to environmental variation. Because individual seeds and seedlings cannot move to their preferred environments, phenotypic adjustments of morphology and physiology are important in both avoidance and tolerance of undesirable conditions, as well as maximization of carbon gain in preferred environments. Reader *et al.* (1993) quantified increases of maximum rooting depth of unwatered seedlings for 42 temperate herbaceous and woody species and found that species with the highest plasticity in rooting depth sustained shoot growth in unwatered soil. Likewise, seedlings exhibit plasticity in response to soil nutrient regimes within the constraints imposed by ontogeny (e.g. Gedroc *et al.*, 1996; Baraloto *et al.*, 2006). Among environmental factors that are known to cause plastic responses in seedlings, light is perhaps the best studied. Hence, our discussion will focus primarily on plasticity in response to light environments that are extremely heterogeneous at small spatial and temporal scales within vegetation.

It is useful to distinguish qualitative and quantitative responses in phenotypic plasticity. Photoblastic germination and etiolation are examples of all-or-nothing qualitative responses that occur when seedlings sense signals above species- and genotype-specific thresholds. Qualitative responses are often restricted to certain ecological and phylogenetic groups. For example, only seedlings of light-demanding species, but not those of shade-tolerant species, exhibit hypocotyl elongation in response to low red:far-red light ratios (Kitajima, 1994; Khurana & Singh 2006). Such a response makes sense only for opportunistic species and genotypes that specialize in vegetation gaps where shade escape is possible by rapid height growth (Schmitt et al., 2003). Signals other than red:far-red light ratio, such as soil temperature fluctuation, high temperature, flush of nitrate, and smoke, are also used by seeds to detect gaps of different sizes and types (e.g. Thompson et al., 1977; Keeley, 1991; Hilhorst & Karssen, 2000; Pearson et al., 2002; Daws et al., 2006). However, compared to seeds, much less is known about these non-light signals for seedling responses to gap environments.

In seedlings, quantitative responses to light quantity, water availability, and soil nutrient availability are perhaps more ubiquitous than qualitative responses. For example, leaves of both shade-tolerant and shade-intolerant seedlings exhibit acclimation response to light environments in both morphological and photosynthetic traits (e.g. Björkman & Holmgrem, 1966; Loach, 1967; Fetcher et al., 1983; Walters et al., 1993a,b; Kitajima, 1994; Lee et al., 1996). From these and many other studies on photosynthetic acclimation, we can generalize that leaves developed under higher light availability tend to have greater thickness, lower SLA, higher nitrogen and carboxylation enzyme concentrations per unit area, higher chlorophyll a/b ratio, and higher rates of respiration and photosynthesis per unit area. These acclimation responses to high light also interact with soil nitrogen availability (Poortsmuth & Niinemets, 2007); under low nitrogen availability, leaves must be protected against potential photoinjury through adjustment of nitrogen allocations to different photosynthetic components (Niinemets et al., 1998; Kitajima & Hogan, 2003).

Levels of phenotypic plasticity are under genetic control, which creates large variations among species, ecotypes, and genotypes (Schlichting, 1986). At the whole seedling level, rates of acclimation should be faster in inherently fast-growing species that produce more leaves per unit time (Newell et al., 1993). Thus, opportunistic species with inherently fast growth rates and shorter leaf life span, including many pioneers and other high resource specialists, should be more plastic (Lortie & Aarssen, 1996). Indeed, many empirical studies demonstrate that early successional species whose seedlings specialize in gaps for regeneration tend to show greater degrees of plasticity than late successional species (e.g. Bazzaz & Carlson, 1982; Straus-Debenedetti & Bazzaz, 1991; Kitajima, 1994; Valladares et al., 2000; Delagrange et al., 2004; Khurana & Singh, 2006; Portsmuth &

Niinemets, 2007). The later successional dominants also have reasons to have genetic potential for high phenotypic plasticity because they may experience greater ontogenetic changes of light availability as juveniles (that typically establish in shade) and adults (exposed to full sun). If phenotypic plasticity allows adaptation to a broader range of environments, then species that experience more variable environments in their lifetime, as well as those that have greater geographical distribution, should show higher degrees of phenotypic plasticity. But, the results may be specific to particular traits and environmental factors. For seedlings of two Australian *Acacia* (Leguminosae) species, Pohlman *et al.* (2006) found only leaf traits, but not whole plant allocation and growth, were more plastic for species with wider geographical distributions.

It is also important to consider whether plasticity is always adaptive, because plasticity may incur costs (DeWitt *et al.*, 1998). It is obvious that stem elongation response is maladaptive if it is not sufficient to escape shade. Thus, in shaded forest understory, seedlings minimize the risk of toppling by not responding to red:far-red light ratios. Likewise, regardless of the current moisture availability, initial biomass allocation to roots is greater for shade-tolerant seedlings (Walters *et al.*, 1993b; Kitajima, 1994; Delagrange *et al.*, 2004). This is a strategy with multiple advantages for stress tolerance. Larger root mass allows not only greater carbohydrate storage (Myers & Kitajima, 2007), but it also serves as a preemptive strategy to cope with droughts. Even though seeds in many communities germinate during the rainy season, dry spells during the rainy season as well as seasonal drought kill many seedlings even in humid forests (Pearson *et al.*, 2003a; Engelbrecht *et al.*, 2006). In some cases, apparent phenotypic similarity within a species may be paradoxically a result of phenotypic plasticity. For example, seedlings of two evergreen Mediterranean *Quercus* (Fagaceae) species exhibited low phenotypic variations that result in conservative resource use in fire- and drought-prone environments with high year-to-year variations in light and soil nutrient availabilities (Valladares *et al.*, 2002).

In summary, phenotypic plasticity of seedlings is widely recognized to have adaptive values for specialization to particular environments by facilitating opportunistic use of resource pulses associated with disturbances, or to ensure phenotypic consistency to minimize maladaptive responses when they exploit a wide range of environments. Yet, debates continue about how to quantify the adaptive nature of observed plastic changes (Valladares *et al.*, 2006).

8.8 | Concluding remarks

The diverse seedling carbon balance strategies discussed here are intimately related to issues addressed in other chapters, including morphological diversity (Chapters 2, 4, 6), life history trade-offs

(Chapter 10), habitat specializations, including specialization to stressful environments (Chapter 3), and population and community dynamics (Chapters 11, 12). Symbiotic associations with mycorrhizal fungi and nitrogen-fixing bacteria also have strong implications for carbon balance strategies of seedlings (Chapter 9). For example, mycorrhizae represent a significant carbon sink that influences photosynthetic response of seedlings to elevated CO_2 (Lovelock *et al.*, 1997).

Carbon balance is a common physiological currency for understanding the evolution of diverse seedling morphology and development patterns. However, a few caveats are worth mentioning. Maximization of short-term carbon gain is not necessarily the target of natural selection because it may be in conflict with maintenance of long-term carbon balance and survival. Likewise, greater phenotypic plasticity is not necessarily adaptive. Multiple traits that influence carbon balance may be simultaneously selected in response to a given environmental factor. Conversely, a given trait, such as high tissue density, may confer multiple advantages including high resistance against herbivory, disease, mechanical damage, and drought, even though it restricts rates of growth and development. Some species are selected to be conservative and slow growing, but this strategy is perhaps adaptive only in combination with large seed size and initial seedling size. In contrast, small-seeded species must compensate for their small size through opportunistic carbon balance strategies that combine efficient rapid growth and high phenotypic plasticity, even if this means survival is limited to the most favorable microsites. Finally, given considerably high phylogenetic constraints on seed size and seedling morphologies, recent advances in phylogenetic comparative analyses (Ackerly, 2000) should be useful in untangling the complicated web of seedling trait evolution.

8.9 | Acknowledgments

Development of key concepts and the results illustrated in Figs. 8.2, 8.4, and 8.5 were supported by NSF-EEP0093303. We appreciate constructive comments from P. Grubb, N. Garwood, J. Phillips, L. Tieszen, the three editors, and an anonymous reviewer.

Chapter 9

The role of symbioses in seedling establishment and survival

Thomas R. Horton and Marcel G. A. van der Heijden

9.1 | Introduction

Seedling establishment is one of the key processes that determines the structure and diversity of natural communities. There are many factors that contribute to seedling establishment as explored in Pickett *et al.* (1987) and this volume. To date, little attention has been paid to the effects of mycorrhizal fungi on seedling establishment. However, there are several compelling reasons to consider these symbiotic fungi. First, the vast majority of all land plant species form symbiotic associations with mycorrhizal fungi and seedlings of most species become colonized by these soil fungi immediately after germination and root formation (Newman, 1988; Read & Birch, 1988; Wang & Qiu, 2006). Second, seedlings usually receive mineral nutrients from mycorrhizal fungi and often show enhanced growth when colonized. In addition, several studies report that mycorrhizal fungi can protect seedlings against drought and the harmful effects of pathogenic soil fungi and heavy metals. Third, many mycorrhizal fungi are not host specific and can colonize a wide range of plant species (Molina *et al.*, 1992; Smith & Read, 1997; Opik *et al.*, 2006). Due to this lack of specificity, seedlings can quickly become integrated into hyphal networks that are usually already present and maintained by the surrounding vegetation. Hence, in this way, seedlings have immediate access to a cheap nutrient absorption machine in which they do not need to invest resources (Newman, 1988). It is unlikely that roots of small seedlings are able to compete effectively with adult plants for nutrients, emphasizing the importance of hyphal networks (Nara, 2006a).

In this chapter, we will show that symbiotic interactions between seedlings and mycorrhizal fungi are essential for seedling establishment of many plant species. We will present several case studies and provide an overview of the impact of mycorrhizal fungi on seedling establishment in different ecosystems and during succession. Additionally, we will briefly discuss the impact of other plant symbionts (e.g. endophytic fungi and nitrogen-fixing bacteria) on seedling establishment.

Plants and mycorrhizal fungi

Fungi were observed in roots in the 1800s, but the mutualistic nature of what would become known as mycorrhizal fungi was not immediately appreciated. Frank was the first to recognize that some fungi found in the roots of plants were beneficial, and coined the term mycorrhiza to describe the root-fungus organ (Frank, 1885, reprinted in English translation, 2005). Since this early work, a tremendous body of literature has developed that documents the typically mutualistic nature of the symbiosis (Smith & Read, 1997). There are also instances of mycorrhizal fungi as parasites (Modjo & Hendrix, 1986; Plattner & Hall, 1995), but most mycorrhizal fungi probably fall along a continuum of mutualistic to less mutualistic. Some fungi reduce fitness of a plant relative to other fungi, but still impart an increased fitness relative to the nonmycorrhizal condition (Johnson, 1993; Johnson et al., 1997; Klironomos, 2003). Further, any plant–fungal pair might fall along different zones of a continuum depending on the environmental condition.

The symbiosis is very old and it has been hypothesized that mycorrhizal fungi contributed to the colonization of land by plants (Stubblefield et al., 1987). Fossil evidence and molecular clock estimates suggest that the symbiosis between plants and one specific group of mycorrhizal fungi, the Glomeromycota (Schüßler et al., 2001), originated some 450 million years ago (Berbee & Taylor, 1993; Simon et al., 1993; Taylor et al., 1995; Redecker et al., 2000).

Although estimates vary, about 80% of the species and 90% of the families of plants worldwide associate with mycorrhizal fungi (Molina et al., 1992; Wang & Qiu, 2006). Many plants obligately depend on the fungi for establishment in nature or even fumigated nursery soil (Molina et al., 1992). Nonmycorrhizal families such as Brassicaceae and Proteaceae have unique root anatomies that contribute to their success despite their nonmycorrhizal status (i.e. well-developed root hairs and cluster roots, respectively).

Mycorrhizal fungi support increased uptake of nutrients [nitrogen (N), phosphorus (P), potassium (K), calcium (C), magnesium (Mg)] and water in plants (Fig. 9.1) (Smith & Read, 1997; Blum et al., 2002). Indeed, plants obtain up to 90% of P and N from mycorrhizal fungi (Li et al., 1991; Hobbie & Hobbie, 2006; van der Heijden et al., 2006b). Mycorrhizal fungi also have abilities to acquire forms of N and P that are not directly available to plants (Smith & Read, 1997; Schimel & Bennett, 2004). The fungi acquire these nutrients with the help of extensive hyphal networks that are formed in soil and with which mycorrhizal fungi can forage actively for nutrients. Once acquired, nutrients are translocated through the network and to plants that are connected (Finlay & Read, 1986; Arnebrant et al., 1993; He et al., 2005). Mycorrhizal fungi are also known to reduce the effects of soil pathogens (Marx, 1969; Marx, 1973; Newsham et al., 1995) and heavy metals (Blaudez et al., 2000; Colpaert et al., 2000; Adriaensen et al.,

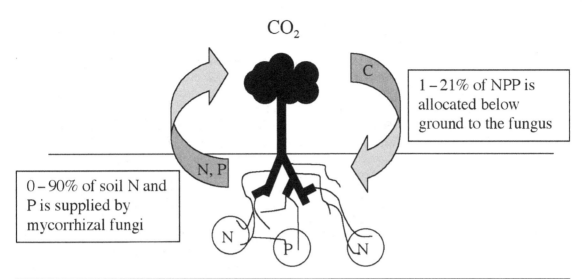

$$CO_2$$

1 – 21% of NPP is allocated below ground to the fungus

0 – 90% of soil N and P is supplied by mycorrhizal fungi

Fig. 9.1 Schematic diagram of a mycorrhizal-seedling interaction. The roots and hyphal networks are highly simplified. Fungi acquire N, P, and other minerals directly from organic and inorganic pools, and allocate these nutrients to the plant. Plants allocate fixed C to mycorrhizal fungi in the form of sugars.

2004). Smith and Read (1997) provide a comprehensive review on these topics.

Several major types of mycorrhizal associations have been distinguished based on fine root anatomy (including the plant and fungal cells) and the taxonomy of the plants and fungi involved (Box 9.1,

---

Box 9.1 | Mycorrhizal types

Mycorrhizae and derivatives refer to anatomical fungus-root structures. AM and EM are abbreviations for *arbuscular mycorrhizal* and *ectomycorrhizal*, respectively (as in AM fungi or EM plants). Brief descriptions of AM, EM, and other mycorrhizal symbioses are introduced below. Table 9.1 provides a summary of the plants and fungi involved in each mycorrhizal type. For more information see Smith and Read (1997) for a comprehensive review of the mycorrhizal literature and Peterson *et al.* (2004) for a general introduction to the various mycorrhizal types, with wonderful microscopic images and drawings focused on anatomy and cell biology.

Arbuscular mycorrhizae
Arbuscular mycorrhizal (AM) roots are named after the tree-like structures that fungi form inside root cell walls known as arbuscules. The arbuscules are separated from the host cell only by the plant's plasma membrane. Importantly, the plant's plasma membrane is not penetrated by the fungus and invaginates around the highly branched arbuscules as they develop, thereby increasing surface area of contact for nutrient transfer. Arbuscules (Fig. 9.2) are not digested or degraded, but rather appear to retract out of the plant cells (for contrast, see monotropoid and orchid mycorrhizae). AM fungi can also form extensive intracellular coils and intercellular hyphae that grow between root cells in the root cortex. Several AM fungi also form

vesicles (Fig. 9.2) that are thought to have a storage function for the fungus. Root colonization by AM fungi is often high, especially in nutrient-poor environments, where over 50% of the roots are colonized. To observe AM fungi in roots, it is necessary to stain the roots and view them under a compound microscope (Brundrett et al. 1996).

AM fungi, a monophyletic group known as the phylum Glomeromycota (Schüßler et al., 2001; James et al., 2006), are present in a wide range of ecosystems (Read, 1991). AM fungi were thought to be rather species poor with only about 200 species described based on morphology. However, molecular data suggests considerable lumping in this estimate. Further, the previous assumption of extremely low host specificity in this group is under investigation (Fitter, 2005).

Ectomycorrhizae

Ectomycorrhizal (EM) roots are so named because the associated hyphae remain external to root cortical cells (Fig. 9.3). EM root anatomy consists of a fungal mantle and a network of hyphae known as the Hartig net wrapped around the cortical cells (Pinaceae) or epidermal cells (angiosperms). While some EM fungi show narrow host ranges, most of the estimated 6000 species do not. In undisturbed plant communities, EM roots are often reported on greater than 99% of the short roots, and root hairs are largely absent, especially in conifer stands.

Mycorrhizae in Ericales: arbuscular, ericoid, arbutoid, and monotropoid types

Plants in Ericales are impressively diverse with respect to mycorrhizal types. Some members, believed to be basal groups, are now known to associate with AM fungi (Kubota et al., 2001). More derived genera in the subfamily Ericoideae form ericoid mycorrhizae. These plants produce extremely fine roots called hair roots (not root hairs) that consist of a stele, one or two layers of cortical cells, and a layer of epidermal cells. It is in the epidermal cells where hyphal coils are located, with nutrient transfer abilities similar to arbuscules. Early culture-based work suggested that only a few species of Ascomycetes formed ericoid mycorrhizae. However, molecular studies have revealed a greater diversity that includes Basidiomycetes (Berch et al., 2002).

Arbutoid and monotropoid mycorrhizae are unique root morphologies on Arbutoideae members of Ericaceae, formed by typical EM fungi with low specificity and high diversity in the fungal associates. The root anatomy consists of a fungal mantle and Hartig net typical of EM associations, but also densely packed intracellular hyphal coils found in the epidermal cells. Like the arbuscular anatomy, the hyphae penetrate only the plant cell wall, not the plant plasma membrane.

Members of the achlorophyllous subfamily Monotropoideae (Ericaceae, with close affinity to Arbutoideae), such as Monotropa, Sarcodes, and Pterospora, form monotropoid mycorrhizae. Here the root anatomy consists of a mantle and Hartig net, but instead of hyphal coils in the epidermal cells, penetration pegs are formed. These reduced structures eventually burst, which may be the mechanism for these achlorophyllous plants to gain C and other nutrients from the fungi. Plant species of the Monotropoideae appear to specialize on specific EM fungi, supporting the idea that monotropoid plants are parasitic in nature with most EM fungi rejecting the plant and driving specificity with a limited group of fungi for each monotropoid plant species (Thompson, 1994).

Orchid mycorrhizae

Orchid mycorrhizae can be grouped into two main types, those associated on green, chlorophyllous orchids and those on achlorophyllous orchids (Taylor *et al.*, 2002). However, all orchids are mycoheterotrophic (Leake, 1994) as seedlings, relying on fungi for establishment in nature. Achlorophyllous orchids remain nonphotosynthetic as they mature, relying on their fungi throughout their lives to provide carbohydrates and other nutrients. The fungi penetrate the orchid root cells of both green and achlorophyllous orchids, forming hyphal coils known as pelotons. The pelotons appear to be digested by the plants, perhaps a mechanism for the orchids to access C and other nutrients similar to the situation in monotropoid mycorrhizae. The achlorophyllous orchids are also like monotropoid plants in that they are highly specific and appear to be parasitic in their associations with EM fungi.

Table 9.1 | Mycorrhizal types and their symbionts

Mycorrhizal types	Plant groups	Fungal groups
Arbuscular mycorrhizae	Some bryophytes, nonvascular plants and ferns; some gymnosperms, such as members of Cupressaceae; most angiosperms	Glomeromycota
Ectomycorrhizae	Pinaceae, Fagaceae, Betulaceae, Dipterocarpaceae; some members of Ericaceae, Myrtaceae, Salicaceae	Basidiomycota, Ascomycota *Endogone* (affinities with Zygomycota)
Ericoid mycorrhizae	Most members of Ericaceae except some ancestral groups that associate with AM fungi and the subfamilies Arbutoideae and Monotropoideae	*Hymenoscyphus ericae* and *Oidiodendron* sp. (Ascomycota), *Sebacina* (Basidiomycota)
Arbutoid mycorrhizae	Arbutoideae subfamily of Ericaceae	Typical ectomycorrhizal fungi
Monotropoid mycorrhizae	Monotropoideae subfamily of Ericaceae	Typical ectomycorrhizal fungi; the plant species are highly specific to fungal genera or species groups
Orchid mycorrhizae – achlorophyllous	Orchids that do not produce green parts	Typical ectomycorrhizal fungi; the plant species are highly specific to fungal genera or species groups
Orchid mycorrhizae – chlorophyllous	Orchids that produce green parts as adults	Basidiomycota such as species of *Tulasnella, Ceratobasidium*, and *Sebacina*

Table 9.1). Associations between plants and arbuscular mycorrhizal (AM) fungi (Fig. 9.2) and ectomycorrhizal (EM) fungi (Fig. 9.3) are most common and widespread. Grasses, herbs, most tropical trees, and

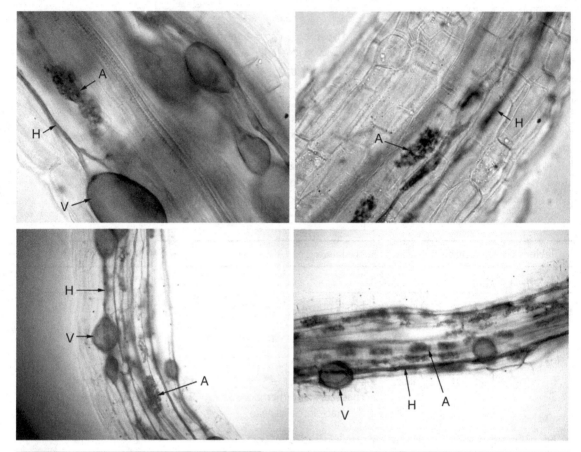

Fig. 9.2 Arbuscular mycorrhizal fungal structures in *Plantago lanceolata* (Plantaginaceae) roots. Hyphae (H) colonize the root cortex and arbuscules (A) and vesicles (V) are formed inside cells of the root cortex. Photos by T. Scheublin.

many temperate trees typically associate with AM fungi, while most trees from temperate and boreal forest usually (not always) form partnerships with EM fungi (Smith & Read, 1997).

Costs of the association

Estimates for net primary productivity allocated by plants belowground range from 5–40% (Smith & Read, 1997). Hobbie (2006) compared a number of laboratory studies and found that allocation to EM fungi ranged from 1–21% of the total NPP. Similar values were found in field studies. As one might predict, allocation to the fungi was high when soil nutrients were limited. The heterotrophic fungus uses this energy for metabolic activity, respiration, nutrient acquisition and transport, vegetative growth, and reproduction. Considering that the size of EM fungal genets vary and that some may grow at the scale of meters belowground (Bonello *et al.*, 1998; Zhou *et al.*, 1999; Redecker *et al.*, 2001; Dunham *et al.*, 2003; Kretzer *et al.*, 2004; Kretzer *et al.*, 2005; Riviere *et al.*, 2006), the cost of maintaining the fungi is probably higher for some species than for others. Similar

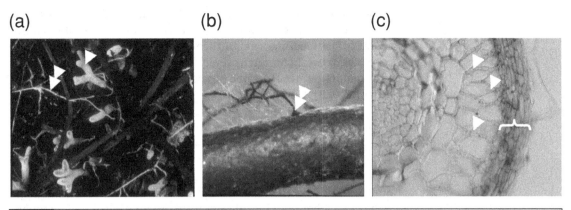

Fig. 9.3 Ectomycorrhizal (EM) roots. (a) EM root tips of *Pinus* colonized by *Rhizopogon* (single arrow). Note the hyphal aggregates or rhizomorphs (double arrow) and the contrast between the fungal covered root tips (white) and the rest of the root system. Root hairs are not produced. The EM tips are the primary location for nutrient acquisition for the plant. (b) EM root of *Pseudotsuga menziesii* colonized by a thelephoroid fungus. Note the rhizomorph (double arrow) attached directly to the fungal mantle. Individual hyphae can also be seen. No root cells are visible; they are beneath the fungal mantle. (c) A cross section of the EM root tip shown in (b) with the mantle (bracket) on the outside of the root and the Hartig net (single arrows) inside the root composed of hyphae packed in the apoplast of the outer cortical cells. Nutrient transfers between plant and fungus occur in the Hartig net region. Photos (a) by J. Trappe, (b) and (c) by T. Horton.

values of NPP allocated to the fungi have been derived for AM fungi even though the fungi do not colonize roots or soils to the same extent as EM fungi (Smith & Read, 1997; Jakobsen, 1999).

The energy demand of initiating and maintaining the mycorrhizal symbiosis has been analyzed in only a few species pairs under fairly narrow conditions. The widely varying estimates of NPP allocated to the fungus shown above are likely a result of the species and age of the symbionts, the age of the mycorrhizal root tip (Jones *et al.*, 1991; Cairney & Alexander, 1992), and the environmental conditions during the experiment. Fungal species and even strains of species vary in their C cost to the plant along a continuum and individual strains may vary under different environmental conditions. For example, Jones *et al.* (1991) observed that C allocation to mycorrhizal roots of *Salix* (Salicaceae) was 1.75 and 3.3 times greater than to non-mycorrhizal roots during the first 50 days and second 50 days of an experiment, respectively. Moreover, it has been suggested that C is a luxury resource for the plant, especially at low nutrient availability, and allocation of surplus C to mycorrhizal fungi does not necessarily reduce plant growth or fitness (Kiers & van der Heijden, 2006). There is a need for more work on the cost of maintaining mycorrhizal fungi to plants, particularly under field conditions and with mature hosts.

Germinating seedlings usually become quickly integrated into mycorrhizal fungal networks that are already established by the surrounding vegetation (Newman, 1988; Read & Birch, 1988). Seedlings that become colonized by fungi already present in soils may benefit from other plants that maintain these fungi. It is still unclear to what extent seedlings benefit from other plants in a community. Seedlings are green, photosynthesize and, thus, have carbon (C) to offer as soon as a functional association has been established. Hence, after initial colonization, seedlings can, in theory, maintain a fraction of the mycorrhizal network. An unanswered question is whether seedlings and adult plants receive the same amount of nutrients from mycorrhizal fungi for every C unit invested.

Specificity interactions between plants and fungi

Many EM and AM fungi lack specificity and can associate with multiple plant hosts. Molina *et al.* (1992) grouped EM fungi into three host ranges, mostly based on the occurrence of mushrooms in pure plant stands and laboratory inoculation trials. Narrow host range fungi are those that associate with plants in only one *genus* such as *Pinus*. Intermediate host range fungi are those that can associate with plants in multiple families, but within the angiosperms or the gymnosperms. Broad host range fungi associate with both angiosperms and gymnosperms. In most cases, only a few fungal host combinations have been investigated, so the actual range of fungi any particular host can associate with is unknown. From the plant perspective, three notable cases highlight the range of specificity to EM fungi. First, *Pseudotsuga menziesii* (Pinaceae) can associate with over 2 000 species of fungi (Trappe, 1977) and other plant species may have a similar number of associates to increase the opportunity to benefit from mycorrhizal fungi present in a community. Second, *Alnus* (Betulaceae) species have only a few fungal symbionts, which may relate to the high nitrogen environment generated by N-fixing root nodules on *Alnus* (Molina *et al.*, 1992). Third, the various achlorophyllous plants are each associated with one or a few closely related species (Taylor *et al.*, 2002; Selosse *et al.*, 2006), which appears to be a function of their parasitic nature. Recent approaches using molecular identification methods have shown that green plants associate with many, if not most, EM fungi that are present in a community, and therefore, each fungal species colonizes multiple hosts (Horton & Bruns, 1998; Horton *et al.*, 1999; Cullings *et al.*, 2000; Kennedy *et al.*, 2003; Dickie *et al.*, 2004; Nara & Hogetsu, 2004; Horton *et al.*, 2005; Dulmer, 2006; Ishida *et al.*, 2006; Lian *et al.*, 2006; Nara, 2006b).

Similar observations have been made for AM fungi. Culture-based studies have shown that most AM fungi that can be isolated lack specificity and different plant species can simultaneously become colonized by the same AM fungi. Several AM fungi (e.g. *Glomus intraradices*) have a worldwide distribution and are present in almost any plant species or ecosystem investigated (Opik *et al.*, 2006). It is likely that seedlings of many species can become integrated in hyphal networks formed by such widespread fungi. However, Fitter (2005) has recently proposed that only a fraction of AM fungal species have been identified and that host-specific associations between particular plant–fungus combinations might exist. Seedling establishment might be influenced by such host-specific fungi through both positive and negative interactions (van der Heijden *et al.*, 1998; Bever, 2003; Klironomos, 2003).

The lack of specificity in mycorrhizal associations makes a great deal of sense in that a rejection of the association by either a fungus or plant may reduce the fitness of the partner (Van der Plank, 1978; Harley & Smith, 1983). However, different mycorrhizal fungi vary in their ability to stimulate seedling growth and nutrition (Klironomos, 2003; van der Heijden, 2004; Nara, 2006a) and it might be beneficial for a plant to associate with specific mycorrhizal fungi (Molina *et al.*, 1992). Mycorrhizal fungal communities vary between plant species (Vandenkoornhuyse *et al.*, 2002; Gollotte *et al.*, 2004; Scheublin *et al.*, 2004) and it has been shown that seedlings can harbor different mycorrhizal communities compared to mature plants (Husband *et al.*, 2002). Hence, this suggests that plants and fungi can discriminate between partners, and a number of mechanisms have been proposed (see Kiers & van der Heijden, 2006). Fungal species that show host specificity may be critical for seedling establishment under some ecological conditions. For example, in the earliest stages of plant community development in pine systems, *Suillus* and *Rhizopogon*, two genera highly specific to members of Pinaceae (Molina & Trappe, 1994; Kretzer *et al.*, 1996; Molina *et al.*, 1999) are particularly abundant and presumably important for seedling establishment (Baar *et al.*, 1999; Ashkannejhad & Horton, 2006).

Mycorrhizal networks

The most reduced form of a mycorrhizal network is a fungal genet and its association with a single plant. The concept can be expanded to include multiple genets of a fungus, multiple plants, and in its most complex form, multiple species of both plants and fungi, all interacting belowground. Mycorrhizal fungi form extensive mycelial networks in soil in which individual hyphae forage for nutrients and new host roots (Olsson *et al.*, 2002; Selosse *et al.*, 2006). Each fungal species, and in fact, each individual of each species forms its own network and can simultaneously colonize and interlink different plant individuals. Using microsatellite markers designed for *Tricholoma matsutake* and *Pinus densiflora* (Pinaceae), Lian *et al.* (2006) demonstrated that matsutake genets typically colonize multiple pine trees. This means that material flow can occur within the fungal individual to the roots of multiple plant hosts that are colonized. The predominant mode of root colonization in forests and grasslands is believed to be through mycelial spread from established fungal individuals.

Some mycorrhizal fungi can form large genets belowground (Bonello *et al.*, 1998; Dunham *et al.*, 2003; Kretzer *et al.*, 2005; Sbrana *et al.*, 2007; Stuckenbrock & Rosendahl, 2005), suggesting a perennial life history and long-lived vegetative structures. Even for fungal species that do not appear to form extensive perennial genets (Redecker *et al.*, 2001; Guidot *et al.*, 2003), hyphal colonization still likely predominates as an inoculum source during the growing season once a new individual is established from spores.

A handful of studies have demonstrated that hyphae of mycorrhizal fungi from the same strain can fuse, or anastomose, to form a continuous network (Giovannetti *et al.*, 1999; De Souza & Declerck, 2003; de la Providencia *et al.*, 2005; Sbrana *et al.*, 2007; Voets *et al.*, 2006). However, fungi have mechanisms for recognizing nonself tissue and hyphae of genetically distinct individuals do not fuse. Thus, the concept of a network must be viewed as a complex system of individual networks operating autonomously and competitively. Further, direct competition between plants for soil nutrients may not be as important as fungal competition for soil nutrients and the level of compatibility between symbiont pairs.

Carbon dynamics in mycorrhizal networks

Many mycorrhizal fungi are not host specific and evidence suggests that plant individuals of different species can be interlinked by common mycorrhizal networks (Simard *et al.*, 2002; Simard & Durall, 2004; Selosse *et al.*, 2006). A consequence of this lack of specificity is that C can be fixed by one plant species, move into the fungus, and picked up by a second plant species via the hyphal network. For example, Read *et al.* (1985) and Grime *et al.* (1987) showed that C moved from canopy dominants to subordinate shaded plants and seedlings via mycelial networks. In a field study, Simard *et al.* (1997) showed that shaded *Pseudotsuga menziesii* received C from *Betula pendula* (Betulaceae) that was growing in full sunlight. Both of these species are EM associates. *Thuja plicata* (Cupressaceae) was used as a control plant (AM host) and did not receive as much labeled C in the experiment as *Pseudotsuga*, suggesting the EM network played an important role in C movement. Another field study has demonstrated the movement of C from the donor herbaceous trout lily (*Erythronium americanum* Liliaceae) into the shoots of the receiver tree seedlings (*Acer* sp. Aceraceae) via AM fungal networks (Lerat *et al.*, 2002). These studies suggest that seedlings can subsidize their C budget from mycorrhizal fungi that acquired C from other green plants in the community through mycorrhizal interactions.

Further evidence for the potential of plants to access C from a fungus is found in about 400 species of mycoheterotrophic plants that parasitize mycorrhizal fungi (Bidartondo *et al.*, 2002; Selosse *et al.*, 2006). Mycoheterotrophic plants lack chlorophyll and receive C and nutrients from AM and EM fungi (Björkman, 1960; Taylor & Bruns, 1997, 1999b; McKendrick *et al.*, 2000a; Bidartondo & Bruns, 2001, 2002; Bidartondo *et al.*, 2002; McKendrick *et al.*, 2002; Trudell *et al.*, 2003; Leake *et al.*, 2004). Mycoheterotrophic plant lineages have evolved in 11 families of plants in 5 orders (Leake, 1994) suggesting that there is a widespread potential for C flow from mycorrhizal fungi to their host plants. Moreover, Girlanda *et al.* (2006) report that an orchid with minimal photosynthetic capacity subsidized its C budget via EM fungi.

Translocation of C to mycoheterotrophic plants is not disputed. However, net transfer of C from one autotrophic plant to another autotrophic plant has been questioned (Fitter *et al.*, 1998). For example, Graves *et al.* (1997) have shown that most, if not all, C remains in fungal structures in the roots of AM plants. Similarly, Wu *et al.* (2001) reported that EM root tips were strong sinks for C, but that the C did not move into the roots or shoots of the receiver plants. Even if C moves into green seedling shoots, the amount of energy may not be sufficient to support seedling establishment. Further, it is possible that the source of C is not associated with a carbohydrate but rather an amino acid that is a source of N. How this might play into the energy budget of a seedling is largely unknown. Clearly, we still need to learn about the forms of C and the C budgets to understand C dynamics in mycorrhizal networks of even single fungal genets, let alone complex communities of plants and fungi. However, the fact that seedlings become colonized immediately after germination by fungi that are already associated with other plants is, in itself, of primary importance (Newman, 1988). This is especially important because a nutrient depletion zone develops in the rhizosphere of a nonmycorrhizal root system, and mycorrhizal fungi can deliver nutrients that are otherwise unavailable to seedlings (Smith & Read, 1997).

9.2 | Ectomycorrhizal fungi and seedling establishment

Tapping into the ectomycorrhizal networks

We now focus on studies of seedling establishment in soils with pre-existing EM networks. Most studies indicate better establishment when seedlings tap into pre-existing mycorrhizal networks (Table 9.2). In three cases, negative effects of mycorrhizal networks were observed (Horton *et al.*, 1999; Dickie *et al.*, 2002; Booth, 2004) and in two cases no difference in survival associated with mature EM trees was reported (Newberry *et al.*, 2000; Booth, 2004). Various measures were used to assess the benefit to seedlings, but in all positive cases the seedlings benefited through their association with the EM network (Table 9.2). It is evident that fungi facilitate establishment of seedlings directly, and pre-established plants facilitate seedling establishment, in part through their mycorrhizal fungi (Nara & Hogetsu, 2004; Nara, 2006a,b).

The absence of compatible mycorrhizal fungi may explain why plant establishment was inhibited in several cases. For example, seedlings of the EM tree, *Pseudotsuga menziesii* that germinated in AM-dominated chaparral patches did not survive the first year of growth, whereas 20% of the seedlings survived in EM-dominated chaparral (Horton *et al.*, 1999). Similarly, soil bioassays revealed that *Thuja plicata* (AM plant) did not form arbuscular mycorrhizae when grown in soils collected in *Tsuga heterophylla* (Pinaceae) (EM plant)-dominated stand and this was believed to play a major role in the lack of *Thuja* establishment in *Tsuga*-dominated stands (Weber *et al.*, 2005). Booth

Table 9.2 Recent studies investigating seedling response to mycorrhizal inoculum

Study	Description	Results
Horton et al., 1999	Mature host species	Arctostaphylos glandulosa (EM) or Adenostoma fasciculatum (Rosaceae) (AM/EM)
	Seedling species	Pseudotsuga menziesii (EM)
	Vegetation type	Chaparral, ≈40 years postfire
	Location	California (USA)
	Experimental set-up	Seed planted beneath canopy of chaparral, harvested with intermingling mature host roots
	Seedling response	Establishment with Arctostaphylos, no survival with Adenostoma
	Mycorrhizal community	High overlap in fungi found on seedlings and Arctostaphylos roots, few EM types observed on seedlings grown with Adenostoma
Newberry et al., 2000	Mature host species	Caesalpinaceae (EM)
	Seedling species	Caesalpinaceae
	Vegetation type	Tropical forest
	Location	Cameroon
	Experimental set-up	Seedlings planted in forest communities
	Seedling response	Variable
	Mycorrhizal community	Not determined
Dickie et al., 2002	Mature host species	Quercus montana (EM/AM) or Acer (AM)
	Seedling species	Quercus rubra
	Vegetation type	Oak savanna
	Location	Pennsylvania (USA)
	Experimental set-up	Seedlings planted near stumps of mature host species
	Seedling response	Positive near Quercus, negative near Acer
	Mycorrhizal community	Not determined

	Booth, 2004	Nara & Hogetsu, 2004	Dickie et al., 2005	Dickie & Reich, 2005
Mature host species	*Pinus strobus, Tsuga canadensis, Betula alleghaniensis, B. lenta, Fagus grandifolia, Acer rubrum* (only Acer is AM, all others are EM)	*Salix reinii* (EM in this location)	*Quercus ellipsoides* and *Q. macrocarpa* (EM/AM)	*Quercus ellipsoides*, AM herbs
Seedling species	*Pinus strobus, Tsuga canadensis, Betula alleghaniensis, Acer rubrum*	*Salix reinii*	*Quercus macrocarpa*	*Quercus macrocarpa*
Vegetation type	Mixed woods	Volcanic desert, primary succession	Old field	Old field
Location	Connecticut (USA)	Mt. Fuji (Japan)	Minnesota (USA)	Minnesota (USA)
Experimental set-up	Seedlings of each species planted with and without access to overstory mycorrhizal networks	Seedlings harvested near isolated mature plants	Seedlings planted at increasing distances from mature hosts	Planted seed at increasing distance from forest edge
Seedling response	EM seedlings did equally well with or without access to networks, AM seedlings did better in the absence of EM networks	Positive	Positive near forest edges, near trees, and with increasing canopy openness and mycorrhizal infection	Not determined
Mycorrhizal community	Not determined	High overlap in fungi between mature plants and seedlings	Greater richness of EM fungi on seedlings grown near mature hosts	Decreasing mycorrhizal infection with increasing distance from EM forest

(cont.)

Table 9.2 (cont.)

Study	Description	Results
Horton et al., 2005	Mature host species	*Pseudotsuga menziesii* (EM)
	Seedling species	*Tsuga heterophylla*
	Vegetation type	40-year-old *P. menziesii* forest
	Location	Oregon (USA)
	Experimental set-up	Naturally established seedlings harvested with intermingling mature host roots
	Seedling response	Not determined
	Mycorrhizal community	High overlap in fungi found on seedling and mature host roots
Weber et al., 2005	Mature host species	*Thuja plicata/Tsuga heterophylla* (AM dominated) and *T. heterophylla/Abies amabilis* (Pinaceae) (EM dominated)
	Seedling species	*Thuja plicata* and *Tsuga heterophylla*
	Vegetation type	Mature conifer forest; *Thuja* regeneration in AM dominated stands, but not EM dominated stands
	Location	Vancouver Island (Canada)
	Experimental set-up	Seeds were planted in soils harvested from the AM or EM dominated vegetation types and seedlings were harvested after 6 months of growth
	Seedling response	*Thuja* survival and growth was significantly greater in AM soils than EM soils; *Tsuga* survival and growth did not differ between soil types
	Mycorrhizal community	AM colonization present in *Thuja* grown in AM soils, but absent when grown in EM soils; EM colonization on *Tsuga* grown in either soil
Ashkannejhad & Horton, 2006	Mature host species	*Pinus contorta*, with *Pseudotsuga menziesii*, *Picea sitchensis*, *Tsuga heterophylla* (all EM),
	Seedling species	*Pinus contorta*
	Vegetation type	Primary successional sand dunes
	Location	Coastal Oregon (USA)
	Experimental set-up	Naturally established seedlings harvested in isolated areas; bioassays with soil or deer fecal pellets used for inoculum
	Seedling response	Not determined
	Mycorrhizal community	Dominated by *Suillus* and *Rhizopogon*

Dulmer, 2006

Mature host species	*Quercus rubra, Q. alba, Betula alleghaniensis, B. lenta, Pinus strobus, Tsuga canadensis, Ostrya virginiana* (Betulaceae), *Fagus grandifolia* (all hosts predominantly EM),
Seedling species	*Castanea dentata*
Vegetation type	Mixed conifer hardwood forest
Location	New York (USA)
Experimental set-up	Seed planted into stands, seedlings harvested with intermingling mature host roots
Seedling response	Not determined
Mycorrhizal community	High overlap in fungi found on *Castanea* seedlings and intermingling roots of dominant tree, *Q. rubra*

Nara, 2006a

Mature host species	*Salix reinii*
Seedling species	*Salix reinii*
Vegetation type	Volcanic desert, primary succession
Location	Mt. Fuji (Japan)
Experimental set-up	Seedlings planted with and without EM networks
Seedling response	Positive, with some fungus species-level variation
Mycorrhizal community	Networked with known fungi

Nara, 2006b

Mature host species	*Salix reinii*
Seedling species	*Betula ermanii, Larix kaempferi* (both EM)
Vegetation type	Volcanic desert, primary succession
Location	Mt. Fuji (Japan)
Experimental set-up	Naturally established seedlings harvested near mature hosts
Seedling response	Not determined
Mycorrhizal community	High overlap in fungi found on seedlings and mature host

(2004) reported a similar pattern in a field study, where AM *Acer* seedlings showed poor establishment and growth when EM networks with or without EM roots were present. The best *Acer* establishment was observed when EM roots and networks were severed and not allowed to re-colonize the soil.

Janzen (1970) and Connell (1971) hypothesized that high species richness in plant communities may be driven by host-specific parasites and soil pathogens associated with mother trees that negatively impact seedling establishment of conspecifics. Because facilitated establishment via common mycorrhizal fungi occurs, their hypothesis should be reevaluated (e.g. Onguene & Kuyper, 2002). This is particularly relevant if belowground interactions with mycorrhizal fungi affect aboveground interactions with herbivores as shown in Gehring & Whitham (1991).

The importance of EM fungi for tree seedling establishment has been recognized and used for commercial seedling production systems. Thus, reforestation without the reintroduction of mycorrhizal inoculum is difficult on sites where the soil microbial community is largely destroyed (Perry *et al.*, 1989). Under conditions where soil microbial communities are not destroyed, Molina and Trappe (1982) hypothesize that EM communities established by one plant species can be used by another. Field evidence under natural conditions support this hypotheses. Horton *et al.* (1999) showed that most of the fungi-associated *Pseudotsuga menziesii* seedlings were also associated with mature *Arctostaphylos* (Ericaceae) shrubs. Nara (2006b) showed a similar pattern where *Betula ermanii* (Betulaceae) and *Larix kaempferi* (Pinaceae) established near previously established *Salix reinii* (Salicaceae), with a high degree of overlap in mycorrhizal fungi between the plant species. Based on work in the Pacific Northwest (USA), Perry *et al.* (1989) hypothesized that plant communities can alternate through time between early successional and late successional plant species, with soil microbes persisting and supporting the establishment of the plants in each stand type even when the plant communities include angiosperms and gymnosperms. Recent evidence provides support for the Perry hypothesis in mixed forests of eastern North America (USA), where 37 of 38 EM fungi associated with *Castanea dentata* (Fagaceae) seedlings in *Quercus rubra* (Fagaceae)-dominated stands were also associated with *Q. rubra* (Dulmer, 2006). The once dominant American chestnut (*Castanea dentata*) was largely replaced by *Q. rubra* after the introduction of chestnut blight fungus (*Cyrophonectria parasitica*) and the current mycorrhizal fungi in the soils appear to be highly compatible with both *Quercus* and *Castanea*. Dulmer (2006) hypothesized that the fungi associated with *Q. rubra* can be utilized to support restoration efforts of American chestnut.

Seedling establishment in the absence of ectomycorrhizal networks

Sites that lack mycorrhizal inoculum are usually first colonized by nonmycorrhizal plants, such as members of Brassicaceae or facultative mycorrhizal plants that become mycorrhizal over time.

Eventually, some inoculum will disperse onto a site and facultative and obligate mycorrhizal plants will out-compete the nonmycorrhizal plants (Janos, 1980a; Allen, 1991). Once inoculum is available, the establishment of seedlings in uncolonized habitats may be as much a function of compatible mycorrhizal inoculum as other factors such as seed availability, herbivory, and soil nutrient levels (Baylis, 1980; Ozinga et al., 1997; Simberloff et al., 2002).

In the absence of existing mycelial networks, spores become the dominant inoculum source. Although seeds from EM trees may germinate in a site otherwise suitable for establishment, without EM inoculum, germinating seedlings will often not survive (Mikola, 1970). Some EM hosts, such as Quercus (Egerton-Warburton & Allen, 2001), Eucalyptus (Myrtaceae) (Lapeyrie & Chilvers, 1985), and members of Salicaceae (Smith & Read, 1997), associate with AM fungi under conditions of low EM inoculum or high nutrient availability. Indeed, even Betula, Pinus, Pseudotsuga, and Tsuga will associate with AM fungi in soils that are low in EM inoculum (Cázares & Smith, 1996; Horton et al., 1998; Smith et al., 1998). The ecological role of AM fungi in otherwise EM hosts has not been fully explored, but Smith et al. (1998) reported increased P uptake in Pseudotsuga menziesii when colonized by only AM fungi. Dickie and Reich (2005) reported that distance from established EM vegetation represents an important gradient for EM inoculum and should be considered a factor potentially influencing seedling establishment. Presumably, increased inoculum potential is associated with hyphal networks or soil spore banks close to the established hosts. This last study corroborates earlier work by Borchers and Perry (1990) who observed better growth and EM formation in Pseudotsuga seedlings grown in soils collected near EM hardwood vegetation than in soils collected away from the EM vegetation.

When soils are taken from undisturbed forests and used to grow local EM hosts in laboratory bioassays, the development of EM roots can be expected despite the lack of a mycorrhizal network. While the fungal species assemblage may show some overlap with that observed in the forest, there is usually a shift in dominance. This was observed by Taylor and Bruns (1999a), who reported that the fungi associated with Pinus muricata roots harvested from a mature forest stand were different than those on bioassay seedlings grown in the mature forest soils under laboratory conditions. The seedlings grown in the laboratory were colonized by Rhizopogon and Tuber spp., fungi that were largely absent on root tips or as fruiting bodies in the mature stands (Gardes & Bruns, 1996; Horton & Bruns, 1998). These authors suggested that the EM fungal species that were characteristic for undisturbed forest could not survive soil disturbance. In contrast, the fungi that colonized the laboratory seedlings were likely present as a dormant spore bank, analogous to a dormant seed bank. These fungi became active and colonized the seedlings when the bioassay plants were grown in the field soils in the lab. When a fire killed all of the mature trees in the field plot, regenerating pine seedlings were colonized by the same fungi observed in the laboratory experiment, presumably from resistant spores (Baar et al., 1999; Taylor &

Bruns, 1999a). These fungi respond to disturbance and are good pioneer species, analogous to r-selected plant species (Grime, 1977). Evidence is accumulating that the same kinds of fungi contribute to plant establishment under primary succession.

Primary succession studies involving EM fungi have been conducted in plant communities established after glacial retreat, sand dunes, and volcanic activity. Under primary succession, spores must disperse to areas without any previous vegetation. Jumpponen (2003) investigated the presence of EM fungi in soils of nonvegetated areas along a chronosequence at a retreating glacier in the Cascade Mountains (Washington, USA). Young substrates at the glacial forefront yielded a diverse group of fungi that included both pathogenic and mutualistic genera. However, there were no living hosts in the forefront area, so the species observed were probably present in the spore bank. The EM genus *Laccaria* was observed in these soils, as is frequently the case in disturbed or primary successional habitats and soil bioassays (Deacon & Flemming, 1992; Helm *et al.*, 1996; Stendell *et al.*, 1999; Nara & Hogetsu, 2004; Dickie & Reich, 2005; Ashkannejhad & Horton, 2006; Dulmer, 2006). *Laccaria* is a wind-dispersed species, suggesting that the spore banks accumulated in the forefront area via wind dispersal from nearby vegetated areas. Other work at the site suggested that small mammals disperse some fungi long distances, particularly fungi that fruit belowground (Cázares & Trappe, 1994).

Large mammals likely disperse EM fungi long distances (>100 m) from forest edges, thereby leading to establishment of *Pinus contorta* seedlings in a primary successional sand dune system (Ashkannejhad & Horton, 2006). During peak fruiting season, deer fecal pellets contained millions of spores of two related genera, *Suillus* and *Rhizopogon*. These same fungi were observed on all naturally established seedlings collected from the field site and in bioassays using soil from the site or deer fecal pellets for inoculum. EM fungi observed on seedlings collected in close proximity to forest edges included *Suillus* and *Rhizopogon* but also a species-rich group of fungi typical of established forests. This last pattern suggests that the forest edge seedlings benefited from hyphal inoculum associated with the established trees, but the influence of an increase in spore rain from the fruiting activity in this zone could not be ruled out.

Dickie *et al.* (2002) showed that *Quercus* seedlings were well colonized by EM fungi within 38 days when planted near *Quercus* stumps (EM networks likely present), but not until 56 days after planting near *Acer* stumps (no EM networks). Pine seedlings naturally establishing in locations dominated by pine before a catastrophic fire were colonized only by dark septate endophytes after 1 month of growth, and by AM and EM fungi after 3 months of growth (Horton *et al.*, 1999). In this same study, pine seedlings establishing in locations dominated by resprouting AM shrubs before the fire were colonized by both AM fungi and dark septate endophytes after 1 month of growth, but not by EM fungi until the third month.

Once seedlings become established under primary successional conditions, then other seedlings of the same and different species can benefit from the nascent mycorrhizal network. Mycelial networks associated with *Salix reinii* facilitate further establishment of *S. reinii* seedlings on Mt. Fuji (Japan) (Nara, 2006a). Further, mycorrhizal networks associated with *S. reinii* also facilitated later successional species *Betula ermanii* and *Larix kaemperi* (Pinaceae), with most fungi observed on *Salix* roots also colonizing *Betula* and *Larix* (Nara & Hogetsu, 2004; Nara, 2006b). Importantly, compared to nonmycorrhizal control seedlings, seedlings connected to the mycorrhizal network showed improved health and nutrient status (Nara, 2006a).

9.3 | Arbuscular mycorrhizal fungi and seedling establishment

Tapping into arbuscular mycorrhizal networks

The majority of plants form symbiotic associations with arbuscular mycorrhizal (AM) fungi (Smith & Read, 1997; Wang & Qiu, 2006), and AM fungi are abundant in a wide range of ecosystems, including grasslands, savannahs, and tropical forests (Read, 1991). Seedlings that germinate in these ecosystems usually become quickly colonized by AM fungi. For example, Birch (1986) observed that seedlings of four herbaceous plant species in calcareous grassland became colonized 3–4 days after radicle emergence. It is thought that mycelial networks are responsible for fast colonization in perennial vegetation because hyphal networks are already present and because the fungi can detect specific substances (strigolactones) that are exuded by roots (Akiyama *et al.*, 2005; Besserer *et al.*, 2006), and the hyphae subsequently can grow toward the roots (Buee *et al.*, 2000). Due to this ability to tap into existing networks, young seedlings have immediate access to nutrients acquired by the fungus.

The presence of hyphal networks is of pivotal importance for seedling establishment of many plant species, especially for plants with poorly developed root systems (Baylis, 1975; Hetrick *et al.*, 1992; Merryweather & Fitter, 1996), woody perennials, and a wide range of plants that are characteristically found in late successional communities (Janos, 1980a). For example, Grime *et al.* (1987) showed that seedlings of 10 out of 18 plant species grew much better when mycorrhizal mycelial networks were present in experimental calcareous grassland microcosms. Moreover, van der Heijden (2004) added seeds of four plant species to 1-year-old grasslands microcosms and monitored whether seedling establishment was higher in microcosms that were inoculated with AM fungi. Seedlings were indeed larger and contained more P when AM fungi were present, indicating that AM fungi also promote seedling recruitment when seedlings establish between adult plants and when they compete with adult plants for mycorrhizal-mediated nutrient resources. Johnson *et al.* (2001)

developed a new method with soil cores to assess effects of AM fungi and hyphal networks on seedling establishment under field conditions. They found that severing the hyphal network reduced P inflow up to tenfold (Johnson *et al.*, 2001).

AM fungi are not always beneficial and seedling establishment of some plant species is either not influenced or negatively influenced by AM fungi. For example, AM fungi often have no positive effect on seedling establishment and plant growth when nutrient availability is high or when P availability is not limiting plant growth (Smith & Read, 1997; Titus & del Moral, 1998; Fujiyoshi *et al.*, 2006; Kiers & van der Heijden, 2006). Furthermore, seedlings of some plant species are negatively influenced by AM fungi, even though their roots are heavily colonized by AM fungi (Francis & Read, 1995). This indicates that effects of AM fungi on host plants are not necessarily positive. Experiments with *Festuca ovina* (*Poaceae*) showed that this plant species only benefits from AM fungi when it is grown in isolation from other plants (Scheublin *et al.*, 2007) and not when it is grown in competition with other species (Grime *et al.*, 1987; van der Heijden *et al.*, 1998; Scheublin *et al.*, 2007). Possibly, *Festuca* is a poor competitor for mycorrhizal resources. Moreover, Kyoviita *et al.* (2003) observed that young seedlings growing in the neighborhood of large plants performed poorly when AM fungi were present.

Plant roots are usually colonized by several AM fungi (Helgason *et al.*, 1998; Bever *et al.*, 2001; Opik *et al.*, 2006) and coexisting plant species harbor different AM fungi communities (Vandenkoornhuyse *et al.*, 2002; Scheublin *et al.*, 2004). The identity of the AM fungi that colonize roots is important because seedling performance differs with different AM fungi (Bever, 2002; van der Heijden, 2004) and plants obtain different amounts of nutrients from different AM fungi. For example, seedlings colonized by AM fungi that produce large amounts of external mycelium usually acquire more phosphorus compared to plants that are colonized by AM fungi that produce low amounts of hyphae (Jakobsen *et al.*, 1992). The impact of AM fungal communities on plant growth and seedling establishment is still poorly understood and recently a number of functional traits have been proposed that could be helpful for predicting how AM fungal communities affect plant growth (van der Heijden & Scheublin, 2007).

Seedling establishment in the absence of arbuscular mycorrhizal networks

AM fungi are often absent or low in abundance in the very early phases of succession or after major soil disturbances such as tillage. Disturbed soils are usually colonized by weedy annuals, especially Chenopodiaceae and Brassicaceae. Members of these families are typically nonmycorrhizal and it has repeatedly been shown that AM fungi have a negative impact on their growth. For example, Francis & Read (1995) established hyphal networks in pots and added seeds of eight species, including the nonmycorrhizal ruderals *Arabis hirsuta* (Brassicaceae) and *Chenopodium album* (Chenopodiaceae). The relative growth rates and the survival rates of these two nonmycorrhizal plant species were much higher in the absence of AM fungi that acted as antagonists (Francis & Read, 1995). The precise mechanisms for adverse effects of AM fungi on seedling establishment are unclear, but allelopathic effects (Francis & Read, 1994) or overexpression of plant defense mechanisms are possibilities. The relative abundance of plant species that host AM fungi usually increases during succession and when nutrient availability declines. Mycorrhizal plant species can out-compete nonmycorrhizal plant species (Allen & Allen, 1984) or plants that are weakly dependent on mycorrhiza (Hartnett et al., 1993; Scheublin et al., 2007) because mycorrhiza provide a nutritional advantage to their host plants. This suggests that there is a role of AM fungi for facilitating succession. Moreover, nonmycorrhizal plant species that establish between mycorrhizal plants often have a disadvantage and the presence of AM fungi can reduce their abundance as has been shown for Carex flacca (Cyperaceae), a nonmycorrhizal sedge typical for nutrient-poor grassland (Grime et al., 1987; van der Heijden et al., 1998).

An interesting situation can develop when nonmycorrhizal species invade new habitats, as demonstrated by Alliaria petiolata (Brassicaceae), a European invader of North American forests. Extracts of this plant have antifungal effects, reducing the abundance of AM fungi and stopping AM fungi spore germination (Stinson et al., 2006). Stinson et al. (2006) proposed that Alliaria petiolata suppresses the growth of mycorrhizal-dependent tree species by disrupting their mutualistic association with AM fungi. This observation also suggests that the relative abundance of hosts versus nonhosts is important in determining the abundance of mycelial networks and, hence, seedling establishment (Booth, 2004). Thus, it appears that interactions between plants and AM fungi are highly diverse, ranging from antagonistic to mutualistic (Fig. 9.4).

9.4 | Other plant symbionts and seedling establishment

Plants not only form symbiotic associations with mycorrhizal fungi, but they form associations with a wide range of other microbial symbionts including nitrogen-fixing bacteria, other endophytic bacteria, and fungi. Some of these microbial symbionts can also promote seedling survival and establishment. Moreover, it is also well known that some microbes are pathogenic and reduce seedling establishment (Packer & Clay, 2000), especially of native plant species (Klironomos, 2002).

Fungal and bacterial endophytes
Almost all plant species, from hepatics to large trees, host diverse communities of endophytic fungi that live in shoots and leaves. Many

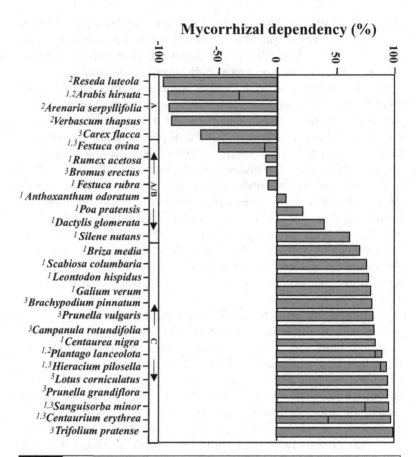

Fig. 9.4 Interactions between plants and arbuscular mycorrhizal fungi range from mutualistic to antagonistic. Mutualistic interactions occur when plant growth is enhanced by AM fungi and plants have a positive mycorrhizal dependency. Antagonistic interactions occur when seedlings grow less with AM fungi and have a negative mycorrhizal dependency. This graph shows the mycorrhizal dependency of 28 plant species characteristic for European grassland (from [1]Grime et al., 1987; [2]Francis & Read, 1995; [3]van der Heijden et al., 1998). The mycorrhizal dependency is calculated after van der Heijden (2002). Plant species that occurred in two studies have two mycorrhizal dependencies: one is shown by the bar and the other, by the black line in the bar. Plants are classified into three groups according to their mycorrhizal dependency (see left bar): Group A comprises ruderals or non-mycorrhizal plants. Group B are grasses and group C mainly comprises forbs. After van der Heijden (2002), published with permission from Springer Science and Business Media.

of these fungal endophytes remain small and colonize only a few leaves (Arnold et al., 2000). It is thought that these endophytes contribute to decomposition of fallen leaves and the endophytic life style is a strategy to have quick access to the leaves once they have fallen, rather than a true symbiosis. However, other endophytes have a large impact on seedling establishment and on the physiology and ecology of their host plants (e.g. Saikkonen et al., 1998). The mutualistic interaction between filamentous fungi belonging to the Clavicipitaceae

(Ascomycota) and many grasses are best studied (Clay, 1990; Schardl *et al.*, 2004). About 20–30% of all grass species associate with fungal endophytes belonging to fungal genera such as *Epichloe*, *Acremonium*, and *Neophytum* (Leuchtmann, 1992). These fungi, which obtain carbohydrates from their plant hosts, systematically colonize the intercellular spaces of leaf primordia, leaf sheaths of blades and tillers, and the inflorescence tissues of reproductive tillers. The plants benefit from this symbiosis by increased resistance to a number of biotic and abiotic stresses such as herbivory, pathogen attack, and drought. The fungi produce several bioprotective secondary metabolites (alkaloids) that provide protection against herbivory. Genes for the biosynthesis of four main classes of alkaloids have now been cloned including genes for the biosynthesis of ergot alkaloids (Wang *et al.* 2004), lolines (Spiering *et al.*, 2005), peramine (Tanaka *et al.*, 2005), and indolediterpenes (Young *et al.*, 2005).

Fungal endophytes can promote seedling survival and growth of several grass species (Clay, 1990). For example, Rudgers *et al.* (2005) showed that survival of *Lolium arundinaceum* (Poaceae) seedlings in diverse plant communities was about 60% higher when the endophyte was present. Moreover, Clay and Holah (1999) showed that the fungal endophyte of tall fescue reduced plant diversity and changed plant community structure in fields by improving the growth of its host.

Nitrogen-fixing symbionts

The 15 000 legumes (Leguminosae) and nitrogen-fixing bacteria belonging to the Rhizobiaceae form another important symbiosis (Sprent, 2001). The bacteria are located inside root nodules, specialized plant organs, that form after complex molecular dialogue (Schultze & Kondorosi, 1998). About 400 tree and shrub species in eight different plant families associate with nitrogen-fixing actinomycetes (Bond, 1983; Benson & Silvester, 1993), about 150 cycad and 65 *Gunnera* (Gunneraceae) species associate with cyanobacteria that can fix nitrogen (Rai *et al.*, 2000), and an unknown number of plant species, including rice (*Oryza sativa*) and sugar cane (*Saccharum officinarum*) (Poaceae), harbor endophytic nitrogen-fixing bacteria such as *Azoarcus* sp. and *Acetobacter* sp. (Boddey *et al.*, 1995; Krause *et al.*, 2006). Rhizobia, actinomycetes, cyanobacteria, and some other endophytic bacteria can fix atmospheric nitrogen into ammonium and by doing so enhance nitrogen availability to the plant. These symbioses are important to seedlings and adult plants because nitrogen is a principal element limiting plant productivity and because plants are not capable of biological nitrogen fixation, a trait restricted to prokaryotes. Under some conditions, nitrogen-fixing symbionts are able to meet the total nitrogen requirements of their hosts (Rai *et al.*, 2000) and this enables the plants to colonize otherwise uninhabitable N-poor environments. The invasion of nitrogen-fixing plants in nutrient-poor ecosystems can alter plant community composition and productivity (Walker & del Moral, 2003). For example, many hectares of the unique and fragile fynbos vegetation in South Africa (one of the biodiversity hotspots on Earth) has been ruined by nitrogen-fixing *Acacia* (Sprent & Parsons, 2000). Moreover, expansion of *Myrica*, which hosts nitrogen-fixing actinomycetes, facilitated succession and promoted plant productivity on Hawaii (Vitousek & Walker, 1989).

Several recent studies have shown that the presence of rhizobia is essential for legume seedling establishment. Survival rates of four grassland legumes increased 13.0-fold, 1.25-fold, 2.75-fold, and 8.5-fold in the presence of rhizobia, for *Trifolium repens*, *Lotus corniculatus*, *Ononis repens*, and *Medicago lupulina* (all Leguminosae), respectively (van der Heijden *et al.*, 2006a). Further, legume seedlings were larger when rhizobia were present and rhizobia enabled the legumes to coexist with other plant species. Another study showed that the presence of rhizobia is also essential for range expansion of the invasive legume, *Cytisus scoparius* (Parker *et al.*, 2006).

Rhizobia are often host specific, meaning that they can only colonize a limited number of host plants. Thus, composition of rhizobia bacteria influences growth and survival of legume species. Many legumes also associate with AM fungi and tripartite symbiotic associations are formed. AM fungi have even been found to colonize root nodules (Scheublin & van der Heijden, 2006) and specific AM fungal species have been observed in several legumes (Scheublin *et al.*, 2004). The relative importance of each symbiont for legume growth and survival is still unclear, but probably both AM fungi and rhizobia are important.

9.5 | Conclusions

It is clear that plants do not establish in isolation from other organisms in the community and that seedlings are involved in multitrophic interactions with a range of parasites, herbivores, and mutualists. Here we focus on plant symbionts and we show that interactions with mycorrhizal fungi and nitrogen-fixing bacteria play a key role in regulating seedling establishment. We provide many case studies, both from field and laboratory experiments, that show mycorrhizal fungi promote seedling establishment through increased access to soil resources. In addition, we also present examples of antagonism between plants and mycorrhizal fungi, including reducing seedling establishment. Other symbionts, such as nitrogen-fixing bacteria or endophytic fungi, may also contribute to seedling establishment.

The recent application of molecular tools and stable isotopes has provided a glimpse into the black box of belowground interactions (Horton & Bruns, 2001), revealing fascinating examples of how mycorrhizal fungal networks interact with seedlings and mature plants to determine the structure and diversity of plant communities. Future work needs to determine how resource fluxes in mycorrhizal networks are regulated and to what extent mycorrhizal diversity regulates seedling establishment and persistence in natural communities.

Importantly, it is far from clear the degree to which mycorrhizal interactions impact seedling establishment versus other mechanisms covered in this book (see also Dickie *et al.*, 2007).

9.6 | Acknowledgments

We would like to thank Kazuhide Nara and Ian Dickie for many helpful comments and suggestions. We also thank Tanja Scheublin and Jim Trappe for providing photographs. Financial support was provided to T.R.H. by the National Research Initiative Competitive Grant No. 99-35107-7843 from the USDA Cooperative State Research, Education, and Extension Service, the USDA Forest Service PNW Research Station, and the US National Science Foundation award DEB-0614384. M.v.d.H. was supported by NWO Vernieuwingsimpuls Grant No. 016.001.023, provided by the Netherlands Organisation for Scientific Research and by funding from the Swiss Federal Government. Further information can be obtained from Tom Horton (trhorton@esf.edu) and Marcel van der Heijden (marcel.vanderheijden@art.admin.ch).

Part IV

Life history implications

Chapter 10

The seedling as part of a plant's life history strategy

Angela T. Moles and Michelle R. Leishman

10.1 | Introduction

In this chapter, we will describe the intricate links between seedling ecology and life history traits such as seed mass, time to maturity, adult size, and reproductive life span. We will pay particular attention to seed mass, as this is the trait most closely linked to seedling ecology. Seed mass affects the initial size of the seedlings, the amount of reserves seedlings have for establishment, the sites to which seeds are dispersed, and the time seeds spend in the soil before germinating.

Much of our understanding of seed and seedling ecology has been based on the idea that plants face a trade-off between producing a few large seeds, each with high rates of survival as seedlings, versus producing many small seeds, each with lower rates of survival as seedlings. We, therefore, begin by reviewing the evidence for this trade-off. Our review shows that a full understanding of seed and seedling ecology requires consideration of life history variables such as plant height, reproductive life span, and the length of the juvenile period. Then we present a new framework for understanding seed and seedling traits as part of an overall life history strategy. Next we outline relationships between seed and seedling traits and other aspects of plant ecological strategy, such as seed dispersal syndrome, the capability to form soil seed banks, tissue density, and adult plant traits. These data complement previous results, and tend to support the idea that seed and seedling traits can be usefully understood as part of a larger spectrum of life history traits ranging from small, short-lived plants with small seeds, fast growth, low tissue density, and low rates of seedling survival to large, long-lived plants with large seeds, slower growth, denser tissues, and higher rates of survival.

Our focus in this chapter is mostly on between-species variation in seed and seedling strategies. This is because the vast majority of the variation in these traits lies at the between-species level. Across all of the species on earth, seed mass ranges over 11.5 orders of magnitude

(Moles *et al.*, 2005b), while within-species variation in seed mass is typically in the range of twofold to fourfold (Leishman *et al.*, 2000). However, variation in seed and seedling traits at the within-species level can have important ecological consequences (Pizo *et al.*, 2006). Thus, we explore variation in seed and seedling traits at the within-species level. We conclude with a discussion of the implications of these data for our understanding of plant regeneration, outlining some areas that we think are ripe for future development. We think that placing seedling ecology in a broader life history context will lead us to new ways of understanding the ecology of plant regeneration.

Some definitions

Many of the terms used in this chapter can be applied to subtly different units/stages. We, therefore, devote this section to clarifying what we mean by seed mass, seedling, juvenile period, and reproductive life span.

By *seed mass*, we mean the dry mass of the unit comprising the seed coat, embryo, and endosperm. In some situations, such as relationships between seed mass and seedling morphology or survival rate, it would be best to know the seed reserve mass (just the embryo and endosperm). In other situations, such as dispersal and seed production, it would be best to know the mass of the whole seed, or even the whole fruit. Fortunately, the various measures of seed mass are tightly correlated. Relationships among seed reserve mass, seed coat mass, and the mass of external protective and/or dispersal structures are approximately isometric across species (Moles *et al.*, 2003; Martínez *et al.*, 2007).

We use the term *seedling* to refer to plants that have germinated but which are still likely to be at least partially dependent on maternal resources. However, there are practical problems in determining whether a young plant is still a true seedling (Fenner & Kitajima, 2000). We, therefore, use the term *juvenile* wherever there is doubt. This term refers to plants at all stages between germination and the time of first reproduction (encompassing seedlings and saplings). Thus, the *juvenile period* is the time between germination and first reproduction. Finally, the *reproductive life span* of a plant is the mean or maximum time between the onset of reproduction and plant death.

10.2 | The trade-off between offspring production and seedling survival

A model for the optimal balance between the size and number of offspring (Fig. 10.1) has been formulated by Smith and Fretwell (1974). This model begins from two observations. First, an organism with a finite amount of energy available for reproduction can either produce a few large offspring or a large number of small offspring. Second, the

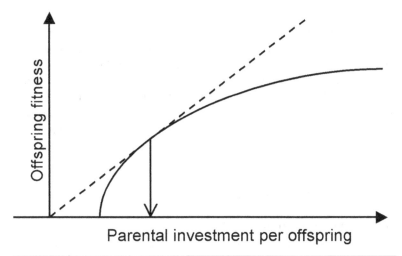

Fig. 10.1 The Smith and Fretwell (1974) model for the optimal size of offspring. This model formalized the idea of a trade-off between producing a few, well-provisioned offspring, versus producing many poorly provisioned offspring.

The curved line represents the relationship between the amount of energy invested in each offspring and the lifetime fitness of that offspring. This line will be curved if there is a diminishing return on parental investment, e.g. if a given increase in allocation increases the fitness of small offspring more than it increases the fitness of large offspring.

The optimum allocation strategy for the parent occurs at the point where the steepest possible straight line from the origin just touches the fitness curve (represented by the arrow). Increasing the amount of energy allocated to each offspring from this point would increase each offspring's fitness, but this increase would not be sufficient to balance the reduction in the number of offspring that could be made for a given amount of energy. Decreasing the amount of energy allocated to each offspring from this point would result in decreases in offspring fitness and this decrease in fitness would outweigh the increase in the number of offspring that can be made for a given amount of energy. Reproduced with permission from the University of Chicago Press.

more energy the organism allocates to each individual offspring, the greater their expected fitness becomes. Thus, organisms face a fundamental trade-off between using their available energy to produce a few well-provisioned offspring, each with high expected fitness, versus producing more poorly provisioned offspring, each with lower expected fitness.

Much of our research and theory on seedling ecology and plant reproductive strategies has been based on the idea that plants face a trade-off between producing either a large number of small seeds, each with low rates of survival as seedlings, or a few large seeds, each with a high chance of survival through seedlinghood. This seed number/seedling survival trade-off idea is clearly derived from the Smith–Fretwell theory. However, the Smith–Fretwell theory and the seed number/seedling survival trade-off idea do differ in an important way: the seed number/seedling survival trade-off idea focuses on survival through seedlinghood, while the Smith–Fretwell theory considers the offspring's lifetime fitness.

The focus on the seedling stage in the plant literature is a reflection of the perceived importance of early mortality in shaping plant population and community ecology (Kitajima & Fenner, 2000). A huge proportion of seedlings die within the first few weeks after germination (De Steven, 1991; Alvarez-Buylla & Martinez-Ramos, 1992; Vander Wall, 1994; Horvitz & Schemske, 1995; Bowers & Pierson, 2001) and thus this life history stage is under strong selection pressure. Further, the advantages of large-seededness seem to be short-lived (Leishman et al., 2000). If individuals of large- and small-seeded species are equally likely to survive from the end of the seedling stage to adulthood, then the Smith–Fretwell theory and the seed number/seedling survival idea will be functionally equivalent.

The seed mass/number trade-off

Several field studies have confirmed that there are strong negative relationships between seed mass and the number of seeds a plant produces per year, and between seed mass and the number of seeds produced per unit canopy per year (Shipley & Dion, 1992; Greene & Johnson, 1994; Jakobsson & Eriksson, 2000; Aarssen & Jordan, 2001; Henery & Westoby, 2001). The trade-off between the size and number of offspring produced for a given amount of energy gives small-seeded species an initial advantage over large-seeded species. However on average, all adult plants must produce exactly one surviving offspring during their lifetime if the population is to remain stable (neither increasing nor decreasing). Thus, the numerical advantage that small-seeded species gain during seed production must be counterbalanced somewhere else in the life cycle. One of the main advantages of producing large seeds is that they produce larger seedlings with higher rates of survival.

The relationship between seed mass and seedling survival

Seed mass affects seedling survival in many ways. First, seedlings from large seeds tend to be taller and/or heavier than seedlings from small seeds, both within and across species (Jurado & Westoby, 1992; Cornelissen, 1999; Jakobsson & Eriksson, 2000; Vaughton & Ramsey, 2001; Baraloto et al., 2005a; Zanne et al., 2005). This larger initial size might give seedlings from large seeds better access to water and light, as well as a competitive advantage (Leishman et al., 2000). Second, seedlings from larger seeds tend to have a greater proportion of their reserves uncommitted, and thus available to support the seedling in times of carbon stress (Kitajima, 1996a; Green & Juniper, 2004b; Zanne et al., 2005). Third, the negative relationship between seed mass and the relative growth rate of seedlings (Maranon & Grubb, 1993; Kitajima, 1994; Grubb et al., 1996; Wright & Westoby, 1999; Fenner & Thompson, 2005; Paz et al., 2005; Poorter & Rose, 2005) might reflect a slower metabolic rate in large-seeded species (Green & Juniper, 2004b). This slower metabolic rate would allow seedlings to persist for longer on a given amount of reserves in times of carbon stress. As is often the case, it is likely that all of these mechanisms contribute to the greater survival rates associated with large-seededness.

Numerous studies have demonstrated that seedlings from large seeds are better able to withstand many of the different stresses that seedlings face in natural environments, including competition from established vegetation, competition with other seedlings, prolonged periods in deep shade, defoliation, nutrient shortage, and burial under soil or litter (Westoby et al., 2002). The survival advantage of seedlings from large-seeded species is also seen under natural conditions (Moles & Westoby, 2004a; Baraloto et al., 2005a), where seedlings are exposed to the full suite of natural hazards, and often face multiple hazards simultaneously.

The survival advantage for seedlings from large-seeded species plus the evidence for a seed mass/number trade-off seem to support the seed number/seedling survival trade-off theory. However, for this theory to work, the survival advantage of large-seeded species would have to exactly counterbalance the seed production advantage of the small-seeded species. For example, if species A produces seeds 10 times as big as species B, then species A will only make $\frac{1}{10}$th as many seeds for a given amount of energy. To recoup this disadvantage, species A's seedlings would have to be 10 times more likely to survive than the seedlings of species B. Is this the case?

Moles and Westoby (2004a) analyzed the slope of the relationship between seed mass and seedling survival to determine whether the advantage that large-seeded species accrue during seedling establishment is sufficient to counterbalance the negative relationship between seed mass and the number of seeds produced per unit of energy. First, they looked at the slope of the relationship between seed mass and seedling survival through the first week after emergence. The slope of this relationship was nowhere near steep enough to counterbalance the seed production advantage of small-seeded species (Fig. 10.2). However, seedlings typically take much more than a week to establish. If the survival advantage of large-seeded species accrued for long enough, they would reach the point where they have as many surviving seedlings remaining as do the small-seeded species.

Moles and Westoby (2004a) also calculated how long the survival advantage of seedlings from large-seeded species would have to persist for them to reach the point where they have as many surviving offspring remaining as do small-seeded species (Fig. 10.2). A curve in the relationship between seed mass and seedling survival means that the answer to this question depends on the species' seed mass. At the small-seeded end of the seed mass spectrum, a larger-seeded species would break even with a species with 10-fold smaller seeds after 8.8 weeks. This seems quite plausible. Many species take several weeks to establish, and the survival advantage of larger-seeded species could easily persist for this long. However, at the large-seeded end of the seed mass spectrum, the survival advantage of the larger-seeded species would have to continue at the same rate as in the first

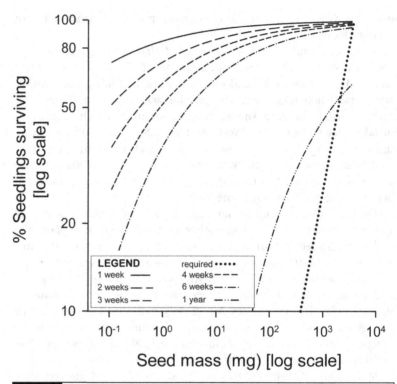

Fig. 10.2 The relationship between seed mass and seedling survival. The solid line represents the relationship between seed mass and seedling survival through 1 week (data shown in Fig. 10.4a). The other lines show the predicted relationship between seed mass and seedling survival through various amounts of time, assuming that mortality proceeds at the same rate as in the first week (see Moles & Westoby, 2004a). The dotted line represents the slope of the relationship required for the survival advantage of large-seeded species to counterbalance the greater seed production (per unit energy invested) of small-seeded species. When the survival lines reach the same slope as the *required* line, the small-seeded species and the large-seeded species produce an equal number of surviving offspring. Modified from Moles and Westoby (2004a), published with permission from Blackwell Publishing, Ltd.

week after emergence for 4.2 years before the larger-seeded species had as many surviving offspring as a species with 10-fold smaller seeds. This is not so plausible because most of the available data show no relationship between seed mass and survival rates of later-stage seedlings or saplings (Saverimuttu & Westoby, 1996a; George & Bazzaz, 1999; Walters & Reich, 2000; Dalling & Hubbell, 2002; Baraloto *et al.*, 2005a; but see Moles & Westoby, 2004a). That is, the advantage of large-seededness is generally temporary, probably expiring when the seed reserves have all been deployed. The time taken for this to happen does vary among species, and there may be correlations between seed mass and aspects of the growth strategy of juvenile and adult plants (Poorter, 2007). However, it seems unlikely that any large-seeded species retains a survival advantage for as long as 4 years. Thus, the available evidence strongly suggests that the survival advantage of large-seeded species is not sufficient to counter the seed production advantage of small-seeded species.

The gains in seedling survival associated with large seededness do not seem to be great enough to counter the decreases in seed production. Thus, the seed number/seedling survival trade-off idea must be missing some important elements. In the following sections, we describe a new understanding of how seed mass and seedling strategy are part of a much larger correlated suite of life history characteristics.

Lifetime seed production

Most cross-species investigations of seed production have focused on annual seed production or on the number of seeds that can be made for a given amount of canopy per year. However, the critical measure of evolutionary success is the total number of surviving offspring produced by an individual throughout its lifetime.

The slope of the relationship between seed mass and the number of seeds that a species can produce per unit canopy area per year is −1 (Henery & Westoby, 2001; Fig. 10.3a). That is, a 10-fold increase in seed mass is associated with a 10-fold decrease in the number of seeds made from a given amount of canopy each year. The difference between this relationship and the relationship between seed mass and lifetime seed production will depend on the relationships between seed mass and canopy area, and between seed mass and plant life span.

Several studies have reported positive correlations between plant size and seed size or seed mass (Levin, 1974; Leishman et al., 1995; Aarssen, 2005; Moles et al., 2005c), and between seed mass and canopy area (Fig. 10.3b). The reason for this correlation between adult size and offspring size has been the subject of much recent debate (Moles et al., 2004; Aarssen, 2005; Grubb et al., 2005; Moles et al., 2005c). Regardless of the mechanism underlying the observed relationship, we would expect that the larger canopies of large-seeded species would mean that they have more photosynthate to allocate to reproduction. Because of this, we would expect the slope of the relationship between seed mass and the number of seeds produced per individual plant per year to be less steep than the −1 slope of the relationship between seed mass and the number of seeds produced per square meter of canopy per year. As predicted, a compilation of data from the literature showed a slope of −0.3 (Fig. 10.3c) (Moles et al., 2004).

Not only do small-seeded species tend to have smaller canopies than do large-seeded species, they also tend to have shorter reproductive life spans (Fig. 10.3d) (Moles et al., 2004; also see Baker, 1972; Silvertown, 1981; Leishman et al., 1995, but note exceptions in Thompson, 1984; Jurado et al., 1991). This further erodes the apparent advantage that small-seeded species have during seed production. In fact, there is no relationship between seed mass and lifetime seed production (Fig. 10.3e) (Moles et al., 2004). Although small-seeded species do make more seeds for a given amount of energy, their smaller canopies

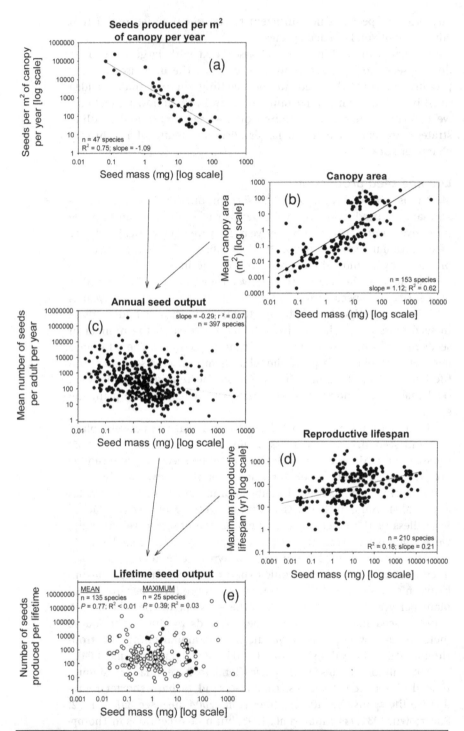

Fig. 10.3 Correlations between seed mass and (a) the number of seeds produced per square meter of canopy per year, (b) canopy area, (c) annual seed production per adult plant, (d) reproductive lifespan, and (e) the total number of seeds produced over a lifetime. Each dot represents the geometric mean value for one species, except in plot (e), where open circles represent geometric means and filled circles represent maxima. Lines show the results of model I regressions. Data are from (a) Henery and Westoby (2001), (b–d) from Moles and Westoby (2006), and (e) from Moles et al. (2004), published with permission from Blackwell Publishing, Ltd.

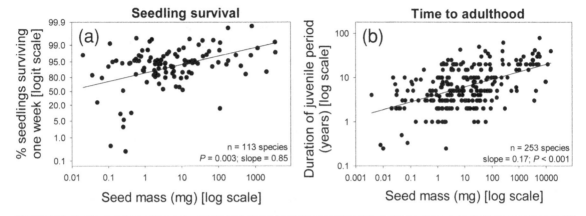

Fig. 10.4 Correlations between seed mass and (a) the percentage of seedlings surviving through their first week after emergence and (b) the time taken for each species to reach reproductive maturity. Each dot represents the geometric mean value for one species. The line in (a) was fit using random effects logistic regression; the line in (b) was fit using model I regression. Data for (a) are from Moles and Westoby (2004a) and (b) from Moles and Westoby (2006), published with permission from Blackwell Publishing, Ltd.

and shorter reproductive lifespans mean that they do not produce more seeds than large-seeded species over a lifetime.

Seed mass and survival to adulthood

Large-seeded species have a survival advantage over small-seeded species during early establishment (Fig. 10.4a) (Moles & Westoby, 2004a). However, it is not survival through the first week, the first season, or the first year after germination (the time frames most commonly used in studies of juvenile survival in plants) that determines a plant's evolutionary success. Rather, it is the number of offspring that survive to reproductive maturity. Two main factors affect the proportion of a plant's offspring that survive to reproductive maturity, namely the probability of survival through a given amount of time and the amount of time elapsed between emergence and maturity.

A compilation of data from the literature showed that small-seeded species tend to reach reproductive maturity much earlier than do larger-seeded species (Fig. 10.4b) (Moles et al., 2004). This finding is consistent with the relationship between seed mass and plant size (Fig. 10.3b) (Moles et al., 2005c), and the positive relationship between plant height and time to reproductive maturity (Moles et al., 2004). That is, small-seeded species tend to have small adult stature and it does not take as long to grow into a small herb as it takes to grow into a large tree.

The longer juvenile period of large-seeded species exposes them to a longer period of juvenile mortality and thus a higher risk of dying before they reach adulthood. This disadvantage counters the higher survival per unit time of seedlings from large-seeded species (full explanation in Box 10.1). Evidence from transition matrix models

Box 10.1 | The effects of seed mass, early rates of mortality, and time to maturity on seedling survival to adulthood

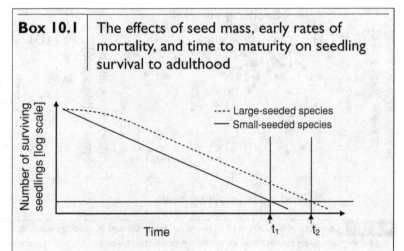

Figure modified from Moles & Westoby, 2006a.

Consider a large-seeded species (dashed line) and a small-seeded species (solid line). The small-seeded species is shown here with exponential rate of mortality, the most common pattern found in empirical studies of seedling mortality (Moles & Westoby, 2004a). Field data suggest that the large-seeded species will have a lower initial rate of mortality than the smaller-seeded species (Moles & Westoby, 2004a). Thus, the number of surviving offspring does not drop as rapidly for the large-seeded species as for the small-seeded species. However, most of the available data suggest that after the seedling from the large-seeded species has deployed all the maternal resources, it will cease to have a survival advantage over the small-seeded species (Westoby *et al.*, 2002). At this point, seedlings from large- and small-seeded species die at approximately the same rate, and the lines on the figure become parallel.

If the large-seeded species and the small-seeded species reached maturity at the same time (say, time t_1) then the lower early mortality of the large-seeded species would translate to a higher number of individuals surviving to adulthood. However, large-seeded species tend to take longer to reach adulthood than do small-seeded species (Moles *et al.*, 2004). Thus, they are exposed to juvenile mortality longer than are the small-seeded species (say, to t_2). This longer period of mortality decreases the total number of individuals surviving to adulthood. It is possible that this additional time to maturity could completely negate the initial survival advantage of the large-seeded species.

Implications for the relationship between seed mass and plant size

If species with large canopies and long reproductive lifetimes produced very small seeds, they would be able to produce a stupendous number of offspring throughout their lives. However, collation of data for 2589 species from all around the world revealed a positive relationship between seed mass and plant height, and an absence of species with very large adult size that produce very small seeds, despite the inclusion of data for several *Ficus* and *Eucalyptus* species (Moles *et al.*, 2005c). The figure above suggests an explanation for the absence of this *large adult, small seeds* strategy: A species with large adults will necessarily have a lengthy juvenile period.

To have sufficient offspring survive to adulthood, the juveniles of large adults will need to have a relatively low weekly rate of juvenile mortality. This can be achieved by producing large, well-provisioned offspring. A species with large adults and very small seeds would not have enough offspring survive to maturity to replace itself and the strategy would become extinct.

also supports the idea that later stages of regeneration can be extremely important determinants of population dynamics, especially in long-lived species (Chapter 11). There are not enough data to be certain about the shape of the relationship between seed mass and survival to adulthood. However, the available evidence suggests that there is either no relationship or perhaps a weak negative relationship between seed mass and survival from seedling emergence through to reproductive maturity (Moles & Westoby, 2006a).

In summary, although small-seeded species produce more seeds per square meter of canopy per year, large-seeded species have larger canopies and longer reproductive life spans. Overall, there is no relationship between seed mass and the total number of offspring produced over a lifetime. Large-seeded species do have higher rates of survival per unit time (at least during the early stages of establishment), but the longer juvenile period of large-seeded species increases the duration of exposure to juvenile mortality. Data are too scarce to make a firm conclusion; however, it does seem unlikely that there is a positive relationship between seed mass and survival from seedling emergence to maturity (the available evidence shows that there is no relationship, or perhaps a weak negative relationship between seed mass and survival from seedling emergence to maturity, Moles & Westoby, 2006a).

Our old interpretation of the trade-off between producing a few large seeds, each with high rates of survival as seedlings, versus producing many small seeds, each with low chances of surviving the juvenile period, is confounded (at least at the cross-species level) by relationships with other plant life history traits. To understand seed and seedling ecology, we need to consider correlations among seedling survival, seed mass, adult size, time to maturity, and reproductive life span.

10.3 | Understanding seed and seedling ecology as parts of a plant's life history strategy

The network of correlations among life history traits can be understood as factors that contribute to one essential variable, the average number of surviving offspring produced during a lifetime. If the population is stable (neither going extinct nor rapidly increasing), then every adult individual will leave, on average, one offspring that survives to reproductive maturity. The number of offspring

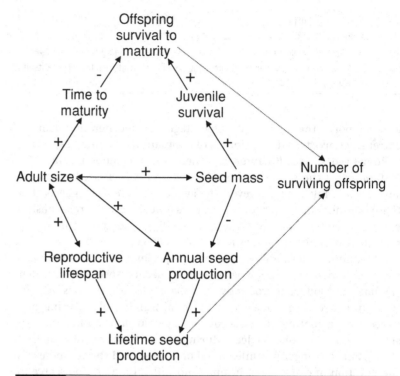

Fig. 10.5 Correlations among the major plant life history traits. Single-headed arrows represent causal relationships; double-headed arrows represent correlations. Plus/minus signs show the direction of correlation.

that survive to maturity is a product of lifetime seed production and the proportion of offspring that survive to adulthood (Fig. 10.5). Seed mass, seedling survival, adult plant size, time to maturity, and length of the reproductive period together determine the level of lifetime seed production and offspring survival to maturity. These variables can be usefully organized into two paths, one stemming from correlations with seed mass and the other from correlations with plant size (Fig. 10.5).

The path through seed mass is the original seed number/seedling survival trade-off. Increases in seed mass decrease the number of seeds that a plant can produce in a year, while simultaneously increasing the rate of seedling survival. This increase in juvenile survivorship (per unit time) increases the probability of a given seed surviving to adulthood. Thus, decreases in annual seed production are balanced by increases in the proportion of seeds surviving to adulthood.

The path through plant height contains variables not usually considered in studies of seedling ecology. However, this path is just as important to the overall life history outcome as the seed mass path. Increasing adult size increases the number of seeds an adult plant can produce because larger plants have larger canopies and, therefore, higher annual photosynthetic incomes, giving the plant more energy to allocate to seed production. However, increases in adult size are associated with increases in time to adulthood and longer juvenile periods carry an increased risk of juvenile mortality. Thus, increases in plant size simultaneously increase seed production and decrease offspring survival to maturity, exactly the opposite effects of increases in seed mass.

10.4 | Correlations between seed and seedling strategy and other aspects of plant ecology

In this section, we briefly outline correlations between seed and seedling strategy and other aspects of the species' ecology. The various correlations are presented in the order in which they are encountered during a typical life cycle, beginning with seed dispersal and continuing through seed survival in the soil, germination and seedling emergence, seedling morphology, seedling growth, and ending with the traits of mature plants.

Seed dispersal

As one might expect, there are strong correlations between seed mass and dispersal syndromes (Leishman *et al.*, 1995; Moles *et al.*, 2005a). Species with unassisted dispersal, species that disperse by adhesion to the outside of animals, and species with wind dispersal tend to have very small seeds, while species that are dispersed by water or via ingestion by animals can be quite large (Moles *et al.*, 2005a). There appears to be a relationship between seed mass and the size of the animals responsible for ingesting/carrying the seeds. Seeds dispersed by mammals tend to be larger than seeds dispersed by birds, which are larger than seeds dispersed by invertebrates.

Differences in dispersal syndrome are likely to affect the shape of the dispersal kernel and the types of environment to which seeds are dispersed (Garcia-Castaño *et al.*, 2006; Russo *et al.*, 2006). These differences in dispersal can affect (1) the relative importance of sibling–sibling competition, intraspecific competition, and interspecific competition among seedlings (Jordano & Godoy, 2002), (2) susceptibility to pathogen attack (Augspurger & Kelly, 1984), and (3) the probability of surviving seed predation (Garcia-Castaño *et al.*, 2006). All of these factors affect a seed's chances of successfully establishing as a seedling. One interesting possibility is that seed and seedling strategies might be affected in the future by selection for dispersal syndromes that enable plants to migrate rapidly in response to climate change.

Seed survival in the soil

In Europe, there is a tendency for species with small, rounded seeds to have seeds that persist for some years in the soil seed bank (Thompson *et al.*, 1993; Cerabolini *et al.*, 2003). This pattern has also been found in Argentina (Funes *et al.*, 1999), but not in Australia (Leishman &

Westoby, 1998), or in the fynbos of South Africa (Holmes & Newton, 2004). In New Zealand (Moles *et al.*, 2000), Iran (Thompson *et al.*, 2001), and central Spain (Peco *et al.*, 2003), there are negative relationships between seed mass and persistence, but no relationships between seed shape and persistence in the soil.

There has been some confusion in the literature regarding seed persistence in the soil and seed dormancy. Some authors use the terms almost as though they are synonyms. However, there is no strong relationship between dormancy and persistence (Thompson *et al.*, 2003). So, is the tendency for smaller-seeded species to have more persistent seed banks matched by a tendency for smaller-seeded species to have some form of dormancy? It seems so; a study of 1795 species showed that species with some form of dormancy have smaller seeds than do species without dormancy (Jurado & Flores, 2005). However, the magnitude of the difference in seed mass was so small that its ecological significance is doubtful.

The idea that small-seeded species are generally more likely to have persistent seed banks is consistent with the finding that short-lived species, which tend to have smaller seeds, are more likely to have persistent seed banks than are long-lived species (Thompson *et al.*, 1998). There is also some evidence that species with high levels of seedling survival tend to have low levels of seed dormancy (Kiviniemi, 2001). Both of these are examples of situations where seed traits can be understood as simple components of the broader life-history strategy of each species. This is consistent with the work of Venable and Brown (1988), who considered the interaction of dispersal, dormancy, and seed size in reducing the impact of environmental variation. They recognized that associations between seed size and other plant attributes are an important component of a species' life history and that there may be trade-offs between these attributes.

Germination and seedling emergence

The tendency for small-seeded species to have persistent seed banks suggests that these species are well placed to have their seedlings rapidly emerge and take advantage of favorable establishment conditions. This idea is consistent with the results of a study of seedling establishment after fire in Australia (Moles & Westoby, 2004b). This study showed a positive relationship between seed mass and the amount of time elapsed between the time at which conditions first became favorable for germination and the first emergence of seedlings. There is also a strong correlation between the size of seeds and the depth of burial from which seedlings can emerge (Maun & Lapierre, 1986; Gulmon, 1992; Jurado & Westoby, 1992; Jurik *et al.*, 1994; Bond *et al.*, 1999). One hypothesis is that this relationship is a simple consequence of the redistribution of mass from the seed to the hypocotyl (Bond *et al.*, 1999).

Seedling morphology

There are strong negative relationships between seed mass and the photosynthetic capacity of a species' cotyledons (Ibarra-Manríquez et al., 2001; Zanne et al., 2005), and between cotyledon thickness and photosynthetic rate (Kitajima, 1992b). Seedlings from large-seeded species also tend to have hypogeal cotyledons that remain below the soil surface or cryptocotylar germination (the cotyledons remain within the fruit wall or seed coat) (Ng, 1978; Garwood, 1996; Wright et al., 2000; Ibarra-Manríquez et al., 2001; Zanne et al., 2005). The correlations between seedling morphology and seed mass may be related to the seedling's ability to resprout after sustaining damage from herbivores. Small-seeded species generally deploy all of their reserves during germination. However, larger-seeded species, with greater reserves, are able to hold some resources in reserve. Seedlings with epigeal cotyledons are usually unable to resprout if their cotyledons and apical bud are removed (Moles & Westoby, 2004b). This is no disadvantage for small-seeded species, which are unlikely to have enough reserves available to resprout, regardless of the availability of buds. However, it would be advantageous for large-seeded species to have their reserves and some meristematic tissue safely stored below-ground, so that they can resprout after sustaining herbivore damage (Harms & Dalling, 1997).

Seedling growth strategies

Many studies have shown negative relationships between seed mass or initial seedling mass and the relative growth rate of young seedlings (Maranon & Grubb, 1993; Kitajima, 1994; Grubb et al., 1996; Wright & Westoby, 1999; Fenner & Thompson, 2005; Paz et al., 2005; Poorter & Rose, 2005). The higher relative growth rate of small-seeded species seems to be associated with low density tissues and a strategy of maximizing surface area. For instance, a study of 33 Australian species showed a positive correlation between relative growth rate and specific root length, and a negative correlation between relative growth rate and leaf mass per area (Wright & Westoby, 1999). In addition, a study of 52 woody species from Europe showed that species whose seedlings had high leaf mass per area had denser, but not thicker leaves (Castro-Diez et al., 2000). These dense-leaved seedlings tended to have more sclerified tissue in the leaf lamina, smaller cells, and lower water and nitrogen concentrations than did species with low leaf mass per area (Castro-Diez et al., 2000). Consistent with this, Cornelissen et al. (1997) found a positive correlation between seedling relative growth rate and foliar nitrogen concentrations. Finally, a study of 80 species from Europe showed that seedlings with high relative growth rates tend to have low stem tissue density and wide xylem vessels (Castro-Diez et al., 1998).

The correlations between seed mass, seedling morphology, and seedling growth strategy undoubtedly contribute to the low survival rates of seedlings from small-seeded species (Fig. 10.4a). One might also expect fast-growing, low tissue density seedlings to have low levels of chemical defense. However, experimental data for 19 species of Asteraceae do not appear to support this prediction (Almeida-Cortez & Shipley, 2002).

Traits of mature plants

Seed and seedling traits are often correlated with the traits of adult plants. We have already described correlations among seed size, seedling size, maximum plant height, time to maturity, and longevity. As one might expect, there are also correlations between a species' morphology as a seedling and as an adult. For example, seedling leaf size is correlated with adult leaf size (Cornelissen, 1999).

As the seedling stage makes up a very small fraction of the life span of a plant, one might think that plant traits would be largely determined by the environment in which the adult plants occur. However, seedlinghood is a time of high mortality, and thus exerts a disproportionately strong selective effect on plant traits. Evidence for the importance of the regeneration phase was recently found in a study of 58 plant species from a rain forest in Bolivia (Poorter, 2007). Poorter showed that several ecologically important leaf traits of adult plants, including leaf mass per area, maximum photosynthetic assimilation rate, and leaf nitrogen content, were more closely related to the light conditions the species experienced during its regeneration phase than to the light conditions experienced by the adult plants on which the traits were measured.

Section summary

There is a spectrum of variation of seed and seedling strategies from species with large seeds and slow-growing seedlings with dense tissues and slow tissue turnover to species with small seeds and fast-growing seedlings with lower-density tissues and fast tissue turnover. The fast-growing, small-seeded species also tend to have the ability to persist for long periods in the soil and then rapidly emerge and grow when conditions become favorable. These data fit well with the idea that plant life histories can be seen as a spectrum from tall, long-lived plant species with large seeds and high rates of seedling survival to small, short-lived species with small seeds and low rates of seedling survival (Fig. 10.5). In the future, it would be good to formally incorporate variables such as tissue density and relative growth rate into the scheme depicted in Fig. 10.5.

10.5 | Seed and seedling strategies within species

The focus of this chapter so far has been on interspecific variation in seed and seedling traits. This is largely because the variation in strategies is so much greater between than within species. However, selection acts within species, so understanding trait variation within species contributes to understanding of the evolution of life history strategies. In this section, we will summarize the state of the literature on seed mass variation, seed production, and seedling survival at the within-species level, and compare this literature to the literature on cross-species relationships.

Seed mass variation

Seed mass typically varies two- to fourfold within species (Michaels *et al.*, 1988; Jacquemyn *et al.*, 2001), compared to seed mass variation within communities of five to six orders of magnitude (Leishman *et al.*, 2000) and 11.5 orders across all species (Moles *et al.*, 2005b). Numerous studies have shown that seed size is a heritable trait (e.g. Schaal, 1980; Byers *et al.*, 1997; Wheelwright, 2004; but see Silvertown, 1989; Wolfe, 1995).

Seed mass variation within species mostly occurs within plants rather than among plants or populations and is largely related to seed position within pods and fruits (Wolf *et al.*, 1986; Wulff, 1986a; Michaels *et al.*, 1988; Winn, 1991; Obeso, 1993; Mendez, 1997; Vaughton & Ramsey, 1997; Willis & Hulme, 2004; Pizo *et al.*, 2006). This suggests that much of within-species seed mass variation is due to physiological or morphological constraints on optimum resource allocation to seeds. Sakai and Sakai (2005) argue that greater allocation to reproductive output should result in increases in seed number rather than individual seed mass because of greater resource use efficiency (the terminal stream limitation model). Seed mass variation among individual plants has also been shown to increase in response to fungal infection in some species of *Hydrophyllum* (Hydrophyllaceae) (Marr & Marshall, 2006). Variation in seed mass among populations has also been attributed to population size (Kery *et al.*, 2000; Jacquemyn *et al.*, 2001; Hensen & Oberprieler, 2005) and resource availability (Willis & Hulme, 2004). Finally, there is some evidence that seed mass varies according to pollen source. Self-pollination is associated with decreased seed mass in *Swertia perennis* (Gentianaceae) (Lienert & Fischer, 2004), *Asclepias incarnata* (Asclepiadaceae) (Lipow & Wyatt, 2000), *Chamaecrista fasciculata* (Leguminosae) (Fenster, 1991), and *Banksia spinulosa* (Proteaceae) (Vaughton & Ramsey, 1997). However, a few studies have found no difference in mass of seeds from self- and cross-pollination (Winn, 1991; Ashworth & Galetto, 2001; Leimu, 2004).

The heritability of seed mass and the relatively small variation in seed mass within species suggests that seed size is determined via a process of stabilizing selection. This can operate through selection via trade-offs in seedling survival, dormancy, dispersal, predation, and seed number (Venable & Brown, 1988). For example, in *Quercus ilex* (Fagaceae), selection pressure on fitness components relating to establishment success favors larger seeds while selection pressure relating to predation pressure favors small seeds (Gómez, 2004). Much of the literature focuses on the central role of the seed number/seedling survival trade-off in determining an optimum seed mass for plant fitness.

The Smith–Fretwell model that formalizes the seed number/seedling survival trade-off produces a curve relating fitness to offspring size that predicts a single optimal seed mass (Fig. 10.1). A key prediction of the Smith–Fretwell model is that if a mother plant can allocate more resources to reproductive output, it should produce more seeds of the same size. The small variation within species in seed mass that is observed is attributed to limitations of a plant's ability to produce completely standardized seed mass or to variability in the shape of the Smith–Fretwell function within species (Leishman et al., 2000). Variations in seed mass among individuals or populations of a species would provide evidence for the role of variability in the seedling environment in producing variation in seed mass. In contrast, variation in seed mass within individuals would provide evidence for the role of the limitation in a plant's capacity to produce completely standardized seeds. Venable (1992) modified the Smith–Fretwell model to account for variation in resource availability and relaxed the assumption of linear fitness returns from offspring number. He showed that within-species trade-offs between seed mass and number may be masked by variation in resource availability and plant size. Further work by Geritz (1995) extended the Smith–Fretwell model into a game-theoretical context. He showed that variation in seedling density favors the evolution of seed mass variation within individuals, based on the idea that seedlings from large seeds win in competition with seedlings from small seeds, but small seeds are able to colonize more sites because of their numerical advantage.

Intraspecific relationships between seed mass and seed production

What evidence is there for intraspecific seed mass/number trade-offs? In studies where plant size is accounted for, the predicted negative relationship between seed mass and number is apparent (e.g. Harper et al., 1970; Ågren, 1989). That is, small-seeded individuals produce more seeds per unit biomass per year. However, when seed mass and number relationships are examined at the whole-plant level, the relationships are either negative (e.g. Aniszewski et al., 2001) or neutral (e.g. Schaal, 1980; Wulff, 1986a; Winn & Werner, 1987). Some authors have argued that these neutral patterns at the whole-plant level can be understood in the context of variation in plant size, where larger plants produce more seeds (Henery & Westoby, 2001). This argument is consistent with Venable's (1992) model. The patterns found in within-species studies are relatively similar to the results of the cross-species studies described in section 9.2. However, no within-species studies have taken the approach of quantifying the slope of the seed mass/number relationship or considered seed mass variation in relation to lifetime seed production.

Intraspecific relationships between seed mass and seedling survival

Many studies have shown that seed mass is positively related to initial seedling size within species (e.g. Dolan, 1984; Wulff, 1986b; Zhang & Maun, 1991; Moegenburg, 1996) and that seedlings from larger seeds can emerge from greater soil depths (Wulff, 1986b). Within particular establishment sites, larger seeds have been shown to have better seedling survival (Stanton, 1984; Morse & Schmitt, 1985; Stanton,

1985; Winn, 1988; Tripathi & Khan, 1990; Wood & Morris, 1990) and greater reproductive output (Stanton, 1985). There is evidence that larger seeds provide an advantage in drier habitats (Schimpf, 1977; Sorenson & Miles, 1978; Wulff, 1986b; Stromberg & Patten, 1990; Parciak, 2002; but see Hendrix, 1991) and that seedlings from large seeds survive longer in conditions of nutrient deprivation (Krannitz et al., 1991) or are more competitive in low fertility conditions (Tungate et al., 2006). Early experiments by Black (1958) showed that large seeds were more successful in seedling–seedling competition. These studies are generally in relation to initial seedling success rather than to success through to reproductive maturity (with the exception of Stanton, 1985).

Cross-species studies show that the greater survival per unit time associated with large-seeded species is countered by the greater time to maturity (section 10.2). We wondered whether this confounding might also occur at the within-species level. We therefore searched BIOSIS (http://www.biosis.org/) to find studies that reported seed mass and plant size, seed mass and plant longevity, and seed mass and time to maturity relationships within species, excluding crop cultivars. The relationship between seed mass and maternal plant size is reported as either neutral (Klinkhamer & De Jong, 1987; Hendrix & Sun, 1989; Dudash, 1991; Stocklin & Favre, 1994; Vaughton & Ramsey, 1997) or positive (Hendrix & Sun, 1989; Klinkhamer & De Jong, 1993; Stocklin & Favre, 1994; see Moles et al., 2004, Appendix 4). Some of these apparently neutral relationships might be due to a lack of statistical power. We were not able to find any information about intraspecific relationships between seed mass and time to maturity. However, variation in time to maturity is probably quite small within species, so may not be an important consideration at this scale.

10.6 | Implications of a holistic understanding of plant life history strategies

Merging theory for plants and other taxa

The idea that seed and seedling strategies are just part of a broader life history syndrome is a departure from most recent thinking about seed mass/offspring survival strategies. However, it is not really a very novel idea. In fact, we seem to have returned to the r–K spectrum of ecological strategies (Pianka, 1970). Species with small-seeds, high rates of juvenile mortality, short life spans, and small adult size are clearly r-selected species, while species with large seeds, low rates of juvenile mortality, long life spans, and large adult size are K-selected species. The present theory (Fig. 10.5) also has much in common with Charnov's (1993) theories on life history, originally developed for understanding animal life history strategies.

The parallels between our present understanding of seed and seedling ecology and these older theory lineages have not been fully explored. It seems certain that applying ideas developed in these other areas of biology will lead to some interesting advances in our understanding of plant life history strategies.

Reassessing theory on the coexistence of multiple recruitment strategies

In the past, theories on plant life history strategy have sought to relate suites of attributes to particular habitats. These habitats may be defined in terms of disturbance, competition, adversity, or demographic factors such as growth rates and mortality schedules (Pianka, 1970; Grime, 1979; Begon, 1985; Sibly & Calow, 1985; Tilman, 1988). However, much of this body of theory fails to account for the wide variation in seed and seedling traits found within communities. The fact that differences in seed size are much greater between species within a community than between very different communities (Leishman et al., 2000) suggests that seed size is more strongly related to other plant attributes than to establishment conditions. Consistent with this, a study of 12 987 species showed that evolutionary divergences in seed mass have been much more consistently associated with divergences in plant traits such as seed dispersal syndrome and growth form than with divergences in environmental variables such as latitude, net primary productivity, temperature, precipitation, and leaf area index (Moles et al., 2005a).

A game-theoretical approach has been used to explain the coexistence of different seed size strategies, both within and across species (Geritz, 1995; Rees & Westoby, 1997; Geritz et al., 1999). In these models, the landscape is seen as a grid of spaces available for seedling establishment. If a large seed and a small seed land in the same grid square, then the seedling from the large seed will outcompete the seedling from the small seed because of its greater resources and larger size. However, the large-seeded strategy can never exclude the small-seeded strategy because the small-seeded strategy is able to produce more seeds than the large-seeded strategy and will, therefore, reach some sites that are not reached by the large-seeded strategy.

Competition/colonization models seemed for some time to be the most plausible explanation for the wide differences in seed mass among coexisting species. Experiments showed that seedlings from large seeds usually have a competitive advantage over seedlings from small seeds (Leishman, 2001; Coomes & Grubb, 2003) and observational studies have supported the idea that small-seeded species produce more seeds per individual per year, and per square meter of canopy per year (Shipley & Dion, 1992; Greene & Johnson, 1994; Jakobsson & Eriksson, 2000; Aarssen & Jordan, 2001; Henery & Westoby, 2001). However, several lines of evidence have recently cast doubt on the relevance of the competition/colonization theory to real-world situations. Leishman (2001) showed that the expected relationship between seed mass and colonization ability was obscured in the field by differences in abundance between species. Further, a compilation of field data showed that (1) competition between seedlings is a relatively minor cause of death compared to drought, herbivory, and pathogen attack and (2) seedlings seldom germinate at high enough densities for interspecific competition to be a major determinant of seedling mortality until after the seedlings have passed the stage at which seed mass might affect the outcome of competition (Moles & Westoby, 2004c). Finally, data from two detailed field studies strongly suggest that the competition/colonization trade-off cannot explain the coexistence of different seed mass strategies. Eriksson (2005) studied plant regeneration in Swedish grassland. He found that recruitment was not related to seed mass, and that larger seeded species did not generally win in competition with small-seeded species. In fact, there was no case in which a larger-seeded species had negative effects on a smaller-seeded species, even though the seeds had been sown at densities at least as great as those found under natural conditions. Thus, Eriksson (2005) soundly rejected the idea that game-theoretical competition/colonization models explained the coexistence of multiple seed-size strategies in this ecosystem. Turnbull *et al.* (2004) studied recruitment in annual plants in limestone grassland in South Wales. Although these authors found some evidence for a competition/colonization trade-off, the trade-off was not strong enough to be the sole explanation for species coexistence (Turnbull *et al.*, 2004). Thus, there is growing consensus that the competition/colonization theory does not provide a general explanation for seed mass variation within communities (Coomes & Grubb, 2003; Eriksson, 2005).

All this leaves us wondering how the vast range of different seed and seedling strategies coexist and how such different strategies were ever selected for in the first place. One possible explanation is that different species utilize different regeneration niches within each environment (Grubb, 1977; Coomes & Grubb, 2003; Turnbull *et al.*, 2004). There is some evidence that different species do perform better under subtly different establishment conditions. However, it is difficult to imagine how subtle differences between microsites within ecosystems could account for the vast differences in coexisting seed and seedling strategies (Eriksson, 2005).

Perhaps understanding seed and seedling strategies as components of a larger life history strategy will help us to resolve this mystery of strategy coexistence. For instance, incorporating information about adult niches along with information about regeneration niches might help us to understand how different plant strategies can coexist. In any case, this strategy coexistence problem is certainly an interesting direction for future research.

10.7 | Conclusions

Our past focus on the trade-off between producing many small seeds, each with low levels of survival as seedlings, versus producing a few large seeds, each with higher levels of survival, was missing several important elements. It appears that seed and seedling ecology are better understood as components of a larger life history strategy, at least at the cross-species level. Plant strategies seem to fall largely along a spectrum from small-seeded species with low rates of seedling survival, short life spans, small stature, rapid growth, and the ability to disperse through time and space to take advantage of short-lived recruitment opportunities (r-selected species) through to large-seeded species with high rates of seedling survival, long life spans, large adult size, slower growth, and seedlings that can tolerate a wide range of conditions during establishment (K-selected species). Of course, there are many exceptions to this generalization. Much of the variation can be understood as species responses to particular combinations of biotic and abiotic stress. However, we think that viewing seed and seedling ecology as part of a broader life history strategy will give us new tools that might help to answer some of the outstanding questions regarding the ecology of plant regeneration.

Chapter 11

Seedling recruitment and population ecology

Ove Eriksson and Johan Ehrlén

11.1 | Introduction

In plant population ecology, *recruitment* refers to the process by which new individuals found a population or are added to an existing population. Although recruitment may refer to clonal offspring, by far the most common means of recruitment is by seedlings. Seedling recruitment includes three basic processes: seed germination, seedling survivorship, and seedling growth. Seedlings represent the interface between the seed and the developing plant, and as a transitional life cycle stage, it has been difficult to define unambiguously when a seedling stops being a seedling (Kitajima & Fenner, 2000). Population studies often define seedlings somewhat arbitrarily, implying also that population processes such as seedling recruitment may be assessed arbitrarily. Irrespective of the definition, it is clear that the seedling stage represents the most sensitive part of the plant life cycle (Harper, 1977; Silvertown & Charlesworth, 2001). Seedlings are usually small and vulnerable to various abiotic and biotic agents. They often have only a tiny supply of resources to consume before they must exploit external resources in competition with other plants. Thus, seedlings are commonly subject to the highest mortality rates of any stage in the plant life cycle.

Seedling recruitment varies widely across species and communities. A stable population is maintained if one reproductive individual is replaced, on average, by one successfully recruited offspring. Therefore, we expect that there is a relationship between the life span of the plants and the temporal pattern of seedling recruitment. Ehrlén and Lehtilä (2002) assessed life span of perennial plant species, using matrix models, and found that the majority of species have a life span exceeding 35 years. The life span of woody plant species was on average 98.6 years, and for nonwoody species, life span was on average 22.2 years. Given such long life spans, it is expected that successful seedling recruitment may occur only sporadically over periods of several tens of years. In one of the longest studies of perennial plant populations extending from 1943 to 1981, Inghe and Tamm (1985) reported seedling recruitment in several forest and meadow herbs. They found, for example, that periods of about a decade could elapse between successful seedling recruitment events in the forest herb *Hepatica nobilis* (Ranunculaceae). Even in short-lived plants such as annuals, recruitment patterns may sometimes be erratic. Symonides (1988) suggested that there are three principal strategies for annuals. Some species appear as temporally ephemeral populations, exploiting occasional bursts of resource availability (e.g. desert annuals), but remain as dormant seeds in the periods between these opportunities. A second group of annuals respond similarly but exploit spatially unpredictable resources, thus behaving as ruderals. In contrast to these two groups that have drastic population fluctuations, there is a third group of annuals typically occurring as rather stable populations, where recruitment occurs yearly (e.g. Symonides, 1979).

Variation in seedling recruitment patterns may also be related to seedling size. Tremendous variation exists in the size and shape of seedlings (Chapter 2). To a great extent, this reflects variation in seed size, which is over 10 orders of magnitude across the plant kingdom (Harper *et al.*, 1970), and often 4 to 5 orders of magnitude within plant communities (Leishman *et al.*, 1995, 2000). How seed size variation mediates seedling recruitment is illustrated by the following two examples, selected from the tails of the seed size distribution.

The smallest seeds, often termed dust seeds, occur in a few families of plants, most notably in the Orchidaceae. These seeds may weigh less than 1 μg and contain only a few cells (4–200) (McKendrick *et al.*, 2000a; Chapter 4). For their germination and development as seedlings, they are completely dependent on being colonized by fungi, suggesting that extremely specific conditions are required for successful seedling recruitment. Indeed, dust seeds are well suited to locate even very specific sites for recruitment. They are produced in vast numbers, often in the magnitude of 10^4–10^6 per plant (Arditti & Ghani, 2000), and they are easily transported by wind. For a considerable period of time, in some cases several years, the seedling development takes place belowground. Seedling recruitment in species with dust seeds have previously been virtually impossible to study in the field, but due to an innovative method developed by Rasmussen and Whigham (1993), there are now several studies providing remarkable insights. Some of the best examples concern mycoheterotrophic species (Leake, 1994), that is, species of the Orchidaeceae and Ericaceae lacking photosynthesis (McKendrick *et al.*, 2000a,b; 2002; Leake *et al.*, 2004). The contrast could not be more pronounced than if we compare recruitment of the tiny dust seeds with the double coconut (*Lodoicea maldivica*, Palmae) famous for having the largest seeds among plants, on average 9.8 kg (Edwards *et al.*, 2002). The seedling has a curious shape, with an extended cotyledonary axis that first buries the embryo at 30–60 cm deep in the soil, and then extends horizontally so that the seedling establishes at a distance of up to 10 m from the location of the seed (Edwards *et al.*, 2002). The seedling of the double coconut is remarkably big, able to expand the first leaf to a height of 1.5 m.

These extremes in seedling recruitment biology illustrate that generalizations may be difficult. There is nothing like a typical seedling. Species-specific idiosyncrasies must always be considered. However, there are some useful approaches for obtaining generalizations for seedling recruitment patterns and their implications for plant population dynamics. In this chapter, we will first review major causes of seedling mortality. We will then place seedling recruitment in the context of recruitment limitation, loosely defined as the degree to which population size is constrained by recruitment. This is a key issue for questions of population and community dynamics in general, as well as for conservation biology and management of habitats and landscapes. Despite progress in the study of plant recruitment over the past several decades, many challenges exist. Seed sowing experiments are easily carried out, but following recruitment often takes several decades. Accordingly, few studies have translated the effects of seedling recruitment to population growth. We will present some of these studies and provide a framework for advancing studies in this relatively unexplored area of plant recruitment ecology. Finally, we will extend the discussion of recruitment to selection processes in seedling populations.

11.2 | The causes of seedling mortality

Much effort has been expended assessing the causes of seedling mortality. A recent review concluded that across species the most common seedling mortality causes were herbivory (38%), drought (35%), and fungal attack (20%) (Moles & Westoby, 2004c). Less frequent were physical damage (4.6%), seedling–seedling competition (1.6%), and competition from established vegetation (1.3%).

Herbivory is often fatal. For example, slugs are responsible for considerable mortality of seedlings of the forest herb *Lathyrus vernus* (Leguminosae) (Fig. 11.1) (Ehrlen, 2003). In contrast, in grasslands, rodents generally killed seedlings, but molluscs often did not, despite removing approximately 75% of seedling tissue (Hulme, 1994). In fact, non-lethal seedling herbivory is probably quite common (e.g. Hoshizaki *et al.*, 1997; Green & Juniper, 2004a). Armstrong and Westoby (1993) found that experimental removal of 95% of photosynthetic tissue in seedlings still left more than 40% survivors in 37 of 44 examined species. However, there may be sublethal effects, with adults suffering from reduced growth and reproduction if subjected to herbivores as seedlings (Hanley & May, 2006).

It is likely that the impact of herbivores is related to the size of seedlings. A large seedling size may increase the risk of attack from

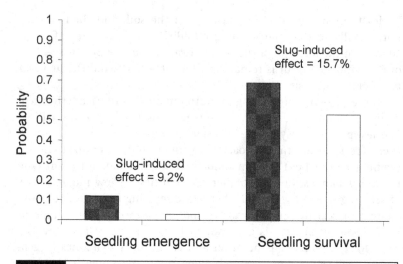

Fig. 11.1 The effect of slug exclusion (mostly *Arion fasciatus*) on seedling emergence and seedling survival in the forest herb *Lathyrus vernus*. Shaded bars represent plots with slug removal and open bars represent control plots. Based on Ehrlén (2003).

herbivores, but also may increase the chance of surviving an attack. It is, therefore, not clear whether increasing seed and seedling size reduces or increases the risk of mortality. Available evidence seems to suggest that larger-seeded species generally experience a lower mortality from herbivory (Moles & Westoby, 2004a), possibly because seedlings developing from larger seeds have a generally higher survival after defoliation than smaller-seeded species (Armstrong & Westoby, 1993).

Chemical defense is another seedling trait directly related to seedling herbivory. Studies of Australian Proteaceae species (Hanley & Lamont, 2001, 2002; Rafferty *et al.*, 2005) provide evidence that high leaf phenolic content deters herbivores. Interestingly, leaf phenolic content was negatively correlated with spininess, suggesting that there may be a trade-off between chemical defense and protection by morphological means. Moreover, leaf phenolic content was positively correlated with specific leaf area (SLA) and relative growth rate (RGR). Because it is likely that fast-growing seedlings are particularly attractive to herbivores (Dalling & Hubbell, 2002), seedling herbivory may promote evolution of chemical defense. Many herbivores are known to be very selective in their choice of food, and several studies suggest that selective grazing on seedlings affects community composition (Hanley, 1998, 2004; Howe *et al.*, 2002; Burt-Smith *et al.*, 2003; Buschmann *et al.*, 2005).

Interactions with pathogens (e.g. fungi) are similar to interactions with herbivores in that the killing agent is biotic and consumes the seedling. Another similarity with herbivores is that chemical defense seems to be the most appropriate strategy for seedlings to avoid damage by pathogens. Interactions between seedlings and pathogens were highlighted by the seminal tests of the Janzen–Connell model for maintaining tree species diversity in the tropics (e.g. Augspurger, 1983, 1984b; Clark & Clark, 1984). These suggested that recruitment was promoted some distance from parent trees due to a decreasing risk of infection. A similar distance-related effect of adults on seedlings has been found in the temperate tree *Prunus serotina* (Rosaceae), where soil pathogens close to adults had a considerable negative impact on recruitment (Packer & Clay, 2000, 2003).

Drought was the second most common mortality cause for seedlings (e.g. Mack & Pyke, 1984; Maun, 1994; Moles & Westoby, 2004c). More generally, it falls into a class of causes that can be termed resource shortage. Seedlings have limited storage and are naturally very sensitive to the accessibility of resources such as water, light, and nutrients (Fenner & Thompson, 2005). Although there are exceptions (e.g. Metcalfe *et al.*, 1998), larger-seeded species recruit better than smaller-seeded species under conditions where water, nutrients, and light are in short supply (Leishman *et al.*, 2000; Moles & Westoby, 2004a).

Several studies have reported complex interspecific differences in responses to resource availability. For example, during recruitment, species responded differently to combinations of shade and drought (Sack & Grubb, 2002), light and water (Battaglia *et al.*, 2000), and combinations of light and N availability (Catovsky *et al.*, 2002), enabling a multitude of niche differentiation among species. In tropical trees and lianas, there is a trade-off between seedling survival and growth in relation to light. Seedlings of light-demanding pioneer species have high growth rate but generally low survival, whereas seedlings of shade-tolerant species are characterized by high survival but slow growth (Wright, 2002; Gilbert *et al.*, 2006). During development from seedlings to juveniles, responses to combinations of resources may shift. Such ontogenetic shifts have been found in, for example, old-field annuals (Parrish & Bazzaz, 1985), ericaceaous shrubs (Eriksson, 2002), and tropical trees (Gilbert *et al.*, 2006).

Although not as important as other causes of seedling death, a common cause is physical damage. Such damage may be due to trampling by animals, but more common is probably death due to falling litter (e.g. Clark & Clark, 1991) or failure of seedlings to emerge through the litter layer (e.g. Dalling & Hubbell, 2002). Litter has, however, more complex effects on recruitment than just preventing seedling emergence (Facelli & Pickett, 1991a; Xiong & Nilsson, 1999; Suding & Goldberg, 1999), ranging from negative to positive, depending on context. Recent studies even suggest that litter can promote seedling recruitment (Quested & Eriksson, 2006). Recruitment of forest plants may be strongly promoted by germination on woody debris, for example, ericaceaous shrubs in boreal forests (Eriksson & Fröborg, 1996), orchids in deciduous forests (Rasmussen & Whigham, 1998b), and trees in *Nothofagus* (Fagaceae) forests (Heinemann & Kitzberger, 2006).

Due to a strong focus in ecological theory on negative species interactions, including competition, herbivory, and predation, facilitation, that is, positive interactions between species, has previously been largely overlooked in recruitment studies (Callaway, 1995). However, it has long been known that interactions with microbes during recruitment are essential for plants dependent on mycorrhiza (McKendrick et al., 2000a,b, 2002) and for legumes dependent on *Rhizobium* infection (Valladares et al., 2002; Chapter 9). There are also numerous examples of how nurse plants promote seedling recruitment (Callaway, 1995). Although positive interactions undoubtedly are more common in harsh environments (Callaway et al., 2002; Chapter 15), recruitment facilitation seems to occur in most vegetation types. The primary mechanisms are amelioration of adverse conditions, resource modification, substrate modification, and protection from herbivores (Callaway, 1995). Shading from nurse plants ameliorated effects of high salinity, thereby promoting seedling recruitment in marsh plants (Bertness & Yeh, 1994). Several species of forbs were strongly favored by neighboring plants during recruitment in limestone grassland, probably due to preservation of moisture (Ryser, 1993). The nurse plant *Honckenya peploides* (Caryophyllaceae) facilitated recruitment of *Leymus mollis* (Poaceae), a key species in dune succession, probably due to substrate modification (Gagne & Houle, 2001). Recruitment of *Taxus baccata* (Taxaceae) depended on protection from herbivores by shrubs such as *Ilex aquifolium* (Aquifoliaceae) (Garcia & Obeso, 2003). This local effect translated to the landscape level produced a distributional association between *Taxus baccata* and its nurse plants. In *Quercus* (Fagaceae) woodlands, recruitment facilitation contributes to maintainance of woodland borders (Weltzin & McPherson, 1999) and development of tree patches (Callaway & Davis, 1998). The invasive shrub *Pyracantha angustifolia* (Rosaceae) promotes recruitment of other invading species, thus enhancing further species invasion (Tecco et al., 2006). Importantly, facilitation has practical implications for conservation. The use of nurse plants seems a promising way to promote restoration of vegetation, especially in arid and semiarid environments (Padilla & Pugnaire, 2006).

Even if seedlings escape mortality factors such as herbivory and pathogens, they may still die before reaching the juvenile stage because of competition from other recruits in the vicinity (Harper, 1977). Specifically, intraspecific density may function as a sieve but the mortality agents acting before this sieve may not affect population growth. Note, however, that these factors may still be important selective agents, determining which seedlings in a cohort can ultimately pass the density sieve. Thus, it is necessary to go beyond listing mortality factors when considering plant population dynamics or plant community composition. For example, what are the implications to a species of a 78% seedling loss from herbivory? Does this imply anything except that most seedlings became food for herbivores? Potentially, such a high mortality rate may affect population size and distribution of the affected species and it may affect community structure, such as species coexistence. However, for such effects to take place, it is necessary that populations are limited by recruitment.

11.3 | Recruitment limitation

Seed versus microsite limitation

Microsites, used interchangeably with safe sites (e.g. Harper, 1965; Grubb, 1977), are small gaps in vegetation where conditions for successful recruitment are present. For many plants, some kind of gap in the vegetation is necessary for seedling recruitment (Bullock, 2000). Considering the gap as competitor-free space, that is, small openings in the vegetation where the composite effects of competition are reduced (Bullock, 2000), suggests that plant–plant interactions are a dominant cause of recruitment failure rather than other potential causes, such as herbivory or drought. However, although there are examples of both intraspecific (e.g. Taylor & Aarssen, 1989) and combined intra- and interspecific (e.g. Harms et al., 2000; Silvertown & Bullock, 2003) density effects on seedling recruitment, other causes of mortality are generally more common. As described previously, there are a plethora of factors that affect recruitment, and a focus just on competitor-free space is likely to be too simplistic. The quality of microsites may be a function of the presence of competitors, but other factors may also be important. Light, herbivory, pathogens, litter, and presence of nurse plants also contribute to determine the quality of microsites for recruitment.

Recruitment may be limited by seed availability, that is, seeds fail to reach sites suitable for recruitment, or by the availability of microsites. Although Harper (1977) believed that microsite limitation was most common in seedling recruitment, a number of studies now suggest that seed limitation is common (e.g. Eriksson & Ehrlén, 1992; Tilman, 1994, 1997). Seed limitation was found in 50% of seed sowing studies performed within local populations and in 53% of the studies in which seeds were introduced at sites where the sown species did not occur at the time of sowing (Turnbull et al., 2000). Thus, seed limitation is likely to affect both local dynamics of populations and the distribution of populations across landscapes. It is therefore useful to distinguish between seed limitation operating at the local scale, that is, the area covered by the habitat of the target species, and the regional scale, that is, incorporating a set of habitat sites of the target species, embedded in a matrix of less suitable habitats (Münzbergová & Herben, 2005).

Most studies have used seed limitation to mean that an experimental increase in seed density leads to increased seedling recruitment (Turnbull et al., 2000). Muller-Landau et al. (2002) distinguished among a limited production of seeds, source limitation, and a limited ability to disperse, dispersal limitation. Furthermore, they remarked that it is necessary to define the life cycle stage actually limited by seed availability. Most sowing studies end their observations at the seedling stage or at an early juvenile stage (Turnbull et al., 2000). However, many plants take several years before reaching adulthood

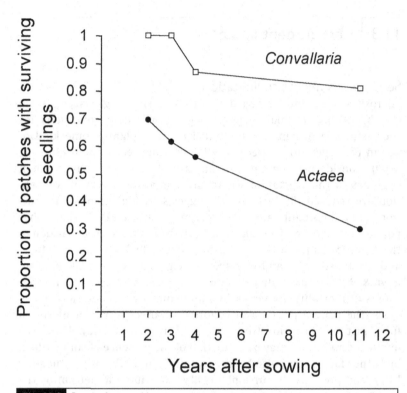

Fig. 11.2 Results from an 11-year sowing experiment concerning recruitment in the two forest herbs *Actaea spicata* (Ranunculaceae) and *Convallaria majalis* (Convallariaceae). The graphs show the fraction of initially unoccupied sites with surviving recruits. Based on Ehrlén *et al.* (2006).

and it is far from clear whether increased seedling recruitment may translate into an increased number of adults.

In an effort to determine seed limitations at life stages beyond juveniles, Ehrlén *et al.* (2006) followed recruitment of six forest herbs for 11 years after experimental seed sowing (Fig. 11.2). The abundance and distribution of the forest herbs were limited by seed availability. This pattern was qualitatively consistent over the study period, although the number of initially unoccupied patches with successful recruitment decreased with time. Across species an average of 57% of the unoccupied patches had surviving experimental populations after 3 years, but only 35% of sites remained occupied after 11 years. A similar trend of recruitment success was found in initially occupied patches, suggesting that the decrease in recruitment success over time was not solely attributable to habitat suitability. The results also suggested that identification of suitable but unoccupied habitats, based on environmental factors, including soil nutrients and vegetation composition, may be difficult.

To assess the importance of seed limitation, experimental design must address several factors (Ehrlén *et al.*, 2006). First, it is important that experiments last for a sufficiently long period of time to encompass the bottlenecks associated with establishment and growth to reproductive adult. Second, because the probability of seedling establishment varies considerably between sites and years, it is important to carry out spatially and temporally replicated sowing experiments that include experimental sowings at reference sites with adult plants and non-sown controls to estimate recruitment from natural seed deposits. Third, demographic stochasticity can affect establishment and, thus, there is a high risk of failure even at suitable sites if sowing densities are too low.

Implications of seed limitation for maintenance of diversity

Seed limitation on a local or regional scale implies that species are not present at all sites where they potentially can grow, and that populations do not fully exploit their resource base. Because recruitment is limited by seed availability (Turnbull et al., 2000), seed production (source limitation; Muller-Landau et al., 2002) and/or dispersal are key processes in the large-scale dynamics of plants. If most species do not fully exploit their resources locally and regionally, then interactions between species due to species traits, such as interspecific competition, are weaker than would be the case without seed limitation. Accordingly, seed limitation is a potentially powerful mechanism promoting coexistence between species and trait diversity in plant communities (Hurtt & Pacala, 1995).

If competitively superior species are more seed limited, competitively inferior species have the opportunity to recruit at sites where the competition for space is lower than if strong competitors were present. For example, Hubbell et al. (1999) found that seed limitation in a tropical forest acted to decouple gap disturbance regime from the control of species richness. Species present in the forest are to some extent dependent on gaps. However, because not all species are present in all gaps, many species that otherwise would be inferior competitors, may be successful, winning by forfeit (Hurtt & Pacala, 1995). Thus, seed limitation contributes to the decoupling of recruitment from competitive ability. This mechanism is the basis for the competition/colonization trade-off, which hypothesizes that coexistence is promoted when competitively inferior species compensate by being better colonizers (Tilman, 1994, 1997). Coexistence in a guild of fugitive prairie plants provides a nice example of this mechanism (Platt & Weis, 1977). If certain life history traits are favored under conditions where recruitment is limited by seed availability, then trait variation may be maintained just as seed limitation promotes species coexistence. Although seed limitation seems to be more common in early successional species (Turnbull et al., 2000), it has been documented in trees, herbs, and graminoids and there is no strong evidence suggesting that the degree of seed limitation differs among these life forms.

The seed size/seed number trade-off may also be involved in competition for successful microsites. Due to this trade-off, smaller seeds would compensate for a competitive disadvantage by reaching more microsites, whereas seedlings from larger seeds would win in direct contest with seedlings from smaller seeds (Rees & Westoby, 1997). If these conditions occur, maintenance of seed size variation within communities would be promoted. Studies aimed at examining this hypothesis have yielded equivocal results. Guo *et al.* (2000) found that possession of smaller seeds improved species distribution, whereas Leishman (2001) and Jakobsson *et al.* (2006) did not find strong evidence that smaller-seeded species had any dispersal advantage, at least not on a local scale. Other studies support the notion that recruitment depends on the presence of seedlings of other species, that is, frequency-dependent seedling mortality (Turnbull *et al.*, 1999; Silvertown & Bullock, 2003). In contrast, Eriksson (2005) found that presence of seedlings from larger-seeded species did not affect recruitment in smaller-seeded species. Seedling–seedling competition does not seem to be a common mortality cause for seedlings (Moles & Westoby, 2004c), although species with smaller seeds are likely to be more dependent on gaps (Jakobsson & Eriksson, 2000), suggesting that seed size broadens the space available for recruitment. If increasing seed size makes species less restricted to microsite conditions, but at the same time, because of the seed size/number trade-off, makes them less likely to reach the microsites, we would predict a positive relationship between seed size and seed limitation. Available evidence supports this prediction (Ehrlén & Eriksson, 2000; Moles & Westoby, 2002; Muller-Landau *et al.*, 2002; Chapter 10).

11.4 | Seedling recruitment and population dynamics

The contribution of recruitment to population growth

Population dynamics is the net outcome of different vital rates, including survival, growth, and reproduction. The influence on population dynamics of a factor that affects a vital rate, such as recruitment, depends both on how much the vital rate is changed and how sensitive the population growth rate is to this change. Vital rates vary among individuals depending on developmental stage or size. As a consequence, the effects of changing recruitment on population growth rate depends both on the magnitude of the actual change and on the pattern of vital rate sensitivities, both of which may vary among populations.

For long-lived plants with complex life cycles, the importance of different vital rates, including survival, growth, and recruitment, are not easily computed. A useful approach to assess the effects of recruitment on population growth is to use transition matrix models. For such a model, individuals are assigned to one of several different stages. Population growth rate can then be calculated by a transition matrix model of the form $n(t+1) = A \times n(t)$ (Caswell, 2001). The matrix A describes how individuals of each stage class in the vector $n(t)$ contribute to the stage classes in $n(t+1)$. Then, to examine how sensitive population growth rate is to changes in various life cycle transitions, computation of elasticity values are useful. Elasticity measures the proportional changes in population growth rate resulting from a proportional infinitesimal change in a matrix transition (de Kroon *et al.*, 1986, 2000). Elasticity can be interpreted as the relative contribution of a stage transition to population growth rate and used to identify those life history stages that contribute most to fitness. Moreover, the total variation in population growth rate among populations or years can be decomposed into contributions from the variation in each of the life cycle transitions by using life table response experiment analysis (LTRE) (Horvitz *et al.*, 1997; Caswell, 2001). The results of such an analysis show the influence of population, year, or treatment on population growth rate through effects on each respective transition in the life cycle.

Comparative studies using transition matrix models have shown that the relative importance of recruitment, defined as the summed contributions of seed production, seed survival, germination, and seedling survival, differs considerably among species (Silvertown *et al.*, 1993, 1996; Franco & Silvertown, 2004). In an examination of 84 species, Silvertown *et al.* (1996) found that the importance of recruitment varied systematically, decreasing in order of importance from semelparous herbs, iteroparous herbs of open habitats, forest herbs, shrubs to trees. Assuming that these groups represent a gradient of increasing life span, the results are consistent with the notion that the importance of recruitment is inversely proportional to life span.

The importance of recruitment in two long-lived herbs

The matrix model is useful because it identifies the potential impact of factors influencing recruitment. From an evolutionary perspective, it also indicates the strength of selection acting during the recruitment phase, compared to factors acting during other stages of the life cycle. It does not indicate, however, if recruitment is limiting long-term population growth or if changes in recruitment will lead to changes in equilibrium population sizes. Whether an increase in seed production or the number of recruited seedlings will translate to an increase in population size depends on whether they lead to compensating increases in mortality during later stages of the life cycle.

Two examples illustrate the use of transition matrix models combined with experiments on recruitment. The long-lived herb *Lathyrus vernus* (Leguminosae), which inhabits deciduous or rich coniferous forests, has been the subject of detailed population studies in forests in southeastern Sweden (see Ehrlén, 2002). This species lacks specialized organs for vegetative spread and survival rates of large flowering individuals are usually over 95%. The average life span of individuals that reach maturity has been estimated to be 58 years (Ehrlén & Lehtilä, 2002). Flowering individuals produce relatively few seeds, usually 5–25, that are dispersed up to 4–5 meters when the pods dehisce explosively.

On average, 9.0% of seeds emerged as seedlings and their 1-year survival rates were on average 79.3%, 82.3%, and 87.9%, respectively, in three consecutive years (Ehrlén & Eriksson, 1996). The estimated contribution of recruitment to population growth, defined as the summed values of elasticities for seed production, seed survival, germination, and seedling survival, was on average 6.2%. Transitions involving the survival and growth of flowering and large vegetative individuals had the largest influence on population growth rate. The value for recruitment elasticities in *Lathyrus vernus* was considerably lower than the average for 43 herbs (21.8%) and also lower than for woody species (9.3%) (Silvertown *et al.*, 1993). However, the contribution of recruitment to population growth varied considerably between populations and years, from 0 to 20.4%. The summed elasticities for fecundity and recruitment were positively correlated to population growth ($r = 0.86$, $P < 0.001$, $N = 32$ patch-year combinations), indicating that recruitment is relatively more important in populations that are increasing while it plays a minor role in decreasing populations. Similarly, Silvertown *et al.* (1996) found that the importance of elasticities associated with recruitment increased with increased population growth rates for *Cirsium vulgare* (Asteraceae) and *Pedicularis furbishiae* (Scrophulariaceae).

In *Lathyrus vernus*, experimental addition of seeds increased both the number of plots with recruitment and the number of seedlings within plots (Ehrlén & Eriksson, 1996, 2000; Ehrlén *et al.*, 2006). The probabilities of seedling emergence and seedling survival were not correlated with seed or seedling density, and germination rates and seedling survival did not decrease with increased densities of established plants (Ehrlén & Eriksson, 1996). Thus, there is no compensating density-dependent mortality during germination or seedling survival. Moreover, seed sowing plots established in 1988–1990 still had higher densities of individuals than control plots 15 years after the sowings (J. Ehrlén, unpubl. data). This strongly suggests that population growth rates and population sizes of *Lathyrus vernus* at current densities are limited by availability of seeds and recruitment.

In a study of *Primula veris* (Primulaceae), a long-lived grassland herb with an average life span of individuals that reach maturity estimated to be 52 years (Ehrlén & Lehtilä, 2002), demography differed considerably between habitat types representing different degrees of canopy closure (Lehtilä *et al.*, 2006). Most study populations had positive growth and only populations in late phases of forest succession consistently showed negative trends. The populations in open habitats had the highest seedling recruitment, whereas the populations of early and middle forest succession had the highest seed production. Elasticity of survival of large individuals increased with increasing canopy closure. In contrast, elasticity of recruitment was lowest in closed habitats. Overall, elasticity analysis suggested that survival of the largest individuals was most important to population growth. However, the life table response experiment analysis (LTRE) showed that survival of the largest individuals contributed little to differences in population growth rates of different habitats, while recruitment had larger effects. The authors concluded that restoration of recruitment is required for positive population development in late-successional populations of *Primula veris*, although the elasticities of recruitment transitions are low.

Another study with *Primula veris* by Ehrlén *et al.* (2005) applied four management treatments to abandoned grasslands and recorded demographic response. Fitness components of established individuals were not affected by experimental treatments, but seedling emergence and juvenile growth increased significantly in plots with disturbance treatment. Likewise, seedling recruitment had higher elasticity in disturbed plots. LTRE analysis showed that a large increase in population growth rate after disturbance treatments was primarily due to increased growth of the smallest rosettes and increased seedling recruitment. These results demonstrate that examination of existing demographic variation, using LTRE analyses, is an important and necessary complement to methods focusing on the average contribution of vital rates to population growth, such as elasticities, in identifying management targets. *Primula veris* sowing experiments demonstrated that seedling emergence was density dependent but the effects were undercompensating (Ehrlén *et al.*, 2005). The proportion of seeds emerging as seedlings the first year after sowing was significantly higher at low sowing densities. Although a smaller proportion of seeds emerged at high sowing densities, the number of seedlings was still higher. There were no effects of sowing density on the proportion of seedlings that survived to their second year or on their size, suggesting that higher mortality during later stages of the life cycle may not further compensate higher seed densities.

These two examples show that recruitment in long-lived plants may have a rather small direct impact on population growth rate, compared with growth and survival of adults plants, particularly in populations that are not increasing. Still, the long-term effect of seedling recruitment may be considerable and decisive for population persistence. Density effects on recruitment are likely to be relatively minor, as judged from the studies on *Lathyrus vernus* and *Primula veris*. The examples show that several tools are needed to assess recruitment limitation and its effects of population dynamics, including sowing experiments and long-term records of recruitment and population development, followed by elasticity and LTRE analyses.

11.5 | Genetic structure and selection in seedling populations

Irrespective of whether variation in seedling recruitment has an impact on population growth rate, mortality during the seedling stage may be selective in the sense that seedlings with certain characteristics experience a survivorship advantage. Because genetic variation in seedling traits affects the performance and survival of the seedlings, it is a fundamental basis for selection on seedling traits. If these traits are correlated with traits expressed later during development, selection during the seedling stage may also mediate changes in the distribution of adult traits in the population. In a study of genetic structure of *Pinus banksiana* (Pinaceae), Saenz-Romero and Guries (2002) found that seedling size was, among a suite of traits, differentiated at the landscape scale in relation to soil conditions. For the dioecious saltmarsh grass *Distichlis spicata* (Poaceae), sex-dependent seedling survival differed between female- and male-dominated populations (Eppley, 2001). In female-biased populations that dominated at lower elevation in the marsh, female seedlings had higher survivorship than male seedlings when subjected to high tide. Thus, variation in population sex ratio was probably mediated by differential seedling survivorship.

The effects of genetic variation and selection on seedling traits have been the focus of study especially in agriculture and forestry and are summarized in Table 11.1. These studies provide substantial evidence for additive genetic variation in many seedling traits related to establishment capacity in general (e.g. Biere, 1991; Casler & Undersander, 2006; Loha *et al.*, 2006) and under specific conditions such as seedling emergence in hard and encrusted soils (Rebetzke *et al.*, 2004) and survivorship when seedlings are subjected to drought (Verhoeven *et al.*, 2004), waterlogging (Marcar *et al.*, 2002), and herbivory (Zas *et al.*, 2005).

Despite the many studies of genetic variation in seedling traits and performance, surprisingly few have focused on selection for seedling traits. This may be understandable for traits that are strongly canalized such as epigeal vs. hypogeal germination. For traits that are more variable, such as cotyledon shape and size, and developmental phenology of roots and leaves, combining studies of additive genetic variation with assessments of seedling survivorship under a range of conditions would enlighten our understanding of seedling evolution. Because seedling features are likely to be constrained by seed size, it is probable that such studies would further add to our understanding of seed size evolution.

The prospects for developing adaptations to local recruitment conditions depend not only on additive genetic variation in seedling traits, but also on the number of sessions during which selection acts on seedlings at a site. Populations where seedling recruitment occurs repeatedly are more likely to develop local adaptations. In contrast, species where seedling recruitment largely occurs when the population is founded are only subject to one session of selection for seedling traits at that particular site. The latter pattern may occur in fugitive annuals that exploit spatially unpredictable, temporally brief windows of opportunity for available resources, but occurs also in many clonal plants where seedling recruitment generally takes place when populations are founded. Eriksson (1989) concluded that seedling recruitment within existing populations of clonal plants had

Table 11.1 | Examples of seedling features, presumed adaptive under certain conditions, for which genetic variation has been shown

Seedling trait	Species	Family	Reference
Seedling size and vigor (establishment capacity)	*Cordia africana*	Boraginaceae	Loha *et al.* (2006)
	Lactuca sativa	Asteraceae	Argyris *et al.* (2005)
	Pithecellobium pedicellare	Mimosaceae	Kang *et al.* (1992)
	Lychnis flos-cuculi	Caryophyllaceae	Biere (1991)
	Pinus banksiana	Pinaceae	Saenz-Romero & Guries (2002)
	Phalaris arundinacea	Poaceae	Casler & Undersander (2006)
	Sorghum bicolor	Poaceae	Cisse & Ejeta (2003)
Developmental plasticity	*Picea omorika*	Pinaceae	Tucic *et al.* (2005)
Susceptibility to herbivory	*Pinus pinaster*	Pinaceae	Zas *et al.* (2005)
Resprouting ability	*Eucalyptus obliqua*	Myrtaceae	Walters *et al.* (2005)
Primary root growth	*Triticum aestivum*	Poaceae	Carnago & Ferreira (2005)
Soil penetration ability of roots	*Triticum aestivum*	Poaceae	Rebetzke *et al.* (2004); Kubo *et al.* (2006)
Allelopathic activity	*Triticum aestivum*	Poaceae	Wu *et al.* (2000)
Cold tolerance	*Oryza sativa*	Poaceae	Z. H. Zhang *et al.* (2005)
Zn toxicity tolerance	*Oryza sativa*	Poaceae	Dong *et al.* (2006)
Desiccation tolerance	*Hordeum spontaneum*	Poaceae	Verhoeven *et al.* (2004)
Response to waterlogging	*Eucalyptus globulus*	Myrtaceae	Marcar *et al.* (2002)

been observed in only 40% of the species for which this kind of data existed. However, because seedling recruitment in long-lived species may be very difficult to observe, we suspect that seedling recruitment in clonal plants generally has been underestimated. Indeed, studies using molecular methods have found that seedling recruitment most likely has occurred repeatedly within populations of clonal plants (e.g. Verburg et al., 2000; Auge et al., 2001; Pluess & Stöcklin, 2004). Initial seedling recruitment (Eriksson, 1993) seems to prevail mainly in exceptionally long-lived clonal plants such as *Populus tremuloides* (Salicaceae) (Romme et al., 2005) and *Spartina alterniflora* (Poaceae) (Travis & Hester, 2005), where individual genets are likely to reach ages of more than 1000 years.

11.6 | Concluding remarks

Despite a rich literature on seedling recruitment, there are still many important issues that need further study. While studies of factors limiting recruitment, for example, seed production, dispersal, various mortality agents, and resource shortage, are always needed, we suggest that it is time to reach beyond the mere listing of limiting factors and turn attention to the actual implications for population dynamics and evolution. Because recruitment is by its very nature variable in time and space, both experimental and observational studies should incorporate variation in study sites and time. To be meaningful, sowing studies should preferably follow the recruitment process over the entire life cycle of the study plant. This will necessitate long-term studies of recruitment. Moreover, the evolutionary implications of recruitment variation is rather unexplored. If we were to suggest a top-priority research program in seedling recruitment ecology, we would strongly emphasize work in the interface between analysis of genetic variation in seedling traits and assessment of how recruitment variation in time and space contributes to population growth rate.

11.7 | Acknowledgments

We are grateful to A. Moles, M. Öster, C. Hatfield, and one anonymous reviewer for comments on the manuscript. Financial support was received from the Swedish Research Council and the Swedish Council for Environment, Agricultural Sciences, and Spatial Planning (grants to O.E. and J.E.).

Chapter 12

Seedling communities

Jon E. Keeley and Phillip J. van Mantgem

12.1 | Introduction

This chapter considers the internal and external processes that affect seedling communities. Internal or endogenous drivers include the density dependence of seedling populations, as well as the relationship of parent to offspring and the competitive relationships affecting seedling populations. There are many external or exogenous drivers, but we will focus on climate, predation, and fire. We will integrate these internal and external drivers of community composition to address the questions: To what extent do seedling recruitment strategies relate to community assembly rules and do these rules dictate the potential combinations of regeneration niches to be found in any given community? This chapter will focus on long-lived woody species because the differences in life history stages and factors affecting them appear much more prominent than in some other growth forms.

12.2 | Internal drivers

Communities of regeneration niches

Plant recruitment strategies have received a great deal of attention generally to determine the environmental conditions that favor one strategy over another (Chapter 11). Although communities comprise an assemblage of different seedling strategies, relatively little attention has been paid to the community combinations or rules that limit possible combinations. A useful concept for understanding seedling communities is that of safe sites (Harper, 1977), which is a species-specific phenomenon driven by unique aspects of ecology and phylogeny. It is important to recognize the diversity of potential safe sites and how they are distributed in space and time. Also of importance is how species reach safe sites and the role of metapopulations.

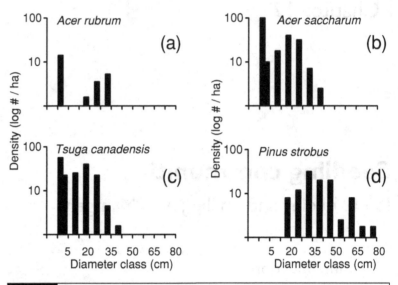

Fig. 12.1 Size structure of dominant tree species in a mid-western North American (Wisconsin, USA) *Acer saccharum* forest. Redrawn from Goff & Zedler (1968).

Plants differ in three important ways in reaching safe sites, which they accomplish by dispersal, seed dormancy, and delayed growth. All three of these may be utilized by the same species, but here we will focus on three functional types that represent modes in a continuum. *Spatial dispersers* reach safe sites that are widely distributed in space although potentially available every year. *Temporal dispersers* have innate or enforced seed dormancy that delays germination until conditions are suitable for generating the appropriate safe sites. These sites may occur annually in the appropriate season for some species or occur periodically following disturbance. *Persistent seedling banks* represent safe sites for seedling recruitment, but require a change in conditions for recruitment into the adult population, something foresters often refer to as *advanced regeneration*. An example in forests would be a pool of seedlings and saplings that opportunistically recruit into the overstory as canopy gaps occur. This is explicit in gap models of forest dynamics (JABOWA and its descendants), where recruitment is wholly dependent on canopy gap availability, size, and arrangement (Bugmann, 2001). Seedling banks that persist in the understory grow slowly, but this provides sufficient advantage for successful recruitment into the canopy when gaps finally occur (e.g. Brown & Whitmore, 1992).

Plant population biology has been concerned with discerning those environments that select for one strategy or another. Here we focus on the seedling community and ask to what extent these strategies coexist and whether there are patterns in the combinations of strategies and their temporal and spatial dispersion.

In North American deciduous forests, seedling communities comprise temporally separated recruitment patterns that illustrate these strategies (Fig. 12.1). *Pinus strobus* (Pinaceae), although a dominant in this forest, lacks seedling recruitment in the understory as it is dependent on disturbances, which apparently have not occurred for some time. Large stand-replacing fires favor recruitment of this species if a nearby seed source exists; however, lighter understory burns may favor recruitment as well (Holla & Knowles, 1988). To recruit, *Pinus strobus* must disperse spatially until a propagule lands on a suitable open site. *Tsuga canadensis* (Pinaceae) and *Acer saccharum* (Aceraceae) recruit continuously, and these are successfully recruited into larger size classes (Hett & Loucks, 1971). *Acer rubrum* exhibits a very different strategy with recruitment occurring continuously, but in the absence of canopy gaps, seedlings fail to develop further. Seedling banks in this species accumulate because shaded understory conditions produce a favorable environment for drought-sensitive seedlings, but recruitment into the canopy requires higher light (Lambers & Clark, 2005). However, not all gaps are suitable for recruitment (Royo & Carson, 2006). Beckage *et al.* (2005) found that dense understory shrubs limit recruitment in small canopy gaps for *Acer rubrum*. They further found that variability in seedling survivorship was seven times greater across time than space, that is, temporal variability was much greater than the effect of light gaps, suggesting the difficulty in precisely defining safe sites for all populations of a species.

Although these seedling/sapling establishment patterns observed for this deciduous forest (Fig. 12.1) imply species-specific differences in recruitment, this is based on the community structure only at a single point in time and misses much complexity. In contrast, if we examine annual patterns of seedling community dynamics over an extended period, we find that recruitment strategies may vary from year to year. Seedling communities for the dominant trees in a mixed-conifer forest in western North America (California, USA) over a period of 8 years illustrates marked annual changes (Fig. 12.2). In the year 2000, the major seedling recruitment was for *Abies concolor* (Pinaceae) and *Calocedrus decurrens* (Cupressaceae), with little contribution from *Pinus ponderosa*. However, if we were to examine this community in 2004, we would see a different picture with substantial recruitment by *P. ponderosa* and relatively few new seedlings of the other two species. These pine seedlings apparently did not persist into older age classes, unlike *A. concolor* and *C. decurrens*. *Quercus kelloggii* (Fagaceae) has modest recruitment each year and these recruits persist well into the older age classes (≥ 3 years).

Other forest types also exhibit spatial and temporal variation in seedling communities (e.g. Houlé, 1992, 1994). For example, based on size structure, *Acer saccharinum* in midwestern North American forests (Wisconsin, USA) appears to recruit successfully into the canopy of undisturbed forests (Fig. 12.1) (Goff & Zedler, 1968), yet in eastern North American forests, this species maintains seedling banks dependent on canopy gaps (Marks & Gardescu, 1998). Likewise, *Tsuga* species recruit in the forest understory as suggested by Fig. 12.1, but may also exhibit increased recruitment around the edges of gaps not exposed to direct solar radiation (Gray & Spies, 1996). *Tsuga canadensis* and

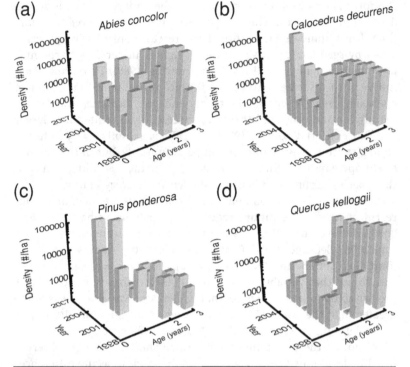

Fig. 12.2 Seedling populations for 1-, 2-, and ≥3-year-old seedlings of the dominant trees in a western North American mixed conifer forest in the southern Sierra Nevada (California, USA). Keeley, van Mantgem, & Stephenson, unpubl. data.

Acer saccharum with similar temporal patterns of seedling recruitment also demonstrate similar microhabitat preferences (Goff & Zedler, 1968).

Within communities seedling recruitment often exhibits marked spatial variation in response to microhabitat gradients. For example, in the oak–pine forest of the New Jersey Pine Barrens (USA), first-year seedling recruitment by six dominant species exhibited significant spatial variation in their preferences for litter depth, moss cover, lichen cover, percentage canopy cover, and nearest neighbor distance (Collins & Good, 1987). Also, as resources change along gradients, competition may change and alter the favorability of sites for seedling establishment (Boerner & Brinkman, 1996; Davis *et al.*, 1998; Catovsky & Bazzaz, 2002). In communities where there is marked temporal and spatial variation in patch favorableness, it has been suggested that this hinders populations from reaching a stable spatial distribution and a stable age distribution with a fixed schedule of recruitment and mortality within a patch (Fowler, 1988).

In many forest types, fire has a profound effect on seedling recruitment and the timing of such events can have important impacts on species that commonly have periodic mast years of seed production. In mixed conifer forests of the southern Sierra Nevada (California, USA), *Abies concolor* typically produces a mast year every 2–3 years (Fowells,

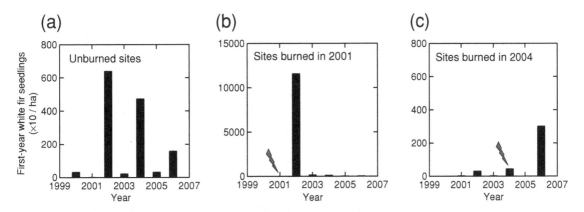

Fig. 12.3 First year *Abies concolor* seedlings in (a) unburned sites, n = 11 1- or 2-ha sites with average density of 376 trees per ha; (b) sites burned in 2001, 1 year prior to a mast year, n = 3 sites with average density of 314 trees per ha; and (c) sites burned in 2004, 2 years prior to a mast year, n = 2 sites with average density of 61 trees per ha. Sites were in *Abies concolor* dominated forests in Sequoia National Park (California, USA). Arrow represents fire year. Keeley, Stephenson, & van Mantgem, unpubl. data.

1965) and historically these forests have been prone to surface fires every decade or two (Caprio & Lineback, 2002). The timing of mast year events relative to fire events is likely to affect seedling recruitment and there is some evidence of this effect in Fig. 12.3. Burning followed immediately by a mast year of seed production (Fig. 12.3b) provides more resources for seedling recruitment than a mast year delayed two years after burning (Fig. 12.3c). In the former case, seedling recruitment is many times greater than in the latter case. This appears to be a widespread phenomenon occurring in different forest types and this interaction between fire and masting cycles has been shown to have long-lasting impacts on forest demography in northern (Alberta, Canada) *Picea glauca* forests (Peters *et al.*, 2005).

Internal dynamics

Obviously, persistence of ecological communities requires reproduction; however, there has not been widespread agreement on the role of regeneration dynamics in determining community composition and structure. Influenced by equilibrium concepts of communities, some see regeneration as the mechanism that maintains community composition and structure, and this is ultimately determined by traits of mature plants. In contrast, others, influenced by the growing appreciation for disequilibrium processes, see community composition and structure as determined heavily by regeneration processes.

Some of the differences in these two approaches center on a different evaluation of the importance of competition as a factor in structuring communities (Howard & Goldberg, 2001). Where competition is a key force, it is often most intense in the adult stage and of lesser importance in seedling communities (Poore, 1964). In forests, competition has long been regarded as a primary driver determining community structure and regeneration dynamics has not been generally seen as critical to determining either composition or structure (size and spatial arrangements of stems) of the overstory community (Swaine & Hall, 1988). Instead, regeneration processes are often thought to be dependent on the availability of sites suitable for overstory recruitment, usually canopy gaps, that provide sufficient resources for recruitment into the canopy. This view is supported by the fact that forest gap models are able to recreate forest communities successfully, although many of the mechanisms proposed by these models are difficult to test independently (Prentice et al., 1993; Bugmann, 2001). As long as processes such as regeneration are not well understood, these models may produce the expected forest communities, but for the wrong reasons.

There is an emerging view, however, that regeneration processes, that is, seed production (fecundity), seed dispersal, seed germination, and seedling survival and growth, may limit populations of many, if not most, forest tree species (Veblen, 1986; Clark et al., 1999). Thus, community composition and structure may, in part, be determined by regeneration processes. This is suggested by lowland tropical forest diversity patterns that show seedling diversity is a subset of diversity in the adult community, setting the community on a new trajectory (Comita et al., 2007). A multitude of factors control the regeneration process, but at the seedling stage, competition with other growth forms, such as herbaceous vegetation effects on tree seedlings, often greatly impact recruitment success (Maguire & Forman, 1983; Meiners & Handel, 2000). In some environments, coexistence of competitively similar species may result from differences in timing and success of seedling recruitment and are predicted by lottery models of recruitment (Bond & van Wilgen, 1996). Of particular importance are recruitment limitations that lessen the effect of competitive asymmetries between dominants and this may contribute to coexistence and greater community diversity (Hurtt & Pacala, 1995).

Thus, a critical question in community ecology is, what life history stage is most critical in determining community structure and distribution? Put another way, is selection operating more on the regeneration niche (sensu Grubb, 1977) or the adult niche (Reich, 2000)? This has been studied by examining the relationship of species occurrence along environmental gradients for seedlings, saplings, and adults (Stohlgren et al., 1998; Collins & Carson, 2004; Chapter 10). In some cases, seedling distribution shows a much broader distribution than adults and in other cases it does not (Fig. 12.4). Interpreting these patterns, however, requires consideration of site history and past climates. For example, in the absence of such information, one might interpret that the distribution of Quercus rubra arises from limited tolerance of adult trees and this restricts them to a narrow subset of sites, despite the larger tolerances of seedlings and saplings (Fig. 12.4a). However, past site history of logging or climate changes could account for their present restricted distribution while the wider distribution of seedlings could indicate a new trajectory for the future distribution of this species (Collins & Carson, 2004). The

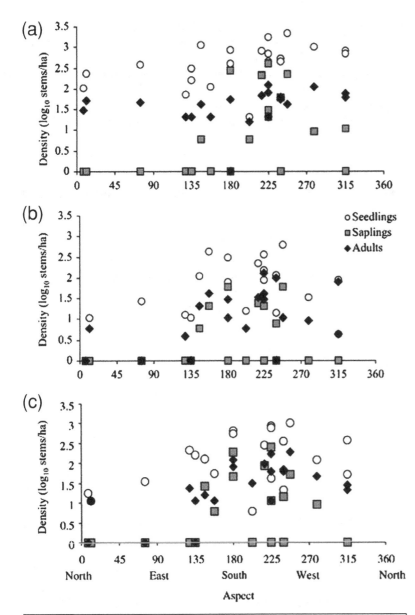

Fig. 12.4 Stem density for seedlings, saplings, and adults across different slope aspects for (a) *Quercus rubra*, (b) *Q. alba*, and (c) *Q. prinus* in an eastern North American (USA) deciduous forest. From Collins & Carson (2004), reprinted with permission from Elsevier.

opposite is possible as well, where adult trees are more widely distributed than seedlings and saplings due to land use changes such as fire exclusion from fire-prone environments, which alters site favorability for recruitment (Parsons & DeBeneditti, 1979; Rooney *et al.*, 2000).

The recruitment strategies of certain chaparral shrubs provide an example of where distributions are more a function of adult traits and seedling recruitment is, thus, largely controlled by adult traits. Species differ in their mechanisms for dealing with drought stress in the semi-arid region of coastal southwestern North America (California, USA). Some species have physiological tolerances to drought and others morphological traits to help avoid drought stress. These mechanisms constrain options available to seedlings and enforce a very narrow range of conditions under which recruitment is successful (see Box 12.1).

Box 12.1 | **Factors determining seedling dynamics in the fire-prone chaparral shrublands**

Several dominant species recruit only on open sites in the first postfire year, whereas others only recruit under the canopy during fire-free periods. In postfire recruiters such as *Adenostoma fasciculatum*, soil-stored seeds are dormant and seedling cohorts occur in a single pulse after fire. *Rhamnus crocea* has transient seed banks leading to recruitment in the understory during fire-free periods and age structure studies show these successfully recruit into the overstory canopy (Keeley, 1992). In others, such as *Quercus berberidifolia*, seed banks are also transient, but seedlings are rare except in the long-term absence of fire (several decades), apparently because of the requirement for shade and deep organic soil. Herbivory of these seedlings or saplings is intense, but they persist in a suppressed form due to resprouting from the base, and successful recruitment into the canopy often requires a fire that provides better growing conditions for the resprouting saplings.

This temporal separation of seedling populations does not appear to be tied to selection for avoidance of competitive interactions in seedling communities, but rather to a suite of traits that are the necessary outcome of differences in how adult populations survive in these semi-arid environments (Keeley, 1998). They are correlated with character syndromes involving physiology, morphology, and other reproductive traits. The postfire recruiting species are capable of tolerating drought with physiological and morphological traits that allow them to persist under extreme water stress. These traits confer on seedlings the ability to tolerate the intense drought stress characteristic of open burned sites and thus the ability to recruit after fire when resource conditions of light and nutrients are highly favorable. Mature shrubs of the fire-free understory recruiters have a much weaker tolerance of water stress, but have deep root systems that can access underground water sources unavailable to the more shallow rooted fire recruiter species. This avoidance strategy works well for established shrubs, but is not an option for their seedlings that require many years' growth to develop root systems sufficient to tolerate conditions on open burned sites. As a consequence, seedling recruitment is restricted to the deep shade under the shrub canopy and to old growth stands with deep litter and duff, which retain moisture longer into the summer drought. In this respect, these shrub species have a similar seedling recruitment pattern to *Tsuga* spp. that recruit in old growth forests and are often dependent on substrates with sufficient decaying wood, which have greater moisture-holding capacity than mineral soil (e.g. Turner & Franz, 1985; Marx & Walters, 2006). Other less known examples of species with seedling recruitment restricted to moist and dense litter microhabitats also have been noted (Williams et al., 1990). Similar seedling avoidance of canopy gaps and restriction to shaded mesic understory sites is seen in some tree species in semi-arid forests of the Mediterranean Basin (Espelta et al., 1995; Gómez-Aparicio et al., 2005).

Correlated with these seedling recruitment modes are very different fruit and seed characteristics. The postfire seeding species are temporal dispersers with weakly developed dispersal and deeply dormant seeds requiring fire cues for germination, a sort of sit and wait strategy. The fire-free understory recruiters have fleshy fruits that are animal dispersed and these transient seed banks fail to accumulate in the soil. These character syndromes also have a biogeographic and phylogenetic component. Fire recruiters appear to represent more restricted taxa that have radiated relatively recently, perhaps in response to the increasing importance of fire in the late Tertiary (Keeley & Rundel, 2005). The fire-free recruiters are widespread in the Northern Hemisphere and appear to be much older lineages that persist in micro-habitats that have been present throughout the late Tertiary (Keeley, 1998).

Thus, we interpret the seedling community dynamics of this shrubland as a reflection of very different strategies for handling water stress and the marked temporal separation in recruitment patterns to be a necessary part of the character syndrome of drought avoiders and drought tolerators. Similarly, one might argue that the Acer saccharum forest community of seedling recruitment patterns is tied to life history trade-offs in shade-tolerance strategies for the adult population (Curtis, 1959), and these appear to be tied to different patterns of seed dispersal and persistence (Houlé, 1991, 1994).

On the other hand, one of the clearest examples where seedling recruitment requirements dictate community composition is in riparian trees. Adult *Salix nigra* (Salicaceae) distribution, for example, is unaffected by factors such as soil moisture, aeration, and nutrients, but seedling recruitment is strictly limited by these factors (McLeod & McPherson, 1973). Thus, it appears that mature trees could survive in a much broader range of environmental conditions than those to which they are restricted because of seedling requirements, and this may generally be the case with riparian species (Sacchi & Price, 1992). Similarly, the distribution of savanna and forest species is apparently constrained by seedling characteristics that differ in tolerance to frequent burning (Hoffmann, 2000). Tropical rain forest communities provide a very compelling case for the importance of the regeneration niche. Across a range of species, the physiological and morphological leaf traits of seedlings, saplings, and adults were most strongly correlated with the crown exposure at the regeneration stage (Poorter, 2007).

Thus in some cases, community composition and structure is determined by characteristics of adult populations and in other cases by characteristics of seedling populations. A third pattern is the potential for adult populations to modify their environment in ways that promote seedling recruitment, a process termed *niche construction* (Odling-Smee et al., 2003). One such example is in mast flowering bamboos, which have the unusual characteristic of dying *en masse* after setting seed. It has been hypothesized that in these high light-demanding forest bamboo species mortality alters the environment by creating a massive fuel load that promotes burning and enhances conditions for seedling recruitment (Keeley & Bond, 1999, 2001; but see Saha & Howe, 2001).

Impacts at larger scales

If recruitment limitation has the potential to influence forest communities, it raises the question of how these processes might affect communities on a broader scale. Specifically, are there patterns in the effective size of recruitment limitations that could give rise to diversity gradients across temperate and tropical forests? The possibility that this might be the case was raised independently by Janzen (1970) and Connell (1971), who argued that density-dependent forces, particularly at the seed and seedling stages, are stronger in tropical than temperate forests, reducing the ability for a particular species to dominate in tropical forests, thereby promoting diversity. Stronger density-dependent effects would arise in tropical forests due to plant enemies, related to the greater abundance of specialist predators and pathogens in the tropics and because adult trees would serve as foci for these enemies. Thus with increasing distance from the parental trees, seeds and seedlings would be further from the enemy population sources and less available to these enemies.

These ideas, known as the Janzen-Connell hypothesis, have been the object of considerable research since they were first proposed. Testing has been concentrated in two areas, finding evidence for density-dependent survival in tropical forests and documenting differences in density-dependent effects in tropical compared to temperate forests. Substantial evidence, both in favor of and against strong density-dependent effects in tropical forests, was recently synthesized in a meta-analysis by Hyatt *et al.* (2003). They found no consistent evidence for density dependence for seeds and weak affirmative evidence for seedlings. The search for density dependence in tropical forests remains an active area of research (e.g. Peters, 2003; Bell *et al.*, 2006; Webb *et al.*, 2006). Comparisons of the effect of density dependence in tropical and temperate forests have received relatively little work, although recent papers have demonstrated the presence of strong density dependence of seedlings in temperate forests (Lambers *et al.*, 2002; Reinhart *et al.*, 2005b). Interestingly, there has been little work investigating the mechanisms driving the Janzen–Connell hypothesis. Moreover, there have been few demonstrations that there is a greater proportion of specialist seed or seedling enemies in tropical than in temperate forests, or that the presence of adult trees modulates visibility to specialist predators (but see Leigh *et al.*, 2004; Freckleton & Lewis, 2006). Furthermore, not all density-dependent effects are negative. For example, oak regeneration may be enhanced by greater mycorrhizal infection when recruiting near rather than far from adult trees (Dickie *et al.*, 2007).

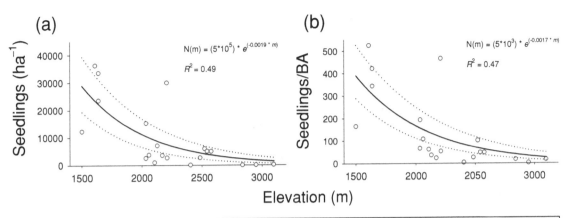

Fig. 12.5 The relationship between elevation and seedling densities across 20 long-term monitoring plots in coniferous forests of the Sierra Nevada (California, USA). (a) Average number of seedlings (ha^{-1}) and (b) average number of seedlings relative to the basal area (m^2 ha^{-1}) of potential parent trees defined as trees within 25 m of the sample plot. The solid line is the nonlinear regression estimate, with the dotted lines showing the 95% confidence intervals calculated from 5000 bootstrapped estimates. N(m) is the number of seedlings at a given elevation and m is the elevation (in meters). From van Mantgem *et al.* (2006), reprinted with permission from Elsevier.

12.3 | External drivers affecting seedling communities

Climate

The distribution and dynamics of forest trees are strongly influenced by climate (Stephenson, 1990, 1998; Stephenson & van Mantgem, 2005). These patterns may arise at least in part by the effects of climate on reproduction, known to be sensitive to climate (Fowells & Stark, 1965; Stohlgren *et al.*, 1998; Collins & Carson, 2004; Gworek *et al.*, 2007). For seedlings, climate may determine patterns of seedling recruitment and growth (Olszyk *et al.*, 1998; Castro *et al.*, 2004a). Climate may interact with other factors, such as fire, and with species-specific recruitment responses that vary in response to different combinations of factors (North *et al.*, 2005). In general, the limited spatial and temporal scale of most seedling studies (Clark *et al.*, 1999) restricts our understanding of how climate may control their dynamics. This is a particularly critical problem in deserts where recruitment may be a rare event (Cody, 2000).

Using a network of long-term seedling monitoring plots, van Mantgem *et al.* (2006) recently showed that seedling densities changed significantly over an elevation (climatic) gradient in the Sierra Nevada Mountains (California, USA) (Fig. 12.5a). The relationship between seedling density and elevation held when potential parent tree basal area was considered (Fig. 12.5b). In this study, *Abies* and *Pinus* species, the two dominant genera in these forests, both showed decreasing seedling densities with elevation, but with a stronger response in *Pinus*. Thus, the seedling response to climate was widespread, but with measurable differences among taxa (Kern, 1996; Green, 2005).

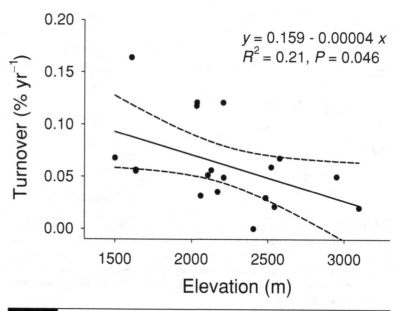

$$y = 0.159 - 0.00004\,x$$
$$R^2 = 0.21,\ P = 0.046$$

Fig. 12.6 The relationship between elevation and turnover rates (the average of mortality and recruitment rates) of tagged seedlings (>10 cm tall) from the same forests as in Fig. 12.5. The solid line is the regression estimate with the dashed lines showing the 95% confidence interval. From van Mantgem et al. (2006), reprinted with permission of Elsevier.

Additionally, van Mantgem *et al.* (2006) found evidence that turnover rates (the average of recruitment and mortality rates) for tagged seedlings more than 10 cm tall declined with elevation (Fig. 12.6). Because climatic data indicate that site potential for primary production may decrease with elevation in the forested zone of the Sierra Nevada (Stephenson, 1988), a negative relationship between seedling turnover and elevation is consistent with broader patterns of forest overstory productivity (Stephenson & van Mantgem, 2005). However, van Mantgem *et al.* (2006) found that not all phases of reproduction were related to climate, with seed production apparently independent of elevation. The response of seedling densities to elevation may be mediated by a greater proportion of seeds that become successfully established at lower elevations (Fig. 12.7). This finding suggested that spatial variation in early phases of recruitment (e.g. proportion of sound seeds, seed predation, germination, and first-year mortality) may strongly influence patterns of seedling community structure.

Herbivores, predators, and pathogens

Herbivores and granivores are known to affect the composition of seedling communities (Guo *et al.*, 1995; Hanley, 1998) and much work has been done to document their effects as a possible mechanism causing density dependence in the context of the Janzen–Connell hypothesis (see above). However, one of the clearest illustrations of predation pressure on seeds and seedlings is the presence of mast flowering behavior in many tree species (Kelly & Sork, 2002). Selection for masting cycles of high and low seed years is widely considered to

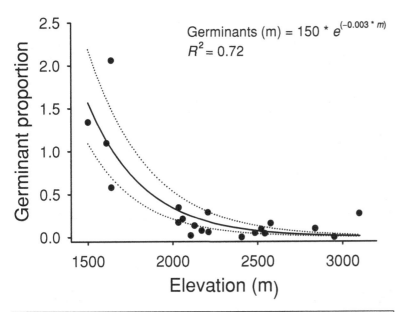

Fig. 12.7 The relationship between elevation and the average proportion of germinants (1st year seedlings at year t+1/seeds at year t) from the same forests as in Fig. 12.5. The solid line is the nonlinear regression estimate, with the dashed lines showing the 95% confidence intervals calculated from 5000 bootstrapped estimates. This pattern may help explain the relationship between elevation and seedling densities (Fig. 12.5). Note that germinant proportions above 1.0 implies the presence of seed banks, but more likely represents measurement error from combining seed trap counts of seed densities versus plot-based seedling counts. Van Mantgem, Keeley, & Stephenson, unpubl. data.

have been driven by seed and seedling predation. The theory argues that a hiatus of one to several years in which seeds and seedlings are in short supply acts to reduce predator populations and mast years satiate predators (Brown & Venable, 1991). One factor working against such a mechanism would be a lack of synchrony with other species within the community. There is some evidence that congeneric oak species in the same community often are synchronized in seed production (Koenig & Knops, 2002). However, more often communities comprise unrelated taxa that are not synchronous in annual seed and seedling production (e.g. Fig. 12.8). This appears to be the case not only in temperate forests but tropical forests as well (De Steven, 1994).

Escape from pathogens is also sometimes considered as a driver of the evolution of masting behavior (Pearson et al., 1994). In tropical forests, pathogens are known to play a major role in seedling mortality because their presence increases near adult trees (Augspurger, 1984b), and is one of the mechanisms favoring the Janzen–Connell model discussed above.

Herbivores affect seedling communities through direct herbivory and indirectly by predation of flowers and seeds. In many cases, seedlings are far more palatable than adults (Fenner et al., 1999). Species replacements along gradients are often interpreted as due to competitive interactions, but seed predation can also be the driver behind species replacements along different resource gradients

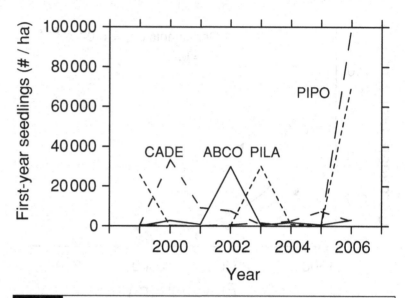

Fig. 12.8 First year seedling populations from 1999 to 2006 for the *Abies concolor* (ABCO), *Calocedrus decurrens* (CADE), *Pinus lambertiana* (PILA), and *P. ponderosa* (PIPO) in a mixed conifer forest in the central Sierra Nevada (California, USA). Keeley, van Mantgem, & Stephenson, unpubl. data.

(Louda, 1989). In addition, density-dependent effects of herbivory on seedling populations can play a significant role in promoting community coexistence (Hulme, 1996). Landscape patterns may also affect the level of impact. For example, small prescribed burns that are surrounded by unburned vegetation may provide patches of rodent habitat in close proximity to patches of seedling recruitment in burn areas, resulting in much more intense herbivory than observed after large wildfires (Bullock, 1991; Keeley, 2000).

Herbivory and granivory need not always be associated with negative effects on seedling communities. For example, in southwestern North American *Pinus ponderosa* forests, frequent fires are a major consumer of tree seedlings, and grazers represent a potential competitor with fire (Bond & Keeley, 2005). When grazing pressure is intense, fuels fail to accumulate sufficiently to carry fire and establishment success of tree seedlings increases.

Fire

The effect of fire on seedling communities must be considered in the context of fire regimes. Different fire regimes have selected for different seedling recruitment patterns that are affected very differently by burning patterns at the community and landscape levels. A quick overview of woody plant regimes is illustrated by comparing four North American plant communities (USA): southeastern long leaf pine, Sierra Nevada mixed conifer, Great Basin pinyon-juniper woodland, and California chaparral.

Long leaf pine ecosystems in the southeastern USA have one of the highest fire frequencies of any ecosystem due in part to the extraordinary density of summer lightning storms (Glizenstein *et al.*, 1995). The dominant tree, *Pinus palustris*, has bark and self-pruning characteristics that make it relatively resistant to frequent fires. The frequency of fires leads to low fuel loads that result in light surface fires and high survival of dominant overstory trees. *Pinus palustris* seedling populations are dependent on fires to clear surface litter and herbaceous competition and on the overstory survival for seed production. There is little evidence that landscape patterns of burning play a significant role in seedling recruitment.

On the other hand, in some western mixed conifer forests, pattern of burning plays a significant role in seedling recruitment. On these sites, productivity is substantially higher and fuel loads heavier and, as a result, fire regimes comprise a mix of low intensity surface fires that burn in the forest understory and high intensity crown fires that create gaps within the community. These gaps favor seedling recruitment (Mutch & Parsons, 1998; Keeley & Stephenson, 2000) and the distance of gaps from parent seed trees is critical to establishing seedling populations for most of the dominant species.

California chaparral shrublands typically burn in high intensity crown fires that often cover thousands of hectares. Dominants in this ecosystem either resprout from the base or have dormant seed banks that establish even-aged cohorts after fires (Box 12.1). As a consequence, landscape patterns of burning are relatively unimportant to their recovery.

Pinyon-juniper woodland characteristically burns at long intervals in large, high intensity crown fires (Wright & Bailey, 1982). None of the dominants, pinyon (*Pinus* spp), juniper (*Juniperus* spp. Cupressaceae), or sagebrush (*Artemisia tridentata* Asteraceae), have much of a dormant seed bank and, thus, the landscape pattern of burning is critical to recovery because seedling recruitment is dependent upon colonization from unburned patches left on the landscape.

12.4 | Seedling community assembly rules

Community assembly rules arise from the concept that if different communities are repeatedly assembled from a common species pool, then there must be rules that constrain community composition. These rules are restrictions based on the presence or abundance of other species and not simply the species-specific response to the environment (Diamond, 1975). The rules are modified by different environmental filters to generate different community composition and different assembly histories may generate multiple stable equilibria (Chase, 2003). Some define assembly rules narrowly as just those patterns due to species interactions (Wilson, 1999). However, because interactions vary as environments change, others consider assembly rules as those patterns resulting from the combination of environmental filters and species interactions (Weiher & Keddy, 1995; Holdaway & Sparrow, 2006).

Because seedlings represent the initial stage of assembly for plant communities, we ask the question, are there assembly rules that determine seedling communities? California (USA) sage scrub provides a useful model system for examining the factors determining seedling community assembly because recruitment is largely centered around fire events and is limited after canopy closure. This semideciduous shrubland is dominated by a variety of growth forms resilient to fires that periodically kill all aboveground biomass. Many herbaceous and woody species maintain deeply dormant seed banks that are fire-stimulated and most woody species are capable of resprouting from underground vegetative parts. In general, seedling communities are restricted to a narrow window of time after fire, but with some variation in the exact timing of recruitment (Table 12.1). For many species, seedling establishment occurs largely as a single pulse in the first growing season after fire. These include *obligate seeding* species that are entirely dependent on this seedling pulse for persistence and *facultative seeding* species that also resprout. However, not all resprouting species have dormant seed banks. In some species, seedling recruitment is dependent on the flowering of resprouts in the first year that produce a transient seed bank, which germinates *en masse* in the second postfire year. Other species produce multiple pulses of seedlings during early succession, as well as exploiting gaps in mature shrublands (e.g. DeSimone & Zedler, 1999). Possible explanations for these different patterns are that the different timing of seedling populations in the sage scrub community is due to selection to avoid competitive interactions, environmental filtering effects, or to phylogenetic constraints.

The sage scrub sites shown in Table 12.1 had a total of 21 woody species in common and there was evidence that these communities were structured due to interactions such as competition. This is indicated by the highly significant co-occurrence pattern. Using EcoSim null model tests (Gotelli & Entsminger, 2006), there was less co-occurrence of all possible pairs of species than predicted if species were distributed randomly ($P < 0.001$). One means of investigating the extent to which resource competition might structure seedling communities is with a trait-based approach that considers seedling recruitment patterns as guilds or functional types, for example, those with a first-year pulse of seedlings, those with second-year pulse, and those with multi-year recruitment (Table 12.1). Species in the same guild might be expected to be similar in their resource use and to compete more with each other than with species from other guilds. EcoSim tests whether the co-occurrence index among these guilds differs from that expected by chance alone. In this analysis, there was a probability of $P = 0.09$ that co-occurrence of guilds is entirely by chance, suggesting only a slight tendency toward structuring of these communities based on combinations of seedling recruitment modes.

However, null models assume all taxa assort at random and this assumption may not hold if there were significant environmental filters at work. In this sage scrub community, there are environmental factors tied to recruitment guilds (Table 12.2). Species with dormant seed banks that recruit largely in the first year after fire comprise a larger proportion of the community on hot interior sites with sandy

Table 12.1 | Woody species seedling recruitment in California (USA) sage scrub communities during the first 5 years after fire (from Keeley et al., 2006). Included are all woody species present at 10 or more of the 50 sites in this study. The chaparral shrub *Adenostoma fasciculatum* (Rosaceae) was a minor component at 11 of these sites, but is not included here because it is not generally considered part of the sage scrub community.

Species	Plant family	Fire response[a]	Resprout (%)[b]	Life form[c]	# sites	Total for years 1–5 (#/ha) x̄	Percentage by year 1	2	3	4	5
First-year seedling pulse:											
Calystegia macrostegia	Convolvulaceae	OS[d]	0	su	34	38 630	92	5	2	0	1
Eriogonum fasciculatum	Polygonaceae	FS	10	ss	32	24 500	71	21	5	0	3
Lotus scoparius	Leguminosae	OS	0	su	43	77 600	72	9	2	1	16
Malacothamnus fasciculatus	Malvaceae	FS	≈15	ss	25	54 400	93	1	5	1	0
Malosma laurina	Anacardiaceae	FS	95	s	27	3 900	90	8	1	1	0
Rhus integrifolia	Anacardiaceae	FS	74	s	13	800	50	20	0	0	30
Ribes spp.	Grossulariaceae	FS	98	ss	11	20 100	85	6	4	0	5
Salvia mellifera	Lamiaceae	FS	14	ss	25	15 900	73	12	3	3	9
Second-year seedling pulse:											
Encelia californica	Asteraceae	OR	83	ss	11	172 100	0	67	15	0	18
Galium angustifolium	Rubiaceae	OR	99	su	23	40 600	4	55	10	17	14
Hazardia squarrosa	Asteraceae	OR	100	ss	22	49 500	0	54	34	6	6
Mimulus aurantiacus	Scrophulariaceae	FS	46	ss	24	142 800	4	60	16	4	16
Mirabilis californica	Nyctaginaceae	OR	≈99	su	19	1 900	0	70	29	0	1

(cont.)

Table 12.1 (cont.)

Species	Plant family	Fire response[a]	Resprout (%)[b]	Life form[c]	Total for years 1–5 (#/ha)		Percentage by year				
					# sites	x̄	1	2	3	4	5
Multi-year recruitment:											
Artemisia californica	Asteraceae	FS	20	ss	41	31 500	41	42	13	1	4
Eriophyllum confertiflorum	Asteraceae	FS	≈99	su	30	24 000	20	33	24	10	13
Galium nuttallii	Rubiaceae	FS	≈99	su	27	27 700	3	32	21	12	32
Lessingia filaginifolia	Asteraceae	OR	≈99	su	10	47 400	0	41	18	12	29
Rhamnus crocea	Rhamnaceae	OR	100	s	14	300	4	0	15	4	77
Salvia apiana	Lamiaceae	FS	63	ss	14	22 900	23	23	42	10	2
Salvia leucophylla	Lamiaceae	FS	87	ss	10	10 000	46	12	31	1	10
Solanum spp.	Solanaceae	FS	≈99	su	23	4 400	40	21	20	4	15

[a]First-year fire response of seeding and vegetative resprouting: OS = obligate seeder, FS = facultative seeder, OR = obligate resprouter.

[b]s = percentage of prefire population resprouting except for those indicated with ≈ that were estimated as a proportion of the 5th-year adult population due to the lack of postfire skeletons.

[c]s = shrub, ss = subshrub, su = suffrutescent, woody at base, li = liana.

[d]This species has the capacity to resprout but failed to survive fire and thus functions as an obligate seeder.

Table 12.2 | Regression analysis, relating the proportion (total 5-year seedling density) for each seedling guild to environmental factors, for the communities in Table 12.1, * $P < 0.05$, ** $P < 0.01$, *** $P < 0.001$, n = 48 sites

			r		
	Coast	Insolation	Sand	Soil P	Fire severity
First-year pulse	+ 0.44 **	+ 0.31 *	+ 0.48 ***	− 0.35 *	+ 0.38 **
Second-year pulse	− 0.51 ***	ns	+ 0.32 *	ns	ns
Multi-year recruit	ns	− 0.36 *	− 0.46 ***	+ 0.42 **	− 0.38 **

soils and low phosphorous content, whereas the multi-year recruiting guild preferred the opposite conditions. In addition, fire severity provides an additional filter by favoring the first-year recruiting guild. These analyses were made at the site level, but it is likely that within site microhabitat differences for some of these factors might also be a force in structuring seedling communities.

In light of the strong effect of distance from the coast on climate, these data were grouped by region (coastal vs. interior) rather than by guild and in this analysis with EcoSim there was a significant ($P < 0.001$) departure from randomness, indicating that co-occurrence structure was significantly different in the two regions. Analyzing these regions separately revealed that at least for the interior sites, there was a significant ($P = 0.018$) chance that the combinations of different recruitment patterns across sites was not the result of the random distribution of taxa, rather there were consistent ratios. It would appear that these seedling communities are the result of both environmental filters that control different seedling recruitment guilds as well as interactions between different guilds that select for particular combinations of guilds.

12.5 | Conclusions

Seedling communities are structured by multiple factors that vary with the vegetation type and landscape position. Here we have focused on woody plant seedlings because the life stages are more distinct than in fast-growing herbaceous species. However, it seems likely that much of this discussion is applicable to herbaceous life-forms as well. The relative contribution of seedling dynamics to later community structure and function can be substantial. In some communities, this regeneration niche seems to be the primary focus of selection whereas in other communities selection is most intense on the adult stage. Seedling community structure may be driven by selection to avoid competitive interactions, environmental filtering effects, or phylogenetic constraints and sorting out the relevant strengths of each of these is a challenge. Testing these ideas across a range of community types will help forge a new perspective on the ecology of seedling communities.

Chapter 13

Spatial variation in seedling emergence and establishment – functional groups among and within habitats?

Johannes Kollmann

13.1 | Introduction

Based on over two centuries of biogeographical research starting with Schimper (1898) and supported by intensive ecological studies during the past decades, a synthesis of large-scale plant distribution patterns and plant functional types may now be possible (Box, 1996; Westoby & Wright, 2006). Plant functional types are nonphylogenetic species groups that show close similarities in their response to ecological factors (Duckworth *et al.*, 2000). Functional types are derived from morphological, ecophysiological, and life history traits, often cutting across taxonomic groupings. One advantage of this approach is that some traits can be readily measured in the field and may act as surrogates for others, which require time-consuming laboratory measurements or experiments. Plant functional types are the ones suggested by Grime (1979, 2001), the three main traits determining plant performance as proposed by Westoby (1998), hierarchical classifications (Lavorel *et al.*, 1997), or the seedling strategy types described by Keddy *et al.* (1998); the latter were experimentally tested by Carlyle and Fraser (2006). These conceptual frameworks and some older systems like the life-forms by Raunkiær (1934) or the scheme devised by Hallé and Oldeman (1975), based on the reiteration of plant modules, may now be integrated into the biogeographical types of the world vegetation (Whittaker, 1970; Holdridge *et al.*, 1971; Breckle, 2002), as done for seed dormancy and germination by Baskin and Baskin (2001).

Seedling dynamics are core issues for plant population ecology (Harper, 1977), and Grubb (1977) advocated that differences in emergence, growth, and survival of seedlings contribute significantly to the regeneration niche of plant species. Recently, ecological research has focused on functional groups at the seedling stage. Cornelissen *et al.* (1997), for example, identified functional correlates of leaf nutrient content among seedlings of woody species and then tried to characterize functional species groups based on leaf nutrient attributes. Seedlings of 81 woody species from the cool-temperate zone of western Europe were grown in an experiment with near-optimal nutrient availability. Mass-based leaf nitrogen content (N-mass) was positively correlated with relative growth rate (RGR) of the seedlings, but the correlation with RGR became tighter when leaf N was expressed on a whole-plant basis, that is, leaf nitrogen mass ratio (LNMR). Functional groups of species and genera could be distinguished with respect to seedling leaf nutrient attributes. Deciduous woody climbers and scramblers had consistently higher leaf N-mass and LNMR than other deciduous species or genera, and seedlings of tall shrubs had higher values than trees. These differences seemed at least partly due to variation in specific leaf area. Seedlings of evergreen species had consistently higher area-based leaf N content than that of deciduous species, but there was no significant difference in N-mass between these two groups. Nitrogen-fixing species were another functional seedling group that had consistently high leaf N-mass compared to non-N-fixers. The emphasis of this study was on leaf traits, although biomass allocation and stem traits may also be important. A complementary paper by Castro-Diez *et al.* (1998) emphasized effects of stem anatomy on seedling RGR, while Cornelissen *et al.* (1996) looked at both whole-plant biomass allocation and SLA as determinants of interspecific variation in RGR of woody seedlings.

Another study on functional seedling groups investigated seedlings of 34 North American grassland and savanna species, representing five functional groups, in a glasshouse on nutrient-poor soil with or without N fertilization (Reich *et al.*, 2003). Here, forbs and seedlings of C_3 and C_4 grasses had similar RGR, followed by legumes and woody species, but the results varied greatly among species within functional groups. All groups, except the legumes, had significantly greater photosynthetic and respiration rates at higher N supply. However, principal component and cluster analyses yielded groupings that corresponded only weakly to the *a priori* functional groups. This study indicated that RGR and related traits may differ among functional seedling groups in significant ways, but in a complex pattern that does not allow simple generalizations about relative plant performance or response to resource supply rates.

The ecological significance of the seedling traits of functional groups needs to be investigated in natural habitats. Zanne *et al.* (2005), for example, found that species with large seeds, large seedlings, thick storage cotyledons, slow germination, large-stature adults, and dispersal by large animals were common in forest and gap habitats, in marked contrast to species from open habitats, that is, grassland and edge. Analyses incorporating phylogeny revealed that these groups of traits showed correlated evolution. However, large-scale comparisons of seedling traits among regional or continental floras are still rare. There was minimal focus on seedling morphology and performance in the handbook by Grime *et al.* (1988) and none in most of the current Internet databases on plant traits, for example, Anderberg and Anderberg (2006), FloraWeb (2006), and LEDA (Knevel *et al.*, 2003). Detailed seedling information is, nevertheless, available for those species included in the series *Biological Flora of the British Isles (Journal of Ecology)* and *Biological Flora of Central Europe (Flora)*, and morphological descriptions of seedlings have been published for some regional floras (Duke, 1965; Csapody, 1968; Bokdam, 1977; Muller, 1978; de Vogel, 1980; Hanf, 1990). There is much literature on seedling types and ecology in relation to forest dynamics, including classifications of *gap dependent, gap independent,* and *facultative gap species* (e.g. Whitmore & Swaine, 1988; Platt & Strong, 1989; Popma *et al.*, 1992; Cornelissen *et al.*, 1994). There is also a series of papers about functional groups of seedlings in terms of adaptations underlying calcifuge vs. calcicolous behavior (Tyler & Strom, 1995). However, few functional concepts of general seedling types have been developed so far (Keddy *et al.*, 1998), although the seedling stage is particularly sensitive to abiotic stress, disturbance, and herbivory, and numerous adaptations have been described to cope with these ecological factors. The aim of this chapter is to give an overview of current knowledge on functional seedling groups among and within habitats using different spatial scales.

13.2 | Description of the seedling stage

In this chapter, *seedling* is used loosely to cover very young plants because there is still no practical and convincing solution of defining the transition from the seedling to the juvenile stage (Fenner & Thompson, 2005, p. 145). Two other terms follow Bullock (2000): *seedling germination* describes the protrusion of some part of the embryo from the seed coat, easily recorded in the laboratory, but difficult to detect in the field, whereas aboveground *seedling emergence* with hypocotyl or cotyledons is the usual measure in the field, that is, the product of germination and early seedling survival. *Seedling establishment* is used for the period until independence from maternal resources.

Although the seedling stage may exhibit less plasticity than observed for adult plants (Harper, 1977), there is considerable variation in seedling size and morphology among and within species (Fig. 13.1; Chapter 2). Among 210 randomly selected species from northwestern Europe (Muller, 1978), average hypocotyl length at the seedling stage varied from 0.15–6.0 cm (CV 82%) and cotyledon length spanned 0.08–12.0 cm (CV 107%). Within species, average variation in hypocotyl length was ±30% and variation in cotyledon length ±26%. This result is not surprising because hypocotyl extension and root:shoot ratio are known to show considerable plasticity, especially in response to light quantity and quality (e.g. Lee *et al.*, 1997). Belowground traits have been less used to describe seedling functional types, although the root system is markedly different among plant species (Fig. 13.2), and responds strongly to variation in light and soil nutrient conditions (e.g. Grubb *et al.*, 1996). Reader *et al.* (1993) tested

Low-light species

High-light species

Agrimonia procera Wallr.

Agrimonia eupatoria L.

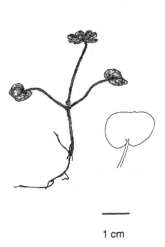

Geranium dissectum L.

1 cm

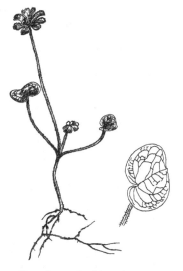

Geranium columbinum L.

Fig. 13.1 Variation in seedling size among closely related species with contrasting shade tolerance (*Agrimonia procera* [Ellenberg light indicator rank value: 5] and *A. eupatoria* [7], Rosaceae; *Geranium dissectum* [6] and *G. columbinum* [7], Geraniaceae). The indicator values for light span from '1' tolerates dark shade, <1% light) to '9' (only found under > 50% light; Ellenberg *et al.*, 1991). Hypocotyl-plus-cotyledon length was 33 mm in the more shade-tolerant *A. procera* and 25 mm in *A. eupatoria*; the species pair *G. dissectum* and *G. columbinum* showed a contrasting pattern (38 mm vs. 49 mm). Drawings modified after Muller (1978), reproduced with kind permission from Springer Science and Business Media, Heidelberg.

Rosa canina L.   Sambucus nigra L.   Cornus sanguinea L.

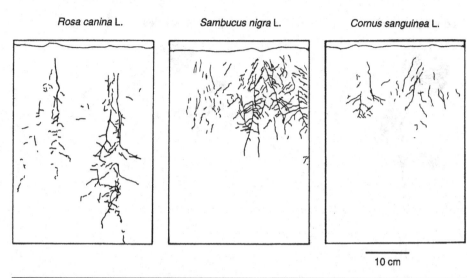

10 cm

Fig. 13.2 Contrasting root systems in co-occurring seedlings of three tall-shrub species (*Rosa canina*, Rosaceae; *Sambucus nigra*, Caprifoliaceae; and *Cornus sanguinea*, Cornaceae) from central Europe; pairs of seedlings were grown in rhizotrons (43 × 29 × 5 cm³), filled with a mixture of sterilized compost material, sand and vermiculite (3:1.5:1), and buried with an inclination of 30° from the vertical in a common garden. Rooting patterns were registered through a glass window after 9 weeks. From J. Kollmann, unpubl. data.

seedlings of 42 species for response to drought and found that the species that showed the greatest plasticity in rooting depth had a greater ability to sustain shoot growth when subjected to drought. Some of these differences are explained by phylogenetic constraints and constraints due to the life-form of the respective species, whereas phenotypic plasticity of seedlings may be an adaptation to cope with a diversity of environmental conditions. The approach of the present chapter is to use variation in aboveground seedling size and morphology during emergence and establishment to identify functional seedling types among and within habitats.

Numerous experiments have shown that plasticity can be essential for seedling survival and growth under unpredictable ecological factors (Fenner & Thompson, 2005). However, ecological factors act at different scales. Short-term microsite effects, for example, might be overruled by regional differences in the long term (Kollmann, 2000). Thus, adaptations of functional seedling groups result from selection pressures involving different scales. Unfortunately, few studies have compared the relative importance of ecological factors during seedling establishment in a multi-scaled analysis (e.g. Bischoff *et al.*, 2006), although the significance of spatial scale in ecological studies was recognized at least a decade ago (O'Neill, 1989; Hoekstra *et al.*, 1991; Levin, 1992; Schneider, 1994).

Spatial scale is also important for understanding contrasting patterns in seedling morphology, emergence, and establishment; vegetative regeneration has been deliberately excluded (but see Chapter 2). The approach is similar to that used by Petersen *et al.* (1999) to review implicit scaling in the design of experimental aquatic ecosystems. The extensive literature on seedling dynamics should allow us to answer the following questions: (1) At which spatial scale(s) are seedling morphology, emergence, and establishment most strongly differentiated and which scales are less relevant? (2) Is the scale dependence different for the various processes associated with the seedling stage? (3) Can the results be used to identify functional seedling groups among and within habitats?

13.3 | Definition of spatial scales

A review of the literature on seed dispersal, seed predation, germination, and seedling establishment reveals that *microhabitat, habitat, landscape, region,* and *biome* are the most commonly used spatial units (Kollmann, 2000; Hulme & Kollmann, 2005). The terminology of scale units in terrestrial vegetation is similar in most ecological publications, and the definitions used within this chapter are in accordance with Wiens *et al.* (1986), Grubb (1987), Schaefer (1992), and Forman (1995). For a definition of these units, it is convenient to delineate them on the basis of vegetation structures because plants are structurally dominant in most ecosystems and determine, to a large extent, the biotic and abiotic interactions. Additionally, a rough estimate of the size of the five spatial categories is needed, although the exact areas of each depends on vegetation type.

The smallest scale used in this chapter is the *microhabitat*, representing, for example, fallen trunks in a forest gap, patches of moss in heathland, or molehills in a grassland. Microhabitats cannot be larger than the largest individual plants in a given vegetation type and thus they can range from a fraction of a square meter in grassland to 100 m^2 in forests. Large forest gaps, on the other hand, develop distinct plant communities that are rather different from the surrounding forest and, thus, should be characterized as a habitat and not as a microhabitat. Studies, which focus, for example, on a comparison of seedling dynamics under different tree species or examine differences between edge and interior forest, are judged as choosing microhabitats as the unit of observation. A *habitat* can be defined as a patch of a specific plant community with a minimum area of about 1 m^2 to 1 ha, depending on the grain of the particular vegetation and the size of the constituting plants. The next scale level is the *landscape*, which can be described as a mosaic of plant communities and abiotic habitat structures repeated in similar form over a km^2-wide area (Forman, 1995). Within a landscape, several sets of attributes tend to be similar across the whole area, including landforms, soil and vegetation types, local faunas, and disturbance regimes. Another spatial level in the scale hierarchy is the biogeographical *region*, for example, the North German lowlands or the Eastern Alps, that is determined by common topographic and climatic traits and a more or less distinctive species composition. The same is true on an even larger scale for the *biome*, for example, the cool-temperate biome or the arctic biome, following Walter's terminology of so-called zonobiomes (Breckle, 2002).

13.4 | Microhabitat effects on seedling dynamics

Various small-scale ecological factors affect seedling emergence and establishment, including microclimate, soil conditions, herbivores, pathogens, and competition with other plants (Crawley, 1997b), and the period between seed germination and establishment of an independent juvenile plant is one of the most vulnerable stages in the life cycle (Kitajima & Fenner, 2000). The main factors causing seedling mortality in contrasting microhabitats are desiccation, heat, herbivory, and fungal diseases (Harper, 1977). Safe sites for seedling establishment, Harper's (1977) "environmental sieve," can be stone cracks, pockets of fine sediments, places with reduced litter cover, or canopy shade. Prominent examples are nurse plants in desert environments or conifer seedling establishment on decaying wood. Another interesting system involves the facilitation of plant establishment by other plants in grazed woodlands (Olff *et al.*, 1999). Harper's (1977) classical experiments on effects of soil preparation on seedling emergence in various species of *Plantago* (Plantaginaceae) and *Ranunculus* (Ranunculaceae) are very instructive. Seed germination and seedling emergence often depend on depth of the seed relative to the soil surface (e.g. Tobe *et al.*, 2006). However, the mechanistic effects of soil depth are still not fully understood and more experiments are needed to relate this dimension of the microhabitat to differential emergence of functional seedling groups. Small-scale soil disturbance can promote seedling emergence and establishment, as observed in a factorial disturbance-seeding experiment with the invasive *Rosa rugosa* (Rosaceae) in five dune habitats (Fig. 13.3) (Kollmann *et al.*, 2006).

Microhabitat factors are highly important for seedling emergence, but less for seedling growth and survival. In a related review on the regeneration niche of fleshy-fruited plants (Kollmann, 2000), most studies (15) dealing with seedling establishment found differences among microhabitats, while 10 others compared habitats and none larger spatial scales. Some studies at the microhabitat scale focused on the effects of the identity of the canopy plants on seedling establishment. Higher seedling mortality occurs under parent plants compared with more distant microsites (Herrera *et al.*, 1994; Böhning-Gaese *et al.*, 1999; Kollmann & Grubb, 1999), supporting the Janzen–Connell hypothesis. In contrast, Schupp (1988) found no such difference associated with proximity to the parent plant. Distance from trunks or to forest edges have been reported to affect seedling establishment (Gleadow & Ashton, 1981; Debussche *et al.*, 1982; Restrepo & Vargas, 1999).

Several studies have compared microhabitats with respect to the degree of canopy cover, for example, old fields or successional shrublands (Gill & Marks, 1991; Guevara *et al.*, 1992; Burton & Bazzaz, 1995; Auge & Brandl, 1997; Cowling *et al.*, 1997). Emergence is often less frequent in canopy gaps, whereas establishment tends to be more

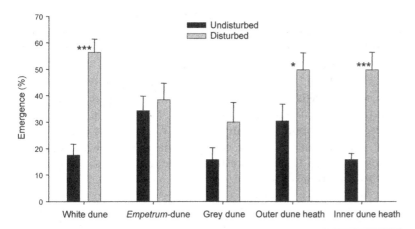

Fig. 13.3 Seedling emergence of the invasive non-native *Rosa rugosa* (Rosaceae) in a factorial disturbance seeding experiment within 5 habitats in a coastal dune system after 2 growing seasons (means $\pm$ 1SE). Fifty seeds were added to 10 disturbed and 10 undisturbed subplots (20×20 cm^2) per habitat. There were significant overall effects of habitat type (two-way ANOVA; $F_{4,90} = 3.4$, $P = 0.013$), disturbance ($F_{1,90} = 35.2$, $P < 0.0001$), and their interaction ($F_{4,90} = 2.9$, $P < 0.0001$). For within-habitat comparisons, effects of disturbance were significant for white dune (Tukey *t*-test; $t_{1,18} = -5.8$, $P < 0.0001$), outer dune heath ($t_{1,18} = -5.0$, $P < 0.0001$), and inner dune heath ($t_{1,18} = -2.2$, $P < 0.05$). Modified from Kollmann *et al.* (2006), reproduced with kind permission from Springer Science and Business Media, Heidelberg.

successful in open microhabitats, but these differences strongly depend on the species involved and on gap size. Microhabitat heterogeneity within gaps may be more important than gap size *per se* (Kitajima & Fenner, 2000). For a discussion of the significance of aboveground and belowground gaps for seedling colonization in various vegetation types see Bullock (2000) and Chapters 8 and 16. The importance of gaps for the invasibility of grassland communities was elegantly demonstrated by Thompson *et al.* (2001) (Chapter 14). Another experiment at the microhabitat scale showed that survival of transplanted seedlings of *Rhamnus cathartica* (Rhamnaceae) in old fields was most dependent on microhabitats where seedling predation by rodents, frost heaving, and desiccation were low (Gill & Marks, 1991). Survival of seedlings after 1 year was 3–6 times higher in experimental aboveground gaps than in closed grassland. Similar effects of competition on establishment of tree seedlings, including *Prunus serotina* (Rosaceae) and *Quercus petraea* (Fagaceae), in old field vegetation and successional grassland have been reported (Burton & Bazzaz, 1995; Kollmann & Schill, 1996). Descriptive studies have often found more frequent establishment of bird-dispersed species under isolated trees and shrubs in grassland due to higher seed dispersal to these microhabitats (Kollmann, 1995; Cowling *et al.*, 1997; Milton *et al.*, 1997; García-Fayos & Verdú, 1998).

Long-term differences between adjacent microhabitats can lead to genetic differentiation of plant communities at a fine scale. In some plant species, genetically determined differences have been observed that exist despite local gene flow. Seedlings of *Veronica peregrina* (Scrophulariaceae) from the periphery of vernal pools in the Central Valley of California (USA) were smaller than those in the center (Linhart & Baker, 1973; Linhart, 1988). Peripheral seedlings result from smaller seeds germinating over a longer time period compared with the subpopulation in the pool center. These seedling differences between microhabitats were interpreted as adaptations to lower and more unpredictable moisture conditions in the periphery. It is less certain, however, that these results can be used to describe generalized seedling functional types.

13.5 | Habitat effects on seedling dynamics

General patterns in seedling establishment

Strong spatial patterns in seedling emergence and establishment among habitats are expected due to differences in competition, for example, in forest patches compared to grassland vegetation. When species are not seed-limited, seedling establishment is more frequent in open habitats. This pattern is determined primarily by the light and nutrient demands of most seedling plants (Grubb *et al.*, 1996), but habitat-specific differences in litter cover and thickness can also control seedling emergence and establishment (e.g. Xiong *et al.*, 2003; Chapter 3). Examples include higher seedling survival in grassland compared with successional scrub (Myster, 1993; Kollmann & Reiner, 1996), and higher seedling densities in gaps compared with the forest interior (Dirzo & Domínguez, 1986; Fisher *et al.*, 1991; Ellison *et al.*, 1993; Restrepo & Vargas, 1999). The opposite pattern has been described in a Panamanian lowland tropical rain forest with eight times higher rates of seedling predation in forest gaps than in closed forest (Schupp, 1988), and in persistent seedling banks from mature temperate and tropical forests (Grime, 2001).

The greater seedling mortality under a canopy might be explained by the less favorable carbon balance of seedlings in shade, which makes it harder to maintain defense traits and to recover from herbivory and pathogen attack (Kitajima & Fenner, 2000). Among neotropical tree species, root allocation and root morphology of seedlings reflect plant adjustments to water or nutrient availability between gaps and closed forest (Paz, 2003). Moreover, life history specialization to light environments is suggested by differences among species groups in allocation to roots and root morphology. For a Chinese forest example of root/shoot responses to light environment see Cornelissen *et al.* (1994). In five contrasting dune communities, seedling emergence and establishment of *Rosa rugosa* (Rosaceae) was positively correlated with soil nutrients and moisture (Fig. 13.3). The most exposed grey dunes had the lowest values for emergence and establishment (Kollmann *et al.*, 2006).

Functional types and seedling size

Species with little shade tolerance often have smaller seeds with relatively high RGR at the seedling stage (Fenner & Thompson, 2005). However, seed mass might be a less useful predictor of species RGR under low light conditions ($< 5\%$ of full light) when the seedlings cannot attain their full growth potential (Poorter & Rose, 2005). Moreover, the strength of the correlation between seed mass and seedling growth declined over time, disappearing in 1–4 years in several tropical tree species studied by Poorter and Rose (2005). Mortality in shade has been shown to be greater in small-seeded species (Leishman & Westoby, 1994b; Grubb et al., 1996), although some small-seeded tropical trees seem to be an exception (Metcalfe & Grubb, 1997). Another example is the failure of Bloor and Grubb (2003) to find any correlation between seed size and seedling survival over a year in 0.2% daylight for 15 species of tropical rain forest tree species, all of them "shade"-tolerant. When comparing six biennial herbs, emergence of the small-seeded species in old-field communities was increased by canopy gaps, whereas the large-seeded species showed no positive response to gaps (Gross, 1984).

The hypothesis that more shade-tolerant plants produce relatively large seedlings was tested for 210 dicotyledonous species from the northwest European lowlands using the drawings and quantitative descriptions in Muller (1978; see Fig. 13.1). The seedlings in this handbook had one to three primary leaves. Differences in seedling performance from contrasting habitats may reflect the ecological niche of the species. Few comparable data exist on the light demands at the seedling stage for a large range of species. A simple approach to characterize the niche of adult plants is the use of Ellenberg indicator values for temperature, moisture, soil reaction, and nutrients, ranging from one to nine for terrestrial species (Ellenberg et al., 1991; Diekmann, 2003). The Ellenberg numbers are based on expert judgement. Ellenberg light values were used to investigate the correlation between shade tolerance and seedling size. To reduce phylogenetic effects, species pairs were chosen from the same genus, having contrasting Ellenberg light values. If, within a given genus, more than two suitable species were available, those with the most similar Ellenberg indicator values for soil moisture, soil pH, and nutrients were chosen. Only species pairs with identical life-form were included to avoid any bias due to life-form. Seedling size was measured as hypocotyl length plus full length of one cotyledon. To achieve normality of the residuals, all seedling size values were log-transformed.

The results of this analysis did not support the hypothesis that shade-tolerant plants have larger seedlings. Seedling size was not significantly different among the light indicator value classes (4–9; ANOVA, $P = 0.65$), although the largest means of hypocotyl-plus-cotyledon length were found in the low light classes (24–31 mm for Ellenberg light values 4–6 vs. 22–23 mm for 7–9). Moreover, there was no significant difference in seedling size when comparing the congeneric species pairs with contrasting light requirements (chi-square

Table 13.1 Effects of (a) phylogeny and (b) life-form on seedling size in 210 northwest European plant species randomly selected from Muller (1978). Seedling size was defined as hypocotyl length plus length of one cotyledon leaf. All analyses were done on log_{10}-transformed data (back-transformed means $\pm$ SE); more information on the data set is given in Section 13.5. Statistical analyses were restricted to families and to life-forms with more than nine species (ANOVA; family effect, F7, 116 $= 2.6$, $P = 0.017$; life-form effect, F3, 204 $= 9.9$, $P < 0.0001$); means with different superscript letters are significantly different (Tukey–Kramer HSD test; $P < 0.05$).

(a) Plant family	n	Seedling size (mm)
Apiaceae	14	33 $\pm$ 12[a]
Asteraceae	36	22 $\pm$ 11[a]
Rosaceae	14	22 $\pm$ 12[ab]
Scrophulariaceae	10	21 $\pm$ 12[ab]
Lamiaceae	16	18 $\pm$ 12[ab]
Fabaceae	12	17 $\pm$ 12[ab]
Brassicaceae	14	16 $\pm$ 12[ab]
Caryophyllaceae	18	11 $\pm$ 12[b]

(b) Life-form	n	Seedling size (mm)
Phanerophytes	20	33 $\pm$ 12[a]
Therophytes	60	22 $\pm$ 11[ab]
Hemicryptophytes	117	17 $\pm$ 11[bc]
Chamaephytes	11	10 $\pm$ 12[c]

contingency table, $P > 0.05$). In 50 pairs, the sister species from a more shaded habitat had longer hypocotyl-plus-cotyledon length, whereas in 53 other pairs the opposite was found, and in two pairs no difference occurred. A similar result was found when analyzing hypocotyl and cotyledon length separately. The lack of support for the hypothesis of larger seedlings in shaded habitats might be explained by higher RGR in seedlings from more open habitats, which would at least partly compensate for some of the initial differences in seed size. The drawings in Muller (1978) give no information about actual seedling age, but it is unlikely that they were collected at the same age. Some differences in size among species might be due to plasticity as the seedlings were not grown under uniform conditions. Another limitation of the analysis is that the Ellenberg light values are based on the light requirements of the adult plant and it might be that seedlings prefer different light conditions. Indeed, ontogenetic change in ecophysiological traits has been shown in several species (Kitajima & Fenner, 2000).

Interestingly, the same data set revealed marked differences in seedling size among the eight most common families (136 spp.) and life-forms (208 spp.). Average seedling size of Apiaceae and Asteraceae was significantly greater than that of the Caryophyllaceae (Table 13.1).

Trees and tall shrubs had significantly larger seedlings than dwarf shrubs, whereas annual plants and hemicryptophytes were intermediate. Thus, effects of phylogeny and life-form seem to be stronger than contrasting light availabilities of the typical habitats in which the species occur. It would be interesting to analyze this data set using the three main traits, *sensu* Westoby (1998), that seem to determine plant performance, that is, seed size, RGR, and specific leaf area (SLA) of seedlings.

13.6 | Landscape effects on seedling dynamics

Relatively little research has been done on the effects of landscape factors on plant regeneration, although sometimes the spatial configuration of habitats seems to be decisive for seedling emergence and establishment. The patterns of seedling establishment of woody species in floodplain ecosystems are a good example of landscape effects (Auble & Scott, 1998; Cooper *et al.*, 1999; Karrenberg *et al.*, 2002, 2003). In the forest and scrub vegetation of these ecosystems, Salicaceae dominate, and their regeneration seems to be adapted to regular disturbance by flooding (Karrenberg *et al.*, 2002). Seed dispersal in *Salix* and *Populus* is achieved by the production of abundant small seeds, which are dispersed by wind and water in spring and early summer. The tiny seeds are short-lived and germinate immediately on moist floodplain sediments. Seedling establishment is only possible if sediments stay moist and undisturbed for a sufficient period of time. Flowering and seed dispersal of Salicaceae are staggered (Karrenberg & Suter, 2003). Thus, the landscape configuration of adult stands of Salicaceae in relation to moist sediments determines the patterns of regeneration, which often lead to monospecific seedling patches (van Splunder *et al.*, 1995). Because of differences in floodplain dynamics, the patterns of regeneration are markedly different in braided and meandering reaches that constitute contrasting landscape types (Karrenberg *et al.*, 2002). However, we have not observed any trend in seedling size of the different Salicaceae species depending on height above water level.

Landscape effects have also been observed for dispersal and regeneration of fleshy-fruited woody species (Kollmann, 2000). Seed dispersal in these species is controlled by the habitat and landscape preferences of frugivorous passerine birds. In this group of plants, at a microhabitat scale, seed rain is poorly structured because chance effects prevail. However, no predictable patterns emerge at a regional scale. At a habitat or landscape scale, seed rain is predictably controlled by the number of available fruit, by specific vegetation structures, and by landscape traits. In mixed scrub or old fields, birds prefer dense woody vegetation compared with open areas (Izhaki *et al.*, 1991; Kollmann, 1995), and postfeeding movements are directed toward structurally rich areas (Herrera & Jordano, 1981; Jordano & Schupp, 2000). Movements of birds within a larger habitat mosaic follow hedgerows or similar habitat corridors (Dmowski & Koza-kiewicz, 1990; Haas, 1995). Frugivorous birds are also abundant along forest edges, causing the highest seed deposition in areas with high densities of forest edges (Kollmann & Schneider, 1997). Here, emergence of seedlings is markedly higher, leading to higher densities of fleshy-fruited species in diverse landscape sections over the long term (Kollmann & Schneider, 1999).

13.7 | Region effects on seedling dynamics

One example for regional effects on seedling emergence and establishment is gradients in seedling performance when transplanted among populations within biomes. The germination characteristics of weed and grassland species (*Daucus carota*, Apiaceae; *Centaurea cyanus* and *Leucanthemum vulgare*, Asteraceae; *Silene alba*, Caryophyllaceae) from Europe along a W–E gradient (England, Germany, Switzerland, Hungary) were studied in growth chambers (Keller & Kollmann, 1999). Germination increment with temperature was different among species and provenances tested. Consistently, the English provenances were least responsive. The results suggest provenance-specific adaptations that were partly correlated with the climate along the European gradient tested. In a related common garden experiment in northern Switzerland, *Daucus carota*, *Leucanthemum vulgare*, and *Silene alba*, but not *Centaurea cyanus*, revealed a trend toward lower germination with increasing distance to the provenance. However, there may be cases where germination and seedling survival show no significant differences among regional populations, although there are large-scale gradients in some of the following life stages (Kollmann & Bañuelos, 2004).

In addition, seedling palatability of *Daucus carota*, *Leucanthemum vulgare*, and *Silene alba* of different European origins to the slugs *Deroceras reticulatum* and *Arion lusitanicus* (both collected locally in northern Switzerland) showed that the amount of seedling herbivory was origin dependent (Keller *et al.*, 1999). The higher losses for all four species were from German and Hungarian provenances compared with English and Swiss plants. The main trend was similar for both slug species except in the case of *Daucus carota*. We found strong correlations between provenance-specific herbivory and certain climatic characteristics of the corresponding regions, that is, winter minimum temperatures and dryness in spring and late summer, which are crucial for the development of slugs. The results were interpreted in terms of a SW–NE European climatic gradient and may be the consequence of differences in the need for plant defenses against herbivory by slugs. The differentiation among the populations used in these studies were so strong that significant heterosis effects were observed in F-1 hybrids and negative epistatic effects in F-2 backcrosses (Keller *et al.*, 2000).

13.8 | Biome effects on seedling dynamics: seed size and seedling survival

Seed and seedling traits vary markedly across biomes to cope with the variation in distribution and amount of rainfall, light, temperature, soil nutrient regimes, and the intensity of predation and disturbance. The mean size of seeds produced by plants near the equator is two to three orders of magnitude greater than the size of seeds produced by plants at 60° north or south (Moles & Westoby, 2003). The larger seed size of tropical plants might be related to more shaded habitats with high leaf area index and high net primary productivity, more woody species, prevailing animal dispersal, and longer growing period. In addition, protection against higher seedling losses due to herbivory, more frequent pathogen infection, and greater competition for light may play a role. The proportion of species producing dormant seeds increases along a gradient of dryness and unpredictability of rainfall, with seed longevity generally higher for dry tropical species than for wet tropical species (Khurana & Singh, 2001a).

Larger seed size in the tropics might lead to a gradient of decreasing seedling size, increasing relative growth rates, and maybe decreasing survival with increasing latitude. Moles *et al.* (2004) investigated whether or not the latitudinal gradient in seed size is correlated with increased seedling survival in the tropics, using the data presented in Moles and Westoby (2004a). They found no relationship between latitude and seedling survival after 1 week in 53 field studies involving 113 plant species. It seems that species growing in the tropics and in the Mediterranean biome need to produce larger seeds to achieve the same levels of seedling survival as species growing farther from the equator. The authors conclude that seeds of a given size are not associated with significantly different rates of seedling survival at different latitudes. However, the assumption of a (linear) correlation between plant traits and latitude is too simple. It is well known that the world climate zones and the resulting major vegetation types do not always follow a latitudinal gradient (Breckle, 2002). The Mediterranean zonobiome, for example, is usually described to occur between 30° and 40° latitude. Actually, in southern Europe, northern Africa, and in the Middle East, it is found at 26–46°, whereas it is largely restricted to 32 to 34° in southern Africa. Similar latitudinal variation can be found for most other zonobiomes. The reasons are sea currents, asymmetric distribution of landmasses, and effects of biogeographical history modifying the latitudinal patterns in climate. Thus, it seems appropriate to use a nonlinear approach in large-scale biogeographical analyses as suggested by Hulme and Kollmann (2005).

I re-analyzed the data of Moles *et al.* (2004) by stratifying the study locations into seven zonobiomes (ZB) *sensu* Breckle (2002), lowland vs. montane sites (>1000 m a.s.l.) and forest and scrub vs. open habitats

Table 13.2 | The 113 plant species used in a study by Moles *et al.* (2004) for analyses of seed size and seedling survival were stratified for reanalysis using Walter's zonobiomes (Breckle, 2002) as shown below. Within each biome, species were further stratified by lowland and montane (>1000 m) sites, and forest and scrub vs. open habitats.

Zonobiomes (abbreviations)	n	Lowland	Montane	Forest and scrub	Open habitats
Tropical biome (I)	9	8	1	9	0
Arid-humid winter rain biome (IV)	8	7	1	7	1
Warm-temperate humid biome (V)	45	43	2	39	6
Cool-temperate biome (VI)	41	41	0	9	32
Aridcold-temperate biome (VII/VIII)	10	8	2	6	4

(Table 13.2). Sample size was unequal and small in some biomes and no studies were available from subtropical grassland or savannah vegetation (ZB II), hot deserts (ZB III), and arctic habitats (ZB IX). The results show significant differences in seed mass among zonobiomes with highest values in the tropical and Mediterranean zonobiomes, and lowest values in the cool-temperate zone (Fig. 13.4a; ANOVA; $F_{4108} = 20.8$, $P = 0.0015$). There were no differences in seed size from lowland and montane sites ($F_{1111} = 3.3$, $P = 0.072$), but forest species had significantly heavier seeds compared with plants from open habitats ($F_{1111} = 27.7$, $P < 0.0001$). The latter was mostly due to heavier seeds in woody species compared with herbs ($F_{1111} = 23.2$, $P < 0.0001$).

Seedling survival after 1 week was highest in the warm-temperate and Mediterranean biomes and lowest in cool-temperate climate (Fig. 13.4b; $F_{4108} = 3.7$, $P = 0.0070$), whereas no altitudinal pattern was observed ($P = 0.90$). Species from open habitats had lower survival than forest species ($F_{1111} = 8.7$, $P = 0.0039$) and the differences among biomes disappeared when the open habitat species were removed from the analysis. High seedling survival in Mediterranean habitats is somehow surprising, because the average climate and soil conditions in these regions tend to be more stressful than in the cool-temperate region. However, it might be that higher seedling predation by mollusks in the cool-temperate zonobiome reverses this pattern.

Moles *et al.* (2004) were guided by the hypothesis that large seed size in the tropics is an adaptation to cope with higher seedling mortality (see Moles & Westoby, 2004a). However, seedling mortality was not particularly high in the tropics and the real challenge seems to be an explanation for the high seedling mortality in cool-temperate regions. Interestingly, the rates of post-dispersal seed predation were also markedly lower in the tropics compared to the temperate zone (Hulme & Kollmann, 2005). This is in contrast to the common idea of more intense herbivory in the tropics (Coley & Barone, 1996). The assumption that biotic interactions are stronger in the tropics is central to a number of hypotheses regarding a positive correlation

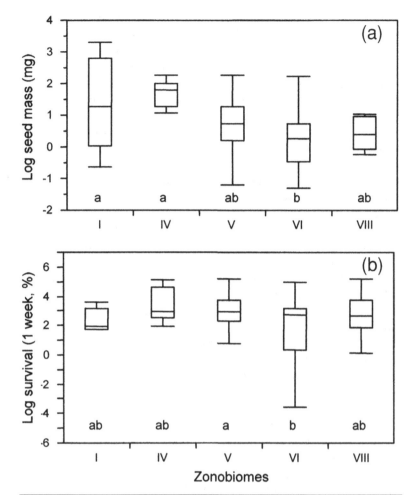

Fig. 13.4 Large-scale patterns in (a) seed mass and (b) seedling survival after 1 week. Re-analysis of data on 113 plant species reviewed by Moles *et al.* (2004). The studies were stratified using Walter's zonobiomes as described in Breckle (2002) (I, tropical biome; IV, arid-humid winter rain biome; V, warm-temperate humid biome; VI, cool-temperate biome; and VII/VIII, arid- and cold-temperate biomes). For statistical results see text; sample size is explained in Table 13.2.

between latitudinal gradient and species richness (Rohde, 1992; Willig *et al.*, 2003), but Moles *et al.* (2004) found no significant relationship between latitude and the proportion of seedling mortality due to biotic factors.

13.9 Synthesis

Functional seedling types – spatial patterns
This chapter has demonstrated considerable variation in seedling emergence and establishment at different spatial scales. The results support the suggestion that separating spatial scales is a useful

approach, although one could argue that the smaller scales are inherent components of the larger units. There is evidence for higher seedling survival (1) in microhabitats with some distance to the parent plant, (2) in open habitats compared with forest patches, (3) in certain types of floodplain landscapes, (4) for populations close to the transplant site in regional comparisons, and (5) in Mediterranean and warm-temperate biomes compared with the cool-temperate biome. Because of the diversity of species, ecological factors, and scales, it may not be possible to translate these differences into simple functional types. Moreover, even in cases with strong theoretical foundation and considerable experimental work, as in the hypothesis of large seed–seedling size for shade-tolerant species, the evidence for functional–seedling types seems rather weak. Clearly, a simple dichotomy of shade-tolerant vs. shade-intolerant seedlings is not appropriate because a *continuum* of types exists corresponding to the wide variation in gap size and in the depths of shade achieved in different community types, both as a proportion of total irradiance and in terms of absolute quanta (Bullock, 2000).

A serious limitation for drawing any conclusions is that there is a strong bias toward the microhabitat and habitat scale in the existing studies. The relatively large number of such studies might also reflect that these two scales are most important for seedling emergence and establishment. However, more studies and preferentially multiscale experiments are needed to substantiate this trend.

Functional seedling types – morphological patterns

Hypogeal and epigeal seedlings are the main morphological types recognized today. Ng (1978) described two other seedling types among tropical forest trees, that is, semihypogeal seedlings, where the hypocotyl is suppressed but the cotyledons exposed at ground level, and durian with extended hypocotyl but cotyledons not exposed and often shed. De Vogel (1980) distinguished 16 seedling types among 150 woody taxa studied in Malaysia using both developmental characteristics as well as morphological traits of cotyledons and hypocotyl. More important functionally may be the distinction between photosynthetic and nonphotosynthetic organs, regardless of their morphological position (Kitajima & Fenner, 2000; Fenner & Thompson, 2005). Although many morphological traits of seedlings appear to be conservative through evolutionary time (Garwood, 1996), the frequency of functional seedling types might vary among life-forms and habitats. The percentage of epigeal species, for example, declined steadily with seed size in 209 Malaysian tropical tree species (Ng, 1978). This might be due to higher dependence on early photosynthesis in small-seeded species but may also be due to a reserve effect, with larger-seeded species holding back reserves in nonphotosynthetic cotyledons against damage to first leaves (Green & Juniper, 2004a). Another characteristic trait of seedlings is the form of the cotyledons, where great variation in length-to-width ratio and SLA can be observed (Fig. 13.1). Variation in SLA can be related to the degree of storage function of the cotyledons and to the shade tolerance of the seedlings (Fenner & Thompson, 2005). However, little is known about the functional importance of cotyledon length. Long cotyledons might be an advantage in dense, short vegetation with a steep vertical light gradient. Future analyses should focus on comparisons in seedling morphology among habitats, regions, and biomes.

Analysis of the data set of Muller (1978) showed surprisingly little difference in aboveground seedling size for plants from contrasting light habitats and aboveground seedling size may be mostly controlled by seed size (Kitajima & Fenner, 2000; Coomes & Grubb, 2003). Not only seedling size but also morphology shows some correlations with seed size. Moreover, allometric relationships are very important here and can explain interspecific relations between seed size and seedling leaf size (Cornelissen, 1999). Small-seeded species tend to have photosynthetically active cotyledons, where the opposite is found in the thicker cotyledons of large-seeded species (Garwood, 1996). For a comparison of the role of cotyledons in five temperate tree species see Ampofo *et al.* (1976). In addition, seeds do vary in the content and concentration of nutrients and, thus, the length of independence from external sources (Kitajima & Fenner, 2000). In contrast, knowledge on belowground seedling size is limited, although high variation among species and habitats is to be expected due to the importance of the root system for seedling survival and growth. New research initiatives should include root traits for delineating seedling functional types.

Seedling RGR is mostly determined by morphological traits as leaf-area ratio (LAR) or SLA, and not by leaf physiology (net assimilation rate, NAR, or unit leaf ratio, ULR). These traits look most promising for future studies of seedling functional types (see Cornelissen *et al.*, 1997; Reich *et al.*, 2003). These traits are very important for seedling establishment in a spatial context. Another interesting set of traits is the nutrient uptake strategy, for instance, hemiparasitism and mycorrhizal dependency (van der Heijden, 2004). Studies of differences in seedling attributes among and within habitats should also include chemical and physiological data, because allocation-based trade-offs are to be expected. Hanley and Lamont (2002), for example, reported a trade-off between physical and chemical defense against herbivores. Because of more effective regeneration (Kitajima & Fenner, 2000), large-seeded species with thick cotyledons might tolerate higher levels of frost damage (Aizen & Woodcock, 1996) and herbivory (Armstrong & Westoby, 1993; Harms & Dalling, 1997).

Functional seedling types – phylogenetic patterns

There has been a strong belief that phylogenetic patterns in comparative plant ecology need to be removed to detect ecological patterns and functional species groups (e.g. Moles *et al.*, 2004). However, it seems that the most marked functional groups determining ecosystem processes, for example, nitrogen-fixing plants (Leguminosae), dwarf shrubs (Ericaceae) of the cold-temperate zone, or coniferous forests (Pinaceae), are confounded by the occurrence of certain systematic groups. This chapter also shows that effects of phylogeny and life-form seem to be stronger determinants of seedling traits than contrasting light availabilities of the typical habitats in which the species occur.

To conclude, the debate between Grime (1979, 2001) and Grubb (1980, 1985, 1998) on the delineation of functional types and plant strategies can be easily transferred to the seedling stage, and thus need not be repeated. Given the diversity of environmental factors among and within habitats resulting in species-dependent safe sites for emergence and establishment, a small number of functional seedling types is not likely. Even the postulate by Grime (2001) that the number of functional types decreases with increasing spatial scale probably needs to be revised with adequate multiscale analyses. More and more, continuous trait spectra (e.g. Wright *et al.*, 2004) rather than distinct functional types are used for the same purposes. We are still a long way from a system of seedling functional types.

13.10 | Acknowledgments

I am most grateful to Angela Moles for generously supplying the original data used in Moles *et al.* (2004). Some of the research described in this chapter was supported by the Landesanstalt für Umweltschutz, Karlsruhe (PAÖ 209118.01), and the Danish Botanical Society. Earlier versions of the manuscript benefited from comments by Hans Cornelissen, Dan Metcalfe, and Jacob Weiner.

Part V

Applications

Chapter 14

Does seedling ecology matter for biological invasions?

Laura A. Hyatt

14.1 | Introduction

Biological invasions constitute an environmental problem of growing global concern. The explosive growth of exotic, invading species
is second only to habitat loss as a factor threatening endangered
plants and animals worldwide (Cronk & Fuller, 1995; Hobbs & Mooney,
1998). Because invasive species did not evolve within their current
ecological contexts, they often process resources and energy differently than natives. Thus, successful invasions often result in changes
to ecosystem dynamics and interspecific interactions. Invasive species
have been shown to alter disturbance regimes (Cronk & Fuller, 1995;
Hobbs & Huenneke, 1992) and resource cycling (Mooney & Drake,
1989). These changes result in reduced biological diversity (Baskin,
1998; D'Antonio & Vitousek, 1992; Sala *et al.*, 2000; Meiners *et al.*,
2001) and contribute to reductions in ecosystem function and productivity. Biological invasions are also estimated to incur costs of nearly
$137 billion per year in the United States alone (Pimentel *et al.*, 2000).

Because invasions lead to so many undesirable ecological and
economic changes, predicting and preventing them have become
research priorities. Ecologists have used historical data on existing
invasions to ask which species are likely to adversely affect newly
recipient ecosystems (e.g. Mack, 1996; Sutherland, 2004) and which
communities are likely to be especially susceptible to alterations by
such invasives (e.g. Elton, 1958; Drake *et al.*, 1989). Universal, reliable
predictive tools have eluded investigators because the mechanisms
leading to changes in biological diversity are varied, as are the traits
of ecosystems or species that predispose successful invasive behavior
(Sher & Hyatt, 1999; Davis *et al.*, 2000).

Although general species traits predisposing invasiveness have
received considerable attention, few seedling-specific traits are targeted. For example, Elton (1958) in his book on biological invasions
suggested that successful invasive plants are likely to be broadly
dispersed, grow rapidly, and be highly competitive. Others have

suggested that successful invasives have phenologies that are broader than those of natives, reproduce in abundance, or have high selfing reproductive capacity (Roy, 1990). In a treatise about weeds, Baker (1965) suggested that successful weeds reach reproductive maturity quickly, make small, easily dispersed seeds, and have a great degree of phenotypic plasticity. It is difficult to know what to make of the absence of seedling traits on these lists.

Thus, several unexamined questions bear investigation: What is the role of seedlings in the process of invasion? Are invasive species competitive as seedlings? Do their demographic roles change as they are relocated from their native habitat to an exotic one? How do they respond to biotic resistance offered by native communities of adult plants?

The literature reviewed in this chapter will reveal a wide variety of seedling strategies used by invading species. While some seedlings have strong impact on recipient communities from the moment they germinate, others have been shown to be easily suppressed by native species at the seedling stage. As seedlings, many of these less competitive species appear to capitalize on resource pulses provided by disturbances, growing rapidly in large numbers. Other inferior competitors receive benefits from growing in the vicinity of adult conspecifics. These adults modify the environment in ways that favor their own seedlings while putting natives at a disadvantage. This chapter will explore variations in the seedling ecology of invasive species as well as how these species, as adults, impact the recruitment and success of seedlings of natives and thereby their communities.

14.2 | Invasive seedlings

Because invasive species have such widespread and substantial effects, one might expect their seedlings to be especially vigorous, and in some cases they are. However, the strategies of successful invasive seedlings are quite variable and many experimental tests of their interactions with native seedlings reveal surprisingly inferior competitive abilities. The wide range of invader seedling strategies means that, demographically speaking, seedlings are of broadly variable importance in facilitating invasion. Intact native adult plant communities are surprisingly resistant to the seedlings of invasives, although global biogeochemical changes might impact these interactions. Capitalizing on the apparent weaknesses in invasive seedling ecology, conservation managers are using several different strategies to protect native plant communities.

Vigorous invasive seedlings?

Some invasives exhibit superior growth as seedlings under a wider range of environmental conditions than natives. One way to evaluate the ecology of invasive seedlings is to compare the ecology of seedlings of exotic, invading plants with functionally similar native species. Comparisons of exotic *Larix* (Pinaceae) and native *Betula* (Betulaceae) species performance at different elevations and exposure conditions revealed that *Larix* seedling survival and biomass accumulation was consistently higher than for *Betula*, especially at higher altitudes (Akasaka & Tsuyuzaki, 2005). This pattern explains *Larix* invasions of habitats above treeline. The success of *Larix* is attributed to its morphological plasticity. It is able to reallocate resources to roots or shoots in a context-appropriate manner far better then the broad-leaved *Betula*. Further, as an adult, *Larix* appears to create microenvironments in its vicinity that facilitate the recruitment of other herbaceous and woody species, furthering the establishment of high-altitude treelines. Thus, the success of *Larix* invasions begins at the seedling stage and can be partly attributed to its evolutionary history.

For some species, seedling invasion is often fostered by microenvironments created by existing adults. This scenario creates positive ecological feedback loops and promotes further invasion. In Florida wetlands, *Sapium sebiferum* (Euphorbiaceae) reproduces by seed and vegetative sprouting much more prolifically than local native species (Jubinsky & Anderson, 1996). Thus, it can quickly respond to natural disturbances that remove competitors (Smith *et al.*, 1997). Seedling recruitment is preferentially enhanced in the immediate vicinity of adult *Sapium* trees through increased N content of litter, which decomposes into forms that *Sapium* seedlings can access, but native C_4 vegetation cannot (Cameron & Spencer, 1989; Siemann & Rogers, 1993; Conway *et al.*, 2002). However, the shade provided by *Sapium* does suppress seedlings and they grow more rapidly when shade is removed (Siemann & Rogers, 1993). Thus, *Sapium* seedlings receive mixed benefits from recruitment in the vicinity of adults, but can colonize new habitats as long as they arrive in them before native species are established.

Other invasive species do not have particularly hardy seedlings, but they mature quickly and escape the vulnerable seedling stage rapidly. Adults of the invasive *Ailanthus altissima* (Simaroubaceae) suppress native vegetation through allelopathic compounds produced in root bark (Lawrence *et al.*, 1991). The success of *Ailanthus* can also be attributed to its extraordinarily precocious seedlings, which can begin flowering within 6 weeks of germination (Feret, 1973). It forms long taproots within 3 months of germination, and can grow a meter in height per year for the first 4 years of life (Adamik & Brauns, 1957). In *Ailanthus*, natural selection has abbreviated the seedling stage and favored rapid transitions to the adult stage where phytochemicals assure dominance.

Other invasives are able to tolerate suboptimal conditions sometimes created by conspecific adults for long periods of time, forming vast seedling banks that can readily capitalize on increased light following disturbances. For example, as an adult, *Acer platanoides* (Aceraceae) casts dense shade, a trait contributing to its deliberate introduction as a street tree. Despite low light conditions created by adults, dense seedling banks form under adult canopies. These seedlings can survive for long periods of time with suppressed growth rates (Martin & Marks, 2006). *Acer platanoides* seedlings have also been shown to perform better than natives under adult *A. platanoides* canopies while other dominant species grow significantly better under native species canopies (Reinhart *et al.*, 2005a). This implies that the facilitation of conspecific seedling growth by adults combined with suppression of natives gives this plant an advantage in invading forest habitats.

Competitive ability of invader seedlings

The few direct quantitative examinations of competitive interactions between seedlings of exotic invasives and natives rarely reveal that invasives are competitively superior. Studies investigating competitive interactions between exotic, invasive *Tamarix ramosissima* (Tamaricaceae) and native *Populus deltoides* (Salicaceae) seedlings show that native trees are easily able to suppress exotic *Tamarix* in a wide range of growing conditions, except when they are vastly outnumbered (Sher *et al.*, 2000, 2002). These authors attribute the success of *Tamarix* to anthropogenic changes in riparian habitats. Disturbances to regular flooding cycles allow *Tamarix* seedlings to recruit in the absence of competition and environmental modifications caused by adult shrubs create unfavorable conditions for natives.

Tamarix is not alone. Several invading species are poor competitors and only successfully recruit in competition-free environments. For example, *Lythrum salicaria* (Lythraceae) suppresses native vegetation as an adult, but its seedlings cannot establish successfully in the midst of intact native vegetation (Hager, 2004). Other species have seedlings that may not be inferior competitors, but are at least equivalent to natives. In controlled greenhouse settings, invasive *Lonicera tartarica* (Caprifoliaceae) seedlings exert competitive effects on *Quercus alba* (Fagaceae) seedlings that are no different from those of native *Cornus* spp. (Cornaceae) (Brudvig & Evans, 2006).

Competitive interactions can be modified by abiotic conditions. In Borneo, competition between exotic *Acacia mangium* (Leguminosae) and the native *Melastoma beccarianum* (Melastomataceae) varies in different light environments (Osunkoya *et al.*, 2005). Under high light conditions, as might occur under treefall or limited disturbances associated with logging, *Acacia* clearly outcompetes the native, while under low light conditions *Melastoma* suppresses *Acacia* seedlings. As for *Tamarix* and *Lythrum*, these varying competitive interactions suggest that although native species can maintain dominance under historical disturbance conditions, large changes in disturbance regimes may favor invasive species.

An implied trade-off between dispersability and competitive ability (Connell & Slatyer, 1977) may account for a diversity of seedling strategies among invasive species. In a study of invasion by annual grasses in perennial grassland habitats, the competitive ability of invasives was immaterial to their invasion success (Seabloom *et al.*, 2003). Invasive success in this habitat was attributed to the invasives' ability to capitalize on disturbance events through broad dispersal rather than competitive ability.

Demographic models

The demographic importance of seedling survival varies among species and habitats. Especially high seedling survival of an invasive species can play a critical role in facilitating invasion, although for some species, it is high survival in the later juvenile and adult life history stages that drive population explosions. Interest in quantitatively examining the relative roles of various life history stages in plant invasions is gaining. If population growth rates are particularly susceptible to change in response to alterations in survival during a particular life history stage, that vulnerability may actually provide an opportunity to exert some control over invasive ecology, exposing what has been called a demographic Achilles' heel for invasives (McEvoy & Coombs, 1999; Parker, 2000).

How can we discern the importance of seedlings to invasive plant life histories? One method is to model population dynamics using matrices (Caswell, 2001). Cells in the matrices correspond to the mean probabilities or frequencies of various life history transitions (germination, growth, and reproduction) as measured in the field. These matrices can be used to model population structure and to project population growth rates and stable age structures. Further computations based on these models, including elasticity and sensitivity analyses, can allow explorations of how changes in life history elements might have changed population growth rates. Population growth rates with high elasticities to particular life history transitions can be strongly influenced by small changes in those transitions. Conversely, population growth rates would change very little in response to even large changes in transitions with low elasticities.

Some invasive species have population growth rates that are not very responsive to changes in seedling growth and survival. In an examination of demographic spatial and temporal variation in *Ardisia elliptica* (Myrsinaceae), per capita population growth showed very low elasticity to seedling recruitment, even in a wide variety of habitats over many years (Koop & Horvitz, 2005). These findings echo those of studies of invasive *Cytisus scoparius* (Leguminosae) (Parker, 2000). Elasticity of per capita growth rate to the seedling stage for this species across a variety of habitats was also fairly small. *Alliaria petiolata* (Brassicaceae) also shows very little elasticity to early seedling survival (Davis *et al.*, 2006; Hyatt, unpubl. data). A comparison of the demography of native and invasive *Rubus* species (Rosaceae) revealed limited elasticity in either species to seedling recruitment (Lambrecht-McDowell & Radosevich, 2005). For these species, control strategies targeted at seedlings are not likely to have substantial effects on per capita population growth.

Other species are more regulated by seedling demography. Elasticity analysis of the population dynamics of *Clidemia hirta* (Melastomataceae) in two different Hawaiian (USA) habitats revealed populations that had high seedling survival and growth elasticity (De Walt, 2006). In all populations studied, more than 50% of the seedlings survived and the elasticities of seedling survival and transitions to small adult stages nearly always made up the largest proportion of demographic elasticity. However, simulations revealed that close to 100% of the seedlings would have to be eliminated to drive populations to extinction. For invasions of *Spartina alterniflora* (Poaceae) in California (USA), spatially explicit invasion rates were most sensitive to seedling establishment rates as well as inflorescence density and outcrossed seed production (Taylor *et al.*, 2004).

The demographic importance of seedlings may change with habitat. One reason exotic species become invasive is because environmental conditions that regulate demographic processes differ in the native and novel habitats. *Polygonum perfoliatum* (Polygonaceae), an annual vine, is native to east Asia and was introduced to the eastern United States in the 1930s. It has become a pernicious invader of wetland and old field habitats there. Asian populations have a high elasticity to seedling survival due to extremely low plant survival in seasonally flooded habitats where *P. perfoliatum* is found. In contrast, North American populations have invaded upland and less regularly flooded habitats. As a result, the elasticity of these populations to seedling survival is much lower, rendering control of these populations much more difficult (Hyatt & Araki, 2006).

Community resistance

Undisturbed communities in which mass and resource cycling have not been modified show a great deal of resistance to seedling invasions. Many exotic species have been shown unable to compete with native species, especially in undisturbed grasslands (Corbin & D'Antonio, 2004). Over 4 years, exotic species' seedlings were consistently unable to gain a foothold within stands of existing adult natives. Intact forest and tree fern canopies can also effectively repel invasion by *Metrosideros polymorpha* (Myrtaceae) seedlings, suppressing both survival and growth (Burton & Mueller-Dombois, 1984). Further, *Metrosideros* seedlings emerging in understory environments cannot capitalize on increased light resources when they appear. Because resource pulses are required to support recruitment, this species must disperse widely so that it may germinate in occasional canopy openings.

Native vegetation is not alone in facilitating or defending against invasives. For example, exotic perennial tussock grass *Cortaderia jubata* (Poaceae) can invade California shrublands, but it seems to be strongly limited by the activities of vertebrate seedling predators (Lambrinos, 2006). Forty percent of invasive seedlings survive on the edges of chaparral stands, but nearly none survive in the center of these stands, mirroring a gradient of mammalian predation activity. While intact shrubland can resist invasion, habitat fragmentation will increase edge habitats, facilitating invasion by seedlings of this perennial grass.

Some invasives meet little to no biological resistance, but are thwarted by abiotic conditions. In Australia, recruitment of invasive *Mimosa pigra* (Leguminosae) seedlings is dependent on emergence time and the availability of water (Lonsdale & Abrecht, 1989). Although *M. pigra* seedlings are not particularly competitive against native vegetation, seedling recruitment is highest in years of unusual rainfall, suggesting that their ability to capitalize on abnormal abiotic conditions enables them to displace natives. Similarly, *Holcus lanatus* (Poaceae), a recent European perennial grass invader of California coastal grasslands, is also reliant on unusual rainfall patterns. Its recruitment patterns are most heavily dependent on spring availability of soil water (Thomsen *et al.*, 2006). Thus, *Holcus* is repelled not by a lack of resistance by the biological community, but by the availability of soil water in the spring. In years with high rainfall, *Holcus* is able to recruit extensively, but has been effectively repelled in drought years.

Carpobrotus edulis (Aizoaceae), a South African invasive of Californian coastal communities, represents an interesting example of seedling invasion that is modified by interacting biotic and abiotic factors. *Carpobrotus* requires soil disturbances to invade successfully (D'Antonio, 1993). In grasslands, gopher activity creates critical resource releases that reduce competition from native vegetation and enable invasion by *Carpobrotus* seedlings. However, in backdunes and coastal scrub where gopher activity is reduced, invasion is effectively deterred by mammal herbivory and competitive suppression offered by existing shrubs. Thus, complex biological interactions lead to landscape variation in susceptibility to invasion by *Carpobrotus*.

Anthropogenic change and consequences

Ecosystems material cycling patterns will be altered due to the accumulation of greenhouse gases and land use changes. There is some evidence that the ecological roles of seedlings in invasions may change as a result. At elevated levels of CO_2 (500 μmol), invasive *Prunus laurocerasus* (Rosaceae) seedlings respond by gaining 56% more biomass, while functionally similar *Ilex aquifolium* (Aquifoliaceae) does not respond measurably (Hättenschwiler & Körner, 2003). These effects may enhance the invasion of *Prunus* in Swiss forests. Phosphorus additions have also been shown to change competitive interactions between the invasive *Eucalyptus albens* (Myrtaceae) and native species in White Box woodlands in New South Wales (Australia) (Allcock, 2002). Because phosphorus runoff from agricultural applications is likely to increase as food production expands (Tilman *et al.*, 2001), this effect may magnify other invasion events worldwide.

When species are relocated around the globe, there can also be substantial changes in selective regimes that can contribute to successful seedling invasions. Reduced tissue losses due to herbivores or predators can occur when species are introduced to novel habitats, due to escape from specialized predators in the native range. This phenomenon has been hypothesized to select for individuals that allocate resources towards growth and away from specialized defenses that are no longer needed, known as the Evolution of Increased Competitive Ability (EICA) hypothesis (Blossey & Nötzold, 1995). The idea is supported by observations that invasives in novel habitats appear to be larger than those in their native range and by specialized studies in a few species. For example, invasive ecotypes of *Sapium sebiferum* grow quickly and have poorly defended leaves, but can rapidly compensate for tissue loss compared to native Asian ecotypes (Rogers & Siemann, 2004). When plants of these two ecotypes were grown under varying nutrient, competition, and root predation conditions, they responded very differently. While native ecotypes were adversely affected by root damage measured by stem height and above- and belowground biomass, the growth of the exotic ecotypes was not significantly impacted. Invading strains were able to compensate swiftly for damage, facilitating rapid growth and recovery in novel habitats where damage is relatively rare.

How might the seedling ecology of deliberately introduced species facilitate or impede invasion? A primary concern about genetically modified (GM) organisms is that they may hybridize with closely related native species and escape from cultivation, introgressing with native species that may then become superweeds, resistant to the herbicides or herbivores that the crop species was originally engineered to resist (Kling, 1996). Changes in the seedling ecology of these GM hybrids could have enormous impact on the adoption of molecularly modified crop species. However, the seedlings of GM crop–weed hybrids have been shown to be competitively inferior (Halfhill *et al.*, 2005; Claessen *et al.*, 2005) and when wild-type weeds compete with hybrids under protection from herbivores, seedlings of genetically engineered *Brassica rapa* are substantially less competitive, measured in terms of growth rate as well as seed production (Vacher *et al.*, 2004). Demographically, the per capita population growth rate of GM *Brassica napus* is primarily elastic to seed bank persistence, with seedling establishment and survival are life history parameters that are unlikely to contribute to changes in population growth rate (Claessen *et al.*, 2005).

Capitalizing on inferior competitive ability of invasives

Because many invasive species appear to be poor seedling competitors, it is possible that the ability of native plants to resist invader seedlings can be exploited to defend against invasion. Two field studies suggest that this might be a useful approach. *Centaurea diffusa* (Asteraceae), a pernicious weed of rangelands, produces copious amounts of inedible biomass, as well as toxic secondary compounds that suppress the growth and reproduction of native species (Callaway & Aschehoug, 2000). *Agropyron cristatum* (Poaceae) can suppress *Centaurea* seedlings by germinating early in the season and drawing down soil water before *Centaurea* emerges (Berube & Myers, 1982). In another study, high densities of native *Gaillardia pulchella* (Asteraceae) were sown in seedling populations of annual *Rapistrum rugosum* (Brassicaceae). This native was able to substantially reduce seedling growth of *R. rugosum*, achieving an 83% reduction in seed set at high densities, when compared with control plots (Simmons, 2005). Given the frequently poor competitive ability of exotic invasives, simply planting native species in disturbances might be an especially effective way to control an invasion, requiring no species importation or introductions of toxic chemicals.

14.3 | Invasive effects on native seedlings

In some circumstances, changes in biodiversity associated with invasion have been shown to be due to declines in native species colonization rates rather than increased extinction rates (Yurkonis *et al.*, 2005). This implies that once invasives are established, they effectively reduce the recruitment of native species seedlings, rather than suppress existing adult residents. How do they do this? The mechanisms for native seedling suppression may include competitive interactions for a variety of resources, modifications of resource availability, and allelopathic interactions. However, studies of mechanistic effects of adult invasives on native species are sparse (Parker *et al.*, 1999). There are a broad array of mechanisms that invading adults use to suppress native seedlings. This section will begin with describing ways members of a single genus impact native vegetation, then provide examples of other, more indirect mechanisms that may be quite widespread.

Example: exotic honeysuckles
Exotic *Lonicera* species in the United States have received considerable study. Although northeastern USA forests contain a few native honeysuckles, they are far outnumbered by introduced vining and bush honeysuckles, including *Lonicera japonica*, *L. maackii*, and *L. morrowii*, *L. tartarica*, and their hybrid, *L.* × *bella*. Observational studies reveal that these *Lonicera* spp. negatively impact the survival and growth of native seedlings, ultimately limiting biodiversity in forests they occupy. Suppression of native seedlings involves a variety of mechanisms.

The shrub, *L. maackii*, is observationally associated with reduced plant diversity (Collier *et al.*, 2002) and removing shrubs enhances the recruitment of the seedlings of native tree and subtree species (Hartman & McCarthy, 2004). Annual natives, including *Galium aparine* (Rubiaceae), *Impatiens pallida* (Balsaminaceae), and *Pilea pumila* (Urticaceae), make reduced numbers of seeds underneath stands of *L. maackii* as compared to other shrubs (Gould & Gorchov, 2000). These effects are likely due to shrub-induced shade as well as drought. The mechanism by which *L. japonica* suppresses native species seedlings is quite different. It creates a dense mat of intertwined vines on the forest floor that few tree seedlings can penetrate. Those that do are often quickly overgrown, bent down by the vine, and consequently die (Thomas, 1980). The combined effects of above- and belowground competition by this species can suppress growth or result in direct mortality of seedlings (Whigham, 1984).

Allelopathy

Allelopathic plants produce compounds that can impair the ability of other plants to establish and grow in the vicinity. Often considered a mechanism whereby invasives successfully suppress native vegetation, allelopathy has been notoriously difficult to demonstrate conclusively (Inderjit & Callaway, 2003).

The most compelling evidence for the allelopathic effects of exotic invasions on native species comes from studies of *Centaurea stoebe* (Bais *et al.*, 2003). These workers have found that it exudes (−)-catechins into the soil surrounding its roots. At biologically realistic concentrations, such compounds can lead to cytoplasm condensation and elevated Ca^+ levels in *Arabidopsis thaliana* (Brassicaceae) – a precursor condition to cell death. Catechins do not appear to affect *C. stoebe*, as no effects were seen on its own root cells. Using activated carbon, the authors found evidence to suggest that allelopathy might be a prevalent plant interaction mechanism in the historical range of *C. stoebe*. The lack of allelopathic history in a community might facilitate the invasive behavior of *Centaurea* in North America.

The success of *Ailanthus altissima* has also been attributed to allelopathy. This Asiatic tree is widespread in northeastern North American urban settings, where it crowds out otherwise competitive native vegetation. A compound present in root bark, newly emerged leaves, and even seeds appears to suppress radicle growth (Heisey, 1990). However, exposure of plants to *Ailanthus* compounds appears to trigger a cascade of events that lead to a reduced sensitivity of seedlings from those plants to *Ailanthus* toxicity (Lawrence *et al.*, 1991).

Third-party interference

Invasives interact with multiple partners ranging from microbes to vertebrates. Some of the most interesting recent explorations of the effects of invasives on seedling ecology reveal extensive third-party interference. By modifying the interactions of seedlings with both beneficial and disease organisms, exotics can suppress or modify the success of native species as well as their own seedlings in profound ways.

Aegilops triuncialis (Poaceae) is a western US grassland invader of Eurasian origin. Its establishment is significantly enhanced by fungal infection (*Ulocladium atrum*) that acts to partially decompose the seed head and decrease germination time (Eviner & Chapin, 2003). At the same time, establishment of the invasive is retarded by the activity of gophers who selectively bury stands of seedlings. Thus, *A. triuncialis* merely coexists in native grassland when gophers and the fungus are both present, but when gophers are absent, it aggressively suppresses natives. The negative impacts of gophers on *A. triuncialis* are magnified by the frequent absence of *Ulocladium* when gophers are present. These interactions are further influenced by interactions between gophers and grazers with heavy grazing suppressing gopher activity while significantly enhancing *A. triuncialis* infestations. The ability of *A. triuncialis* to suppress native seedlings is dependent on its interactions with members of three different kingdoms.

Alliaria petiolata appears to exude compounds that indirectly impair the growth of native tree seedlings. It produces an as yet unidentified phytochemical that negatively interacts with arbuscular mycorrhizal (AM) fungi. These fungi modify the ability of native tree seedlings to acquire enough resources to persist and grow. Stinson *et al.* (2006) grew seedlings of *Acer saccharum* (Aceraceae), *Acer rubrum* (Aceraceae), or *Fraxinus americana* (Oleaceae) in soils that were either from *A. petiolata*–invaded habitats or from nearby locations free from *A. petiolata*. They autoclaved (sterilized) soils from each location and quantified percent colonization by AM fungi and total plant biomass. The autoclaving treatment nearly completely eliminated AM colonization, and only *A. rubrum* showed any signs of colonization in unsterilized *A. petiolata*–derived soils. The biomass gain in *A. petiolata* soils was equivalent to that seen in autoclaved soils, far lower than biomass in unautoclaved, uninvaded soils. These findings were consistent across a variety of experiments, leading to the conclusion that this mechanism of suppressing native seedlings may also act to facilitate the success of *A. petiolata* in recipient habitats in the United States.

14.4 | Conclusions

The potential invasiveness of an exotic species would be difficult to predict given only information about seedling ecology because of the wide variation in invasive seedling behavior. While some biological invaders are not particularly competitive as seedlings, as adults they can create environments in their immediate vicinity that facilitate their own seedlings' success while impairing the success of natives. These habits, in conjunction with biotic and abiotic factors, can transform species that are well-regulated in their native communities into dominant ecosystem drivers in novel ones.

Seeking traits that might facilitate predictions about invasives, ordination studies have examined the traits that are associated with successful invasives within a taxonomically or geographically restricted set of species. Although seedling traits are included in these ordinations, they rarely emerge as significant contributors. Neither seedling relative growth rates (RGR) nor seedling survival effectively predicted rate of spread in a wide variety of New Zealand taxa (Bellingham *et al.*, 2004), although other, more taxonomically restricted studies have found higher RGR to be a good predictor of invasiveness in *Pinus* spp. (Grotkopp *et al.*, 2002). A recent examination of *Oenothera* (Onagraceae) showed that successful invasive seedlings were exceptionally poor at competing with native species (Mihulka *et al.*, 2006). These authors implied that successful invasives were likely to deploy a jack-of-all-trades strategy, never committing to one partic-ular strategy, but being moderately successful at many.

Invasives impact the ecology of native species' seedlings in many different ways. Not only do invading adults interfere with the ability of native seedlings to acquire light, nutrients, and water, they also change biotic and abiotic processes that feed back to alter the success of both invasive and native seedlings. Hopefully, the current gap in our mechanistic understanding of invading exotic effects will narrow, as ecologists begin to focus on quantifying impacts.

As we have seen, invasive seedlings often do best in disturbed parts of recipient communities. Human land use changes that alter distur-bance regimes can create environments that either favor or reduce the recruitment of exotic seedlings relative to native species. Exotic species with population growth rates that are elastic to seedling sur-vival will likely benefit from land use changes that increase recruit-ment opportunities, but they may also be controlled by a reduction in those same opportunities.

Nearly all of the studies cited above examine the ecology of invad-ing species in their novel ranges. This is not surprising; as most of these species draw attention in their exotic habitats, little is known about their ecology within their native ranges. This lopsided knowl-edge base contributes to our difficulty in making generalizations, if any can be made, about seedling traits that predispose species toward invasive behavior because the ecology of seedlings may be dramati-cally different in native and exotic ranges. Hierro *et al.* (2005) and oth-ers call for comparative studies of invasives in their native and inva-sive ranges, not only to test major outstanding hypotheses explaining the occurrence of invasions, but also to help develop control plans that are realistic.

In conclusion, the life histories of exotic invasive seedlings are highly variable. Complex biotic and abiotic interactions can facilitate the success of seedlings. Many invasives tend to gain footholds in native communities under conditions that free them from commu-nity resistance. Future detailed quantitative studies of seedling ecol-ogy and adult impacts should yield a wide variety of species-specific insights, leading to substantial advances in control strategies, effec-tive mitigation techniques, and more accurate predictive systems for invasives.

14.5 | Acknowledgments

Useful feedback for improvement of the original manuscript was pro-vided by Anna Sher. This work was supported in part by National Science Foundation Grant DEB 0344218 to L. A. H.

Chapter 15

The role of seedlings in the dynamics of dryland ecosystems – their response to and involvement in dryland heterogeneity, degradation, and restoration

Bertrand Boeken

15.1 | Introduction

To understand the role of seedlings in the dynamics of dryland ecosystems, I consider functions seedlings fulfill in population and community dynamics of desert plants and how these functions affect ecosystem and landscape processes. My main focus is to explore the interactions of seedlings with the structure and function of dryland ecosystems. These include not only the responses of seedlings to changes in their environment, but especially their roles in formation and maintenance of spatial heterogeneity and its degradation. These interactions are also involved in sustainable management of arid and semiarid rangeland and in restoration of landscapes degraded by unsustainable human land use. Understanding the roles that plants in general, and seedlings in particular, play in dryland ecosystem and landscape processes may help in formulating appropriate management and restoration methods for sustainable dryland land use under changing climatic conditions.

Interactions involving seedlings are complex and diverse, concerning both pure ecosystem processes, such as resource dynamics of energy, water, nutrients, and gas flows and fluxes, foodweb relations, and the dynamics of the substrate within which the processes occur, that is, the landscape. Part of the complexity arises from the interrelations between ecosystem functions and landscape structure at various spatial and temporal scales. This is strikingly illustrated by the nature of dryland ecosystems, which are characterized by the temporal variability of resources due to scarce pulses of rainfall and by the spatial heterogeneity of the landscape. Both vary from modest in semiarid areas to extreme in hyperarid zones.

Another aspect of the complex network of relationships between plants and their environment is that resource variability and spatial heterogeneity affect plants, but are also affected by them. Plants respond to their abiotic and biotic environment, but alter it in return, especially as they grow large or dense. However in drylands, individual plants can have a greater effect on their environment than in other climatic zones because larger portions of ground are exposed, solar radiation is higher, air humidity lower, and air and soil temperatures are often more extreme.

Within this context, I will identify the contribution of the seedling stage of the life cycle of dryland plants to the functioning and structure of their environment. I will rely on experience in semiarid and arid zones of the Negev Desert of Israel, augmented with information from other dryland regions. The enormous diversity of plants, biological interactions, and dryland climate and soil types may add further complexity, but may in the end be helpful in reaching a generalized understanding.

15.2 | The importance of the seedling stage

The seedling as an individual within a population

Although there is considerable variation in the definition of *seedling* (Chapter 2), I assume that the seedling stage begins with germination and ends when the plant is well-established (a juvenile), no longer relying on seed reserves. Sometimes, the seedling stage is considered to end with maturation as a reproductive adult, that is, annuals, or after the first growing season in evergreen or ephemeral perennials. In any actual situation, the definition should reflect plant development, but ultimately depends on what question is being asked.

For an individual plant (genet) of any life form, the seedling stage is as crucial for its survival as the seed, juvenile, or adult stages. However, if individuals are ramets produced by vegetative propagation, as in many desert plant species, the plant usually starts its life as a larger juvenile that has been heavily subsidized by resources from the mother plant. In general, the seedling stage is by far the most vulnerable stage of a plant's life cycle because the plant neither enjoys the relative independence of inputs from the environment as seed nor the benefits of large size and a developed root system as a maturing juvenile or adult. In deserts, small, developing plants are especially vulnerable because of the usually extreme conditions of air and soil temperatures and water content. However, the fact that every individual plant ultimately started as a seedling does not imply greater importance from a demographic perspective, nor does the greater vulnerability of a plant as seedling automatically imply strong selection for adaptations during this stage. Demographic theory predicts that this depends on the reproductive value of the seedling stage (Fisher, 1930) and on the elasticities (relative contributions to population growth) of the rates of seedling recruitment and survival

(Caswell, 1989), including costs and benefits of reproductive allocation and trade-offs with other traits.

Adaptations of desert plants can be viewed as a result of interactions between the demographic relationships of the seedling stage with the rest of the life cycle in terms of the value of reproductive allocation to seedlings and its effect on fitness on one hand, and the dryland environment with its low, temporally and spatially highly variable resource availability on the other. From the numerous detailed descriptions (Evenari *et al.*, 1982; Batanouny, 2001; Gutterman, 2002), six patterns of adaptations emerge that are relevant for understanding the role of seedlings in desert ecosystems.

- In most perennials (shrubs, trees, succulents), many adaptations to the desert environment occur during the established stages maximizing adult survival (K-strategy; Harper & Ogden, 1970) by means of tolerance of or resistance to adverse conditions of solar radiation, heat, and atmospheric and soil drought. Morphological or physiological adaptations directly affecting seedling recruitment or survival are rare, except rapid root growth responses to soil water (Franco & Nobel, 1990; Nobel & Linton, 1997; Dubrovsky, 1997b; Dubrovsky & Gómez-Lomelí, 2003), and developing or already operational adult traits such as photosynthetic pathways, succulence, and stomatal structure (Sharifi & Rundel, 1993; Hamerlynck *et al.*, 2000). Also, herbivory defenses (morphological or chemical) are similar to those in adults (Rafferty *et al.*, 2005; Fang *et al.*, 2006).
- Herbaceous perennials evade drought by a seasonal ephemeral habit (geophytes and hemicryptophytes), allowing them to endure adverse periods in a dormant state. Few seedling adaptations are known, and primarily juvenile and adult survival is maximized. Subterranean organs offer a degree of protection against consumption of stored reserves.
- In annuals and short-lived perennials, most adaptations are found in the regeneration phase, as seed traits, which maximize either seedling recruitment or seedling survival (r-strategy, bet-hedging, or risk reduction; Brown & Venable, 1986; Venable & Brown, 1988; Ellner *et al.*, 1998). This helps the plant utilize conditions when and where they are less extreme. Often, additional adaptations occur that may reduce or prevent granivory or herbivory (Gutterman, 2002). Specific seedling traits that aid drought evasion are not known in ephemeral dryland species, but there is mounting evidence that desiccation tolerance may often play a role in annual dryland grasses (Gutterman, 2002; Huang & Gutterman, 2004; Zhang *et al.*, 2005). Thus far, the only perennial in which this has been reported is *Ruschia spinosa* (Aizoaceae), a leaf-succulent of the Karoo (Esler & Phillips, 1994).
- Most desert annuals produce numerous small seeds that disperse easily and that germinate with minimal soil moisture (Went, 1948, 1949), maximizing seedling recruitment in favorable sites. Low-cost dispersal mechanisms are most common, with small wings or pappi

(Gutterman, 2002) for primary wind dispersal or without appendices, aiding dispersal along the ground with wind or runoff water flow (Yeaton & Esler, 1990; Boeken & Shachak, 1994; Esler, 1999).

- Elaborate high-investment dispersal mechanisms, especially by zoochory, are rare in desert annuals (Howe & Smallwood, 1982). Although some species have mechanisms mediating dispersal to favorable sites by animals (Esler, 1999), most involve topochory or atelechory (Zohary, 1937; Ellner & Shmida, 1981; Gutterman, 1994; Esler, 1999) that resist dispersal and granivory.

- Some annuals have seeds with highly elaborate mechanisms of delayed predictive germination (Cohen, 1967; Venable et al., 1993; Gutterman, 1994; Esler & Cowling, 1995; Smith et al., 2000), involving seed release during rainfall, timing of germination, and long-term seed dormancy. These traits maximize seedling survival by cuing germination to conditions that also ensure establishment and maturation. Ungerminated and/or undispersed seeds wait for their chance, relatively protected, beneath or at the soil surface or on the (dead) mother plant. In the leaf-succulent Aizoaceae of the Karoo, such mechanisms are also found among perennials (Esler, 1999).

Seedlings and dryland plant community dynamics

Plant traits before and during the seedling stage determine to a large degree how species respond to the spatially and temporally variable desert environment. However, the importance for community, ecosystem, and landscape dynamics depends on different relationships than those of demographic importance.

Community dynamics only depend on seedlings insofar as the latter critically determine the community's development. This happens when seedlings dominate the vegetation, as in pioneer communities and in annual plant communities, but especially when communities are seedling-limited, that is, if changes in seedling density, by lack of germination, competitive interactions, or seed predation, affect species richness, composition, and/or succession. Annual-dominated plant communities are common in semiarid and arid deserts, and their density and diversity vary significantly in time and space. Seedling recruitment does not occur and survival is low during dry periods and high during resource pulses, but only in some parts of the landscape. Due to high seasonal and interannual variation, the role of seedlings can be decisive for local species richness, and for the fate of the community.

In contrast to mesic plant communities where high vegetation density causes competitive suppression or exclusion (Grime, 1973), seedlings of many dryland plants germinate and survive better in resource-rich patches than in the rest of the landscape, often aided by the effects of vegetation density and height on microclimate. Whether facilitation exceeds competition in resource-rich sites (Silvertown & Wilson, 1994; Callaway et al., 1996; Aguiar & Sala, 1998; Holzapfel & Mahall, 1999; Holmgren, 2000) varies along geographical rainfall gradients, between wet and dry years and among species (Tielbörger & Kadmon, 2000; Lloret et al., 2005), and among climatic regions and continents. Generally in more mesic subhumid regions, competition by fast-growing dominant annuals or established perennials may suppress many species in the seedling stage (Silvertown & Wilson, 1994; Holmgren et al., 1997; Walker et al., 2001), while at greater aridity fewer species may be susceptible. However, because the suppressed plants are numerically subordinate initially, their demise usually does not control community dynamics, except for seedlings of perennials that become dominant by size.

In dryland plant communities, the relationships between individual plants and their communities are strongly affected by landscape heterogeneity of patches with different levels of productivity (Montaña, 1992; Boeken & Shachak, 1994; Silvertown & Wilson, 1994). Direct effects of plants on species richness and species composition also vary among different types of patches, but their indirect effects as ecosystem engineers drive structural changes in some of the patches that may control species richness at least as much.

Seedlings and dryland ecosystem and landscape dynamics

During their development, plants not only respond to their biotic and abiotic environment, but also affect it in many ways. They do this directly by means of typical trophic interactions of resource use, primary production, and provision of food for animals and microorganisms, and indirectly as ecosystem engineers by means of structural changes to the landscape. Seedlings may play crucial roles in trophic interactions, but in general these are limited by their small stature, except temporarily when they are dominant in annual-dominated vegetation.

Seedling growth of drought-evading desert plants (annuals and ephemeral perennials) is very fast, but with low water use efficiency (Sandquist et al., 1993; Angert et al., 2007). Therefore, seedlings take up considerable amounts of water and produce most of the biomass during the early part of rainfall pulses before other plants grow in size. Early season soil moisture depletion can lead to interspecific resource competition, but it is not known how it constrains further vegetation development and possibly succession. Biomass production is a transient role of seedlings and because of the short time they are seedlings, herbivory on annual herb seedlings seems to be rare in deserts. Many wild and domesticated grazers feed on annual vegetation, but seedlings may escape grazing because of their small size. Many evergreen perennials of dryland have thorns or chemical herbivore repellants as seedlings (Rafferty et al., 2005; Fang et al., 2006).

Plants modify local microclimate, soil and soil surface structure, resource availability, and site availability for other organisms, and, thus, give rise to a large portion of spatial heterogeneity. Especially in dryland systems, plants are major contributors to the formation and maintenance of landscape patches (Yeaton & Esler, 1990; Silvertown & Wilson, 1994), primarily due to capture of sediment and resources

(Boeken & Shachak, 1994) and/or stabilization of the substrate, in the case of sandy soils (Singh & Rathod, 2002; Singh *et al.*, 2003). The contributions of seedlings to these processes vary significantly among species and depend on plant form, size, and density. Larger and denser plants sustain more root growth, cast more shade, produce more litter, capture more sediment and water, and provide more protection against herbivores. In addition, the timing and location of seedling development through germination and establishment plays an important role (Silvertown & Wilson, 1994).

In contrast to fitness-related roles of seedlings, their importance to the ecosystem and the landscape is independent of survival. Even unsuccessful individuals that die before they reproduce contribute to environmental change, especially if they have high density. Seedlings contribute to unlignified plant litter, which has a trophic component as it feeds decomposers and a structural side that changes surface flows of water and sediment and the microclimate near the ground (Montaña, 1992; Boeken & Shachak, 1994; Silvertown & Wilson, 1994). As standing dead, seedlings also add structure that affects other organisms (Boeken & Shachak, 2006).

15.3 | Seedlings and spatial heterogeneity of drylands

Vegetation and landscape patchiness

One of the most pronounced features of water-limited ecosystems is the spatial heterogeneity of plant growth among patches with high or low productivity, density, biomass, and species richness (Montaña, 1992; Boeken & Shachak, 1994; Soriano *et al.*, 1994; Boeken *et al.*, 1995). Within watersheds, tops of hills, slopes, and valleys have different vegetation cover. At the smaller scale, patches often vary internally in plant density and size due to position and edge effects. The contrast among patch types increases inversely with rainfall, from patches with more vegetation in semiarid grasslands and shrublands to contracted vegetation only in/along depressions in more extremely arid landscapes (Fig. 15.1).

Spatial heterogeneity of plant recruitment arises due to the interaction of rainfall with landscape patchiness, usually associated with localized accumulation of water. Other factors also vary among patches, like soil depth, water and nutrient content (Zaady *et al.*, 1996a), seed accumulation (Boeken & Shachak, 1994; Prasse & Bornkamm, 2000), and microclimate amelioration (Montaña, 1992; Silvertown & Wilson, 1994). Plants respond in a great variety of ways to all or some of these factors, depending on their local prevalence and on species-specific traits. Therefore, primary production, species richness, and composition vary along with spatial landscape heterogeneity.

Resources are more available in some patches due to source-sink relationships between source patches where soil water infiltration

Fig. 15.1 Spatial heterogeneity of vegetation patches from scattered shrubland to contracted vegetation on loessial soils along a rainfall gradient in the Negev Desert, Israel. (a) Dense shrubland in Lehavim, northern Negev (200–300 mm/yr); (b) scattered shrubland in Sayeret Shaked Park, western Negev (150–200 mm/yr); (c) sparse shrubland near Sede Boqer, central Negev highland (50–100 mm/yr); and (d) shrub vegetation in dry streambeds near Ein Aqrabim, Eastern Negev (<50 mm/yr). Photos by B. Boeken.

rate is lower than rainfall intensity, causing runoff (Montaña, 1992; Boeken & Shachak, 1994; Silvertown & Wilson, 1994; Zaady & Shachak, 1994; Tongway et al., 1996; Shachak et al., 1998; Eldridge et al., 2000), and sink patches where water accumulates (Montaña, 1992; Boeken & Shachak, 1994; Silvertown & Wilson, 1994; Shachak et al., 1998; Prasse & Bornkamm, 2000; Boeken & Orenstein, 2001; Zaady et al., 2004; Belnap et al., 2005; Ludwig et al., 2005). The sinks function as the basis for denser vegetation patches, which exist in a variety of forms, sizes, and configurations in most semiarid and arid systems of the world. In contrast, vegetation cover of source patches is sparse (Montaña, 1992; Boeken & Shachak, 1994; Silvertown & Wilson, 1994; Zaady & Shachak, 1994), both because of resource loss by runoff and because of the dense surface structure that causes runoff. Surface runoff during rainfall is generated on bedrock surface (Yair et al., 1980; Yair & Shachak, 1982; Shachak et al., 1991; Boeken et al., 1995), slopes with compacted soil surface (Schlesinger et al., 1990), or biological crust cover on loess (Boeken & Shachak, 1994; Zaady & Shachak, 1994;

Tongway *et al.*, 1996; Shachak *et al.*, 1998; Eldridge *et al.*, 2000) and stabilized sand (Verrecchia *et al.*, 1995; Kidron *et al.*, 1999).

Sink patches are the result of geomorphologic processes, such as
depressions, runnels, or soil patches within rock outcrops (Boeken
et al., 1995), animal disturbances like burrows, diggings, and wallows
(Gutterman, 1982; Boeken *et al.*, 1995; Gutterman, 1997; Whitford &
Kay, 1999), or patch formation by large plants. Animals, plants, and
crust microorganisms are important ecosystem engineers in water-
limited systems, shaping landscape heterogeneity. The result of their
activity is a mosaic of source and sink patches with low and high vegetation density, productivity, and species richness (Montaña, 1992;
Boeken & Shachak, 1994; Silvertown & Wilson, 1994; Boeken *et al.*,
1995). On sandy and heavier loessial soils in all semiarid landscapes, vegetation patches (islands of fertility) are formed by shrubs,
trees, cacti, and tussock grasses, as herbaceous understory (Boeken &
Shachak, 1994; Franco & Nobel, 1989), grass rings around the perennial canopy (Soriano *et al.*, 1994; Reisman-Berman, 2004), or more
complex patches such as arcs (Montaña, 1992) and bands (Dunkerley,
1997). In more arid systems, geomorphic processes and animal disturbances are often more pronounced (Yair & Danin, 1980; Shachak *et al.*,
1991; Boeken *et al.*, 1995), and plant and crust-based patch formation
less pronounced (Shem-Tov *et al.*, 1999, 2002).

The difference between animal- and plant-driven patch formation processes is that animal disturbances in the form of diggings
arise suddenly and may recover faster to their pre-disturbance state
(Gutterman *et al.*, 1990) than plant-based patches, which grow slowly
due to accumulation of sediment and litter and capture of runoff
water (Montaña, 1992; Soriano *et al.*, 1994; Shachak *et al.*, 1998). Transitions from the former to the latter are also possible when diggings
enable development of large perennials (B. Boeken, unpubl. data).
Trapping runoff and debris in animal disturbances is a direct effect
of topography (basically soil pits), which limits expansion and even
ensures recovery to undisturbed surface (Gutterman *et al.*, 1990). Plant-
based patches can in principle expand until source limitation stops
it or erosion processes reverse the expansion.

Seedling roles in plant responses to landscape patchiness
All life cycle states can be affected in various ways by spatial heterogeneity, as patches differ in structure, resource availability, microclimate, and animal consumers. Plants differentiate between dense
vegetation and open patches, and seedling recruitment and survival
often play an important role. Most plants, especially annuals, are
more abundant in dense vegetation patches (Went, 1942; Montaña,
1992; Boeken & Shachak, 1994; Silvertown and Wilson, 1994), while a
minority is specialized for growing in open, exposed sites. However,
only a few species have adaptations for increased seedling recruitment and survival in specific patch types and these are mostly found
in species with a preference for open patches. Vegetation patch preference is based either on differential seed arrival, retention, and

germination, or on greater seedling survival and maturation. While greater resource availability and conditions for germination and survival in vegetated patches (islands of fertility) are usually cited as the main factors (Nobel, 1989; Van Auken & Bush, 1990; Valiente-Banuet et al., 1991a,b; Valiente-Banuet & Ezcurra, 1991; Bowers & Pierson, 2001; Walker et al., 2001; Nobel & Bobich, 2002; Nobel & Zutta, 2005), seed arrival and retention alone may be sufficient to differentiate between open and vegetated patches (Esler, 1999). Although no quantitative data are available, it appears that for a majority of annual herbs in dryland systems, passive secondary dispersal is responsible for their greater incidence and abundance in vegetated patches. Some adaptations for primary dispersal may occur, but are rarely elaborate or involve high investment costs. Notable exceptions exist, such as in the annual chili pepper, *Capsicum annuum* (Solanaceae), in Arizona (USA) (Tewksbury & Nabham, 2001), whose fruit is consumed by birds that differentially deposit seeds under perch trees, while the capsacin repels mammals, preventing random dispersal. Directed animal dispersal is exceptionally common in the South African Karoo, and includes the mutualism between *Cucumis humifructus* (Cucurbitaceae) and aardvarks (*Orycteropus afer*) (Meeuse, 1963), and the diverse fruit structures for endo- and exozoochoric dispersal in many herbaceous species that inhabit resource-rich Mima mounds (*heuweltjies*) (Esler & Cowling, 1995; Esler, 1999).

Seeds are easily transported by water and wind across flat surfaces to sink patches (Boeken & Shachak, 1994, 1998a,b; Soriano et al., 1994; Aguiar & Sala, 1997), where rough surface, litter, and plants capture the seeds (Prasse & Bornkamm, 2000; Boeken & Orenstein, 2001; Ludwig et al., 2005). Round seeds without much structure are good ground dispersers, and preferentially come to rest in vegetated patches (Boeken & Shachak, 1994; Gutterman, 1994). Interestingly, Ellner and Shmida (1981) interpreted their lack of morphological adaptations as evidence for nondispersal, similar to topochory. After arrival in vegetated patches, seeds are less likely to be removed by physical flows (Ludwig et al., 2005; Prasse & Bornkamm, 2000; Boeken & Orenstein, 2001) or granivory (Reichman 1984; Wilby & Shachak, 2004), and species with dormancy can build up seed banks. Following rainfall, simple minimum moisture requirements for germination combined with greater resource availability (Zaady et al., 1996a,b, 1998; Walker et al., 2001) increase seedling recruitment rate, growth, and survival compared to open patches.

Many perennial dryland plants have a more exclusive preference for vegetation patches, as protégés of nurse plants (Cody, 1993; Holmgren et al., 1997). This is generally attributed to milder microclimate and greater resource availability compared to open patches (Silvertown & Wilson, 1994; Brittingham & Walker, 2000; Hastwell & Facelli, 2003) where they rarely occur as seedlings (Cardel et al., 1997; Bowers & Pierson, 2001; Barchuk et al., 2005). Seed trapping may also play a role in the success of perennials (Flores & Jurado, 2003), but it appears that differential seed arrival at the nurse plant is not explicitly considered as in annuals. The facilitative effects of nurse plants include milder winter and summer soil temperatures (Nobel, 1989; Nobel & Bobich, 2002), lower summer air temperatures and evaporation rates (Valiente-Banuet et al., 1991a,b; Valiente-Banuet & Ezcurra, 1991), higher soil moisture (Bowers & Pierson, 2001), and higher nutrient concentrations (Carrillo-Garcia et al., 2000; Walker et al., 2001), and mycorrhiza presence (Carrillo-Garcia et al., 1999). While most of the nurse plant situations in dryland systems have been described for stem and rosette succulents in North America, especially Agave, Yucca (Agavaceae), and various large Cactaceae, many shrubs and trees grow preferentially as seedlings in resource-rich patches. The high vulnerability of nurse plant protégés during the seedling stage is compensated by very effective resource use and high survival rates as adults (Esler & Phillips, 1994; Bowers et al., 2004), as demographic theory predicts for perennials. Eventually, the survivors grow large and may replace the nurse plant in patch dominance (Yeaton & Esler, 1990; Valiente-Banuet et al., 1991b; Cody, 1993).

Part of the success of desert plants in densely vegetated patches relative to open patches is due to protection against granivores and herbivores (Flores et al., 2004). Harvester ants in semiarid and arid shrubland, for instance, rarely collect seeds in dense vegetation (Gutterman & Shem-Tov, 1996, 1997b; Wilby & Shachak, 2004). Many rodents and birds also forage more in open areas (Reichman, 1984; Kotler & Brown, 1999), although in sandy dryland some granivorous rodents forage more under shrubs due to predation risk in the open (Kotler et al., 1994).

Seedlings in open patches

Exposed patches have a more extreme microclimate with greater variation than vegetated patches (Montaña, 1992; Boeken & Shachak, 1994; Silvertown & Wilson, 1994; Boeken et al., 2004). Because they usually function as sources for resource flow into the more mesic vegetated patches, open patches are poor for establishment and growth. Nevertheless, only a minority of annual and perennial species never appears in open patches, and I expect that most occasionally occur in open patches, but with significantly lower densities and higher mortality. There are also many desert species for which open patches provide preferred sites for seedling recruitment with greater seedling densities. Bare areas between vegetated patches in the Chihuahuan Desert (Mexico) (Montaña, 1992), for example, were colonized by short-lived perennials, which hardly ever appeared in the vegetation patches.

For species preferring open patches, growing in sparse vegetation with few competing neighbors compensates for the lower growth rate and fecundity due to limited soil water (Callaway et al., 1996; Holmgren, 2000; Walker et al., 2001). For this to work, evergreen perennial seedlings require drought resistance or tolerance to survive dry seasons or prolonged drought, and often have rapid vertical and/or horizontal root growth. Ephemeral perennial and annual grasses and forbs actively grow only during rainfall seasons and pulses, but dry and hot spells are very common, causing low growth rates and possibly stress and mortality in open, resource-poor patches.

Timing of germination is crucial, especially for annuals that prefer exposed patches. Indeed, rainfall-triggered dispersal and predictive germination (Ellner & Shmida, 1981; Esler, 1999; Gutterman, 2002), which increase seedling survival and spread germination over time, work best in exposed sites and in more arid environments with extended open areas. In fact, such specialized seed dispersal and germination adaptations (Gutterman, 2002) are primarily associated with preference for open patches, in contrast to vegetation patches where minimal germination thresholds and passive secondary dispersal are adequate. Interestingly, some species with adaptations for dispersal to and retention in open sites, such as *Plantago coronopus* (Plantaginaceae) and *Carrichtera annua* (Brassicaceae), germinate and grow significantly better in vegetated patches (Zaady *et al.*, 1997; Boeken & Shachak, 1998a,b), but recruitment is seed limited due to nondispersal. Many small-statured annual species, such as *Rostraria, Schismus* (Poaceae), *Filago, Ifloga* (Asteraceae), and small Caryophyllaceae and Crassulaceae in semiarid Negev shrubland, are more dense (as seedlings or adults) in open patches covered with biological crusts (Boeken & Shachak, 1994, 1998a,b; Prasse & Bornkamm, 2000; Tielbörger & Kadmon, 2000). Aided by early germination and maturation, their seedlings may avoid competition for light and water, especially during wetter years, as was shown for *Ifloga spicata* (Prasse & Bornkamm, 2000; Tielbörger & Kadmon, 2000). In more arid regions, the larger open areas with less-developed biological crusts (Shem-Tov *et al.*, 2002) are populated by larger annuals with small wind-dispersed seeds accumulating near stones, many of them drought-tolerant annual Chenopodaceae like *Salsola* and *Suaeda* (Gutterman, 2002).

Species with preference for open patches have to overcome site-constrained seedling recruitment and survival due to harsher microclimate and lower resource availability relative to vegetated patches. They also manage to avoid adverse effects of seed-limited seedling recruitment (Boeken & Shachak, 1994, 1998a,b, 2006) due to removal from open patches by transport to vegetated patches (Boeken & Shachak, 1994, 1998a,b; Esler, 1999; Prasse & Bornkamm, 2000) and by seed predation (Gutterman & Shem-Tov, 1996, 1997b; Wilby & Shachak, 2004). Seed removal by transport and predation is closely associated with the surface properties of open patches and directly or indirectly with their source function. This depends on whether the soil is bare or covered with a biological crust (Zaady & Shachak, 1994; Verrecchia *et al.*, 1995). On biological crusts, runoff flow and wind along the ground remove all materials, including dust, plant litter, and seeds, until deposited in structured sink patches, that is, either vegetated patches or recent disturbances (Boeken & Shachak, 1994; Boeken *et al.*, 1995). Because well-developed flat, dense cyanobacterial crusts form a barrier against seed penetration into the soil (Boeken & Shachak, 1994, 1998a,b; Zaady & Shachak, 1994; Shachak *et al.*, 1998), larger seeds are more likely to be transported than smaller ones, which are more likely to be retained near obstacles and in surface cracks (Yeaton & Esler, 1990; Boeken & Shachak, 1994). In addition to passive transport, exposed seeds in open patches have a higher chance to be removed by seed predation (Boeken & Shachak, 1994), which may or may not limit seedling recruitment. Their greater apparency makes open patches more profitable than densely vegetated ones, especially for diurnal foragers using visual cues like birds (Mares & Rosenzweig, 1978; Kerley, 1991; Folgarait & Sala, 2002) and harvester ants (Abramsky, 1983; Gutterman & Shem-Tov, 1996, 1997b; Wilby & Shachak, 2004), although energetic cost-benefit considerations suggest that granivores would forage more in patches with high seed densities (Reichman, 1984; Aguiar & Sala, 1997; Folgarait & Sala, 2002).

Many well-known topochoric and short-distance dispersing annual plant species (Gutterman, 1993; Boeken et al., 1995; Boeken & Shachak, 1994, 2006) are more numerous in open patches because they adhere to or penetrate the soil surface, especially where covered by firm microbial crust (Gutterman & Shem-Tov, 1996, 1997b; Zaady et al., 1997). Some of the most widespread and common dominants of open areas have elaborate mechanisms for early occupation of germination sites. This is illustrated by the hydrochastic soil penetration mechanisms of the annual grass Stipa capensis (Evenari et al., 1982; Gutterman, 1993; Kadmon, 1993; Boeken et al., 2004; Boeken & Shachak, 2006) and hemicryptophytic and annual desert species of Erodium (Geraniaceae) (Stamp, 1984; Gutterman, 1993). Most of these are dominants of open sites in semiarid and arid shrubland, respectively, throughout North Africa, the Middle East, and Central Asia (Feinbrun-Dothan, 1986).

Locally, very common annual forb species of open patches have seeds that are dispersed during rainfall by rain-triggering mechanisms and rain splash, and adhere to soil surface by means of mucilage in the seed coat. In the Middle East and North Africa, such species include Anastatica hierochuntica, Carrichtera annua (Brassicaceae), Plantago coronopus (Plantaginaceae) (Gutterman & Shem-Tov, 1996, 1997b; Gutterman, 2000, 2002), Aizoon hispanicum and Mesembryanthemum nodiflorum (Aizoaeae), and Blepharis (Acanthaceae) (Gutterman, 1972). In the South African Karoo, such mechanisms are known in Augea capensis (Zygophyllaceae), Euryops multifidus (Asteraceae), Lepidium spp. (Brassicaceae), and many Aizoaceae, annuals as well as perennials (Esler, 1999). It is not known whether they are also common in open site species of other deserts, for example, Pectocarya recurvata (Boraginaceae), Plantago patagonica, and Schismus barbatus (Poaceae) in North and South America (Pake & Venable, 1995).

Besides timing of germination, seed retention and surface adherence are especially effective mechanisms for protecting seeds against predispersal and postdispersal predation (Gutterman & Shem-Tov, 1996b; Gutterman, 2002). Soil penetration alone is also effective, as in Stipa capensis in semiarid shrubland of the Middle East (Wilby & Shachak, 2004). Harvester ants (Messor spp.) collect large amounts of *Stipa capensis* seeds, but generally seed predation does not limit seedling recruitment. The collected seeds are apparently the zero-fitness individuals that did not manage to penetrate the soil.

Among the species that prefer open, sparsely vegetated patches are many perennials (clonal grasses and geophytes, shrubs, trees, and succulents), some of which are important landscape modulators as they form vegetation patches (Yeaton & Esler, 1990; Montaña, 1992), often becoming nurse plants themselves. Dominant patch-forming tree and shrub species, such as *Acacia papyrocarpa* (Mimosaceae) of open Australian woodland (Facelli & Brock, 2000), establish in exposed sites, unaided by obstacles, but often having rapid germination and root development during rainy periods. However, in most cases, exact recruitment sites and modes are not known. In contrast, *Attractylis serratuloides* (Asteraceae), a small patch-forming shrub of loessial soils in the semiarid northern Negev of Israel, requires small-scale distur-bances of soil crust such as wider cracks and small pits dug by beetles or larger ones by porcupines (*Hystrix indica*) (B. Boeken, unpubl. data). *Ibex* and *Gazella* diggings and wallows (Gutterman, 1997, 2001) also enable plants to evade the barrier posed by biological crust. Other perennial species of *Stipa* and *Erodium* have crust-penetrating seed mechanisms (Stamp, 1984; Haase *et al.*, 1999; Hensen, 1999).

In contrast to loessial soil where biological crust cover is prevalent in open patches and bare uncrusted soil is transient in disturbance patches, sandy deserts have large areas with mobile sand (dune tops, complete dunes, and dune areas) with restricted areas of stabilized surface covered with biogenic crust and scattered vegetation patches (Verrecchia *et al.*, 1995). Similar to loessial soil, the crusted surface of stabilized sand forms a barrier to seed arrival, seed retention, and seedling recruitment, but mobile sand poses entirely different chal-lenges because of deep-water infiltration and blowing sand. Many of the pioneer species of semiarid and arid shifting sands are peren-nial grasses and woody plants, whose rapid vertical or lateral root development contributes to survival and to stabilization (Huang & Gutterman, 1998; Shirato *et al.*, 2005). In perennial grasses like *Poa bulbosa* and *Stipagrostis* spp. (Poaceae) of Africa, the Middle East, and Central Asia (Bornkamm *et al.*, 1999; Milton & Dean, 2000), layered growth overcomes periodic covering by sand. In addition, seedling establishment may be aided by the sand-binding and water-holding effects of mucilaginous rhizosheaths, which are known in other grasses (Bailey & Scholes, 1997) and cacti (North & Nobel, 1997). Woody pioneers of mobile sand have rapid vertical root growth, some with lateral vegetative propagation. Examples include *Prosopis glandulosa* (Mimosaceae) in SW North America (Campbell, 1929), the chenopod trees *Haloxylon persicum* and *H. aphylla* of sand dunes in central Asian deserts (Nechaeva, 1985), the asteraceous shrubs *Artemisia monosperma* in the Middle East and China (Huang & Gutterman, 1998) and *A. halo-dendron* in China (Shirato *et al.*, 2005), and the shrubs *Hedysarum laeve* (Leguminosae), *Sabina vulgaris* (Cupressaceae), and *Salix psam-mophila* (Salicaceae) of western China and Inner Mongolia (Roels *et al.*,

2001). Possibly, hydraulic lift (Horton & Hart, 1998), as found in *Yucca* (Agavaceae) (Yoder & Nowak, 1999) and *Prosopis* (Leguminosae) species (Hultine *et al.*, 2004), enhances water availability and sand stabilization near the plant.

Seedling involvement in patch formation

The widespread occurrence of stable vegetation patches illustrates that plants not only respond to landscape patchiness, but also play a crucial role in patch development. Here, too, the seedling stage is of importance if a perennial plant establishes in open space or a small disturbance and starts to function as a trap for airborne dust (Offer *et al.*, 1998; Shachak & Lovett, 1998), locally produced litter (Montaña, 1992; Boeken & Shachak, 1994; Silvertown & Wilson, 1994), and animal feces (Zaady *et al.*, 1996b). Many nurse plants are involved in patch formation as seedlings, but their protégés only as adults (Yeaton & Esler, 1990). Under semiarid conditions with loessial soil, the accumulated material forms mounds under shrubs or perennial grass clones. In more arid conditions, the mounds are often not more than a layer of plant litter (Boeken *et al.*, 1995). In addition, mounds may be enhanced as a result of erosion of the interpatch matrix (Schlesinger *et al.*, 1990). When there is runoff from source patches during rainfall, the mound functions as a sink as it absorbs water (Fig. 15.2) because of slower flow (Kosovsky, 1994) and increased infiltration rate on its loose friable surface (Eldridge *et al.*, 2000).

In sandy soils, perennial seedlings established in open sites also form mounds (Carrillo-Garcia *et al.*, 1999). On stabilized sand with biological crusts (Verrecchia *et al.*, 1995), similar dynamics take place as on loess, but differ when the sand is loose (Fig. 15.2). Water loss due to deep infiltration in unstabilized sand can be locally counteracted by the combined effects of litter accumulation on the surface, organic matter in the soil, and by rhizosheaths and hydraulic lift of rapidly developing root systems. This results in greater soil integrity and shallower soil moisture storage around the plant (Su *et al.*, 2005). The improved soil and surface properties enhance colonization of the patch by other plants, producing a vegetation patch. In plants that grow in layers (*Stipagrostis* spp., *Poa bulbosa*, and various shrubs), such patches secondarily develop raised mounds (nebkhas) by sand accumulation on the patch and wind erosion of the intershrub space (El-Bana *et al.*, 2003).

The development of vegetation patches around a single perennial (grass tussock, shrub, tree, or stem of a rosette succulent) or a small group of them involves multiple feedback relations (Fig. 15.2). I hypothesize that interactions among plant size, mound size, and capture of runoff and sediment drive patch formation. If a mound begins to form around a seedling by material accumulation, more runoff water is captured and the plant grows in size, which increases accumulation, and so on. The second positive feedback process takes place as the mound also retains seeds and site quality is improved by runoff capture. Other plants will establish on the mound, forming

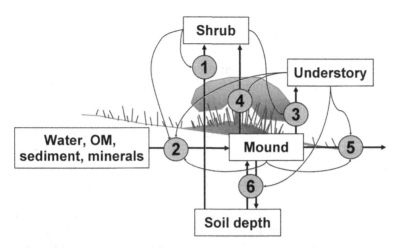

Fig. 15.2 Diagram of resource flows into, within, and out of a shrub patch, illustrating patch formation by a shrub on crusted loessial soil or unstabilized sand. Boxes denote state variables of system components [shrub dimensions, understory density, mound dimensions and surface structure, and pools of water, organic matter (OM), and sediment], thick arrows are flows of resources, and numbered circles and thin curved arrows are controls of flows by system components. On loessial soils horizontal movements predominate and vertical flows establish later, whereas in sand vertical flows initially predominate until stabilization of the intershrub matrix by crust development allows horizontal flows. Major processes are: 1 – Water uptake by the shrub seedling from deeper soil (especially on sand) is affected by root structure and OM, increasing water holding capacity, supporting shrub growth, and greater rooting space. 2 – Accumulation of local and imported litter and sediment under the growing shrub seedling forms a mound that traps runoff water (on loessial and/or crusted surface) and other resources. 3 – The enriched mound supports a colonizing herbaceous understory community, facilitated by microclimate amelioration by the shrub canopy. 4 – The mound contributes to shrub growth, affected by interference and/or facilitation by the understory. 5 – The understory vegetation consolidates and expands the mound, increasing resource supply and decreasing leakage. 6 – resources captured in the mound move to deeper layers, depending on vegetation-induced changes in soil structure.

the understory (Boeken & Shachak, 1994). These often smaller plants add to the mound's surface roughness, friability, and cover, even when they become standing dead or loose litter, further increasing capture and retention of resources, debris, and seeds. Expansion of the vegetated patch will, according to the hypothesis, decelerate to a halt when negative feedback mechanisms cause (1) source limitation due to area reduction of the unvegetated matrix, including interference by other vegetation patches (Boeken et al., 1998), reducing further accumulation of material and runoff capture, or (2) limitation of the sink patch as erosion exceeds material accumulation and water capture becomes insufficient to sustain understory plant growth. Crust encroachment or animal trampling (Yeaton & Esler, 1990; Golodets & Boeken, 2006) can further cause their decay.

When understory individuals of larger species overtop the patch, succession sometimes takes place with dominant species acting as ecosystem engineers. This continues the process of patch expansion from a patch with a small canopy and mound to one with a taller, wider, or denser canopy and a larger mound. The early successional patch formers are nurse plants, while some of the later successional species are obligate nurse plant protégés. This has been observed for shrubs, trees, rosette succulents, and cacti in semiarid and arid drylands worldwide (Yeaton & Esler, 1990). Examples from semiarid Negev shrubland at 150–200 mm of rainfall per year are the dominant chenopod shrub *Noaea mucronata* and the asteraceous sub-shrub *Attractylis serratuloides*, which have higher seedling recruitment and growth in any sink patch with or without a specific nurse plant (B. Boeken, pers. obs.), apparently facilitated by soil moisture more than microclimate. Adult *Attractylis serratuloides* patches do not always contain *Noaea mucronata*, while adults of the latter are almost exclusively found in larger patches associated with *Attractylis serratuloides* or its remains.

Because the processes of patch formation depend on rainfall and runoff, the rates of plant growth and patch expansion vary in time with local annual rainfall, especially when frequent or prolonged droughts occur. Similarly, spatial variation within watersheds occurs in depressions, gullies, and stream beds, where more resources and materials arrive, but also erode faster if channel flows are strong and frequent. Moreover, because of geographic variation along rainfall gradients, patches are highly diverse in size, shape, density, and configuration. Thus, closed shrub or tree canopy with scattered unvegetated patches occur in mesic locations, scattered vegetation patches within an exposed matrix in semiarid zones, and contracted vegetation in arid and hyperarid climates.

15.4 | Seedlings and dryland system degradation

Dryland degradation

Dryland degradation is very common in many semiarid and arid regions of the world (Dregne, 1986), resulting from xerification (West, 1986) by large-scale reduction of resource input, when rainfall patterns change or when water flow within watersheds is blocked or eliminated altogether, and desertification by land-use-related changes that operate at the smaller local scale. The latter anthropogenic impacts include cutting of woody plants for firewood and heavy livestock grazing. They affect the vegetation, ecosystem functions, and landscape structure of patches within watersheds, while their spatial extent often encompasses many whole watersheds or entire geographic regions (Dregne, 1986).

The intricate multiple relationships among plant communities, resource distribution, and landscape heterogeneity outlined above indicate that dryland ecosystem degradation is not simply a matter of resource loss, but a system-wide phenomenon simultaneously involving processes in many components, often linked by feedback interactions. There are many factors driving degradation processes, such as long-term drought, grazing and browsing, trampling, shrub/tree cutting, and fire (Le Houérou, 1992). These factors operate by two primary pathways, direct impacts on plants, increasing mortality, which cascades to other components, and direct impacts on landscape heterogeneity by transforming patch structure, usually their surface properties. Either leads to changes in resource distribution in the form of water leakage out of the system by overland runoff (Montaña, 1992; Shachak *et al.*, 1998; Schlesinger *et al.*, 1999), subterranean water flow (Huxman *et al.*, 2005), or deep infiltration (Berndtsson *et al.*, 1996), and are often accompanied by soil erosion and dust production in heavier soils and sand movement in dune areas (Barth, 1999). During the processes of accelerated decay of vegetated patches, caused either by direct impacts on plants or indirectly by substrate erosion, seedling recruitment and survival decline or do not occur, primarily because of the herbaceous layer. While large perennial adults may survive in suppressed form, their seedlings have an equally low chance of establishing in decaying vegetation patches.

A very common cause of degradation in dryland is heavy livestock grazing (e.g. Dregne, 1986; Le Houérou, 1992). Although low levels have positive effects on diversity in closed-canopy shrub and woodland vegetation of subhumid zones (Osem *et al.*, 2002), grazing suppresses dominance and opens up sites for new colonization, as predicted by the Intermediate Disturbance Hypothesis (Fox, 1979). Whether the Grazing Optimization Hypothesis (Dyer *et al.*, 1993; Noy-Meir, 1993) also pertains to these systems is not known, but positive effects of low grazing intensity on primary production by means of removal of slow-growing species and elimination of self-shading are possible. At lower rainfall, most drylands have patchy perennial vegetation, where opening of sites is irrelevant and grazing in the open patches suppresses the dominants. In open semiarid shrubland of the northern Negev, moderate sheep grazing (Golodets & Boeken, 2006) primarily impacts the surface properties of vegetated and exposed patches more than the herbaceous vegetation itself. Soil and litter mounds under the patch-forming shrubs *Noaea mucronata* and *Thymelaea hirsuta* (Thymelaeaceae) became smaller and lower, while the crust of the intershrub matrix became cracked to a greater degree (Oren, 2000; Golodets & Boeken, 2006). At higher grazing intensity, the grazing of the shrub understory vegetation (unpubl. data), browsing of the shrub canopy, and trampling of mounds together gradually eliminate the mounds and herbaceous understory, often killing the shrubs. Oren (2000) demonstrated this process experimentally in semiarid Negev shrubland; artificial trampling led to erosion of shrub mounds within two rainy seasons. Trampled crust surface also eroded, but this only initially decreased runoff. During degradation of vegetation patches, seedling recruitment and survival is severely limited, reducing mound protection and maintenance.

In semiarid perennial grassland, heavy grazing relative to productivity can easily lead to mortality of ramets or whole clones, especially during dry seasons (Grice & Barchia, 1992). Above a certain level, positive feedback interactions among vegetation, soil accumulation,

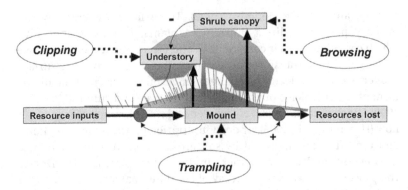

Fig. 15.3 Diagram of the effects of grazing on shrub patches in semiarid Negev shrubland. Boxes denote state variables (shrub canopy dimensions, understory density and height, mound dimensions and surface structure, and amounts of water, organic matter, and sediment), thick arrows are flows of resources, thin lines and circles are controls (positive or negative) of flows, and ovals (and dotted lines) are grazing impacts; browsing reduces the shrub canopy, clipping the herbaceous understory exposes the mound, and trampling disintegrates it. Positive feedback interactions occur among mound, resource inputs, and understory vegetation. Because reduced mounds capture less resources, both mounds and resources offer less support for herbs and shrubs, and sparser vegetation exposes the mound to erosion. Seedling involvement in the degradation of the herbaceous understory vegetation is due to their reduced recruitment in the understory.

and resource capture lead to dwindling tussock or bunch patches, increased resource leakage, and reduced productivity (Schlesinger et al., 1999; Snyman, 1999). Whether changes in diversity occur depends on the facilitative effects of the dominant grasses on other plant species. Decay of perennial grass patches (Mangan et al., 2004) is similar to that of shrub patches (Fig. 15.3) with a single feedback process as the grass clone takes the function of both canopy and understory plants. In North American and South African, semiarid perennial grasslands, the altered spatial heterogeneity enables seedling establishment of unpalatable shrubs, resulting in shrub encroachment (Schlesinger et al., 1999; Roques et al., 2001; Berkeley et al., 2005; Huxman et al., 2005; Ward, 2005).

Seedlings of species that prefer open patches are similarly affected by heavy livestock grazing, because degradation occurs in the expanding and increasingly bare matrix between vegetated patches. The degradation processes of the open interpatch matrix set in motion by heavy livestock grazing are different for heavy, loessial soils and sand, due to the greater vulnerability to trampling of biological crusts of stable sand (Verrecchia et al., 1995) and to greater water infiltration rates of uncrusted sand (Abu-Awwad, 1997; Kidron et al., 1999). Although few details are known about how these processes vary among the drylands of the world, it seems that heavy grazing on sandy soil primarily impacts the crust cover of the interpatch matrix (Verrecchia et al., 1995; Mazon et al., 1996; Karnieli et al., 1999; Kidron et al., 1999; Shirato et al., 2005), and secondarily the vegetated patches. On loessial soil, the primary impact is on shrub patches as described earlier

Fig. 15.4 Degraded landscapes in spring, resulting from heavy livestock grazing. (a) Reduced shrub cover and gully erosion on loessial soil near Beer-Sheva. (b) Absence of soil crusts with moving sand along the Sinai side of the Israel–Egypt border (west, background) and, in the foreground, spring vegetation on stabilized sand on the Negev side. Photos by B. Boeken.

with less pronounced and smaller-scale impacts of crust degradation (Oren, 2000). It should be noted that if biological crust on loess is completely removed by deeper soil disturbance, rainfall water infiltration increases significantly, decreasing runoff (Eldridge *et al.*, 2000). However, such disturbances are not due to livestock trampling, but to localized processes such as animal diggings, which unlike degradation contribute to heterogeneous landscape structure. Grazing and trampling cause sink limitation on loessial soils and source limitation on sandy soil. Either way, both source and sink limitation lead to system-wide decline in productivity and diversity.

On semiarid loessial soil, sink limitation by reduced vegetation patches causes increased runoff flow, resulting in greater energy over larger source areas (Shachak *et al.*, 1998) and decreasing local retention of water, sediment, and seeds (Fig. 15.4). Under these conditions, seedlings may completely fail to establish, including perennials that would otherwise maintain patchiness. As a result of degradation processes in vegetated patches and open areas, productivity and plant species diversity of entire watersheds inadvertently decline due to diminishing heterogeneity and resource availability. The processes of degradation involve a number of positive feedback interactions, which given their natural course, would only stop when no more erodable material is available. Failure of seedling recruitment and survival of patch-forming perennials is part of a positive feedback process at a larger temporal scale, counteracting or preventing recovery of degraded landscapes by shrub reestablishment. In addition, the increased overland water flow causes upward moving gully erosion (Fitzjohn *et al.*, 1998) and floods in the watershed's outflow channel (Domingo *et al.*, 2001), both leading to severe erosion of slopes and channels by removal of large quantities of soil.

On sandy substrate (dunes and ergs), livestock and people cause large-scale breaking of the biological crust (Meir & Tsoar, 1996; James *et al.*, 1999). The process is strongest near human settlements and watering points, spreading outward over large areas (James et al., 1999). Therefore, biological crust cover of sand is rare in many drylands, restricted to stabilized areas where grazing is excluded or greatly reduced (Meir & Tsoar, 1996; Shirato et al., 2005). A well-known consequence of crust cover presence or absence is the albedo contrast of the same sandy landscapes as shown between the brighter soil reflectance in Sinai (Egypt) and the darker-colored stabilized sands of the Negev (Israel) (Fig. 15.4) (Meir & Tsoar, 1996; Tsoar & Karnieli, 1996). While runoff is hardly reduced on loessial soil (Oren, 2000; Zaady et al., 2001), the disappearance of the crust on sandy soil increases water infiltration (Verrecchia et al., 1995; Abu-Awwad, 1997). This affects seedling recruitment and survival in both the vegetated patches and the open interpatch matrix, because vegetated patches receive less runoff, and water is lost by deep infiltration in the matrix. Positive feedback interactions occur as surface sand, unfixed by cyanobacterial crust in the matrix and by the higher plants of the vegetation patches, becomes mobile, further impeding establishment of crust and plants. In less arid sandy systems in Israel (Kutiel, 1998) and South Africa (Milton & Dean, 2000), the positive effects of crust removal on site availability for seedling recruitment outweigh the reduced effects on horizontal and vertical water distribution.

Dryland restoration

Recovery of degraded dryland landscapes without human intervention is at best a very slow process and often unattainable, depending on how strong the positive feedback processes are and to what degree they are constrained by counteractive factors. Even if the direct causes of degradation such as livestock grazing and woody plant removal stop, degradation may go on, only sporadically or rarely mitigated by favorable rainfall periods. During exceptionally wet years with regular but low intensity rainfall, crust recolonization may be accelerated and vegetation heterogeneity may start to reestablish, at least on the soil that has not been eroded. In general, recovery depends on rainfall, the intensity of ongoing impacts, as well as the local severity of the degradation and its spatial extent. In addition, the rate and success of recovery critically depends on whether the degradation is driven by source reduction of runoff-contributing crust surface on sand or by sink reduction of runoff-absorbing vegetation patches on loessial soils. Thus, because even in degraded loessial dryland systems the exposed soil surface produces runoff, recovery is likely to be far more rapid and system functions are easier to restore than sandy systems where horizontal water redistribution only starts if surface stabilization is already occurring. Moreover, sand stabilization may be strongly inhibited by constant covering by blowing sand from large contiguous areas.

Because degradation of loessial landscapes is due to sink limitation, adding sink patches on slopes, flats, and streambeds is a very efficient means to restore landscape heterogeneity. Biological crust cover often recovers by itself if no heavy grazing occurs, probably starting

Fig. 15.5 Restoration of heterogeneity of a degraded loessial slope by means of 30 cm high contour dikes during continuing low intensity grazing, near Beer-Sheva (Israel) (150–200 mm/yr). Photo by B. Boeken.

during the first rainy season (E. Zaady, pers. comm.). Crusts may locally be absent, but because the soil is usually truly bare only at a small scale, sufficient sources for recolonization are available in other parts of the watershed or adjacent ones. Construction of patches, similar to vegetation patches formed by plants (Tongway & Ludwig, 1996) or animal soil disturbances (Shachak *et al.*, 1989; Boeken *et al.*, 1995), at once reduces loss of runoff water, sediment, and organic matter and provides places for the formation of vegetation patches (Boeken & Shachak, 1994, 1998a,b; Tongway & Ludwig, 1996; Shachak *et al.*, 1998). Various techniques and designs are employed on degraded slopes, from 1- × 0.3-m soil pits to 15-m long contour dikes in semiarid and arid parts of the Negev of Israel (Boeken & Shachak, 1994; Shachak *et al.*, 1998) (Fig. 15.5) and similar barriers in the semiarid and arid zones of Chile (B. Boeken, pers. obs.) and Australia (Tongway & Ludwig, 1996). Although most of these restoration methods leave clear signs of human activity, they effectively reduce resource leakage and restore productivity and diversity of the vegetation (Boeken & Shachak, 1994; Zaady *et al.*, 2001) and of various animal groups (Whitehouse *et al.*, 2002).

The involvement of seedlings in restoration of degraded loessial dryland systems is secondary, as runoff into artificial sink patches and lower runoff flow on the interpatch matrix improve seedling recruitment of colonizing plants. In pits and mounds in degraded northern Negev shrubland, numerous herbaceous plant species found under shrubs colonized within 1 to 4 years (Boeken & Shachak, 1998a,b) along with seedlings of the dominant shrubs *Atractylis serratuloides* and *Noaea mucronata* (B. Boeken, unpubl. data), while the larger shrub *Thymelaea hirsuta* successfully colonized the interpatch matrix.

Adding sink patches to degraded sandy dryland by landscape management alone, unlike loessial systems, is ineffective because degradation is driven by source reduction and natural colonization by seedling recruitment from incoming seeds may be very improbable due to unavailability. If only relatively small sand dune areas are degraded, spontaneous recovery is possible during decades of grazing exclosure, as in the Negev desert sands (Meir & Tsoar, 1996). Restoration of system functions requires both source and sink patches, which are absent in degraded moving sands. Restoration of complete crust cover would stabilize the sand surface against wind erosion (Kurtz & Netoff, 2001; Li et al., 2002; Hupy, 2004; Orlovsky et al., 2004), but due to inhibition of seedling establishment on crust (Orlovsky et al., 2004), this does not enhance landscape heterogeneity with productive vegetation patches. Therefore in many sandy dryland regions in the world, early successional psammophytes that stabilize sand are planted in efforts to recover heterogeneity and productivity (Dregne, 1986; FAO, 1997; Singh & Rathod, 2002; Li et al., 2004). Seedlings play a crucial role in restoration, but this is offset by physical processes and ongoing land use. Sand dune degradation often covers large areas (Dregne, 1986), with colonization of local species severely seed limited. Furthermore, survival of naturally recruited and planted seedlings is improbable due to constant sand cover and shear (Singh & Rathod, 2002). However, restoration by means of revegetation with sand-binding perennials simultaneously stabilizes the surface and creates vegetation patches (Nechaeva, 1985; Dregne, 1986; Babaev, 1999; Mukhammedov et al., 1999; Roels et al., 2001; Li et al., 2004; Su et al., 2004, 2005). Locally, planted seedlings start processes of soil and surface improvement, leading to greater water holding capacity, organic matter, nutrient content, and microbial activity, enhancing the development of vegetated patches with higher productivity and herbaceous species richness (Li et al., 2002, 2004; Su et al., 2004, 2005). Often native perennials are planted, such as *Caragana microphylla* (Leguminosae) and *Artemisia halodendron* (Asteraceae) in northern China (Su et al., 2004, 2005); *Calligonum* spp (Polygonaceae), *Haloxylon persicum*, and *Salsola paletzkiana* (Chenopodiaceae) in Turkmenistan (Mukhamraedov et al., 1999); and *Populus simonii* (Li et al., 2003), *Hedysarum laeve* (Leguminosae), *Sabina vulgaris* (Cupressaceae), and *Salix psammophila* (Salicaceae) (Roels et al., 2001) in Inner Mongolia, among others. Nonnatives are also planted, as in Rajastan, India (Singh & Rathod, 2002), with *Acacia tortilis* and *Prosopis juliflora*, but care should be taken not to introduce aggressive invaders. If stabilization by revegetation is successful, biological crust is formed within 3 to 5 years, creating runoff flow between the open areas and the vegetated patches (Li et al., 2004; Shirato et al., 2005).

The success of these efforts at stabilizing moving sand by artificial revegetation is quite variable, from beginning stabilization in a few years to failure for more than a decade of seeding and planting attempts in the large sand deserts of western China and Inner Mongolia (FAO, 1997; Zhou et al., 1997). One of the problems in many degraded sandy dryland systems is that livestock grazing is very difficult to control or stop (Katoh et al., 1998; Shirato et al., 2005), counteracting attempts at sand stabilization and establishment of patchforming perennials and biological crusts. Grazing continues because in transitional states of stabilization in semiarid regions bare sand often still supports exploitable annual plant cover during rainy periods (Katoh *et al.*, 1998; Mukhammedov *et al.*, 1999). If control is possible, grazing during wet seasons can effectively be used to prevent crust inhibition of annual vegetation (FAO, 1997; Orlovsky *et al.*, 2004). However, this kind of management is risky because it heavily depends on ample rainfall and seed availability. Drought and high stocking rates relative to productivity can easily cause local disruption of stabilizing surface and seed depletion of the annual vegetation, while the large scale of land use can prevent recolonization by sand encroachment and seed limitation.

Climate change and dryland systems

Changes in regional climates, especially in rainfall regimes, will undoubtedly affect system structure and function in all types of dryland because of the dependence of the interactions among the vegetation, landscape heterogeneity, and resource dynamics on the amount and temporal variability of water input. Reduced annual precipitation or greater frequency, longevity, and/or intensity of droughts may change these systems, as will higher temperatures and elevated CO_2 levels. The way seedlings respond to the direct and indirect effects of climate change will be crucial for the resilience of the ecosystem. Seedling recruitment and survival will determine population maintenance, community composition, species richness, and the reshaping of landscape heterogeneity.

Higher temperatures, which may be insignificant at the latitudes of most arid and semiarid regions (Le Houérou, 1996), can affect plants during seed germination, seedling establishment, and later survival. For winter annuals with a low temperature requirement (Gutterman, 2002), seedling recruitment may decline, while for others seedling survival may increase in shaded patches, but decrease in the open. Depending on whether important patch-forming plants and others can establish and survive, the result may be either positive or negative for landscape heterogeneity, ecosystem functioning, and species diversity, especially when accompanied by unpredictable changes in levels of herbivory and elevated CO_2 levels that accelerate growth and reduce transpiration (Le Houérou, 1996).

Less rainfall alone could move transitions between subhumid, arid, and semiarid environments, leading to replacement of existing systems by drier ones (Le Houérou, 1996). Whether such a change would be gradual is doubtful, because seedling recruitment and survival of many species, patch forming and others, may be seriously limited by their inability to colonize fast enough. This may cause a delay in replacement and a depauperate landscape with less heterogeneity, productivity, and species richness, which may only slowly and incompletely recover.

However, climatic changes may not always be toward lower productivity and species richness. Longer periods of drought could increase heterogeneity in semiarid dryland with originally high cover of shrub patches, resulting in increased productivity and species richness of herbaceous annuals. In the subhumid to semiarid zone with

200–400 mm/yr in the northern Negev of Israel, for instance, the dominant patch-forming shrub *Sarcopoterium spinosum* (Rosaceae) often forms more or less continuous aggregate patches with scattered open patches (Reisman-Berman, 2004). Annuals grow at high density in a ring around the dense canopy (M. Segoly, pers. comm.). Fragmentation into smaller and fewer patches would increase the patch perimeter and, therefore, seedling establishment of annuals. In more water-limited semiarid landscapes like the 150–200 mm/yr zone in the Negev, scattered shrub patches may become smaller, increasing water and nutrient flows within the watershed, but the decline in productivity and species richness of annuals may be partially compensated by greater water availability and establishment of patch-forming shrubs, like *Thymelaea hirsuta*, along streambeds. More arid desert systems may respond less to drought frequency and more to changes in rainfall intensity. If quantity stays the same but intensity decreases, less or no runoff will occur and the few sink patches near runnels or outcrops will decline, whereas if intensity increases, the additional runoff may not be utilized completely due to sink limitation.

Although semiarid and arid systems of the world will respond differently to changes in their climate, there is little doubt that exploitative land use will not increase ecosystem resilience. Drylands where grazing is reasonably sustainable can easily be degraded and subsequent recovery or restoration may be very difficult.

In humid regions, greater periods of drought may set degradation processes in motion similar to those found in dryland landscapes. Temperate and boreal rangeland with sustainable livestock production may become increasingly more heterogeneous as a result of reduced productivity and recruitment of dominant perennial grasses and forbs. Because these systems are situated far removed from existing water-limited dryland systems with their associated flora, they are less resilient than current dryland systems against drought and heavy livestock grazing. Fewer species are available that are able to cope simultaneously with the more arid conditions, maintain heterogeneity, and reduce the rate of degradation. Recognizing the early signs of changes in spatial heterogeneity may serve the sustainability of water-limited and non-water-limited rangeland systems alike.

15.5 | Conclusions

Concentrating on landscape heterogeneity and its interactions with resource dynamics and the functioning of the various components of the vegetation highlights the central role played by plant seedlings in the functioning of dryland ecosystems. In relation to dryland landscape heterogeneity of vegetated and exposed patches, seedlings perform two distinct roles, first as organisms responding to the environment in different patch types and in different configurations. Second, they serve as landscape modulators creating patches and patterns of patchiness.

Plant responses to landscape patchiness during seedling recruitment and survival are mostly determined by the interaction of plant traits with the structure of the landscape and the movement of material within these relatively open landscapes. Many annual plant species, mostly grasses, respond to landscape patchiness with greater seedling recruitment and survival in vegetated patches where they are preferentially trapped during passive dispersal, avoiding high investment in adaptations for dispersal or dormancy. Elaborate adaptive traits of seeds and seedlings are found mainly in species with a specific preference for exposed patches. In perennials, adaptations for survival in vegetated or exposed patches are mostly found during the established stages, reflecting the lower importance of the seedling stage for longer-lived plants.

Seedlings of many perennial dryland plants play a crucial role in shaping landscape patchiness because of their role in patch formation. While much is known about effects of nurse plants and the islands of fertility they form for other plants, the process of patch formation is rarely documented. The formation of vegetated patches and their effects on other organisms are generally considered related to resource and microclimate amelioration because patches function as sinks for resource flows in a source–sink relationship with their immediate surroundings. While this determines differential survival of seedlings, their recruitment also depends to a large degree on arrival and retention of seeds, which is low in open patches due to removal by water, wind, and predators.

From literature and personal observations of the dynamics of patches formed by shrubs in dryland ecosystems, I have proposed conceptual models for the relationships among patch-forming shrub growth, material accumulation, and resource flows and herbaceous plant colonization and production. These schemes may be the basis for synthesis of the diverse and seemingly unrelated phenomena in various semiarid and arid systems. The positive feedback interactions highlighted are especially promising subjects for comparative study because they may be the key to understanding how landscape structure and ecosystem function are related and how these vary at larger spatial and temporal scales. They are also crucial in understanding how variation in climate and human land use affect landscape patchiness, resource dynamics, productivity, and biodiversity.

Human-driven and natural disruption of these processes creates new interactions between patches as systems become degraded, including strong positive feedback relations leading to continuing deterioration. Although not all cases of dryland degradation are caused by grazing, trampling, or perennial plant removal, the occurrence of feedbacks that make efforts to restore degrading or degraded dryland systems difficult seems universal. Research aimed at conservation of dryland systems and biodiversity and/or sustainability of human land use should identify such feedback mechanisms.

Chapter 16

Anthropogenic disturbance in tropical forests: toward a functional understanding of seedling responses

James W. Dalling and David F. R. P. Burslem

16.1 | Introduction

The last decade has seen rapid growth in research dedicated to seedling ecology (Kitajima, 2007). This reflects an increased recognition of the importance of variation in seedling survival in determining patterns of adult abundance and distribution. Many seedling studies have grown out of larger programs investigating community dynamics in natural systems. Nonetheless, the results of these studies are often directly relevant to the management and conservation of human-altered ecosystems where changing environmental conditions and altered biotic interactions can directly affect seedling recruitment success. In extreme cases, human-mediated disturbances may be sufficient to invoke community-wide recruitment failure, leading to stalled succession or to shifts in vegetation type. Disturbance effects, however, can also have more subtle effects on recruitment success, resulting in changes in forest composition that may take decades or more to become apparent (Dirzo & Miranda, 1990).

Seedlings are particularly vulnerable to disturbance. Most species have no equivalent of the dormancy that facilitates seed survival through periods with adverse environmental conditions. Instead, alterations in light or soil moisture availability often impact seedlings first because their small leaf area and shallow rooting depth limit their ability to integrate resource capture over space. Similarly, only a small fraction of tree species have seed reserves that persist for more than a few months after germination, leaving seedlings vulnerable to damage and to temporal fluctuations in resource supply (Kitajima, 1996a; Harms & Dalling, 1997). Seedlings are also especially susceptible to natural enemies and fire. They may be less well defended or more accessible to herbivores than seeds and adults, and disease outbreaks may spread more rapidly through clustered populations

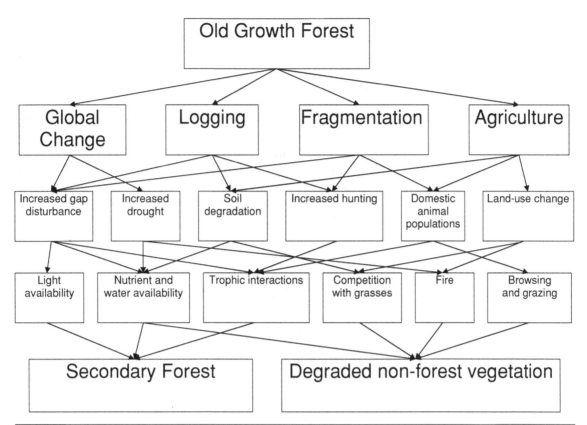

Fig. 16.1 Schematic diagram illustrating the pathways and mechanisms of forest responses to anthropogenic disturbance. We define four initial classes of disturbance that influence old growth tropical forests (level 1), their principal ecological effects (level 2), and mechanisms by which they influence seedling growth and survival (level 3). In turn, these effects collectively determine the variant of disturbed vegetation at a site (level 4). To improve clarity, we have illustrated only the most important linkages between levels, including all those discussed in the chapter. *Trophic interactions* include all biotic interactions (plant–herbivore, plant–pathogen, plant–mycorrhizal, plant–pollinator, and plant–disperser) that contribute to the demography of seedlings in disturbed forests, except competition with grasses, ferns, and shrubs, which is considered separately. *Increased gap disturbance* includes both direct effects of gap creation on abiotic factors plus indirect effects of tree density on tree reproduction (see text for details).

of seedlings than among more widely spaced adults (Gilbert *et al.*, 1994).

In this chapter, we highlight the different pathways by which human disturbances influence seedling recruitment (Fig. 16.1). Here we take a broad view of what constitutes anthropogenic disturbance: localized effects of forest fragmentation, logging, and hunting have clear effects on seedling populations that can often be ascribed to individual processes. However, climate change is a disturbance affecting all forests. Increased tree growth and mortality rates in tropical forests reported over the last few decades should increase the frequency of canopy gap formation, a natural disturbance that influences the species composition of tropical forests. We also recognize that seedlings are not the only life stages affected by disturbance and therefore, we start with a discussion of how disturbance impacts populations. We then address the predominant mechanisms by which disturbance influences seedling recruitment patterns before examining in more detail the potential for species-specific responses. The focus of this chapter is on tropical forest ecosystems where seedling trait variation is high and the conservation implications of a functional understanding of seedling ecology is greatest. Understanding how disturbances associated with logging and with the cultivation and abandonment of agricultural lands affect seedlings will be of critical importance in the developing science of restoration ecology. Although our emphasis is on tropical forest ecosystems, recent work highlighting globally consistent variation among key plant functional traits (e.g. I. J. Wright *et al.*, 2004) implies that our comments should be generally applicable to other forests as well.

16.2 | Significance of the seedling stage for forest management

In practice, the capacity of tropical forests to regenerate following disturbance may be more or less dependent on the seedling stage of the life cycle compared to other stages. In relatively undisturbed tropical forests, seed addition experiments have shown that, for most species, seedling recruitment is limited by the availability of seeds rather than suitable microsites for seedling regeneration (Makana & Thomas, 2004; Svenning & Wright, 2005). This result is also supported by seed trap studies that have shown that common tree species fail to disperse seeds to more than a small fraction of recruitment sites less than 100 m from reproductive adults (Hubbell *et al.*, 1999; Dalling *et al.*, 2002).

By disrupting the reproduction of forest trees, anthropogenic disturbance may exacerbate the relative importance of seed limitation, and thereby, change the relative importance of the seedling stage as a key factor for forest regeneration. Key disturbance effects on early life history stages are of three types, (1) reduced fecundity or seed dispersal, (2) reduced seedling growth or survival, or (3) reduced or expanded number and range of suitable microsites for germination and establishment. Type (1) impacts will exacerbate seed limitation at the population level, type (2) impacts are more or less neutral in terms of seed limitation but may shift the balance toward limitation by seedling survival, while type (3) impacts may increase microsite limitation for some habitat specialist species, while reducing microsite limitation for pioneer species and colonizers of disturbed soil and litter.

The main focus of this chapter is on seedling responses to disturbance and resource availability (types 2 and 3 impacts above), but seed or dispersal limitation may be exacerbated by anthropogenic disturbance and these impacts may represent the primary constraint on regeneration for some species. Logging results in direct loss of reproductively mature trees of commercial timber species and accidental damage to noncommercial species. In addition, these individuals may become more isolated, thus reducing pollination success and outcrossing rates and increasing the frequency of inbreeding (Murawski *et al.*, 1994; Ghazoul, 2005). These impacts may collectively lower seed or fruit production. For example, seed production in logged forest was only 23% that of primary forest for dipterocarps in West Kalimantan, Indonesia (Curran & Webb, 2000), and was lower for individuals in logged compared to unlogged forests in Amazonia one decade after logging (Johns, 1992). Logging may also lower the densities of large vertebrate dispersers by reducing their food supply and/or destroying their habitat (Johns, 1997). This effect is particularly important in tropical forests where 65–90% of species are adapted for vertebrate dispersal (Willson *et al.*, 1989; Jordano, 1992). The increasing isolation imposed by forest fragmentation may also disrupt patterns of gene flow by pollination for facultative and obligatory out-crossing species and reduce genetic diversity of progeny by limiting the number of pollen donors (Dick *et al.*, 2003). Dispersal may also be reduced by fragmentation if the abundance or foraging patterns of animal dispersers are affected (Laurance, 2005). Consequently, seed production by plants may be lower in forest fragments (Aizen & Feinsinger, 1994).

Certain traits confer greater susceptibility to the negative effects of logging or forest fragmentation on seed production. For example, species that have relatively small and/or immobile pollinators (Ghazoul, 2005), have a dioecious breeding system (Mack, 1997), or exist at low population density (Ghazoul *et al.*, 1998) may show reduced pollination success in response to disturbance-induced reduction in the density of large adult trees. Similarly, trees with few, large fruits may be dispersed less effectively following disturbance because these species often depend on relatively few, large-bodied frugivores that are susceptible to hunting (Corlett, 1998; Laurance, 2005). Because most tropical trees exist at low population densities (<5 reproductive-sized individuals/ha), one or more of these traits is exhibited by a large proportion of tropical forest tree species (Turner, 2001). Therefore, although impacts of anthropogenic disturbance on seedling recruitment may contribute to population and community responses, effects on earlier stages in the life cycle may also be important.

16.3 | Effects of human disturbances on seedling regeneration

Human disturbances can influence seedling growth and survival via numerous pathways. Direct effects include altered resource availability (light, moisture, and nutrients), physical damage to plants (fire, trampling), and compaction or removal of soil. Indirect effects on plant communities are equally important and are mediated by changes in the range and frequency of biotic interactions between plants and their natural enemies (herbivores, seed predators, and pathogens) and between plants and their mutualists (pollinators, dispersers, and mycorrhizas).

Disturbance changes to canopy structure and light availability

Mature forests vary greatly in canopy structure and light availability. The amount and seasonal distribution of rainfall influence canopy deciduousness and herbaceous vegetation cover, while soil fertility and rates of natural disturbance (e.g. landslides, cyclones) additionally affect canopy stature and gap frequency (e.g. Hartshorn, 1980; Condit et al., 2000; Stephenson & van Mantgem, 2005). The species composition of forests, to some extent, also reflects variation in structure and disturbance. Forests with low rates of canopy turnover lack fast growing pioneer species whose seedlings recruit exclusively in treefall gaps (Condit et al., 1999). Conversely, when disturbances such as logging and forest fragmentation increase the frequency of gap formation, pioneer or longer lived shade-intolerant species proliferate (Laurance et al., 2006). Molino and Sabatier (2001) showed how increased forest disturbance associated with logging in French Guiana can also influence species diversity. As the frequency of gap disturbance increased, so too did the proportion of recruits that were shade intolerant. However, because shade-intolerant species are a small fraction of species diversity, increasing opportunities for recruitment of these species beyond a threshold where both guilds coexist resulted in reduced stand species richness.

Increasing canopy disturbance may also affect stand species richness by promoting the establishment of lianas (woody vines). Most lianas are light demanding (Webb, 1958; Putz, 1984), and many recruit from seeds that await gap formation in the soil seed bank (e.g. Kennedy & Swaine, 1992; Dalling & Denslow, 1998). Increasing liana abundance potentially affects the recruitment of both shade-tolerant and shade-intolerant species, either by competing with trees for light and soil-borne resources (Stevens, 1987; Perez-Salicrup & Barker, 2000) or by smothering the surface of gaps, preventing seedling recruitment (Schnitzer et al., 2000). Numerous studies have documented increased liana abundance in logged and fragmented forests in Queensland (Australia) (Laurance, 1991), Sabah, Malaysia (Campbell & Newbery, 1993), Atlantic coastal forest in Brazil (Viana et al., 1997), and central Amazonia (Laurance et al., 2001). However, liana abundance is also increasing in some forests with no recent human disturbance (e.g. over 10–20 years at 47 sites in western Amazonia, Phillips et al., 2002; over 18 years in central Panama, S. J. Wright et al., 2004; Wright & Calderón, 2006). These increases have been attributed to global climate change. Lianas may be more abundant because increasing atmospheric CO_2 concentrations have generally enhanced forest growth, resulting in higher stand turnover rates and greater recruitment opportunities for shade-intolerant species. Furthermore, lianas may have an increased competitive advantage associated with a strong growth response to elevated CO_2 (Granados & Körner, 2002; Zotz et al., 2006).

Changes in canopy structure may also affect conditions for seedling recruitment in the forest understory. Laurance et al. (2002) summarized the spatial extent of edge effects associated with fragmenting lowland Amazonian forest. Microclimatic effects were mostly limited to a band of forest less than 50 m from the forest edge and included reduced soil moisture and increased vapor pressure deficit. Surveys conducted in the same fragments showed that seedlings of woody tree species were little affected (Sizer & Tanner, 1999). Seedling recruitment and growth were elevated within 10 m of the edge, whereas seedling mortality was unaffected. However, overall, densities of shade-tolerant tree seedlings surveyed a few years later were lower at fragment edges than in the interior, perhaps resulting from altered biotic interactions (reduced seed output and dispersal and increased seed predation) rather than changes in resource availability (Benitez-Malvido, 1998).

Changes to soil structure and fertility

Logging affects soil conditions as well as light availability. Soil disturbance affects 14–43% of land surface area in conventionally logged forests in Malaysia, Indonesia, and Suriname (Pinard et al., 2000). The most severely impacted areas are the skid trails created during log extraction, and the log landings, where logs are stored prior to transportation (Nussbaum et al., 1995a). These disturbances typically result in reductions in soil nutrient availability, pore volume, and infiltration rates, and increases in soil bulk density and erosion rates (Lal, 1987; Malmer & Grip, 1990; Chauvel et al., 1991; Douglas et al., 1992; Nussbaum et al., 1995a).

The effects of soil disturbance on the abundance and growth of seedlings varies depending on the degree of compaction, the extent of topsoil loss, and changes to soil surface characteristics and competing vegetation. Soil disturbance may inhibit post-logging forest recovery when loss of topsoil eliminates recruitment from seeds present in the soil. Almost all tropical pioneer species that dominate in large canopy gaps maintain seed banks that allow germination from seeds dispersed months or sometimes even decades prior to the disturbance (Dalling, 2005). Topsoil removal also affects residual soil fertility and porosity, reducing seedling growth and survival. Both effects have been reported for logged forest in the Ulu Segama Forest Reserve, Sabah (Malaysia). Skid trails that had lost their topsoil had lower densities of seeds in the seed bank (Howlett & Davidson, 2003) and seedling regeneration was lower in areas where topsoil had been removed by logging equipment compared to areas with intact topsoil (Pinard et al., 2000). Conversely, soil disturbance may sometimes remove surface litter, improving establishment success from the soil seed bank or may improve seedling recruitment success if logging machinery damages existing vegetation (Fredericksen & Pariona, 2002).

Longer-term impacts of skid trails on forest regeneration have been studied in selectively logged forests in Costa Rica and Malaysia. In Costa Rica, there was a reduction in the density and diversity of stems ≥ 1 m tall and ≤ 5 cm dbh in the center of abandoned skid trails compared to plots on the edge of skid trails or in adjacent forest 12 to 17 years after logging (Guariguata & Dupuy, 1997). Similarly, skid trails in Sabah had lower richness and density of small woody stems (>1 m tall, <5 cm dbh) than adjacent unlogged forest 18 years after logging (Pinard et al., 2000). In the Costa Rican study site, recovery of basal area on skid trails to values equivalent to the unlogged forest would take approximately 80 years. However, the responses of tree seedling establishment and growth to soil disturbance are still not well explored, and likely depend on interactions among logging practices, soil properties, and species traits.

Changes to fire regimes

Fires have always impacted tropical forests. However, the scale and frequency of tropical forest fires may be increasing because land use changes, such as logging (Holdsworth & Uhl, 1997; Curran et al., 1999; Nepstad et al., 1999; Siegert et al., 2001) and habitat fragmentation (Cochrane & Laurance, 2002; Gascon et al., 2000), increase susceptibility of forests to burn. Simultaneously, the number of ignition events is rising because of human colonization of formerly forested regions (Laurance, 1998) and because El Niño-related droughts are increasing in severity and frequency (Trenberth & Hoar, 1996; Dunbar, 2000). Surface fires in otherwise intact forest are difficult to detect remotely, but their effects are manifested in increased mortality rates of many components of the biota, including plants, invertebrates, and vertebrates (Laurance, 2003). Burning kills thin-barked (<1 cm thick) tropical forest plants directly, and seedlings are particularly at risk because of their small size (Uhl & Kauffman, 1990; Cochrane & Schulze, 1999). These direct effects may result in the death of one third of all trees greater than 10 cm dbh, and three quarters of all saplings, as well as many lianas and forbs during the first 15 months after fire (Barlow & Peres, 2004). Total tree mortality may increase to nearly 50% after 3 years, which leads to a collapse of forest biomass, increasing canopy openness and an invasion of light-demanding trees, shrubs, bamboos, and herbs at ground level (Barlow & Peres, 2004).

Recovery from fire may be inhibited by a positive feedback of fire on susceptibility to burning (Cochrane & Schulze, 1999). The increased canopy openness and greater quantity of necromass in once burned forests increases the likelihood that they will burn a second time and with greater intensity (Cochrane et al., 1999; Slik & Eichhorn, 2003). In East Kalimantan (Indonesia), the density and diversity of understory trees were lower in forests that had burned twice (1982/1983 and 1997/1998) than in those that had only burned in 1997/1998 (Slik & Eichhorn, 2003). These changes lead to a cycle of degradation that ultimately eliminates forest cover and promotes the expansion of anthropogenic savanna, scrubland, and grassland (Laurance, 2003).

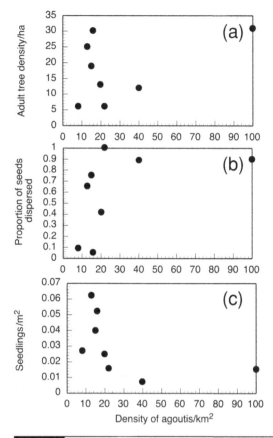

Fig. 16.2 Effects of poaching on recruitment of the canopy palm *Astrocaryum standleyanum* at sites in central Panama. The density of agoutis (*Dasyprocta punctata*), a large rodent that consumes and disperses *Astrocaryum* seeds, is plotted on the abscissa. Sites with low agouti densities also had low densities of other large mammals that consume seeds. (a) Adult *Astrocaryum* density was unrelated to mammal density, but (b) seed dispersal was reduced at low mammal densities, and (c) *Astrocaryum* seedling recruitment increased, reflecting reduced seed predation. Therefore, in the short-term, poaching appears to favor recruitment of this plant species. Redrawn from data in Figs. 2, 5, and 6 in S. J. Wright *et al.* (2000).

Changes to biotic interactions by hunting and forest fragmentation

Hunting can have dramatic effects on seedling communities by eliminating critical biotic interactions (Fig. 16.2; reviewed by Wright, 2003). Unrestricted hunting inside many protected areas in the tropics has removed entire guilds of herbivores, granivores, and frugivores (Robinson *et al.*, 1999; Peres, 2000, 2001; Dirzo, 2001). The effect of this defaunation on seedling communities was first studied in detail by Dirzo & Miranda (1990) in southern Mexico. They showed that a heavily defaunated site (Los Tuxtlas) had twice the seedling density but only a third of the understory species diversity of that found in a site with an intact fauna (Montes Azules). Large effects of defaunation on seedling recruitment patterns have now been reported in several other tropical forests (Wright *et al.*, 2000; Roldán & Simonetti, 2001; Wright & Duber, 2001; Galetti *et al.*, 2006).

Forest fragmentation can have similar and often synergistic effects to hunting. Even large fragments (>1000 ha) may be too small to support populations of large carnivores resulting in elevated populations of herbivores and granivores (Terborgh, 1988). Several experiments have addressed how the trophic structure of tropical food webs influences plant recruitment. In Lake Gatun (Panama) and Lake Guri (Venezuela), islands of different sizes support subsets of the mainland fauna. In Lake Gatun, small islands (<1 ha) that have been isolated from the mainland for greater than 90 years support a highly impoverished tree community when compared to equal-sized patches of forest on the mainland (Leigh *et al.*, 1993). These islands support spiny rats (*Proechimys semispinosus*), but no larger resident mammals. Experimental studies to determine the causes of tree species loss on these islands have shown that seed predation rates and rates of seedling herbivory by mammals are higher on these islands than on the mainland (Asquith *et al.*, 1997; Asquith & Mejia-Chang, 2005). In Lake Guri, islands have been isolated for 20 years and allow a test of the early effects of a loss of large carnivores. Small islands at Guri (0.25–0.9 ha) lack carnivores but often retain large populations of a few generalist herbivores (e.g. tortoises, iguanas, and howler monkeys). On these islands, seedling and sapling densities are only a third of those found on large island and mainland sites. Low juvenile recruitment rates at Guri were attributed mainly to increased seedling herbivory by leaf-cutter ants, whereas seed predation rates appeared to be similar between island and mainland sites (Rao *et al.*, 2001; Terborgh *et al.*, 2006).

The variation observed in biotic responses to these disturbances argues for much more site-specific research to determine the causes of shifts in recruitment patterns. In forests where hunting and fragmentation primarily affect large frugivores, such as primates, fruit characteristics affecting diet choice in these animals may best predict effects on seedling recruitment via altered seed dispersal (Chapman & Onderdonk, 1998; Cordeiro & Howe, 2001). When terrestrial granivores and herbivores are most affected, as occurred in forests in Mexico and Panama, then the potential for small rodents less affected by fragmentation and hunting to compensate for the losses of larger mammals becomes important. For example, Wright *et al.* (2000) have suggested that the increase in seedling recruitment associated with hunting in Los Tuxtlas (Mexico), but not in Panamanian forests, may reflect the prevalence of spiny rats in Panama.

A striking contrast to tropical forests, where hunting has typically reduced densities of large herbivores (but see Ickes, 2001), occurs in boreal and temperate forests. In eastern North America, Europe, and Japan, deer populations have increased dramatically over the last century as a result of reduced hunting pressure, extirpation of natural predators, and increased availability of winter forage from agricultural and sylvicultural activities (Côté *et al.*, 2004). Deer browsing on seedlings and saplings can be sufficient to alter the chemical and physical defenses of plant populations (Vourc'h *et al.*, 2002), reduce species richness (e.g. Horsley *et al.*, 2003), and can lead to dramatic shifts in tree species composition (reviewed in Côté *et al.*, 2004). For example, seedlings and saplings of *Tsuga canadensis* (Pinaceae) are now rare across the upper midwestern USA, while virtually all seedlings greater than 30 cm tall of *Thuja occidentalis* (Cupressaceae) have disappeared (Rooney *et al.*, 2000, 2002). Remaining seedlings of these browse-sensitive species tend to be distributed on tip up mounds, rock faces, and other inaccessible microsites (Comisky *et al.*, 2005; Krueger & Peterson, 2006).

Changes to microbial communities

Disturbances that affect animal communities may also have cascading effects on pathogenic and mutualistic microbial communities. Browsing mammals and insects may be important vectors of plant diseases and can provide wounds that facilitate microbial infection (García-Guzmán & Dirzo, 2001; García-Guzmán & Benítez-Malvido, 2003). Changes to animal communities may also directly affect mycorrhizal communities. Mycorrhizas confer nutritional and other benefits to their hosts that are critical for seedling growth and survival and may potentially influence tree species community composition (Janos, 1980b; Kiers *et al.*, 2000; Chapter 9). In Australia and the neotropics, marsupials and rodents are important dispersers of spores of some mycorrhizas (Janos *et al.*, 1995; Reddell *et al.*, 1997; Mangan & Adler, 1999) and mammal exclosures in Australian rain forest show significant reductions in mycorrhizal diversity and density (Gehring *et al.*, 2002).

Logging damage to soils can also affect mycorrhizas. In a Malaysian forest, Alexander *et al.* (1992) found that logging reduced spore and inoculum density of arbuscular mycorrhizas and lowered the proportion of root length infected for bioassay plants. In Sabah, seedlings of the ectomycorrhizal species *Hopea nervosa* (Dipterocarpaceae) had the same proportion of infection but 40% greater morphospecies diversity when transplanted into unlogged (42) than logged (30) forest (Lee *et al.*, 1996; Alexander & Lee, 2005). Limited infectivity or functional diversity of mycorrhizas in forest soils may be an important potential constraint on forest recovery after logging.

16.4 | Application of seedling functional ecology to tropical forest management and restoration

The expansion of agricultural land, urban development, and logging are leading to the loss of tropical forest at unprecedented rates

(Mayaux *et al.*, 2005; Kirby *et al.*, 2006). As a consequence, greater than 60% of tropical forests are now classified as secondary forest or degraded forest land (Chazdon, 2003). For ecologists, the imperative now is to learn how to manage the secondary forests that remain after logging and that develop after agricultural land has been abandoned. A shift in attention to tropical secondary forests will require an assessment of how ecological processes that determine seedling recruitment success might differ between primary and secondary forests, and further, how plant traits that confer a fitness advantage in mature forest environments might constrain recruitment success in degraded sites.

Research in mature tropical forests has identified three processes that potentially mediate seedling recruitment success: habitat specialization along gradients of resource availability (e.g. Kobe, 1999; Fine *et al.*, 2004), competitive interactions among plants (e.g. Lewis & Tanner, 2000), and frequency-dependent mortality arising from the activity of specialized natural enemies (e.g. Harms *et al.*, 2000). Here, we examine how seedling traits can influence the importance of these processes in determining regeneration success in disturbed forests.

Resource specialization: implications for forest management

A major unresolved question in forest ecology is whether habitat specialization can account for much of the local or landscape scale diversity seen in tree communities. Variation in resource availability (light, nutrients, and water) across habitat types might be expected to lead to selection on traits that maximize plant performance along portions of these resource gradients (Dalling & Burslem, 2005). The resulting performance trade-offs, where higher fitness under one set of conditions reduces fitness under another, may have a dominant effect on community structure (e.g. MacArthur & Levins, 1964; Tilman, 1988). Selection is expected to be strongest on seedling traits, as seedlings experience more variability in supply rates of light, water, and possibly nutrients than do adults or seeds. An understanding of performance trade-offs is of particular importance for forest management. If performance gradients are strong, then disturbed sites with altered resource supply rates may only allow the recruitment of a small subset of species in tree communities. Conversely, if performance gradients are shallow, then management practices may permit rapid recovery of species diversity and ecosystem function.

Seedling responses to light and moisture availability

Numerous studies reveal a performance trade-off between fast growth and low survivorship in high light, versus slow growth and high survivorship in shade (e.g. Hubbell & Foster, 1992; Kitajima, 1994; Kobe, 1999). Intensively logged or clear-cut forest should therefore stimulate the recruitment of the fastest growing pioneer species (e.g. *Trema* spp. Celtidaceae in the neotropics, Africa and Asia, and *Ochroma pyramidale* Malvaceae in the neotropics). These species require elevated or fluctuating soil temperatures for germination from seeds that can

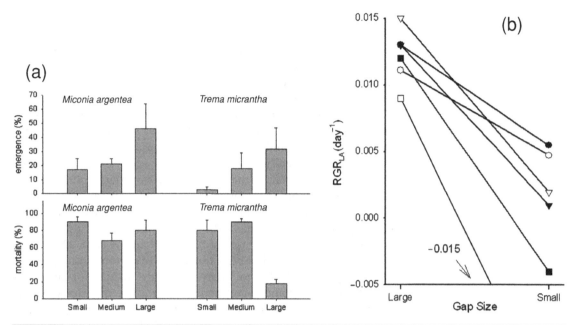

Fig. 16.3 (a) Mean and standard error emergence and seedling mortality after 10 weeks for seeds of *Miconia argentea* and *Trema micrantha* experimentally sown into small (25 m²), medium (64 m²), and large (225 m²) gaps on Buena Vista Peninsula (Panama). (b) Schematic illustration of the change in ranks of species relative growth rate of leaf area (RGR$_{LA}$) across gap sizes (small and large as above) for seedlings growing inside (closed symbols) or outside (open symbols) invertebrate herbivore exclosures over the first year following transplantation. ● – exclosed *Miconia argentea*, ○ – unexclosed *Miconia argentea*, ▼ – exclosed *Cecropia insignis*, □ – unexclosed *Cecropia insignis*, ■ – exclosed *Trema micrantha*, □ – unexclosed *Trema micrantha*. *Trema* showed poor germination, early seedling survival, and growth except in the largest gaps, while *Miconia* and *Cecropia* changed rank in growth rate between small (*Miconia*) and large (*Cecropia*) gaps. Redrawn from Pearson et al. (2003b).

persist in the soil for several decades after dispersal (Pearson *et al.*, 2002; Dalling, 2005; Fig. 16.3a). Smaller gaps, resulting from selective logging, and the partially shaded edges of forest fragments are often also colonized by pioneer species. Pioneers that occupy these sites are usually small seeded (e.g. *Cecropia* spp. Cecropiaceae, *Miconia* spp. Melastomataceae, *Piper* spp. Piperaceae) and detect canopy openings by discriminating small changes in red:far-red light ratio (e.g. Daws *et al.*, 2002). However, small seed size and light-sensitive germination prevents seedlings from successfully emerging from beneath leaf litter. Recruitment of these species may, therefore, also depend on disturbances that expose mineral soil (Vázquez Yanes *et al.*, 1990; Williams-Linera, 1990).

Changes in light availability may continue to affect species composition at the post-establishment phase and for decades after the initial disturbance. Fast growing species may become increasingly restricted to large gaps and clearings by either increased mortality or, in some cases, reduced growth in small gaps and forest edges, relative to competitors (Fig. 16.3b). Even once the canopy has regrown, spatial patterns of variation in light availability can differ from mature forests for several decades. For example, 20-year-old secondary forest in Costa

Rica had a larger fraction of microsites receiving intermediate light levels (2–5% full sun) than mature forest (Nicotra et al., 1999). Changes in the availability of low to intermediate light microsites may strongly impact recruitment patterns of shade-tolerant species by increasing the availability of habitat where positive seedling growth rates can be maintained (Bloor & Grubb, 2003).

In contrast, recruitment of shade-tolerant species into high light microsites may be limited by the ability of seedlings to acclimate to high irradiance and leaf temperatures (Bazzaz & Carlson, 1982). Growth rates of seedlings of shade-tolerant species typically decline above 25–50% full sun (Veenendaal et al., 1996; Poorter, 1999), partly due to photoinhibition, a temporary reduction in the efficiency of photosynthesis due to damage to photosystem II (Langenheim et al., 1984). Species vary in susceptibility to photoinhibition in proportion to their shade tolerance (Krause et al., 2001; Houter & Pons, 2005). However, longer-term experiments have shown that even strongly shade-tolerant species are able to survive exposure to full sun and acclimate to high irradiance conditions within a few months (Clearwater et al., 1999; Krause et al., 2006). Furthermore, after a prolonged period of acclimation, biomass growth of shade-tolerant seedlings may reach that of plants growing in partial shade (Krause et al., 2006).

Photoinhibition, however, may contribute to recruitment failure when combined with water stress. In seasonal tropical forests, plants can experience considerable water stress during the dry season, indicated by low leaf water potentials (e.g. Chiarello et al., 1987; Tobin et al., 1999). These conditions may either be exacerbated in large gaps and clearings because of higher temperatures and vapor pressure deficits (Robichaux et al., 1984), or ameliorated because of reduced root competition for water (Veenendaal et al., 1995). In logged forests, soil compaction may also exacerbate seasonal water shortages by reducing soil pore volume or by affecting moisture release characteristics (Brooks & Spencer, 1997). These changes in soil moisture availability have the potential to exert large effects on plant communities. In a seasonally moist tropical forest in central Panama, experimental drought treatments have shown that seedlings of co-occurring species vary greatly in drought tolerance, with species from wet habitats especially susceptible to drought-induced mortality (Engelbrecht et al., 2005). Even in aseasonal forests, dry spells of only a few days' duration may elevate seedling mortality in large gaps and clearings, with newly emerging seedlings with shallow rooting systems most susceptible (Engelbrecht et al., 2006).

Seedling responses to nutrient availability

Tree species distributions are associated with soil factors at a variety of spatial scales (Clark et al., 1999; Tuomisto et al., 2003; Phillips et al., 2003). In mature tropical forests, up to half the species in a local community may have distributions that are biased in relation to one or more soil nutrients (John et al., 2007). Changes to soil structure and fertility arising from compaction or erosion (Congdon & Harbohn, 1993; Nussbaum *et al.*, 1995a), or from atmospheric deposition of nitrate and sulfate (Driscoll *et al.*, 2003; Fabian *et al.*, 2005), might therefore strongly affect community composition. As yet, however, direct evidence of differential seedling responses to soil conditions in the context of these disturbances is rare. In the Ulu Segama Forest Reserve, Sabah, four tree species (two pioneer and two shade-tolerant dipterocarp species) were transplanted onto skid trails. The addition of either inorganic nutrients or forest topsoil increased the growth of all species (Nussbaum *et al.*, 1995b). This experiment supports the common finding that soil compaction may slow root growth and lower nutrient availability to plants (Greacen & Sands, 1980) and thus delay regeneration or reduce seedling growth (Malmer & Grip, 1990; Jusoff & Majid, 1992; Pinard *et al.*, 1996; Guariguata & Dupuy, 1997; Whitman *et al.*, 1997; van Rheenen *et al.*, 2004).

Predictions about the long-term effects of soil disturbances are also hampered by an inadequate understanding of the mechanisms underlying species differences in soil nutrient requirements. In short-term experiments, fast growing species from nutrient-rich habitats outgrow slow growing species from nutrient-poor habitats, even under oligotrophic conditions (Fichtner & Schulze, 1992; Keddy *et al.*, 1994; Baraloto *et al.*, 2006; dos Santos *et al.*, 2006). Over longer time scales, however, differences in whole-plant nutrient use efficiency might cause rank reversals of seedling growth under differential nutrient supply. Species specialization along soil fertility gradients may reflect a growth–mortality trade-off similar to that described for gradients of light availability. Fine *et al.* (2004, 2006) carried out experiments in which seedlings of species found in relatively nutrient-rich clay soil forests in Amazonia and congeneric species limited to nutrient-poor sand soils were transplanted into both habitat types. They found that the clay-soil specialists grew faster in both soil types, but were less resistant to herbivory, and suffered higher mortality than the better defended white-sand specialists (Fig. 16.4). This result may be exceptional because of the extreme variation in fertility between soil types. However, it has important implications for the restoration of forest cover on degraded soils. Matching species to restoration sites must be based on the soil conditions of their native habitats. In nutrient-poor environments, the selection of species with traits that confer defense against herbivores may be more important than selecting species for rapid growth rate.

On sites with the most highly degraded soils, growth rates of seedlings of pioneer species commonly used for restoration may be extremely low (Aide & Cavalier, 1994) and some amelioration of site conditions may be needed before forest cover can be restored. Fertilization of these sites with nitrogen or phosphorus may increase seedling growth rates (Gehring *et al.*, 1999; Davidson *et al.*, 2004). However, fertilization may also stimulate the growth of competing herbs and grasses with negative effects on woody seedling growth (Pareliussen *et al.*, 2006).

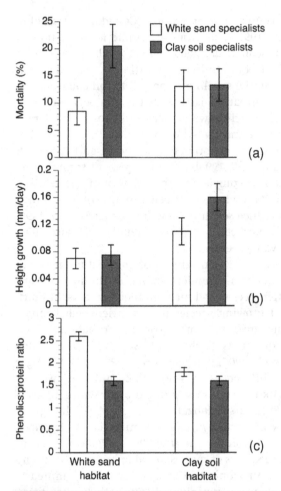

Fig. 16.4 Effects of soil type (fertile clay and infertile sand) on (a) percent mortality, (b) height growth, and (c) defense allocation for seedlings of 10 congeneric pairs of tree species, specialized to either sand or clay soils. Clay soil specialist species grew as fast as sand soil specialists on the nutrient-poor sand soils, but were less well defended from herbivores and suffered higher mortality rates. Redrawn from Fine *et al.* (2004, 2006).

Seedling responses to competition

In mature forests, competitive interactions among plants can strongly affect seedling growth. For example, in low fertility soils in central Amazonia, trenching around seedlings in the understory to remove root competition has a positive effect on seedling growth with a magnitude comparable to that of creating small canopy gaps (Lewis & Tanner, 2000). However, competitive effects on seedlings are mostly exerted by canopy vegetation, which intercepts most of the light and accounts for a large proportion of root biomass. There is little experimental evidence for competition for resources among understory seedlings (Wright, 2002).

In contrast, tree seedlings in degraded forests and abandoned pastures must often compete with lianas, pasture grasses, or ferns. Seedling transplant experiments in logged forest in Ivory Coast show that lianas can dramatically reduce seedling growth, with the strongest competitive effects exerted belowground (Schnitzer *et al.*, 2005). Similarly, establishment of tall tropical grasses, such as *Saccharum*, *Pennisetum*, and *Melinis* (Poaceae) can often strongly inhibit seedling recruitment (Aide *et al.*, 1995; Holl *et al.*, 2000; Posada *et al.*, 2000; Slocum *et al.*, 2006). In abandoned pastures in central Panama, a combination of mowing and shading treatments showed that *Saccharum spontaneum* exerts both above- and belowground competitive effects on seedling recruitment (Hooper *et al.*, 2002). Small-seeded species were unable to tolerate either above- or belowground competition from *Saccharum*. In contrast, large-seeded, shade-tolerant species performed well and were recommended for reforesting these sites (Hooper *et al.*, 2002).

The selective barrier to recruitment imposed by grasses may in part explain the divergent successional trajectories observed between abandoned pastures and natural disturbances. For example, natural gaps in Puerto Rico are dominated by pioneer taxa (*Cecropia schrebiana* Urticaceae, *Schefflera morototoni* Araliaceae), whereas pastures become dominated by Melastomataceae, Rubiaceae, and Myrtaceae (Aide *et al.*, 1995). Similarly, abandoned pastures in Amazonia are colonized by a species-poor community dominated by the pioneer genus *Vismia* (Clusiaceae), while nearby secondary forests not used for pastures are dominated by a more species-rich community in which *Cecropia* is most abundant (Mesquita *et al.*, 2001).

The presence of competition-free microsites may be especially important for stimulating the recruitment of woody species in abandoned pastures. In lowland and montane forests in Costa Rica, small-seeded taxa (e.g. Melastomataceae) are restricted to nurse logs (often the stumps of trees that remained after pasture clearance) or fern patches (Peterson & Haines, 2000a; Slocum, 2000). However, ferns may also compete with tree seedlings on sites abandoned after human use and frequent fire, such as *Dicranopteris linearis* (Gleicheniaceae) in Sri Lanka and *D. pectinata* in the Dominican Republic (Cohen *et al.*, 1995; Slocum *et al.*, 2006). Once trees do become established in pastures, they may catalyze succession by modifying environments for seedling establishment both above- and belowground (Nepstad *et al.*, 1991; Rhoades *et al.*, 1998) and by providing perches for seed dispersers (Toh *et al.*, 1999; Slocum, 2001).

An alternative to stimulating regeneration on competition-free microsites is to circumvent the seedling stage altogether. When dense grass cover slows seedling recruitment or prevents establishment because of increased fire frequency (Cavalier *et al.*, 1998; Cabin *et al.*, 2002; Lwanga, 2003), recruitment can be started from large cuttings. In active pastures in Africa and the neotropics, farmers frequently use large woody shoots of resprouting tree species (e.g. *Gliricidia sepium* Leguminosae) as living fence posts planted directly into the soil (Budowski, 1987). Trials using these species indicate that many species can readily establish canopy cover and stimulate seedling recruitment in grass-dominated degraded pastures (Zahawi, 2005; Zahawi & Augspurger, 2006).

Seedling responses to natural enemies

Human disturbances frequently have large effects on populations of herbivores and granivores. However, impacts on seedling recruitment may be the hardest of all to predict. This is because compensatory effects often accompany the removal of suites of natural enemies and may have either positive or negative effects on focal plant species. For example, an important timber species and food source for mammal communities in Central American forests is *Dipteryx panamensis* (Leguminosae). In mature forests, *Dipteryx* seeds are dispersed by bats and primates, predated by mammals (squirrels and peccaries), and secondarily dispersed or predated by rodents (Bonaccorso *et al.*, 1980; Forget, 2004). Contrasting studies at nearby sites have shown either increased recruitment of *Dipteryx* in forest fragments, suggesting escape from mammalian seed predation (Hanson *et al.*, 2006), or increased seed predation and reduced recruitment (Guariguata *et al.*, 2002), suggesting ecological release of small granivorous rodents at sites where dispersers or large predator populations have been reduced by hunting.

The effects of altered herbivore and granivore communities are also likely to be influenced by seedling traits. Elevated seedling herbivory rates on small islands in Panama suggest that differences among species in tolerance or susceptibility to mammalian herbivory might contribute to the characteristic species composition of these sites (Leigh *et al.*, 1993; Asquith *et al.*, 1997). Similarly, differences in susceptibility to leaf-cutter ant defoliation might account for shifts in species composition on islands in Lake Guri (Rao *et al.*, 2001). Traits that confer resistance to foliar herbivory and browsing damage might include tough leaves, low nitrogen content, rapid leaf expansion, and early stem lignification (Coley, 1983; Kursar & Coley, 1991). Some of these traits may also confer resistance to fungal diseases (Augspurger, 1983; Benitez-Malvido *et al.*, 1999; Benitez-Malvido & Lemus-Albor, 2005), although these interactions remain poorly understood.

In mature forests, density-dependent mortality of seeds and seedlings appears to be an important constraint on adult population density (e.g. Augspurger, 1984b; Clark & Clark, 1984; Webb & Peart, 1999; Harms *et al.*, 2000). Not surprisingly therefore, attempts to improve the regeneration success of several important timber species in secondary forests have been hindered by pathogen and insect outbreaks to planted seedlings and saplings (e.g. *Swietenia* spp., *Khaya* spp. Meliaceae, *Stryphnodendron microstachyum* Leguminosae, *Melicia excelsa* Moraceae; Newton *et al.*, 1993; Folgarait *et al.*, 1995; Nichols *et al.*, 1998). However, the potential for compensatory density-dependent mortality arising from increased pathogen attack when herbivores and granivores are absent has not been studied in tropical secondary forests. In both temperate and tropical forests, damping off pathogens can accumulate in the soil surrounding maternal sources of susceptible species (Augspurger, 1984b; Packer & Clay, 2000) and may develop populations rapidly in secondary forests (Packer & Clay, 2004). Therefore, when dispersers that remove seeds from the vicinity of maternal trees are absent, pathogen infection may further contribute to recruitment failure.

16.5 | Future directions

Forest regeneration requires the successful completion of a sequence of steps between successive generations of reproductive adults. Consequently, the failure of tropical forests to regenerate following anthropogenic disturbance may be caused by disruption of any one or more of these steps, including floral induction, pollination, seed maturation, dispersal, germination, seedling establishment, growth or survival, and onward growth of saplings to reproductive maturity. The mechanisms that determine the transitions between these stages are not equally well understood, but land managers and conservation biologists need to recognize that barriers to forest regeneration may be occurring at any one or more of them. Research is required to understand the relative importance of these processes for a given site and to address potential solutions.

Our review has emphasized the sensitivity to anthropogenic disturbance of the processes that determine seed limitation and seedling recruitment, but this perspective is based on a limited number of case studies and is therefore necessarily anecdotal. In particular, because of the complexity of the biotic and abiotic interactions that impinge on tree reproduction, generalization beyond this level of detail may never be possible without detailed site-specific studies. The current state of knowledge suggests that seed limitation and seedling recruitment failure represent the most significant constraints on regeneration following anthropogenic disturbance; management of these problems is most likely to ameliorate the significant barriers to forest succession. However, further research is still required to address gaps in our understanding of how environmental conditions and biotic interactions affect the fate of individuals during seedling recruitment. The least well understood stages of plant life history are those that occur from primary dispersal to seedling emergence, which may include processes of secondary dispersal, seed burial, seed mortality due to pathogens, germination, and early seedling growth (e.g. Levey & Byrne, 1993; Gallery et al., 2007; Marthews, 2007). Integrating this research with current knowledge of forest regeneration processes will require further development of mechanistic and spatially explicit forest simulation models typified by SORTIE and TROLL (Chave, 1999).

To circumvent seed limitation of forest regeneration and to manipulate species composition in favor of commercially valuable species, land managers often resort to planting nursery-raised seedlings, despite the high costs of this approach (Lamb *et al.*, 2005). However, until recently the selection of species for reforestation or enrichment trials for degraded tropical lands has been biased toward well-known pioneers from a relatively small number of genera (Lamb *et al.*, 2005). The fundamental criterion for species selection has been the higher survival and growth of pioneer species in high light conditions (Poorter, 1999), which may reflect the greater susceptibility of shade-tolerant species to the short-term damaging effects of photoinhibition particularly when combined with water or nutrient shortage (Chiarello *et al.*, 1987; Tobin *et al.*, 1999; Bungard *et al.*, 2000). However, this emphasis on short-term biomass growth as the primary criterion for species selection in restoration programs may need to be reviewed. Recent research on the longer-term responses of tree seedlings to photoinhibition has highlighted how relatively shade-tolerant species can reach maximal rates of biomass accumulation in open conditions (Clearwater *et al.*, 1999; Krause *et al.*, 2006).

More generally, optimizing survival in sites with low nutrient or water availability may require experimentation with species that use these resources conservatively (Craven *et al.*, 2007) and possess an allocation strategy that emphasizes defense against natural enemies and capture of belowground resources over fast growth rates. These traits are more often found in relatively shade-tolerant species, which may, therefore, have greater potential for restoration of degraded tropical lands than previously supposed. A comparison of the growth and survival of seedlings of 18 tree species planted into clearings in fern thickets on infertile post-agricultural soils in Dominica found that the species with the highest survival over 3 years, *Inga fagifolia* (Leguminosae), a late-successional nitrogen-fixer, and *Alchornea latifolia* (Euphorbiaceae), were ranked third and sixth in terms of growth rate over the same interval (Slocum *et al.*, 2006). In the most degraded or exposed sites, such species may require amelioration of site conditions by an established canopy of a species that can tolerate high light combined with low water and nutrient availability. For example, in the lowland wet zone of Sri Lanka, a wide range of relatively shade-tolerant herbaceous and tree species establish successfully as seedling transplants beneath a canopy of thinned *Pinus caribaea* (Pinaceae) plantations on degraded sites previously dominated by pasture grasses (Ashton *et al.*, 1997, 1998). Although we do not advocate the widespread planting of exotic species such as *Pinus caribaea* in Sri Lanka, trials with native species with similar traits may prove fruitful as part of a two-stage process for restoring the most degraded sites.

Our perspective on forest restoration has two fundamental implications for the direction of future research. First, the response of tree seedlings to anthropogenic disturbance may be unpredictable on the basis of current knowledge because different types of disturbance vary in their effects on the biotic and abiotic environment and are highly site specific. Although seedling growth in response to abiotic resource supply should be predictable, survival is less easily predicted when it is determined by the abundance of natural enemies rather than abiotic resources. Second, if natural enemies are important determinants of seedling responses to anthropogenic disturbance, then a diverse seedling community from an early stage may serve as an insurance against the unpredictability in response to interactions at higher trophic levels (Montagnini *et al.*, 1995). Seedling community diversity may also be desirable if it promotes early reestablishment of ecosystem function (Erskine *et al.*, 2006) and gives rise to mixed forests that facilitate regeneration of a diverse plant community (Carnevale & Montagnini, 2002). However, these topics remain poorly explored aspects of tropical forest restoration ecology.

16.6 | Acknowledgments

We thank our collaborators, colleagues, and students who have contributed to our work on disturbance and tree seedling ecology, in particular, Thilanka Gunaratne, Nimal Gunatilleke, Toby Marthews, Chris Mullins, and Michelle Pinard. Our research was funded by the British Ecological Society, Darwin Initiative, European Union, Natural Environment Research Council, National Science Foundation, and Leverhulme Trust.

Chapter 17

Seedling establishment in restored ecosystems

Susan Galatowitsch

17.1 | Introduction

Over the past two decades, ecological restoration has progressed to rely on more refined techniques, to include a greater array of ecosystems, and to attempt larger and more complex problems. Despite this progress, the outcome of many restorations fails to result in ecosystems that are similar to their natural counterparts. Restored ecosystems typically have fewer species and do not accumulate species over time, as expected. A lack of available seeds or suitable microsites for seedling establishment can hinder community development. Not surprisingly, seed availability is more often reported to be the key limitation to higher richness (e.g. Pywell *et al.*, 2002; Martin & Wilsey, 2006; Kettenring, 2006). Most restorations introduce a small subset of the species expected and often at much lower abundances than exist in unaltered sites. To do otherwise seldom has been considered necessary because dispersal has the potential to add species over time. Unfortunately, habitat fragmentation has diminished native species propagule pressure and hinders dispersal in many landscapes (Galatowitsch & van der Valk, 1996; Honnay *et al.*, 2002; Young *et al.*, 2005) leading to increased recognition of the importance of adequate seed introductions for restorations.

When the investment in acquiring native seed for restoration is significant, there needs to be a reasonable likelihood that conditions are suitable for seedling emergence and growth. This can be especially challenging considering that site conditions at the start of a restoration project can be radically different than what might have ever existed in an unaltered community, even after natural disturbances. Newly restored sites are typically more stressful growing environments than their natural counterparts with respect to temperature, moisture, available nutrients, and wind abrasion. Young seedlings are generally more vulnerable to physical hazards than any other life stage (e.g. Grime & Curtis, 1976; Grubb, 1977), and high mortality of newly germinated seeds can make seeding an unreliable restoration practice. Moreover, many restorations, regardless of initial conditions or ecological goals, are expected to be established within a few years (generally <5 years), with minimal resources for follow-up actions available in the long-term. Consequently, later successional species are often introduced to restorations at their onset, rather than after some period of community development.

Restorationists have employed a variety of techniques to promote the establishment of seedlings. Approaches for species selection, placement of seeds, and soil amendments initially relied on production practices from agronomy, horticulture, or forestry, but are increasingly informed by ecological processes. The emphasis on rapid vegetation establishment in restoration affects the choice of species in seed and planting mixes, the genotypes selected by restoration nurseries and seed producers, and perhaps even long-term vegetation dynamics. Practical approaches for vegetation establishment informed by ecological principles are needed to increase the likelihood that restorations are reasonable facsimiles of the natural ecosystems they have replaced, not simply assemblages of a few aggressive colonizers. However, because most ecological principles have emerged from the study of intact, natural systems, knowing how to translate this knowledge into practices that affect vegetation establishment in newly restored ecosystems is not always obvious. Developing practices that promote seedling establishment and the formation of initial plant communities has been a primary research concern for the young field of restoration ecology. This chapter reviews how the conditions in restored ecosystems affect vegetation establishment, what practices are typically pursued to encourage establishment, and the extent to which these practices are linked to restoring processes (Table 17.1).

17.2 | Selecting initial community composition for restoration

Restoration plans typically revolve around reestablishing particular target plant communities situated where environmental conditions are expected to be suitable. Determining which species will be planted often will be based on both commercial availability and likelihood of rapid seedling establishment. Most natural ecosystems experience periodic episodes unfavorable to plant growth and there can be an evolutionary disadvantage for seeds to germinate rapidly and synchronously after dispersal (e.g. Rogers & Montalvo, 2004). However, if seeds sowed in a restoration fail to germinate at high rates and with minimal lags (i.e. a few weeks to a few months), sites can face an increased likelihood of soil erosion or (re)invasion by exotic species. Consequently, predictable, high rates of germination and seedling growth are highly desirable for restoration even though seed production and recruitment are more often episodic in natural systems. Selecting readily colonizing species in seed mixes, selecting strains of particular species that have been increased in a cultivated setting,

Table 17.1 | Commonly encountered limitations to seedling recruitment in newly restored ecosystems and approaches to counteract the limitations

Limitations/problems	Cause(s)	Restoration approaches	Reported potential problems
Episodic or slow recruitment	Germination may be cued by changes in temperature or moisture, or by disturbances, such as flooding or fire, that only occur episodically	Use species strains selected for high germination Use species that germinate under a broad range of conditions Treat seeds to break dormancy (e.g. scarify, stratify, treat with smoke extract)	Trade-off between maximizing initial recruitment and the long-term advantages of adaptations for episodic recruitment; legacy effects of founder community
Relatively low recruitment compared to similar, intact sites with established populations	Ontogenic niche shifts	Select locales to seed based on juvenile not adult occurrences	Information on juvenile habitat suitability often lacking
Little or no recruitment on exposed sites	Presence of physical hazards (e.g. soil instability, abrasion, inadequate moisture) Sites denuded by atmospheric deposition of metals, sulfuric acid	Mulching, installing natural (rock) or artificial shelters or use of nurse plants Lime to neutralize acids and minimize metal uptake by seedlings	Mulch can limit germination; shelters difficult to install on large sites; nurse plants can compete with establishing vegetation
Little or no recruitment on compacted sites in arid environments	Poor infiltration due to chemical sealing of clays on soil surface	Land imprinting, gypsum treatments to replace sodium on clays, increasing permeability	
Little or no recruitment or seedling growth on moisture-limited sites	Low precipitation, poor soil-water retention, minimal opportunity to irrigate new plantings	Treat seeds or planting sites with hydrophilic polymers	Use on sites with too much soil moisture will cause decay

Little or no recruitment on sites with existing vegetation	Flushes of annual growth reduce water & nutrients available to other species Competition from perennial grasses in deforested sites Competition in nutrient-enriched sites	Remove annuals with burning, solarization, herbicides, tillage Install nurse logs, close canopy with fast-growing trees Soil impoverishment, scraping, biomass removal	Seed banks of weedy annual and perennials species will cause reinvasion; trees used to be carefully chosen so they do not impede growth of establishing seedlings; soil impoverishment treatment can have short life span
Reduced seedling growth of later successional species	Lack of symbiotic mycorrhizae	Install nurse plants or barriers to trap windblown spores, inoculate seeds or seedlings, install decaying nurse logs in forests	
High seedling mortality from grazing	Food quantity or quality low relative to herbivore population	Install grazing exclosures, use nurse plants	Competition between nurse plants and target species can develop over time

relying on vegetative propagules or juvenile container stock, and detailed planting design are all commonly used techniques for ensuring high rates of initial revegetation success. Restorationists often assume the initial vegetation established from these selected seedings and plantings will create suitable conditions for other colonists and the legacy of these founders will be minimal in the long run. There is, however, increasing recognition that the initial communities of restorations can be very persistent.

Choosing species to avoid episodic recruitment

Although seedling emergence is cued to climatic events in many ecosystems, this is particularly the case for regions that are chronically moisture limited. In arid and semiarid ecosystems, germination is well known to be episodic, with greatly enhanced recruitment during wet climatic cycles (e.g. Beatley, 1974; Noy-Meir, 1974). Repairing damage from uses such as overgrazing, oil and gas development, and minerals extraction, must happen regularly, resulting in a reliance on very few species that do have predictable recruitment under a wide range of conditions. For several decades, the United States government relied on exotic species, especially the bunchgrass *Agropyron desertorum* (Poaceae), to restore semiarid grasslands because it was considered more reliable than native species (Johnson, 1986). Unfortunately, colonization of native plants in monotypic stands of *A. desertorum* is rare, and these restorations do not develop over time into typical grasslands of the region (e.g. Marlette & Anderson, 1986). Because one of the goals of managing these federal lands is to conserve native diversity, public land restorations have, over the last 10 years, attempted restorations with mixes of native species matched to the expected community for each locale. In a study of oil and gas road restorations in the Little Missouri Badlands (North Dakota, USA), Simmers (2006) found that of the 30 species seeded in this national grassland on abandoned roads over the past 2 decades, only 8 reliably recruited (including three nonnative species). In response to poor recruitment, managers sometimes simplify seed mixes so they are comprised of a few species that can be broadly specified for use across a broad array of soils and landforms. Unfortunately, communities along these road corridors are largely comprised of seed mix species, even after more than a decade (Simmers, 2006). These legacy effects of seeding are stronger than edaphic differences in this region. Selecting seed mixes to prefer species that offer reliable, rapid recruitment is not confined to semiarid grasslands, and is a nearly universal practice on projects that must be completed rapidly, because of regulatory oversight. Establishing vegetation in several phases is relatively uncommon, although it offers the advantage of initially focusing on the establishment of colonizing species and later including those species that will be more likely to succeed with some existing vegetation (Betz, 1986; Pywell et al., 2003).

Selecting plant strains based on rapid recruitment

Increased interest in using native plants in restorations combined with their lack of recruitment reliability has likely resulted in the selection of strains of particular species, either deliberately or inadvertently, with reduced seed dormancy (Roundy & Call, 1988). The common practice of increasing seed availability in an agricultural setting can cause large genetic shifts within a single generation based on harvesting patterns (McKay *et al.*, 2005). In a study that compared the establishment of populations of a rare species, *Gaura neomexicana* subsp. *coloradensis* (Onagraceae), from wild and nursery sources, Burgess *et al.* (2005) found evidence that the maternal plant growing conditions impacted dormancy. Nursery production favored genotypes with lower dormancy that could pose problems for long-term population maintenance after reintroductions into natural settings. The evolutionary benefit of dormancy in later successional species to ensure all seeds do not germinate under unfavorable environmental conditions is of relatively little value in new restorations, which are more suited to rapidly establishing pioneers with a strategy of high fecundity and dispersal. On restoration sites after sowing, plants with dormant seeds may be slow to establish, unable to compete with earlier establishing vegetation. These species may be labeled poor performers, candidates for improvement through deliberate selection or for avoidance in restoration mixes. In many ecosystems, the genetic composition of the original planting will have long-term consequences to the genetic diversity of the restored populations (Rogers & Montalvo, 2004; Gustafson *et al.*, 2005). Restoring a population by deliberately introducing a large number of propagules at one time is unlikely to yield the same genetic diversity as natural immigration of seed over an extended period (Rice & Emery, 2003) and runs the risk of contributing to genetic bottlenecks.

Choosing between juvenile and adult stock for restoration

Because seedlings are typically more vulnerable to unfavorable environmental conditions than are larger plants, transplanting nursery-reared seedlings or using divisions of adult vegetative propagules (e.g. rhizomes) are often assumed to assure more predictable survival and vegetation establishment than direct seeding. In forest restorations, however, plants establishing from sowed seed tend to perform better in the long term than those from container stock, although seeds may suffer some initial losses (Young & Evans, 2000). In reintroductions of *Quercus lobata* (Fagaceae) in California (USA), oaks grown from field-sown acorns had greater first-year survivorship and growth than container stock, suggesting they were more drought tolerant, had less root deformation, or avoided mortality due to transplant shock. Welch (1997) reported similar findings for *Artemisia tridentata* (Asteraceae). If these effects are due to transitory shock, the long-term effect on the restoration may not be significant. However, Halter *et al.* (1993) observed that after 11 years, *Pinus contorta* (Pinaceae) transplanted at 1 year of age grew more slowly than naturally regenerated seeds at the same time. The individuals transplanted as container stock had major root deformation, lacking deep taproots. The loss of tap roots in container culture may be a common phenomenon, with long-term consequences for restorations that indicate significant advantages to establishing woody plants in restorations by direct seeding. Damage to root systems has also posed problems for transplanted wetland seedlings in the Everglades (Florida, USA). *Cladium jamaicense* (Cyperaceae) nursery-reared seedlings that were grown at lower densities incurred greater root damage during transplanting that caused higher mortality than plugs with more seedlings. After establishment, growth rates of the highest density plugs were limited presumably due to intraspecific competition (Miao *et al.*, 1997). The authors observed that avoiding seedling root disturbance during nursery production was a major factor in maximizing survivorship in the field. In wetland plants, transplanting seedlings may offer advantages for flooding tolerance over transplanted rhizomes. Experiments in restored prairie potholes (Yetka & Galatowitsch, 1998; Budelsky & Galatowitsch, 2000) demonstrated that transplanted seedlings of two dominant sedge meadow species, *Carex lacustris* and *C. stricta* (Cyperaceae), were initially more successful than transplanted rhizomes. For *Carex stricta*, seedlings had a rate of establishment across a wide range of water regimes and depth (generally greater than 80%), whereas rhizome survivorship never exceeded 50% for any hydrologic treatment and was most often less than 10%. Although establishment from rhizomes was higher for *Carex lacustris* than *C. stricta*, they still were more sensitive than seedlings to flooding stress.

Designing plantings to maximize establishment

Because quantities of seeds and plants are often limited, restorationists are often precise about locations they choose for particular species or assemblages, informed by vegetation patterns they observe in extant ecosystems. Relying on the occurrences of adult plants to predict optimal survivorship of seedlings can be misleading because juvenile and adult life stages do not necessarily have similar requirements (Grime, 1979; Augspurger, 1983).

The importance of these ontogenic niche shifts has received little attention in restoration ecology, although they could be a major factor in revegetation success (Young *et al.*, 2005). For example, in natural wetlands, some sedges (i.e. *Carex lacustris, Carex stricta*) form large stands in areas of prolonged standing water, yet in restored wetlands, seedling establishment and survivorship is much lower in these settings than at higher elevations (Budelsky & Galatowitsch, 2000, 2004). Juveniles in these newly restored wetlands may be less tolerant of flooding than adults because they lack well-developed rhizomes to aid in oxygen transport (Galatowitsch *et al.*, 1999; Fig. 17.1). In other situations, changes in niche requirements as plants mature could also be a reaction to local hazards (e.g. pathogens, resource competition) created by the presence of adult plants (Augspurger, 1983; Ohlson *et al.*, 2001; Miriti, 2006). Predicting when juveniles will likely have different, or even conflicting, niche requirements than adults could alter how restorations are planned and implemented.

Fig. 17.1 Nursery-reared seedlings of several *Carex* species (e.g. *C. lacustris*, *C. stricta*) were planted at a wetland restoration at elevations comparable to where they typically occur in natural wetlands (Minnesota, USA). Only those at the highest elevations survived; seedlings at lower elevations could not survive prolonged flooding. Photo by S. Galatowitsch.

17.3 | Creating safe sites to promote seedling establishment

A lack of safe sites restricts germination and seedling establishment on many restorations, especially those with extensive, exposed soils and those at the other end of the extreme, with a dense cover of competing vegetation (Urbanska, 1997). Restoration practices promoting plant establishment often target specific safe site requirements of dominant species in the desired community. For example, facilitating regeneration in bauxite-mined Jarrah forests (*Eucalyptus marginata* Myrtaceae) of southwestern Australia depends on handling and respreading removed topsoil and its seed bank so that most seeds are buried, but not more than 5 cm deep (Grant *et al.*, 1996). Most common indigenous species of Jarrah forests are photoinhibited, likely an adaptation to ensure germination occurs where there is adequate moisture and anchorage (Heydecker, 1956) and minimal predation by ants, rodents, and birds (Rokich *et al.*, 2002). However, only large-seeded species can emerge from deep within the soil, so the optimal seed depth is around 1–2 cm, posing a considerable challenge when stripping and respreading topsoil over extensive, uneven terrain. To improve vegetation establishment in a restoration, site preparation and seeding practices are increasingly tailored to specific kinds of ecosystems based on target species adaptations that evolved to avoid prevailing hazards to regeneration.

Creating safe sites in exposed landscapes

In newly restored lands of arid, arctic, alpine, and other barren regions, early seedling growth is often severely limited by physical hazards typical of very exposed settings, like soil instability, abrasion, and inadequate moisture (Chapter 3). Safe sites often in short supply are those that entrap seeds and sediment, provide shelter from desiccating winds, and retain soil moisture. Mulching is perhaps the most widely used technique to add structural complexity to exposed soils to serve as shelters and traps. In addition, mulch can reduce raindrop impact on soils, reduce evaporation, insulate the soil from temperature extremes, and in some cases, contribute to soil fertility. Ensuring the mulch layer is thick enough to provide protection but not so thick as to impede seedling emergence is important, as is determining how to anchor it. In the most extreme settings, depositing rocks or creating artificial shelters (e.g. from wood) are likely to be more effective, although these solutions are usually only feasible to implement in small, strategically selected areas (Fig. 17.2). On reclaimed grasslands in Iceland, colonizing native species preferentially occurred in the vicinity of small rocks that provided shelter and shade and on biological soil crusts that stabilize the soil surface, retain soil moisture, and increase nitrogen fixation from cyanobacteria (Elmarsdottir *et al.*, 2003). Soil moisture limitations in arid lands are often compounded by chemical sealing of the soil surface, which will not be mitigated by mulches and other physical barriers. Hydraulic conductivity of these soils has been improved by treating sites with gypsum, a slow-dissolving calcium salt that replaces some of the sodium bound to clays, causing the soil to be more permeable. Beukes and Cowling (2003) explored approaches for triggering vegetation establishment in historically overgrazed lands in the Succulent Karoo of South Africa, which had exhibited minimal recovery over decades. They determined that regeneration from existing seed banks could be promoted by improving soil surface conditions with both mulch and gypsum. Increasing permeability can also be accomplished by imprinting the soil surface so it is fractured into patterns that promote infiltration (Fig. 17.3).

Fig. 17.2 Restoration crews creating regeneration microsites on an abandoned road in a desert scrubland (California, USA). Holes are bored into soil and dead branches of local woody plants placed vertically in the holes. Seeding is focused in close proximity to these vertical mulch shelters to facilitate germination and seedling establishment. Photo by Paul Brink, Bureau of Land Management, California State Office (USA).

Fig. 17.3 Land imprinting is a commonly used site preparation technique for restoration of arid lands. The microtopography created by the land imprinter increases infiltration in compacted clay soils. Photo by Robert M. Dixon.

Physiological failure is also a significant limitation for seedling establishment in many restored ecosystems, especially those that are chronically water-limited. The dense growth of invasive, annual grasses precludes reestablishment of coastal scrub and perennial grass species by severely reducing water availability near the soil surface, directly impacting juvenile plants with shallow root systems (Dyer & Rice, 1997; Eliason & Allen, 1997; Cione *et al.*, 2002). Restoration strategies such as spring burns and solarization that reduce the competitive advantage of annual grasses during initial establishment should favor both scrub and perennial grass species (e.g. Moyes *et al.*, 2005). Creating bare patches by removing annual weeds and thatch is another potentially useful restoration approach for providing safe sites in these water-limited systems (Eliason & Allen, 1997).

Mitigating the effects of soil alteration and loss on industrial and mined lands

Soil surface conditions can be especially extreme on restorations in industrial and mined sites. In some cases, high concentrations of contaminants pose hazards for establishing vegetation, while in others, total topsoil removal – sometimes reworked, stored, and replaced – can alter physical and chemical soil properties. Atmospheric deposition of pollutants (metals and sulfuric acid) has denuded vast areas of North America and Eurasia that once were boreal or temperate forest. Shelters of stone and wood can provide safe sites for woody plant seedlings (e.g. *Betula* spp. Betulaceae), as long as metal contamination is not too severe (Eranen & Kozlov, 2006). In these cases, incorporating lime is needed to neutralize the acidified soils and minimize metal uptake by the seedlings. On the Kola Peninsula in northwestern Russia, Eranen and Kozlov (2006) found liming doubled the survival of juvenile *Betula pubescens* between the first and fifth years, from 40–80%. Peat mining has also impacted extensive areas of boreal North America and Eurasia, and while this extraction does not concentrate or expose toxins, re-establishing hydrology after mining and a lack of soil fertility on the mined surface create near total barriers for plant establishment. In these settings, mulch is essential for providing safe sites for the establishment of *Sphagnum* and other moss species, while phosphorus additions promote seedling growth of herbaceous and Ericaceous bog plants, as well as some mosses that serve as nurse plants (Rochefort *et al.*, 2003).

Whether seeds occupy restored, potential safe sites is still a matter of chance, and if seeds are in short supply, the odds may not be sufficient to ensure a high rate of seedling establishment. In some cases, particularly those where soil moisture is likely to severely limit germination or seedling emergence, one strategy is to coat seeds or seedling roots with hydrophilic polymers, essentially making some aspects of safe sites portable (e.g. Woodhouse & Johnson, 1991; Specht & Harvey-Jones, 2000). In addition to the polymers which keep the area around seeds and roots moist, the coatings used to cover seeds also can incorporate pesticides, fungicides, nutrients, and growth regulators to promote seed germination and seedling establishment (Venning, 1988). Turner *et al.* (2006) investigated the effectiveness of polymer coatings to enhance recruitment of 11 common species of *Banksia* (Proteaceae) woodlands used to restore sand mines in Australia. Seed coatings increased emergence rates by 17–55%. They speculated that in addition to providing a reliable moisture supply, seed coatings may have improved recruitment by reducing light (most species are photoinhibited) and removal by ants and water erosion.

Creating safe sites for seedlings in restorations with existing vegetation

Because limitations to recruitment are generally responses to multiple hazards, effective restoration strategies designed to provide safe sites often mimic important features of comparable natural systems, rather than treating each limitation with separate prescriptions. Such is the case for forest restoration where woody seedling failure is often very high due to physiological limitations (e.g. inadequate moisture), competition from grasses, and lack of mycorrhizal associates (Hammond, 1995; Slocum, 2000). In tropical, temperate, and boreal forests, decaying logs are preferred sites for woody plant seedling establishment. In cut-over tropical pastures, decaying logs and stumps are most likely to be safe sites because they offer protection from cattle grazing, reduce competition with exotic grasses as the rotting logs are nutrient poor, and increase light availability by allowing seedlings to be situated above the grass canopy (Peterson & Haines, 2000b; Slocum, 2000). These decayed wood seed beds, or nurse logs, also promote recruitment of later successional boreal conifer species, such as *Thuja occidentalis* (Cupressaceae) for reasons speculated to range from providing an accumulation of bacterially fixed nitrogen or ectomycorrhizae to increasing light availability (Cornett *et al.*, 2001). Because decaying logs are often missing from logged secondary forests, germination and seedling establishment is much lower than expected. For many forest restorations, replacing nurse logs is considered an essential restoration step for promoting regeneration (Cornett *et al.*, 2001; Fig. 17.4). In tropical forests, closing the canopy with fast-growing, drought-tolerant plantation species has been pursued as a restoration strategy to minimize the time before understory conditions are amenable for plant establishment, as well as a way to accelerate seed dispersal to these sites by birds (Parrotta, 1992; Khurana & Singh, 2001b; Holl, 2003). Focusing initial planting efforts on fast-growing species that are easy to propagate is likely to be more feasible than attempting to replant a diverse assemblage typical of tropical forests (Holl, 2003). Nonnative tree plantations have facilitated seedling establishment of woody plants in rain forests in Australia, Sri Lanka, and Central America (Ashton *et al.*, 1997; Keenan *et al.*, 1997; Lugo, 1997), although careful selection of plantation trees is critical (Holl, 2003).

(a)

(b)

Fig. 17.4 To restore fire-prone Mediterranean shrublands, seeds must be pre-treated with smoke or smoke extract from local native woody species. A seed smoker consists of (a) a drum in which the fire is situated and a cooling pipe that delivers smoke to seeds, and (b) connected to this pipe, a tented area where seeds are spread out on trays. Seeds are treated for an hour, and then watered to initiate germination. Photo by Trevor Adams, Kirstenbosch Botanic Garden, South Africa.

17.4 | Managing biotic interactions that affect seedling survival and growth

Because many restorations seek to accelerate succession, seed and plant mixes typically include many later successional species. This practice has long been known to be problematic in forest restorations where environmental conditions change dramatically with canopy closure. The difficulty of establishing later successional species in newly created habitats is increasingly understood to be a challenge in grasslands and shrublands, as well as where changes in biotic interactions, such as herbivore pressure, soil microbial associations, and plant competition during community development, influence plant regeneration. For many restorations, the transition from seedling establishment to mature vegetation is often the most labor intensive and prone to failure.

Interactions with neighbors

Competition from species well-adapted to early successional habitats, notably annuals and introduced perennials, can greatly reduce the establishment of later successional species. In many cases, the density of seeds in the seed bank of these early successional weeds greatly exceeds what is seeded. Even under what is considered heavy seeding, the weed seed banks in grasslands and meadows, for example, can be several orders of magnitude greater than the density of deliberately seeded species. Not surprisingly, the synchronous germination of fast growing, often high density species, such as *Bromus tectorum* (Poaceae), can reduce water and nutrients available to other species. Site preparation tactics designed to reduce weed seed banks prior to seeding are commonplace in restorations. After a flush of seedling emergence of weedy species, sites are disked or chemically treated. In some cases, several cycles of emergence followed by control are needed to sufficiently reduce seed bank densities to a level that does not interfere with the establishment of later successional species. While this practice (using herbicides) is effective at reducing *Phalaris arundinacea* (Poaceae) seed densities, even very low densities of this strongly rhizomatous grass can impede the establishment of sedges (*Carex* spp.; Reinhardt, 2004). When soils are amended with sawdust to reduce nitrogen levels, *Phalaris* seedling establishment is limited and *Carex* is the superior competitor (Perry *et al.*, 2004). Reducing soil fertility by scraping, plant harvest, and amendments are increasingly being considered for restoration, especially in North America and western Europe, where intensive agriculture has increased atmospheric and surface inputs. Even though these tactics are only effective for a limited time (especially sawdust amendments), if later successional species can close the canopy while nitrogen is depleted, germination of many light-requiring weedy species will be greatly reduced.

In contrast to inherently productive ecosystems, existing vegetation is widely reported to facilitate the regeneration of other species in severely stressed ecosystems (e.g. Niering *et al.*, 1963; Urbanska & Chambers, 2002). Facilitation in restorations is not surprising because exposure intensifies moisture and temperature stress on many sites. Seedlings establishing near existing vegetation may benefit from favorable canopy and/or soil conditions. An existing canopy of vegetation can lower irradiance, moderate temperature fluctuations, and reduce evapotranspirational demands. In addition, some canopies may reduce herbivory on establishing seedlings by making them less visible or accessible to grazers (Gómez-Aparicio *et al.*, 2005). The soil near existing vegetation may have increased nutrients and organic matter from litter accumulation beneath the canopy, symbiotic associations with nitrogen-fixing organisms, or trapped fine sediments. The advantages of more suitable canopy and soil conditions outweigh potential limitations from resource competition in many systems, especially sites that are moisture limited or experience very high or low temperatures (Milton *et al.*, 1997; Eccles *et al.*, 1999; Urbanska & Chambers, 2002; Gómez-Aparicio *et al.*, 2004). Several restoration strategies have been devised to capitalize on facilitation effects. Seeding near remnant vegetation has proved beneficial in restorations of cloud forests (Alvarez-Aquino *et al.*, 2004), Mediterranean scrub (Gómez-Aparaicio *et al.*, 2005), deserts (Milton *et al.*, 1997), and alpine meadows (Urbanska & Chambers, 2002). Planting stress-tolerant pioneer species to facilitate seedling establishment of a broader array of species is a restoration technique commonly used in mined land restoration (Lunt & Hedger, 2003). In many cases, the suitability of plants within an ecosystem to provide facilitation varies. For example, *Pinus* seedling establishment benefits most from association with a short-statured, shallow-rooted shrub, *Salvia lavandulifolia* (Lamiaceae) because the pines could easily out compete shrubs as they grew larger (Castro *et al.*, 2002). Likewise, not all species within a community are equally likely to benefit from facilitation. In a Great Basin desert restoration, Huber-Sanwald and Pike (2005) found that while sagebrush (*Artemisia tridentata* Asteraceae) could facilitate *Agropyron smithii* (Poaceae) establishment under some conditions, *Agropyron desertorum* (Poaceae) establishment was impaired by reduced irradiance under any level of canopy closure. Because *Agropyron desertorum* is not native to the Great Basin (USA), this difference in facilitation could be a benefit, not a problem to overall restoration success.

Interactions with soil microbes

Early successional soils, typically the product of recent disturbances and alterations, have depauperate microbial communities, including low abundance and diversity of symbiotic mycorrhizal fungi. Allen *et al.* (2002) noted that restorations typically attempt to establish late successional vegetation in early successional soils. For most later successional species, symbiotic mycorrhizae, most commonly arbuscular mycorrhizae (AM), contribute to the efficient acquisition of nutrients and protection against pathogens during seedling establishment (Smith & Read, 1997; Renker *et al.*, 2004). These symbiotic associations often have major effects on nutrient cycling because fungal hyphae bind soil particles and immobilize nutrients (Renker *et al.*, 2004). Surprisingly, few studies have focused on the role of AM fungi in seedling establishment, although pot culture experiments have shown that the competitive balance among species can be affected by the presence or absence of AM fungi (Allen & Allen, 1984; Hetrick *et al.*, 1989). Whether a restoration site will favor early successional, often weedy species, or deliberately planted later successional species may depend on whether a sufficiently high density and diversity of fungal spores occur there. In a study of vegetation succession in sagebrush steppe (USA), Allen *et al.* (1989) found that *Salsola kali* (Chenopodiaceae), the dominant weedy colonizer of disturbed sites, was not only nonmycotrophic, but was actually parasitized by AM fungi. AM fungi increased seedling mortality and inhibited the growth of survivors. The vegetation shifted to facultative mycotrophs (i.e. the grasses *Elymus smithii* and *Bouteloua gracilis* Poaceae), as the density of fungal inoculum increased. *Artemisia tridentata*, an obligate mycotrophic shrub, established after the grasses were well established (Renker *et al.*, 2004). Changes in fungal composition accompanied these vascular plant changes. Early successional *Glomus* spp. with small, wind-dispersed spores, facilitated the growth of grasses. Later successional species, including *Scutellospora* spp. and *Acaulaspora* spp., were likely responsible for enhancing growth of *Artemisia* (Weinbaum *et al.*, 1996).

The restoration of mycorrhizal fungal communities is clearly crucial for revegetation (Chapter 9). Managing soil to preserve inocula ensures an on-site source of spores of both early and later successional species. In cases where this opportunity is lost, restorationists can adopt passive or deliberate tactics to accelerate fungal succession. Early in succession, windblown species (e.g. *Glomus*) can be trapped in the lee of hill slopes, nurse plants, or other barriers (Allen *et al.*, 2002). These locations are logical focal areas for planting or seeding midsuccession species that will help increase inoculum densities further and also attract animal dispersers of later-succession fungi. A more active restoration approach is to inoculate sowing sites, seeds, or seedlings, preferably with local mycorrhizae (Renker *et al.*, 2004). Although using inocula from a mature ecosystem would seem to best support the establishment of later successional species, studies from both arid sagebrush steppes (Weinbaum *et al.*, 1996) and dry tropical forests (Allen *et al.*, 2005) report that applying an inoculum of a few *Glomus* spp. stimulates initial establishment and growth. Ectomycorrhizal fungi, common symbionts of many woody species, can be stimulated on restoration sites by ensuring the soil surface has pockets of decaying organic matter, for instance from bits of coarse woody debris (St. John *et al.*, 1983). On a landfill restoration, *Quercus prinus* and *Q. rubra* (Fagaceae) seedling establishment was higher on soil microsites that included fragments of decayed wood and their associated high rates of mycorrhizal infection (Parson *et al.*, 1998).

Herbivore impacts on newly established vegetation

Seedlings in new restorations often must contend with high herbivore pressure compared to many natural ecosystems; newly created habitat makes excellent foraging sites for granivores and grazers looking for readily available food sources with relatively little difficulty of movement. Seed predation and seedling herbivory are known to be major limitations for establishing vegetation in tropical forest systems (e.g. Hammond *et al.*, 1995; Holl, 2003). In some ecosystems, seedling herbivory may be of comparable intensity in restored and natural sites. The adult populations of natural sites may be the product of occasional rare events or episodic recruitment. Such is the case of *Xanthorrhoea* (Xanthorrhoeaceae) populations in Jarrah forests of southwestern Australia. *Xanthorrhoea gracilis* and *X. preisii* are common understory species in unmined forests, although seedlings are rarely observed, presumably because they are grazed by western gray kangaroos (Koch *et al.*, 2004). Grazing by kangaroos also severely limited the early survival and growth of *Xanthorrhoea* seedlings on mined sites. Because new restorations seldom have the potential for recurrent recruitment attempts (either from natural recolonization or planting), grazing exclosures are frequently used to establish an initial population. For both species of *Xanthorrhoea*, exclosures increased establishment threefold (Koch *et al.*, 2004). Grazing exclosures are also commonly used to protect tree seedlings from deer (*Odocoileus virginianus*) and aquatic plants from muskrats (*Ondatra zibethicus*), among others, and are removed as plants grow large enough to withstand some biomass removal or until self-seeding occurs. Although artificial barriers offer the greatest protection, some nurse plants can also reduce seedling mortality from herbivory, especially if the nurse plants are spiny or provide visual protection (Castro *et al.*, 2004b).

17.5 | Mimicking the effects of disturbances in restoration

In disturbance-prone natural ecosystems, seedling emergence from the seed bank is often cued by some feature of major events. This is also true in ecosystems that experience long periods unfavorable for seedling survival, such as those with frigid or dry seasons. For example, in wetlands and riparian systems, saturated mudflats exposed in the aftermath of flooding episodes stimulate germination (e.g. van der Valk, 1978; Casanova & Brock, 2004). The smoke or heat associated with fires will release canopy-stored seeds or promote emergence in forests, grasslands, and shrublands in regions with pronounced dry seasons prone to ignition (e.g. van Staden et al., 2000; Rokich et al., 2002). At their initiation, many restorations are devegetated, as might be similar to a recently disturbed, natural system. To encourage introduced or remnant seed populations to germinate, however, restorationists often need to mimic the key aspects of the disturbances or climatic episodes that serve as germination cues. Because disturbances also affect vegetation composition, restorationists also frequently use fire or flooding as a management tool to favor the intended community.

Promoting seedling emergence by treating seeds to break dormancy

Priming dormant seeds to trigger germination is a strategy that should minimize the evolutionary liability of selecting for reduced dormancy while allowing for a pulse of initial establishment. Cold stratification and smoke treatments are perhaps the two most widely employed seed priming techniques used in restoration. In cold climate regions of the world, seed dormancy ensures that brief warm spells during the winter will not trigger germination. To prime seeds for germination, restorationists store harvested seed for several months in cold conditions prior to sowing. Without stratification, establishment can be protracted, giving weedy species a head start. Species that evolved to regenerate after fire disturbances can be stimulated to germinate with smoke extract. Restorationists treat seed mixes with aerosol smoke or an extract of smoke water. For example, Rokich et al. (2002) found that smoke treatments of common species of Banksia woodlands in Australia increased germination rates up to 42 times. Aerosol smoke and smoke water treatments are used to maximize germination of seedings in mined land reclamation of Jarrah forests of Western Australia and Mediterranean scrub in South Africa (Brown & van Staden, 1997; Roche et al., 1997).

Recreating disturbance regimes

Vegetation changes in many ecosystems are typically cyclic, triggered by disturbances such as fires or floods that cue seedling recruitment.

Reestablishing disturbance regimes can be a key aspect of restorations, and in some cases, the primary practice. For example, Mediterranean climate shrublands of California chaparral, Australia kwongan, and South African fynbos historically experienced fires every 4–100 years and the community dominants have long-lived soil or canopy stored seeds that germinate after fires (Le Maitre & Midgely, 1992; Holmes & Richardson, 1999). For fire-tolerant species in these systems, seed dormancy may be broken directly by the heat pulse or smoke from fire, or indirectly by increases in diurnal temperature fluctuations as a result of vegetation removal. Where fire prevention has shifted the composition toward nontolerant species, prescribed fires can create the post-fire mineral soil patches that these species depend on for seedling recruitment. In some cases, extensive woody overgrowth needs to be removed before prescribed burning to minimize the likelihood of unnaturally hot fires that can kill even fire-tolerant seeds or make soils hydrophobic. The effectiveness of prescribed fires to facilitate seedling establishment is a function not only of burning the canopy vegetation, but also of the microsite conditions created. Lamont *et al.* (1993) studied the fate of seedlings in a restored Australian shrubland that was treated with prescribed fire, creating a mosaic of sand and litter microsites. Seeds occupying sand and litter patches had comparable levels of survivorship through summer and fall, as well as subsequent winter germination. However, seedling survival and growth were both much lower in litter than sand microsites for many species. While some investigators have reported that litter microsites are especially critical for post-fire regeneration, Lamont *et al.* (1993) counter that more seeds are dispersed into litter patches resulting in greater competition for water and, thus, much lower seedling success rates.

Prescribed disturbances, such as fire, can be counterproductive, if the timing or periodicity are not appropriate for the system or if the disturbance triggers the emergence or spread of invasive species. For example, wet meadows in the midcontinental USA likely burned infrequently (during severe droughts) until agricultural settlement. Within this region, fire is frequently used in tallgrass prairie restoration to reduce the growth of cool season species. However, the exposed gaps created by prescribed burns in meadows allow the invasive grass *Phalaris arundinacea* to emerge from the seed bank. In California chaparral, repeated burning favors aggressive, herbaceous alien species over shrubs. Shrub recovery cannot occur if the time between fires is not adequate (e.g. Parker, 1990; Keeley, 2002; Keeley & Fotheringham, 2006).

17.6 | Conclusions

With an increasing array of restoration techniques, establishing an initial community is possible on a wide variety of sites, including some extreme environments. On these harshest sites, addressing soil degradation and unnatural microsite conditions are understood to be essential for both community establishment and for catalyzing natural processes essential for long-term community development. Without deliberate intervention, the conditions on new restorations seldom support safe sites suitable for seedling establishment of many of the desired species. Formulating restoration approaches to facilitate seedling recruitment for restorations has, to a great extent, been informed by key factors in natural ecosystems that govern regeneration of plant communities. Various seed treatments, such as stratification and smoke extract, mimic important climatic and disturbance cues for seedling recruitment and thus can be used in restorations around the world that share those same kinds of regeneration-triggering events.

For many restorations, though, practices aimed at optimizing initial establishment have little to do with facilitating long-term community development and are primarily short-term remedies for vexing problems. For example, a need to prevent an influx of invasive species can result in implementing restoration approaches solely focused on choosing species with rapid seedling establishment and growth. But if the initial colonists selected to establish quickly and grow aggressively preempt establishment of later arrivals, succession can be arrested, not accelerated. Of equal concern is that optimizing practices for initial success can actually mask environmental problems that will hinder long-term community development. One pervasive example is soil compaction. While an assemblage of a few tough colonists can be established on abandoned roadbeds that have received minimal physical preparation, the likelihood that these conditions will be suitable for the full complement of species expected in many communities is very low. Many restoration projects are pursued in very short time frames (i.e. a few years), often compressing the schedule for site preparation and placing increasing expectations on establishing seedlings in still highly altered settings. The innovative practices developed to encourage seedling recruitment across a great variety of altered ecosystems are certainly one of the major accomplishments of restoration ecology over the past few decades. Understanding the long-term implications of these revegetation practices and developing more approaches for directly addressing environmental stressors across sites and landscapes are important challenges that deserve additional attention.

Part VI

Synthesis

Chapter 18

The seedling in an ecological and evolutionary context

V. Thomas Parker, Robert L. Simpson, and Mary Allessio Leck

18.1 | Introduction

All theories concerning metapopulation, source–sink, and metacommunity dynamics require dispersal, seed-bank dynamics, and seedling establishment to structure populations and communities (Hubbell, 2001; Leibold *et al.*, 2004). Our understanding of dispersal (Howe & Smallwood, 1982; Nathan & Muller-Landau, 2000; Levine & Murrell, 2003), seed germination ecology (Baskin & Baskin, 1998; Fenner & Thompson, 2005), and seed-bank dynamics (Leck *et al.*, 1989a), as well as the structure and dynamics of adult plants and communities, emphasize the need to bring the seedling life history stage to the fore. This is particularly important for the concept of recruitment limitation (Hurtt & Pacala, 1995).

Seedlings are clearly a vulnerable stage, shifting in short time periods from complete dependence on maternal reserves to physiological independence. Variation in seed size, carbon allocation patterns, and seedling structure and physiology has considerable influence on the potential for individual seedlings to survive to establishment. Dispersed across a variable habitat, mortality results from discordance in those characters and the environment, limiting potential establishment of seedlings. Their high mortality results from drought, herbivory, and disease (Moles & Westoby, 2006b; Fenner & Thompson, 2005; Kitajima, 2007), although many other factors also determine seedling success.

Reducing vulnerability may be accomplished by maternal investment in the seed that shepherds the seedling toward independence, or by facilitation by a nurse plant or other microhabitat (rock crack). Typically, the nurse plant is not the parent, raising the question of implications for community associations and long-term dynamics. Vulnerability may be lessened by the ambient disturbance regime, for example, gap formation. However, disturbance can create changes that reduce favorability for one species and enhance it for another.

Where an invasive is involved, disturbance or a new habitat may release it from constraints of its native habitat. Reduction of vulnerability may come from symbioses that enhance nutrient and/or water availability (Chapter 9). The degree of seedling vulnerability, which can change with time and habitat, may contribute to species diversity and the presence of rare species that have low seed production or lack a persistent seed bank.

Plant population recruitment is limited largely by inadequate dispersal or failure of seedlings to establish. At the same time, seedlings are only one stage in the life cycle of plants, and we should expect coordination among life stages resulting from evolution in varying ecological circumstances. An extensive base of knowledge is developing, especially for some environments, although research has focused on a limited array of seedling types and research questions (Chapter 11; Garwood, 1996; Levine & Murrell, 2003), and research questions may have been inaccurately formulated (Moles & Westoby, 2006b). The available evidence suggests that the seedling stage requires consideration in the context of entire life histories and resident communities to further our understanding of population and community dynamics, as well as ecological evolution. Thus, seedling ecology must consider the spectrum of life history traits, including seed size, seed-bank type, seedling longevity, adult size and longevity, and time to reproduction (e.g. Chapters 10, 11, 13; Moles & Westoby, 2006b).

Although recruitment varies greatly in space and time, spatial patterns are understood better than temporal patterns because of experimental accessibility. Temporal patterns govern different scales of dynamics within communities, and given the context of global change, more studies are needed that focus on longer-term processes. Indeed, the differential yearly success of cohorts may influence long-term trajectories of community change (Chapters 12, 16).

18.2 | Dispersal, seed bank dynamics, and seedling banks

Plant populations cope with spatial and temporal heterogeneity in recruitment by dispersal and seed-bank dynamics. Seed dispersal is a key process creating spatial patterns (Harper, 1977; Nathan & Muller-Landau, 2000). Differences in numbers of seeds dispersed, the shape of the seed shadow, and the frequency of long-distance events initiate much of the dynamics within and among communities. Subsequently, other processes, like predation and/or germination and establishment, reinforce or modify the initial spatial pattern established by dispersal. For many species, recruitment success increases with distance from the parent (Janzen, 1970; Connell, 1971; Clark & Clark, 1984; Hammond & Brown, 1998; Harms et al., 2000), with most seedlings not related to nearby adults (Hardesty et al., 2006). In other circumstances, seedlings may differentially be associated with or near adults (Arii & Lechowicz, 2002; Lazaro et al., 2006).

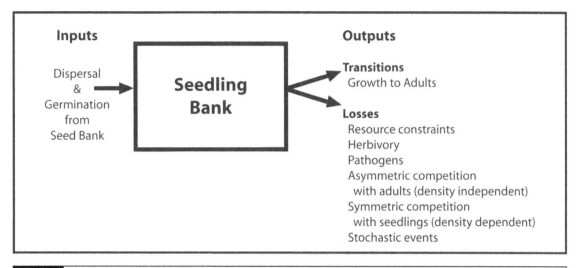

Fig. 18.1 The dynamics of a seedling community reflects different rates and sources of inputs and losses. While inputs are generally dependent on germination from the soil seed bank, outputs from the seedling bank can result from growth into adults or from mortality.

Dispersal patterns are heterogeneous (Nathan & Muller-Landau, 2000; Levine & Murrell, 2003; Clark *et al.*, 2004), as are post-dispersal processes (Chambers *et al.*, 1999; Rey & Alcantara, 2000; Gomez-Aparicio *et al.*, 2005). Seedling recruitment can be highly dependent on directed dispersal to suitable habitats for seedlings (Tewksbury & Nabhan, 2001; Wenny, 2001), or set the stage for subsequent differential patterns of mortality (Rey & Alcantara, 2000; Gomez-Aparicio *et al.*, 2007).

In almost all of these studies, plants exhibited transient seed banks. Dormancy complicates the concordance of dispersal with recruitment because of the temporal dimension (Parker *et al.*, 1989; Baskin & Baskin, 1998; Fenner & Thompson, 2005). In some cases, potential seedling recruitment is spread over a longer period; in others, recruitment from the seed bank may respond to a disturbance event equivalent to a mass dispersal effect (Shmida & Ellner, 1984). Because persistent seed banks uncouple seedling recruitment from current conditions, they can act to ensure higher rates of seedling survival. Environmental conditions following wildland fire, for example, create substantially different abiotic and biotic conditions favoring seedlings arising from persistent soil seed banks or serotinous fruit or cones (Whelan, 1995; Bond & van Wildgen, 1996). In wetlands, climate determines drawdown/inundation cycles that engage different components of the persistent seed bank (e.g. van der Valk, 1981; Keddy & Reznicek, 1982; Brock, 1998).

Seedlings emerging from soil seed banks give rise to spatial and temporal patterns of seedling distributions called *seedling banks*. This stage is transient among annuals and most other herbaceous seedlings and more prolonged in woody species. Inputs to the seedling bank arise from soil seed banks, which in turn are dependent on dispersal patterns (Fig. 18.1). Losses from the seedling bank result

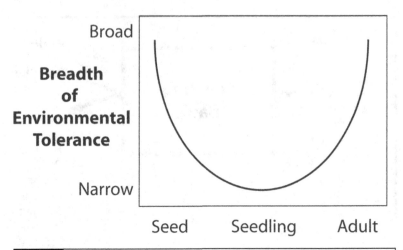

Fig. 18.2 Schematic of environmental (biotic and abiotic) tolerances of seedlings relative to seed and adult life stages.

from successful transition of individuals into adulthood, or to mortality from a variety of factors with resource limitations and herbivory often being of greatest importance. Extremely low rates of seedling success may be sufficient as long as parental populations produce high numbers of seeds, creating a mass effect (e.g. Brown & Fridley, 2003; Thomsen *et al.*, 2006). Even in extreme environments, such as arctic or alpine tundra, adequate numbers of seeds are produced to assure species continuity over the long term (Körner, 2003).

18.3 | Dynamics of individual seedlings

Individual seedlings face a larger number of physiological and environmental constraints than do either seeds or adults (Fig. 18.2). This results from two separate characteristics, their diminutive size and large-scale physiological transitions, as occur during development of the embryo into a seedling and maturation into a juvenile or adult.

Following germination, maternal reserves are mobilized for the construction of the young plant. This complex development results from the interplay of phytohormones and environmental cues (Chapter 7) initiating carbon allocation (Chapter 8). Light cues become the first environmental constraint on many seedlings, controlling seedling emergence and early development (Chapter 8; Kitajima, 2002). The transition to deetiolation, development of leaves, and growth in shifting light environments depends on interactions among phytohormones (Chapter 7; Garcia-Martinez & Gil, 2002; Vandenbussche & Van Der Straeten, 2004; Vandenbussche *et al.*, 2005). In the aerial environment, subtle sifts in red:far-red ratios are detected by phytochrome indicating the presence and direction of neighbors (Vandenbussche *et al.*, 2005). Similarly, root development is critical to avoid fatal water stress and optimize mineral availability. Even in tropical forests, water stress seems to be a critical constraint during seedling establishment (Chapter 16; Engelbrecht *et al.*, 2007). Although equally important, much less is known about developmental patterns and diversity of seedling roots in the soil environment than above-ground plant parts.

Patterns of allocation shift depending on the amount of light, the patchiness of soil minerals, and whether other stresses are present. For example, seedlings exposed to drought develop higher levels of brassinosteroids that appear to confer protection by regulating membrane permeability and structure of proteins (Chapter 7; Krishna, 2003). These physiological changes also appear to influence resistance to other environmental factors. Because carbon reserves are scarce, allocation patterns often become all-or-nothing developmental decisions. As seedlings utilize reserves, dependency on light develops (Chapter 8; Kitajima, 2002), suggesting carbon is the first limiting factor for small seedlings. Relative growth rates of seedlings, which are negatively correlated with seed size in many plant families and habitats, quickly become positive as long as resources are available (Chapter 8; Shipley & Peters, 1990; Poorter & Garnier, 2007).

Contrasts between seedlings in low and high light provide a measure of their adaptive developmental plasticity (Bazzaz, 1990, 1996). Seedlings in low light must balance complex interactions between leaf- and plant-level responses to light, nutrient, and water availability (Givnish, 1988, 1995; Messier *et al.*, 1999). Comparisons of shade to higher light environments at the leaf level show a consistent shift from high to low specific leaf area within species (Bazzaz, 1990; Niinemets & Kull, 1994; Messier *et al.*, 1999). Similarly, adult maximum rates of photosynthesis were lower among species whose seedlings can tolerate shade (Leverenz, 1996; Bassow & Bazzaz, 1997). Among woody plants, those seedlings best tolerating shade seem able to cease height growth in diminished light, and maximize the leaf area relative to nonphotosynthetic tissues (Kitajima, 1994; Messier *et al.*, 1999).

Other factors are also important for seedling success. Individual trees impact the soil environment differently due to characteristics of their litter, modifying soil pH and nutrient combinations (e.g. Gosz *et al.*, 1973; Ellis & Mellor, 1995; Finzi *et al.*, 1998). While light is usually the initial environmental constraint on seedlings in shaded environments (Kitajima, 2002), soil nutrients and/or moisture will later impact seedlings to a similar or greater extent (Burslem *et al.*, 1995; Walters & Reich, 1996; Lusk *et al.*, 1997; Engelbrecht *et al.*, 2007). Likewise, early season temperature constraints can be important determinants of seedling establishment (Wang & Lechowitz, 1998).

These responses to abiotic gradients by seedlings become critical in the context of climate change. Temperature shifts predicted by the IPCC (2007) and the Arctic Climate Impact Assessment (2004) indicate that boreal and arctic areas will differentially warm with climate change (Stewart *et al.*, 1998). Because climate change will not occur uniformly over the globe, changes in processes that affect seedling recruitment will vary greatly among regions. Some species are sensitive to winter temperature extremes, while others are more sensitive to moisture flux. Thus, climate change will not be a simple northward shift of climate patterns, but a complex mosaic of temperature and precipitation shifts. Changes in winter snowpack, for example, will have large impacts on watersheds, leading to earlier summer droughts in large regions including western North America, northern Africa, and southern Europe, with the consequence of threshold changes for many species. These changes (such as fire regimes) will lead to increases in processes that will strongly affect patterns of seedling recruitment and ultimately lead to discontinuous changes in the structure and dynamics of communities.

18.4 | Seedlings in heterogeneous environments

The transition to a larger juvenile or adult often "solves" many of the most critical environmental constraints of fluctuating water supply, insufficient nutrients, and/or limited light availability. Seedlings establishing in the understory of tall vegetation, such as a shrubland or forest, may live for multiple years as suppressed individuals in a persistent seedling bank. These seedlings must tolerate the temporal extremes of their environments, including summer droughts, extreme cold, or the weight of snow in winter. Physiological responses in seedlings must accommodate and anticipate these seasonal environmental fluctuations. Seedlings remain vulnerable unless they can protect their tissues from excessive stress such as ice crystal damage, high temperature denaturation of proteins, or osmotic flux due to drought or cold (Chapter 7). Morphological traits, including contractile roots and/or hypocotyls that place the growing point belowground, may also avoid stresses (e.g. Pate & Dixon, 1982).

Habitats exhibit considerable spatial heterogeneity of resources, with establishment patterns often reflecting the size and physiological constraints of seedlings (Garcia et al., 2005). Nurse plants ameliorate resource extremes that seedlings could not otherwise tolerate (Chapter 15; Turner et al., 1966; Jordan & Nobel, 1979; Nobel, 1984; Bertness & Callaway, 1994; Callaway, 1997; Callaway & Walker, 1997; Chambers et al., 1999). Disturbance regimes may also provide microsites suitable for seedling establishment (Chapters 14–17). Mycorrhizal mutualists are similarly patchily distributed in many habitats (Horton et al., 1999), and their presence or absence determines the success of seedling germination and establishment (Chapter 9; Dunne & Parker, 1999; Horton et al., 1999).

Ecological syndromes of seedlings are generally described as exploitive or opportunistic versus conservative (e.g. Chapters 8, 11; Bazzaz, 1979; Bormann & Likens, 1979). The former have high resource demands and relatively high potential growth rates. Successful seedling establishment is dependent on the availability of higher levels of resources. Conservative syndromes are characterized

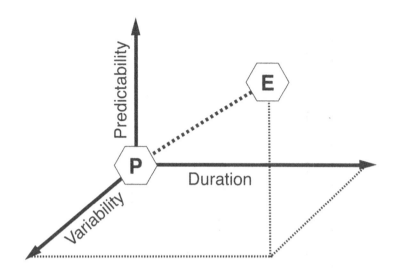

Fig. 18.3 Environmental stressors may range from persistent (P) to ephemeral (E). Stressors vary in predictability, variability, and duration. The predictability and duration axes run from high (near P) to low and the variability axis runs from low to high in the direction of the arrow. For example, most temporary wetlands lie along the dotted line and inundation varies annually or on another time scale. From Boutlon and Brock (1999), reproduced with permission from the authors.

by longer-lived seedlings tolerant of resource limitation. Conservative seedlings also appear to be better defended against herbivores or fungal pathogens. These alternative syndromes are extreme representatives of a gradient pattern of seedling responses (e.g. Hubbell, 2004). Other life history stages clearly interact with these syndromes, often in compensatory ways. Likewise processes, such as herbivory and pathogens, may influence characteristics including relative growth rates and specific leaf area (Hanley & Lamont, 2002).

Processes of seedling mortality or constrained seedling growth vary with habitat (Table 18.1). These may vary spatially and temporally and their influence is determined by their predictability, variability, and duration (Fig. 18.3) (e.g. Canham *et al.*, 1996; Wright & Westoby, 2001; Arii & Lechowicz, 2002; Gomez-Aparicio *et al.*, 2005; Baraza *et al.*, 2006; Lazaro *et al.*, 2006). Multiple stresses act on seedlings (Chapter 3), and many of the coping strategies available to adults (e.g. deep tap roots of desert species) may not be available to seedlings because of size alone. In some cases, benefactors such as nurse plants that facilitate seedling establishment may change roles and cause seedling mortality when they out compete seedlings for limited resources such as moisture during droughts (e.g. Chambers *et al.*, 1998). In the tropics, where the highest seedling diversity occurs and woody dicots and monocots, as well as bromeliads and orchids, abound, establishment occurs despite high biotic stress due to pathogens and herbivory. At the other end of the spectrum are arctic and alpine deserts where low temperatures, lack of water, and wind limit establishment to very infrequent events, with low levels of seed availability compounded by low frequency of establishment niches.

Seedling establishment is dependent on availability of safe sites/regeneration niches. These may be limited by prevailing abiotic and/or biotic factors. Safe sites in many systems are determined by disturbance (Table 18.1), which may either facilitate or obstruct seedling establishment depending on species. For example, in tropical forests,

Table 18.1 | Major constraints to seedling growth and agents of disturbance in selected biomes

Habitat	Seedling richness ranking[a]	Major constraint variables	Agent of disturbance
Tropical forests	1	Shade, litter, low nutrients, pathogens, herbivores	Litter, animal activity, tree fall, tree death
Deserts	2	Drought, heat, high irradiance, salinity	Animal activity, wind-blown sand
Grasslands	3	Fire, grazing	Fire, grazing, animals
Semi-arid shrublands	4	Fire, drought, heat, grazing	Fire, debris flow
Temperate forests	5	Shade, litter, winter temperature, fire, drought	Hurricanes or other storms, tree fall, tree death, frost heaving
Arctic/alpine	6	Cold, irradiance, snow cover, wind	Frost heaving, desiccation

[a] Ranking is speculative.

litter may serve as a deterrent to seedling growth, while in deserts it may enhance establishment (Chapters 3, 16). Maintenance of diversity in temperate deciduous forests or western conifer forests, however, may be related to seed availability, coupled with temporal heterogeneity in recruitment (Chapter 12; Hille Ris Lambers, *et al.*, 2002; Beckage & Clark, 2005) rather than availability of regeneration niches.

If facilitators of establishment rather than limiting factors are examined, a different picture emerges. Phylogenetic heritage endows seedlings with the ability to tolerate or avoid environmental stressors; what may be stressful in one habitat for a given species may be advantageous in another. Seedlings may be seen as opportunistic or conservative, generalists or specialists, avoiders or tolerators. These attributes may be related to their plasticity and to above- and below-ground architecture, with success often the result of physiological and morphological characteristics, determined by the genetic potential of the seedling.

18.5 Alternative strategies

A variety of alternative seedling strategies exist. Usually seedlinghood occurs away from the parent, but in the case of viviparous species, growth is directly supported by the parent (Chapter 2). In some cases of vivipary, the seedling can be well developed before abscission and dispersal occur. In certain other species that lack seeds, bulbils or other asexual reproductive structures substitute for seeds, and in the case of pseudovivipary, plantlets develop on inflorescences in the place of seeds. Young plants from such structures are analogous to seedlings, for example, seedling equivalents. These plant parts may produce new plants that are quite similar in appearance to seedlings. Ultimately, the importance of alternative strategies lies in the ability of plants, facing differing selective factors, to recruit new individuals successfully.

Alternative strategies among seedlings may exist within the same habitat. These may result from seedlings differing in soil resource tolerance (e.g. Arii & Lechowicz, 2002) or in their ability to tolerate shade from established adults. Among woody plants, a gradient of shade tolerance exists (Humbert *et al.*, 2007) that results in many species completely failing to establish in shade, while others produce complex age- and size-class seedling banks. Among temperate forest herbs are species whose seedlings exhibit various dormancy patterns (Chapter 2; Baskin & Baskin, 1998). These alternatives reflect reliance of populations on different life history stages for persistence (Parker *et al.*, 1989). While structural and physiological processes may gradually shift along the gradient of shade tolerance, the ecological impact is quite distinct.

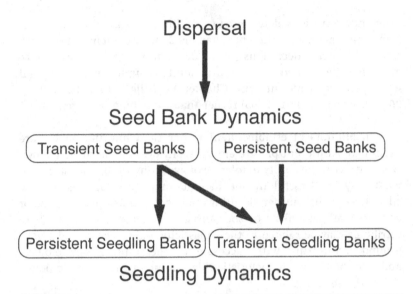

Dispersal

Seed Bank Dynamics

(Transient Seed Banks)   (Persistent Seed Banks)

(Persistent Seedling Banks)(Transient Seedling Banks)

Seedling Dynamics

Fig. 18.4 Types of seedling dynamics in plant communities. Transient seedling banks are those living for less than a year, arising from either transient or persistent seed banks. Persistent seedling banks contain individual seedlings that live for multiple years and generally arise from transient seed banks.

Seedling banks are generally of two types, transient and persistent (Fig. 18.4). Often species arising from long-term persistent seed banks have seedlings that are exploitative of resources and if they are not in an appropriate location, do not survive to the next growing season. This may be seen in post-fire establishment of seedlings in chaparral and other Mediterranean climate systems, and conifers with serotinous cones in other climate regimes. Other plants producing transient seedling banks are long-lived woody plants that produce transient seed banks from which seedlings arise following yearly seed dispersal. These seedlings have similar exploitative resource demands and usually fail due to lack of light, moisture, or other resources. Annuals and many herbaceous perennials are similarly characterized by transient, rapidly growing, exploitative seedlings. The primary advantage of this type of seedling is the ability to fill gaps in vegetation rapidly, exceeding growth rates of competitors and surviving herbivore damage. The difference in seed-bank type indicates the selective pressures on the population as a whole and the integration of all life history stages. For these types of plants, the seedling phase is rapid and transitional; other stages of the life history act as persistent stages.

Persistent seedling banks, in contrast, appear to arise from species producing transient seed banks. These seedlings exhibit tolerance to resource limitations, frequently low light levels, and often water and nutrient stresses. These species represent the conservative end of a seedling syndrome gradient, characterized by slower rates of growth and more investment in defense and wood density. Persistent seedling banks are perennial by definition, of multiple ages and size classes. The primary advantage to these types of seedlings is their preemption of future gaps. In some case, however, persistent seedling banks can be short-lived. For example, many tree species produce seedlings that are shade intolerant, rarely surviving more than 1 year except in the rare circumstance of sufficient light (e.g. *Betula papyrifera* Betulaceae, Perala & Alm, 1990; *Pinus banksiana* Pinaceae, Beland & Bergeron, 1993; *Populus tremuloides* Salicaceae, Messier *et al.*, 1999). These contrast with long-term persistent seedling banks of more shade-tolerant species that produce a higher density of understory seedlings and saplings (e.g. *Abies lasiocarpa* Pinaceae, *Acer saccharum* Aceraceae, *Fagus grandifolia* Fagaceae, *Picea engelmannii* Pinaceae, and others; Aplet *et al.*, 1988; Canham, 1988; Canham *et al.*, 1996; Antos *et al.*, 2000; Walters & Reich, 2000; Casperson & Kobe, 2001).

A tidal freshwater wetland in New Jersey (USA), where climate is seasonally predictable and where water and nutrients are not limiting, illustrates the complexity of the interactions between these strategies. The 13 seed-bank species in the wetland show a range of dispersal, seed bank, and seedling types (Table 18.2). Those with persistent seed banks, including the monocots, *Carex* spp., *Dulichium arundinaceum*, and *Juncus effusus*, do not recruit frequently. Their patchy, deeply buried seed banks require soil disturbance for germination. In contrast, transient seed-bank species, like *Impatiens capensis* and *Polygonum arifolium*, germinate readily in the spring once stratification requirements are met. These early germinating annuals are more likely to establish than perennials (Parker & Leck, 1985) unless canopy gaps occur. Also, the two *Polygonum* species have differing seedling tolerances that result in zonation segregation: *Polygonum punctatum* able to tolerate the inundation stresses, occurs along tidal channels, while *Polygonum arifolium*, able to tolerate biotic stresses (competition), is typically found on the high marsh. Along the slightly elevated levees of the tidal channels, seedlings of the annuals *Amaranthus cannabinus* and *Ambrosia trifida* can be prominent. *Cuscuta gronovii*, a parasite with several hosts, is an important component of the seed bank and vegetation.

This tidal freshwater system shows that many families have varied seed dispersal and size, seed bank, and seedling solutions for the same habitat. Such multiple solutions make generalizations difficult. Seedling coexistence is made possible in part by differing emergence times and tolerances (light, inundation), as well as growth form (upright vs. viney) and tolerance to duration of inundation. Our observations suggest that very high densities ($>4000/m^2$) of large seedlings (*Impatiens capensis, Polygonum arifolium*) can preempt vegetation composition. Smaller seedlings (e.g. *Bidens laevis, Pilea pumila*, and perennials) may not survive unless they are shade tolerant or a gap develops. In this system where annuals are important, their seedling dynamics are crucial to understanding community composition.

Table 18.2 | Seed, seed bank, and seedling characteristics of common tidal freshwater wetland species (New Jersey, USA). Data are from Parker and Leck (1985); Leck et al. (1989b); and Leck and Simpson (1993).

Species	Family	Ranking of emergence[a]	Seed size (mg)	Dispersal	Seed bank type[b]
Amaranthus cannabinus	Amaranthaceae (D, A)[c]	3	1.6	H_2O	III
Ambrosia trifida	Asteraceae (D, A)	3	84.1	H_2O	II
Bidens laevis	Asteraceae (D, A)	2	2.1	H_2O, animal	III
Carex spp.	Cyperaceae (M, P)	5	na[d]	H_2O	IV
Cuscuta gronovii	Convolvulaceae (D, A, H)	4	8.3	H_2O	III
Dulichium arundinaceum	Cyperaceae (M, P)	5	na	H_2O	
Impatiens capensis	Balsaminaceae (D, A)	1	26.7	H_2O, explosive	II
Juncus effusus	Juncaceae (M, P)	5	0.01	H_2O, wind?	IV
Peltandra virginica	Araceae (M, P)	3	seed 530	sink	I
			fruit 880		
Pilea pumila	Urticaceae (D, A)	4	0.7	H_2O	II
Polygonum arifolium	Polygonaceae (D, A)	1	46	H_2O[d]	II?
Polygonum punctatum	Polygonaceae (D, A)	2	3.2	H_2O[d]	II/III
Typha latifolia	Typhaceae (M, P)	5	0.03	H_2O, wind	III
					IV

[a]Ranking is approximate for a growing season.

[b]Types: I - transient, <1 year; II - transient, <1 year, overwinters; III - short-term persistent, some annual turnover; IV - long-term persistent, not present in surface soil layers.

[c]D = dicot; M = monocot; A = annual, P = perennial; H = heterotroph (parasite); na = not available.

[d]Seeds lacking attached perianth sink.

18.6 | Conclusions

The importance of the seedling stage in the life history of a species now demands consideration of molecular ecology, pollination ecology, maternal investment in seeds, seed maturation, phylogeny, morphological and physiological plasticity both above- and belowground, and other relevant processes. The great variety of structures, physiologies, phenologies, and persistence among seedlings challenges our ability to make generalizations. Many characteristics of seedlings are constrained by their evolutionary history, both long-term (phylogenetic) and short-term (ecological). The physiological and structural patterns described above reflect that evolutionary history. Understanding phylogenetic constraints will require evaluating seedlings within and among clades (e.g. Chapter 6). The evolutionary progression in development (Chapter 5) and the types of reduction in monocot embryos (Tillich, 1995, 2000) indicate that environment has had selection impact on embryo, seed, and thus seedling traits. Moreover, better understanding of characteristics of seedlings of functional groups, including C_3 vs. C_4 species, forbs vs. legumes (Chapter 13), and phylogenic dichotomies such as gymnosperms vs. angiosperms, dicots vs. monocots, basal terrestrial vs. aquatic angiosperms (Chapter 6), as well as special groups (orchids, bromeliads, carnivorous, and parasites, Chapter 4), would improve understanding of phylogenetic constraints on seedling establishment priorities. From this might come insights regarding, for example, the significance of epigeal vs. hypogeal seedling forms.

The underlying genetic control of a number of aspects of seedling ecology, such as the relationship of seed size to seedling establishment (Chapter 10), has been well investigated. However, as noted by Harper and White (1974), the length of the seedling stage of development in plants is also under genetic control as demonstrated by plants that have been anthropologically selected to reduce the length of time from germination to flowering. Because the duration of the seedling period is heritable, the extent to which species become adjusted represents an adaptive response. Researchers who have focused on the seedling period have emphasized the processes that result in mortality or patterns of transition to adulthood and have assumed that minimizing the seedling stage is the ecologically optimal solution, but have not examined the length of the seedling stage as an adaptive phenomenon *per se*.

Incorporating the diversity of seedling types and functions into community studies is a well-established tradition in some ecosystems but not others. Within the context of community dynamics, differences in the types of seedling banks have been a starting point for forest studies more than in other systems, because persistent seedling banks are prominent in the forest understory and in community dynamics. Variation among species in tolerance to shade, temperature flux, nutrient conditions, and moisture regimes are all critical to the dynamics of these systems. At the same time, theoretical considerations, that is, Hubbell's neutral theory (Hubbell, 2001), suggest that such species-specific adaptations may be less critical at larger spatial or temporal scales. How these disparate views are reconciled may rely on critical and long-term experimental work on seedlings at multiple scales.

The seedling stage of the plant life history is critical to the long-term success of plant populations and communities. Whether transient or persistent, it is requisite to species success. Seedlings display multiple and overlapping strategies that transcend evolutionary and phylogenetic boundaries. For the seedling to be successful, a suitable suite of environmental conditions must be present to support germination and establishment. Equally important, conditions must be favorable for the seedling to survive the "environmental sieve" and reach adulthood.

There is no typical seedling. Seedlings with a long evolutionary history have adapted to exploit most environments on earth, except the hottest, driest, and coldest places. The range of strategies includes autotrophic (with/without symbionts), mycoheterotrophs, and parasites. There is, at present, no all-encompassing classification of seedlings. While de Vogel (1980) has classified multiple groups of dicots, any future view must incorporate the array of monocot seedling and establishment types (Tomlinson & Esler, 1973; Bell, 1991; Tillich, 1995, 2000).

Multispecies comparisons will continue to inform questions regarding seedling ecology and evolution within and among populations, communities, and habitats. It is also of interest to compare traits of generalist and specialist seedlings, as well as transitions between stages. Other aspects are also significant, including dispersal ecology and seed limitation, that are made more significant by anthropogenic disturbance. Given the extraordinary impact of humans on landscapes, it becomes necessary to shift attention to seedling traits that determine recruitment success in degraded habitats, whether for restoration or to mitigate the impact of invasives. Yet, impact of perturbations is unpredictable, variably affecting abiotic and biotic factors, and is site-specific.

As a defined field of study, seedling ecology and evolution is still far from mature. While further work is needed on all aspects of seedling establishment, growth, and survival, several areas emerge from the chapters in this volume as needing special attention and collectively promise to extend significantly our current understanding of the role seedlings play from ecological and evolutionary perspectives. These are summarized below:

• Much thought has gone into the significance of seed and seedling size. The tools of molecular biology and stable isotopes now permit examining such questions in the context of the evolutionary constraints of basal and modern angiosperm taxa.

- In a similar way, much research has focused on seed germination ecology, which is a precursor to seedling emergence. Now the tools of molecular biology and new techniques in physiology permit greater understanding of both germination and establishment ecophysiology, allowing seedling ecologists to explore previously unresolved questions. Moreover, the development of genetic mutants in a wide array of species should further understanding of seedling responses.

- There is an urgent need to increase our understanding of the mechanisms involved in seedling tolerance to stressful conditions and attendant evolutionary pathways and inherent trade-offs. This will require consideration of seed and seedling ecology in the context of broader life history strategies.

- Phytohormones shape many seedling strategies. Ecologists working in collaboration with geneticists and physiologists have access to many newly developed tools to explore the roles phytohormones play in guiding and constraining establishment.

- Carbon balance is a common physiological currency for understanding the evolution of diverse seedling morphology and development strategies, but no clear adaptive patterns emerge because of the complicated web of seedling trait evolution that requires extensive phylogenetic comparative analysis.

- Incorporation of spatial and temporal scales is essential to furthering knowledge of seedling ecology and vegetation dynamics. These dimensions apply directly to the mechanisms involved in seedling tolerance to stressful conditions and the attendant evolutionary pathways and inherit trade-offs.

- Understanding of seedling recruitment and population growth would greatly benefit from studies at the interface between genetic variation in seedling traits and assessment of recruitment variation in time and space.

- Underground dynamics of seedlings is still largely a black box. Molecular tools and stable isotopes are helping to elucidate the complex network of interactions between seedling and mature plants and the role played by mycorrhizal fungi. Future work must address how resource fluxes in mycorrhizal networks are regulated and to what extent mycorrhizal diversity regulates establishment and persistence in natural communities vs. other mechanisms such as competition, herbivory, and resource constraints.

- Much of our current understanding of evolutionary trends is tentative and subject to reinterpretation as new tools become available. Genetic mapping of plants coupled with emerging evidence from the fossil record promises to enhance significantly our understanding of evolutionary and phylogenetic relationships, as well as the role of seedlings in population and community dynamics. This approach can also help determine if there are suites of traits associated with seedling establishment and growth that are common among species in a given family.

- Seedling communities are structured by multiple factors that vary with vegetation type and landscape position. In some circumstances, the relative contribution of seedling dynamics to later community structure and function can be substantial. Seedling community structure may be driven by selection to avoid competitive interactions, environmental filtering effects, or phylogenetic constraints. Knowledge of the relative importance of each is essential to understanding fully the ecology of seedling communities and community dynamics more generally.
- Given the diversity of environmental factors among and within habitats resulting in species-dependent safe sites for emergence and establishment, a small number of functional seedling types are not likely. Thinking in terms of continuous trait spectra from a multivariate perspective may be more fruitful.
- For many groups, only a few species have been studied in depth, precluding generalizations. Quantitative studies of seedling establishment and survival are essential if we are to understand the ecology of these groups and apply this knowledge to conservation and restoration issues.
- Comparative studies of invasives in the native and invasive ranges will help both with the understanding of invasion and the control thereof. Especially important is the question of how quickly species in novel habitats change to optimize the use of resources there, including interactions with symbiotic fungi and bacteria that facilitate access to resources.
- In systems impacted by human activity, it is essential to identify feedback mechanisms and take these into account when managing ecosystems for biodiversity and/or sustainability of human land use.
- Comparative studies of seedling germination and establishment, especially as related to anthropogenically altered ecosystems, for example climate change, habitat fragmentation, nutrient disturbance, salinization, desertification, and invasive species, would benefit from knowledge gained in agricultural and horticultural work.
- Understanding the long-term implications of revegetation practices is essential for successful restoration, as is development of new approaches for directly addressing environmental stressors across sites and landscapes. Superimposed on this is the need to understand the potential impacts of rapid climate change on contemporary restoration strategies.

Plant recruitment is a multi-staged, temporally and spatially structured process central to the dynamics of plant communities. Understanding the variability in time or space of this process requires a fundamental understanding of the seedling stage. Only with systematic evaluation of seed (size, maternal investment trade-offs, morphology/dispersal) and seed-bank strategies, germination requirements, and seedling tolerances of species from a number of habitats can we evaluate the special relationships of seedlings to the habitats in which they are found, and the likely consequences at the population and community level portended by global climate change. While seedling diversity may be large, only a subset exists within any one community. The great vulnerability of this stage lends itself to experimental studies at many levels, from physiological or life history evolution, to community levels. While the importance of abiotic stressors and herbivory are well known, the role of mutualists like mycorrhizal fungi in moderating abiotic stress or the more diverse nature of nurse plants in which facilitation is due to release from herbivory is only just beginning to be understood. Individual seedling establishment success is strongly dependent on chance. In many habitats, seedling recruitment reflects reproductive and dispersal limitations, with the distribution of adults providing little, if any, insight into seedling establishment. Moreover, the diversity of seedlings is a consequence of the diversity of ecological and evolutionary processes acting on other stages. Although ecologists have long appreciated the importance of the seedling stage, clearly, there is now a surge of interest in integration of seedlings with other life history stages, within their abiotic and biotic context, and with their evolutionary history.

References

Aarssen, L. W. (2005). Why don't bigger plants have proportionately bigger seeds? *Oikos*, **111**, 199–207.

Aarssen, L. W. & Jordan, C. Y. (2001). Between-species patterns of covariation in plant size, seed size and fecundity in monocarpic herbs. *Ecoscience*, **8**, 471–7.

Abramsky, Z. (1983). Experiments on seed preparation by rodents and ants in the Israeli desert. *Oecologia*, **57**, 328–32.

Abu-Awwad, A. M. (1997). Water infiltration and redistribution within soils affected by a surface crust. *Journal of Arid Environments*, **37**, 231–42.

Achard, P., Cheng, H., De Grauwe, L., *et al.* (2006). Integration of plant responses to environmentally activated phytohormonal signals. *Science*, **311**, 91–4.

Ackerly, D. D. (2000). Taxon sampling, correlated evolution, and independent contrasts. *Evolution*, **54**, 1480–92.

Ackerman, J., Sabat, A., & Zimmerman, J. K. (1996). Seedling establishment in an epiphytic orchid: an experimental study of seed limitation. *Oecologia*, **106**, 192–8.

Adamec, L. (1997). Mineral nutrition of carnivorous plants: a review. *Botanical Review*, **63**, 273–99.

Adamec, L. (2000). Rootless aquatic plant *Aldrovanda vesiculosa*: physiological polarity, mineral nutrition, and importance of carnivory. *Biologia Plantarum*, **43**, 113–19.

Adamec, L. (2002). Leaf absorption of mineral nutrients in carnivorous plants stimulates root nutrient uptake. *New Phytologist*, **155**, 89–100.

Adamik, K. J. & Brauns, F. E. (1957). *Ailanthus glanulosa* (tree of heaven) as a pulpwood: part II. *Tappi*, **40**, 522–7.

Adams, W. W. & Martin, C. E. (1986). Morphological changes accompanying the transition from juvenile (atmospheric) to adult (tank) forms in the Mexican epiphyte *Tillandsia deppeana* (Bromeliaceae). *American Journal of Botany*, **73**, 1207–14.

Adams, W. W., III., Zarter, C. R., Ebbert, V., & Demmig-Adams, B. (2004). Photoprotective strategies of overwintering evergreens. *BioScience*, **54**, 41–9.

Adriaensen, K., van der Lelie, D., Van Laere, A., Vangronsveld, J., & Colpaert, J. V. (2004). A zinc-adapted fungus protects pines from zinc stress. *New Phytologist*, **161**, 549–55.

Agrawal, A. A., Conner, J. K., & Stinchcombe, J. R. (2004). Evolution of plant resistance and tolerance to frost damage. *Ecology Letters*, **7**, 1199–1208.

Ågren, J. (1989). Seed size and number in *Rubus chamaemorus*: between-habitat variation, and effects of defoliation and supplemental pollination. *Journal of Ecology*, **77**, 1080–92.

Aguiar, M. R. & Sala, O. E. (1997). Seed distribution constrains the dynamics of the Patagonian steppe. *Ecology*, **78**, 93–100.

Aguiar, M. R. & Sala, O. E. (1998). Interactions among grasses, shrubs, and herbivores in Patagonian grass-shrub steppes. *Ecologa Austral*, **8**, 201–10.

Agyeman, V. K., Swaine, M. D., & Thompson, J. (1999). Responses of tropical forest tree seedlings to irradiance and the derivation of a light response index. *Journal of Ecology*, **87**, 815–27.

Aide, T. M. & Cavalier, J. (1994). Barriers to lowland tropical forest restoration in the Sierra Nevada de Santa Marta, Colombia. *Restoration Ecology*, **4**, 219–29.

Aide, T. M., Zimmerman, J. K., Herrera, L., Rosario, M., & Serrano, M. (1995). Forest recovery in abandoned tropical pastures in Puerto Rico. *Forest Ecology and Management*, **77**, 77–86.

Aiken, S. G. (1986). The distinct morphology and germination of two species of wild rice (*Zizania*, Poaceae). *Canadian Field-Naturalist*, **100**, 237–40.

Aizen, M. A. & Feinsinger, P. (1994). Forest fragmentation, pollination and plant reproduction in a Chaco dry forest. *Ecology*, **75**, 330–51.

Aizen, M. A. & Woodcock, H. (1996). Effects of acorn size on seedling survival and growth in *Quercus rubra* following simulated spring freeze. *Canadian Journal of Botany*, **74**, 308–14.

Akasaka, M. & Tsuyuzaki, S. (2005). Tree seedling performance in microhabitats along an elevational gradient on Mount Koma, Japan. *Journal of Vegetation Science*, **16**, 647–54.

Akiyama, K., Matsuzaki, K., & Hayashi, H. (2005). Plant sesquiterpenes induce hyphal branching in arbuscular mycorrhizal fungi. *Nature*, **435**, 824–7.

Alberdi, M. & Corcuera, L. J. (1991). Cold-acclimation in plants. *Phytochemistry*, **30**, 3177–84.

Alberdi, M. & Rios, D. (1983). Frost resistance of *Embothrium coccineum* Forst. and *Gevuina avellana* Mol. during development and aging. *Acta Oecologia (Oecologia Plantarum)*, **4**, 3–9.

Albert, V. A. (1999). Shoot apical meristems and floral patterning: an evolutionary perspective. *Trends in Plant Sciences*, **4**, 84–6.

Albert, V. A., Williams, S. E., & Chase, M. W. (1992). Carnivorous plants, phylogeny and structural evolution. *Nature*, **257**, 1491–5.

Albrecht, C., Geurts, R., & Bisseling, T. (1999). Legume nodulation and mycorrhizae formation: two extremes in host specificity meet. *EMBO Journal*, **18**, 281–8.

Alexander, I., Ahmad, N., & Lee, S. S. (1992). The role of mycorrhizas in the regeneration of some Malaysian forest trees. *Philosophical Transactions of the Royal Society of London Series B*, **335**, 379–88.

Alexander, I. J. & Lee, S. S. (2005). Mycorrhizas and ecosystem processes in tropical rain forest, implications for diversity. In *Biotic Interactions in the Tropics*, ed. D. F. R. P. Burslem, M. A. Pinard, & S. E. Hartley. Cambridge: Cambridge University Press, pp. 165–203.

Al-Hamdani, S. & Francko, D. A. (1992). Effect of light and temperature on photosynthesis elongation rate and chlorophyll content of *Nelumbo lutea* (Willd.) Pers. seedlings. *Aquatic Botany*, **44**, 51–8.

Allcock, K. G. (2002). Effects of phosphorus on growth and competitive interactions of native and introduced species found in White Box woodlands. *Austral Ecology*, **27**, 638–46.

Allen, A., Chambers, J. L., & Pezeshki, S. R. (1997). Effects of salinity on baldcypress seedlings: physiological responses and their relation to salinity tolerance. *Wetlands*, **17**, 310–20.

Allen, E. B. & Allen, M. F. (1984). Competition between plants of different successional stages: mycorrhizae as regulators. *Canadian Journal of Botany*, **62**, 2625–9.

Allen, M. F. (1991). *The Ecology of Mycorrhizae*. Cambridge: Cambridge University Press.

Allen, M. F., Allen, E. B., & Friese, C. F. (1989). Responses of the non-mycotrophic plant *Salsola kali* to invasion by VA mycorrhizal fungi. *New Phytologist*, **111**, 45–9.

Allen, M. F., Allen, E. B., & Gomez-Pompa, A. (2005). Effects of mycorrhizae and non-target organisms on restoration of a seasonal tropical forest in Quintana Roo, Mexico: factors limiting tree establishment. *Restoration Ecology*, **13**, 325–33.

Allen, M. F., Rincon, E., Allen, E. B., Huante, P., & Dunn, J. J. (1993). Observations of canopy bromeliad roots compared with plants rooted in soils of a seasonal tropical forest, Chamela, Jalisco, Mexico. *Mycorrhiza*, **4**, 27–8.

Allen, M. L., Jasper, D. A., & Zak, J. C. (2002). Micro-organisms. In *Handbook of Ecological Restoration*, vol. 1, ed. M. Perrow & A. Davy. Cambridge: Cambridge University Press, pp. 257–78.

Allen-Diaz, B., Bartolome, J. W., & McClaran, M. P. (1999). California oak savanna. In *Savannas, Barrens, and Rock Outcrop Plant Communities of North America*, ed. R. C. Anderson, J. S. Fralish, & J. M. Baskin. Cambridge: Cambridge University Press, pp. 322–39.

Allessio, M. L. (1967). Observations of seedling development in *Polygonum bistortoides*. *American Journal of Botany*, **54**, 1272–4.

Allison, S. K. & Ehrenfeld, J. G. (1999). The influences of microhabitat variation on seedling recruitment of *Chamaecyparis thyoides* and *Acer rubrum*. *Wetlands*, **19**, 383–93.

Almeida-Cortez, J. S. & Shipley, W. (2002). No significant relationship exists between seedling relative growth rate under nutrient limitation and potential tissue toxicity. *Functional Ecology*, **16**, 122–7.

Alvarez-Aquino, C., Williams-Linera, G., & Newton, A. C. (2004). Experimental native tree seedling establishment for the restoration of a Mexican cloud forest. *Restoration Ecology*, **12**, 412–8.

Alvarez-Buylla, E. R. & Martinez-Ramos, M. (1992). Demography and allometry of *Cecropia obtusifolia* a neotropical pioneer tree, an evaluation of the climax-pioneer paradigm for tropical rain forests. *Journal of Ecology*, **80**, 275–90.

Alvarez-Clare, S. & Kitajima, K. (2007). Physical defense traits enhance seedling survival of neotropical tree species. *Functional Ecology*, **21**, 1044–54.

Alvey, L. & Harberd, N. P. (2005). DELLA proteins: integrators of multiple plant growth regulatory inputs? *Physiologia Plantarum*, **123**, 153–60.

Ampofo, S. T., Moore, K. G., & Lovell, P. H. (1976). Role of cotyledons in 4 *Acer* species and in *Fagus sylvatica* during early seedling development. *New Phytologist*, **76**, 31–9.

Anderberg, A. & Anderberg, A. L. (2006). *Den Virtuella Floran*. Stockholm: Naturhistoriska Riksmuseet (http://linnaeus.nrm.se/flora/).

Anderson, A. B. (1991). Symbiotic and asymbiotic germination and growth of *Spiranthes magnicamporum* (Orchidaceae). *Lindleyana*, **6**, 183–6.

Andresen, E. & Feer, F. (2005). The role of dung beetles as secondary seed dispersers and their effect on plant regeneration in tropical rainforests. In *Seed Fate: Predation, Dispersal and Seedling Establishment*, ed. P.-M. Forget, J. E. Lambert, P. E. Hulme, & S. R. V. Wall. Wallingford: CAB International, pp. 331–49.

Andrews, H. N. (1963). Early seed plants. *Science*, **142**, 925–31.

Andrews, H. N. & Pannel, E. (1942). Contributions to our knowledge of American carboniferous flora. II. *Lepidocarpon*. *Annals of the Missouri Botanical Garden*, **29**, 19–28.

Angert, A., Huxman, T., Barron-Gafford, G., Gerst, K., & Venable, D. (2007). Linking growth strategies to long-term population dynamics in a guild of desert annuals. *Journal of Ecology*, **95**, 321–31.

Aniszewski, T., Kupari, M. H., & Leinonen, A. J. (2001). Seed number, seed size and seed diversity in Washington lupine (*Lupinus polyphyllus* Lindl.). *Annals of Botany*, **87**, 77–82.

Antos, J. A., Guest, H. J., & Parish, R. (2005). The tree seedling bank in an ancient montane forest: stress tolerators in a productive habitat. *Journal of Ecology*, **93**, 536–43.

Antos, J. A., Parish, R. A., & Conley, K. (2000). Age structure and growth of the tree-seedling bank in subalpine spruce-fir forests of South-central British Columbia. *American Midland Naturalist*, **143**, 342–54.

Aplet, G. H., Laven, R. D., & Smith, F. W. (1988). Patterns of community dynamics in Colorado Engelmann spruce–subalpine fir forests. *Ecology*, **69**, 312–19.

Arai, N. & Kamitani, T. (2005). Seed rain and seedling establishment of the dioecious tree *Neolitsea sericea* (Lauraceae): effects of tree sex and density on invasion into a conifer plantation in central Japan. *Canadian Journal of Botany*, **83**, 1144–50.

Arber, A. (1920). *Water Plants*. Cambridge: Cambridge University Press.

Arber, A. (1934). *Gramineae: A Study of Cereal, Bamboo and Grass*. 1965 reprint. Weinheim: J. Cramer.

Arber, E. A. N. & Parkin, J. (1907). On the origin of angiosperms. *Journal of the Linnean Society (Botany)*, **38**, 29–80.

Arctic Climate Impact Assessment. (2004). *Impacts of a Warming Arctic*. Cambridge: Cambridge University Press.

Arditti, J. (1992). *Fundamentals of Orchid Biology*. New York: Wiley.

Arditti, J. & Ghani, A. K. A. (2000). Numerical and physical properties of orchid seeds and their biological implications. *New Phytologist*, **145**, 367–421.

Argyris, J., Truco, M. J., Ochoa, O., *et al.* (2005). Quantitative trait loci associated with seed and seedling traits in *Lactuca*. *Theoretical and Applied Genetics*, **111**, 1364–76.

Arii, K. & Lechowicz, M. J. (2002). The influence of overstory trees and abiotic factors on the sapling community in an old-growth *Fagus-Acer* forest. *Ecoscience*, **9**, 386–96.

Armstrong, D. P. & Westoby, M. (1993). Seedlings from large seeds tolerate defoliation better: a test using phylogenetically independent contrasts. *Ecology*, **74**, 1092–1100.

Arnebrant, K., Ek, H., Finlay, R. D., & Söderström, B. (1993). Nitrogen translocation between *Alnus glutinosa* (L.) Gaertn. seedlings inoculated with *Frankia* sp. and *Pinus contorta* Doug. ex Loud seedlings connected by a common ectomycorrhizal mycelium. *New Phytologist*, **124**, 231–42.

Arnold, A. E., Maynard, Z., Gilbert, G. S., Coley, P. D., & Kursar, T. A. (2000). Are tropical fungal endophytes hyperdiverse? *Ecology Letters*, **3**, 267–74.

Ashkannejhad, S. & Horton, T. R. (2006). Ectomycorrhizal ecology under primary succession on the coastal sand dunes: interactions involving *Pinus contorta*, suilloid fungi and deer. *New Phytologist*, **169**, 345–54.

Ashton, P. M. S., Gamage, S., Gunatilleke, I. A. U. N., & Gunatilleke, C. V. S. (1997). Restoration of a Sri Lankan rainforest: using Caribbean pine *Pinus caribaea* as a nurse for establishing late-successional tree species. *Journal of Applied Ecology*, **34**, 915–25.

Ashton, P. M. S., Gamage, S., Gunatilleke, I. A. U. N., & Gunatilleke, C. V. S. (1998). Using Caribbean pine to establish a mixed plantation: testing effects of pine canopy removal on plantings of rain forest tree species. *Forest Ecology and Management*, **106**, 211–22.

Ashworth, L. & Galetto, L. (2001). Pollinators and reproductive success of the wild cucurbit *Cucurbita maxima* ssp. *andreana* (Cucurbitaceae). *Plant Biology*, 3, 398–404.

Asquith, N. M. & Mejia-Chang, M. (2005). Mammals, edge effects, and the loss of tropical forest diversity. *Ecology*, 86, 379–90.

Asquith, N. M., Wright, S. J., & Claus, M. J. (1997). Does mammal community composition control recruitment in neotropical forests? Evidence from Panama. *Ecology*, 78, 941–6.

Aston, H. I. (1977). *Aquatic Plants of Australia*. Carlton: Melbourne University Press.

Auble, G. T. & Scott, M. L. (1998). Fluvial disturbance patches and cottonwood recruitment along the upper Missouri River, Montana. *Wetlands*, 18, 546–56.

Auge, H. & Brandl, R. (1997). Seedling recruitment in the invasive clonal shrub, *Mahonia aquifolium* Pursh (Nutt.). *Oecologia*, 110, 205–11.

Auge, H., Neuffer, B., Erlinghagen, F., Grupe, R., & Brandl, R. (2001). Demographic and random amplified polymorphic DNA analyses reveal high levels of genetic diversity in a clonal violet. *Molecular Ecology*, 10, 1811–19.

Augspurger, C. K. (1983). Seed dispersal of the tropical tree, *Platypodium elegans*, and the escape of its seedlings from fungal pathogens. *Journal of Ecology*, 71, 759–71.

Augspurger, C. K. (1984a). Light requirements of neotropical tree seedlings – a comparative study of growth and survival. *Journal of Ecology*, 72, 777–95.

Augspurger, C. K. (1984b). Seedling survival of tropical tree species: interactions of dispersal distance, light gaps, and pathogens. *Ecology*, 65, 1705–12.

Augspurger, C. K. & Kelly, C. K. (1984). Pathogen mortality of tropical tree seedlings: experimental studies of the effects of dispersal distance, seedling density, and light conditions. *Oecologia*, 61, 211–17.

Aukema, J. E. (2003). Vectors, viscin, and Viscaceae: mistletoes as parasites, mutualists, and resources. *Frontiers in Ecology and the Environment*, 1, 212–19.

Baar, J., Horton, T. R., Kretzer, A. M., & Bruns, T. D. (1999). Mycorrhizal colonization of *Pinus muricata* from resistant propagules after a stand-replacing wildfire. *New Phytologist*, 143, 409–18.

Babaev, A. G. (1999). *Desert Problems and Desertification in Central Asia: The Researches of the Desert Institute*. Berlin: Springer-Verlag.

Bailey, C. & Scholes, M. (1997). Rhizosheath occurrence in South African grasses. *South African Journal of Botany*, 63, 484–90.

Bais, H. P. & Ravishankar, G. A. (2002). Role of polyamines in the ontogeny of plants and their biotechnological applications. *Plant, Cell, Tissue and Organ Culture*, 69, 1–34.

Bais, H. P., Vepachedu, R., Gilroy, S., Callaway, R. M., & Vivanco, J. M. (2003). Allelopathy and exotic plant invasion: from molecules and genes to species interactions. *Science*, 301, 1377–80.

Baker, H. G. (1965). Characteristics and modes of origin of weeds. In *The Genetics of Colonizing Species*, ed. H. G. Baker & G. L. Stebbins. New York: Academic Press, pp. 147–68.

Baker, H. G. (1972). Seed weight in relation to environmental conditions in California. *Ecology*, 53, 997–1010.

Bakker, C., van Bodegom, P. M., Nelissen, H. J. M., Aerts, R., & Ernst, W. H. O. (2007). Preference of wet dune species for waterlogged conditions can be explained by adaptations and specific recruitment requirements. *Aquatic Botany*, 86, 37–45.

Baldwin, A., Egnotovich, M., Ford, M., & Platt, W. (2001). Regeneration in fringe mangrove forests damaged by Hurricane Andrew. *Plant Ecology*, **157**, 151–64.

Ball, E. (1956a). Growth of the embryo of *Ginkgo biloba* under experimental conditions. I. Origin of the first root of the seedling *in vitro*. *American Journal of Botany*, **43**, 488–95.

Ball, E. (1956b). Growth of the embryo of *Ginkgo biloba* under experimental conditions. II. Effects of a longitudinal split in the tip of the hypocotyl. *American Journal of Botany*, **43**, 802–10.

Ball, M. C. (1988). Salinity tolerance in the mangroves *Aegiceras corniculatum* and *Avicennia marina*. 1. Water-use in relation to growth, carbon partitioning, and salt balance. *Australian Journal of Plant Physiology*, **15**, 447–64.

Ball, M. C., Egerton, J. J. G., Leuning, R., Cunningham, R. B., & Dunne, P. (1997). Microclimate above grass adversely affects spring growth of seedling snow gum (*Eucalyptus pauciflora*). *Plant Cell and Environment*, **20**, 155–66.

Ball, M. C. & Farquhar, G. D. (1984). Photosynthetic and stomatal responses of two mangrove species, *Aegiceras corniculatum* and *Avicennia marina*, to long-term salinity and humidity conditions. *Plant Physiology*, **74**, 1–6.

Ball, V. C., Hodges, V. S., & Laughlin, G. P. (1991). Cold-induced photoinhibition limits regeneration of snow gum at tree line. *Functional Ecology*, **5**, 663–8.

Baltzer, J. L. & Thomas, S. L. (2007). Physiological and morphological correlates of whole-plant light compensation point in temperate deciduous tree seedlings. *Oecologia*, **153**, 209–23.

Baraloto, C., Bonal, D., & Goldberg, D. E. (2006). Differential seedling growth response to soil resource availability among nine neotropical tree species. *Journal of Tropical Ecology*, **22**, 487–97.

Baraloto, C. & Forget, P.-M. (2007). Seed size, seedling morphology, and response to deep shade and damage in neotropical rain forest trees. *American Journal of Botany*, **94**, 901–11.

Baraloto, C., Forget, P.-M., & Goldberg, D. E. (2005a). Seed mass, seedling size and neotropical tree seedling establishment. *Journal of Ecology*, **93**, 1156–66.

Baraloto, C., Goldberg, D. E., & Bonal, D. (2005b). Performance trade-offs among tropical tree seedlings in contrasting microhabitats. *Ecology*, **86**, 2461–72.

Baraza, E., Gomez, J., Hodar, J., & Zamora, R. (2004). Herbivory has a greater impact in shade than in sun: response of *Quercus pyrenaica* seedlings to multifactorial environmental variation. *Canadian Journal of Botany*, **82**, 357–64.

Baraza, E., Zamora, R., & Hodar, J. A. (2006). Conditional outcomes in plant-herbivore interactions: neighbors matter. *Oikos*, **113**, 148–56.

Barchuk, A. H., Valiente-Banuet, A., & Diaz, M. P. (2005). Effect of shrubs and seasonal variability of rainfall on the establishment of *Aspidosperma quebracho-blanco* in two edaphically contrasting environments. *Austral Ecology*, **30**, 695–705.

Barker, N. G. & Williamson, B. G. (1988). Effects of a winter fire on *Sarracenia alata* and *S. psittacina*. *American Journal of Botany*, **75**, 138–43.

Barlow, B. A. (1981). The loranthaceous mistletoes in Australia. In *Ecological Biogeography of Australia*, ed. A. Keast. The Hague: Dr. W. Junk Publishers, pp. 556–74.

Barlow, J. & Peres, C. A. (2004). Ecological responses to El Niño-induced surface fires in central Brazilian Amazonia: management implications for flammable tropical forests. *Philosophical Transactions of the Royal Society of London Series B*, **359**, 367–80.

Barrat-Segretain, M.-H. (1996). Germination and colonization dynamics of *Nuphar lutea* (L.) Sm. in a former river channel. *Aquatic Botany*, **55**, 31–8.

Barth, H. J. (1999). Desertification in the eastern province of Saudi Arabia. *Journal of Arid Environments*, **43**, 399–410.

Barton, L. V. (1961). *Seed Preservation and Longevity*. New York: Plant Science Monograph Series, Interscience.

Baskin, C. C. & Baskin, J. M. (1998; 2001). *Seeds: Ecology, Biogeography, and Evolution of Dormancy and Germination*. San Diego: Academic Press.

Baskin, C. C., Chesson, P. L., & Baskin, J. M. (1993). Annual seed dormancy cycles in two winter annuals. *Journal of Ecology*, **81**, 551–6.

Baskin, C. C., Hawkins, T. S., & Baskin, J. M. (2004). Ecological life cycle of *Chaerophyllum procumbens* variety *shortii* (Apiaceae), a winter annual of the North American eastern deciduous forest. *Journal of the Torrey Botanical Society*, **131**, 126–39.

Baskin, J. M. & Baskin C. C. (1989). Physiology of dormancy and germination in relationship to seed bank ecology. In *Ecology of Soil Seed Banks*, ed. M. A. Leck, V. T. Parker, & R. L. Simpson. San Diego: Academic Press, pp. 54–66.

Baskin, Y. (1998). Winners and losers in a changing world: global changes may promote invasions and alter the fate of invasive species. *BioScience*, **48**, 788–92.

Bassow, S. L. & Bazzaz, F. A. (1997). Intra- and inter-specific variation in canopy photosynthesis in a mixed deciduous forest. *Oecologia*, **109**, 507–15.

Batanouny, K. H. (2001). *Plants in the desert of the Middle East*. Heidelberg: Springer Verlag.

Battaglia, L. L., Foré, S. A., & Sharitz, R. R. (2000). Seedling emergence, survival and size in relation to light and water availability in two bottomland hardwood species. *Journal of Ecology*, **88**, 1041–50.

Batty, A. L., Brundrett, M. C., Dixon, K. W., & Sivasithamparam, K. (2006). New methods to improve symbiotic propagation of temperate terrestrial orchid seedlings from axenic culture to soil. *Australian Journal of Botany*, **54**, 367–74.

Batty, A. L., Dixon, K. W., Brundrett, M., & Sivasithamparam, K. (2001a). Constraints to symbiotic germination of terrestrial orchid seed in a mediterranean bushland. *New Phytologist*, **152**, 511–20.

Batty, A. L., Dixon, K. W., Brundrett, M., & Sivasithamparam, K. (2001b). Long-term storage of mycorrhizal fungi and seed as a tool for the conservation of endangered Western Australian terrestrial orchids. *Australian Journal of Botany*, **49**, 619–28.

Baylis, G. T. S. (1975). The magnolioid mycorrhiza and mycotrophy in root systems derived from it. In *Endomycorrhiza*, ed. F. E. Sanders, B. Moss, & P. B. Tinker. London: Academic Press, pp. 378–89.

Baylis, G. T. S. (1980). Mycorrhizas and the spread of beech. *New Zealand Journal of Ecology*, **3**, 151–2.

Bazzaz, F. A. (1979). Physiological ecology of plant succession. *Annual Review of Ecology and Systematics*, **10**, 351–71.

Bazzaz, F. A. (1990). Plant–plant interaction in successional environments. In *Perspectives on Plant Competition*, ed. J. B. Grace & D. Tilman. San Diego: Academic Press.

Bazzaz, F. A. (1996). *Plants in Changing Environments: Linking Physiological, Population, and Community Ecology*. Cambridge: Cambridge University Press.

Bazzaz, F. A. & Carlson, R. W. (1982). Photosynthetic acclimation to variability in the light environment of early and late successional plants. *Oecologia*, **54**, 313–6.

Beare, P. A. & Zedler, J. B. (1987). Cattail invasion and persistence in a coastal salt-marsh – the role of salinity reduction. *Estuaries*, **10**, 165–70.

Beatley, J. C. (1974). Phenological events and their environmental triggers in Mojave Desert. *Ecosystem Ecology*, **55**, 856–63.

Beckage, B. & Clark, J. S. (2005). Does predation contribute to tree diversity? *Oecologia*, **143**, 458–69.

Beckage, B., Lavine, M., & Clark, J. S. (2005). Survival of tree seedlings across space and time: estimates from long-term count data. *Journal of Ecology*, **93**, 1177–84.

Beentje, H. J. (1993). A new aquatic palm from Madagascar. *Principes*, **37**, 197–202.

Begon, M. (1985). A general theory of life-history variation. In *Behavioural Ecology: Ecological Consequences of Adaptive Behaviour*, ed. M. R. Sibly & R. H. Smith. Oxford: Blackwell Scientific Publications, pp. 91–8.

Beland, M. & Bergeron, Y. (1993). Ecological factors affecting abundance of advance growth in jack pine (*Pinus banksiana* Lamb.) stands of the boreal forest of northwestern Quebec. *Forest Chronicles*, **69**, 561–8.

Bell, A. D. (1991). *Plant Form: An Illustrated Guide to Flowering Plant Morphology*. Oxford: Oxford University Press.

Bell, D. T., Plummer, J. A., & Taylor, S. K. (1993). Seed germination ecology in southwestern Western Australia. *Botanical Review*, **59**, 24–73.

Bell, T., Freckleton, R. P., & Lewis, O. T. (2006). Plant pathogens drive density-dependent seedling mortality in a tropical tree. *Ecology Letters*, **9**, 569–74.

Bellingham, P. J., Duncan, R. P., Lee, W. G., & Buxton, R. P. (2004). Seedling growth rate and survival do not predict invasiveness in naturalized woody plants in New Zealand. *Oikos*, **106**, 308–16.

Bellingham, P. J. & Richardson, S. J. (2006). Tree seedling growth and survival over 6 years across different microsites in a temperate rain forest. *Canadian Journal of Forest Research*, **36**, 910–8.

Belnap, J., Welter, J. R., Grimm, N. B., Barger, N., & Ludwig, J. A. (2005). Linkages between microbial and hydrologic processes in arid and semiarid watersheds. *Ecology*, **86**, 298–307.

Benítez-Malvido, J. (1998). Impact of forest fragmentation on seedling abundance in a tropical rain forest. *Conservation Biology*, **12**, 380–9.

Benítez-Malvido, J., García-Guzman, G., & Kossman-Ferraz, I. D. (1999). Leaf fungal incidence and herbivory on tree seedlings in tropical forest fragments: an experimental study. *Biological Conservation*, **91**, 143–50.

Benítez–Malvido, J. & Lemus-Albor, A. (2005). The seedling community of tropical rain forests and its interaction with herbivores and pathogens. *Biotropica*, **37**, 301–13.

Bennett, J. R. & Mathews, S. (2006). Phylogeny of the parasitic plant family Orobanchaceae inferred from phytochrome A. *American Journal of Botany*, **93**, 1039–51.

Benson, D. R. & Silvester, W. B. (1993). Biology of *Frankia* strains, actinomycete symbionts of actinorhizal plants. *Microbiology and Molecular Biology Reviews*, **57**, 293–319.

Benzing, D. H. (1978). Germination and early establishment of *Tillandsia circinnata* Schlecht. (Bromeliaceae) on some of its hosts and other supports in South Florida. *Selbyana*, **5**, 95–106.

Benzing, D. H. (1980). *The Biology of the Bromeliads*. Eureka: Mad River Press.

Benzing, D. H. (1981). The population dynamics of *Tillandsia circinnata* (Bromeliaceae): cypress crown colonies in southern Florida. *Selbyana*, **5**, 256–63.

Benzing, D. H. (1990). *Vascular Epiphytes: General Biology and Related Biota*. New York: Cambridge University Press.

Benzing, D. H. (2000). *Bromeliaceae: Profile of an Adaptive Radiation*. Cambridge: Cambridge University Press.

Benzing, D. H., Friedman, W. E., Peterson, G., & Renfrow, A. (1983). Shootless, velamentous roots, and the pre-eminence of Orchidaceae in the epiphytic biotope. *American Journal of Botany*, **70**, 121–33.

Benzing, D. H. & Renfrow, A. (1974). The mineral nutrition of Bromeliaceae. *Botanical Gazette*, **135**, 281–8.

Berardini, T. Z., Bollman, K., Sun, H., & Posthig, R. S. (2001). Regulation of vegetative phase change in *Arabidopsis thaliana* by cyclophilin 40. *Science*, **291**, 2405–7.

Berbee, M. L. & Taylor, J. W. (1993). Dating of the evolutionary radiations of the true fungi. *Canadian Journal of Botany*, **71**, 1114–27.

Berch, S. M., Allen, T. R., & Berbee, M. L. (2002). Molecular detection, community structure and phylogeny of ericoid mycorrhizal fungi. *Plant and Soil*, **244**, 55–66.

Berkeley, A., Thomas, A. D., & Dougill, A. J. (2005). Cyanobacterial soil crusts and woody shrub canopies in Kalahari rangelands. *African Journal of Ecology*, **43**, 137–45.

Berndtsson, R., Nodomi, K., Yasuda, H., *et al.* (1996). Soil water and temperature patterns in an arid desert dune sand. *Journal of Hydrology*, **185**, 221–40.

Bertness, M. D. & Callaway, R. M. (1994). Positive interactions in communities. *Trends in Ecology & Evolution*, **9**, 191–3.

Bertness, M. D. & Yeh, S. M. (1994). Cooperative and competitive interactions in the recruitment of marsh elders. *Ecology*, **75**, 2416–29.

Berube, D. E. & Myers, J. H. (1982). Suppression of knapweed invasion by crested wheatgrass in the dry interior of British Columbia. *Journal of Range Management*, **35**, 459–61.

Besserer, A., Puech-Pages, V., Kiefer, P., *et al.* (2006). Strigolactones stimulate arbuscular mycorrhizal fungi by activating mitochondria. *Public Library of Science Biology*, **4**, 1239–47.

Betz, R. F. (1986). One decade of research in prairie restoration at the Fermi National Accelerator Laboratory (Fermilab) Batavia, Illinois. In *Proceedings of the Ninth North American Prairie Conference*, ed. G. K. Clambey & R. H. Pemble. Moorhead & Fargo: Tri-College University Center for Environmental Studies, pp. 179–85.

Beukes, P. C. & Cowling, R. M. (2003). Evaluation of restoration techniques for the Succulent Karoo, South Africa. *Restoration Ecology*, **11**, 308–16.

Bever, J. D. (2002). Negative feedback within a mutualism: host specific growth of mycorrhizal fungi reduces plant benefit. *Proceedings of the Royal Society of London Series B – Biological Sciences*, **269**, 2595–601.

Bever, J. D. (2003). Soil community feedback and the coexistence of competitors: conceptual frameworks and empirical tests. *New Phytologist*, **157**, 465–73.

Bever, J. D., Schultz, P. A., Pringle, A., & Morton, J. B. (2001). Arbuscular mycorrhizal fungi: more diverse than meets the eye, and the ecological tale of why. *BioScience*, **51**, 923–31.

Bews, J. W. (1927). Studies on the ecological evolution of angiosperms. *New Phytologist*, **26**, 1–21.

Beyrle, H. F., Penningsfeld, F., & Hock, B. (1991). The role of nitrogen concentration in determining the outcome of the interaction between *Dactylorhiza incarnata* (L.) Soó and *Rhizoctonia* sp. *New Phytologist*, **117**, 665–72.

Beyrle, H. F., Smith, S. E., Peterson, R. L., & Franco C. M. M. (1995). Colonization of *Orchis morio* protocorms by a mycorrhizal fungus: effects of nitrogen nutrition and glyphosate in modifying the responses. *Canadian Journal of Botany*, **73**, 1129–40.

Bhaskar, V. (2003). *Root Hairs: the 'Gills' of Roots: Development, Structure and Functions*. Enfield: Science Publishers, Inc.

Bidartondo, M. I. (2005). The evolutionary ecology of myco-heterotrophy. *New Phytologist*, **167**, 335–52.

Bidartondo, M. I. & Bruns, T. D. (2001). Extreme specificity in epiparasitic Monotropoideae (Ericaceae): widespread phylogenetic and geographical structure. *Molecular Ecology*, **10**, 2285–95.

Bidartondo, M. I. & Bruns, T. D. (2002). Fine-level mycorrhizal specificity in the Monotropoideae (Ericaceae): specificity for fungal species groups. *Molecular Ecology*, **11**, 557–68.

Bidartondo, M. I. & Bruns, T. D. (2005). On the origins of extreme mycorrhizal specificity in the Monotropoideae (Ericaceae): performance trade-offs during seed germination and seedling development. *Molecular Ecology*, **14**, 1549–60.

Bidartondo, M. I., Redecker, D., Hijri, I., *et al.* (2002). Epiparasitic plants specialized on arbuscular mycorrhizal fungi. *Nature*, **419**, 389–92.

Biere, A. (1991). Parental effects in *Lychnis flos-cuculi*. 1. Seed size, germination and seedling performance in a controlled environment. *Journal of Evolutionary Biology*, **4**, 447–65.

Bierhorst, D. (1971). *Morphology of Vascular Plants*. New York: Macmillan.

Birch, C. P. D. (1986). Development of VA mycorrhizal infection in seedlings in semi-natural grassland turf. In *First European Symposium on Mycorrhizas*, ed. V. Gianinazzi-Pearson & S. Gianinazzi. Paris: INRA, pp. 233–9.

Birschwilks, M., Haupt, S., Hofius, D., & Neumann, S. (2006). Transfer of phloem-mobile substances from the host plants to the holoparasite *Cuscuta* sp. *Journal of Experimental Botany*, **57**, 911–21.

Bischoff, A., Cremieux, L., Smilauerova, M., *et al.* (2006). Detecting local adaptation in widespread grassland species – the importance of scale and local plant community. *Journal of Ecology*, **94**, 1130–42.

Björkman, E. (1960). *Monotropa hypopitys* L. – an epiparasite on tree roots. *Physiological Plantarum*, **13**, 308–27.

Björkman, O. & Holmgren, P. (1966). Photosynthetic adaptation to light intensity in plants native to shaded and exposed habitats. *Physiologia Plantarum*, **19**, 854–9.

Black, J. N. (1958). Competition between plants of different initial seed sizes in swards of subterranean cover (*Trifolium subterraneum* L.) with particular reference to leaf area and the light microclimate. *Australian Journal of Agricultural Research*, **9**, 299–318.

Blanc, P. (1986). Edification d'arbres par croissance d'établissement de type monocotylédon: l'exemple de Chloranthacae. In *Colloque international sur l'arbre 1986*. Montpellier: Nautralia Monspeliensia, numéro hors série, pp. 101–23.

Blaudez, D., Jacob, C., Turnau, K., *et al.* (2000). Differential responses of ectomycorrhizal fungi to heavy metals *in vitro*. *Mycological Research*, **104**, 1366–71.

Blennow, K. & Lindkvist, L. (2000). Models of low temperature and high irradiance and their application to explaining the risk of seedling mortality. *Forest Ecology and Management*, **135**, 289–301.

Bloor, J. M. G. & Grubb, P. J. (2003). Growth and mortality in high and low light: trends among 15 shade-tolerant tropical rain forest tree species. *Journal of Ecology*, **91**, 77–85.

Blossey, B. & Nötzold, R. (1995). Evolution of increased competitive ability in invasive nonindigenous plants: a hypothesis. *Journal of Ecology*, **83**, 887–9.

Blouin, M., Zuily-Fodil, Y., Pham-Thi, A.-T., *et al.* (2005). Belowground organism activities affect plant aboveground phenotype, inducing tolerance to parasites. *Ecology Letters*, **8**, 202–8.

Blum, J. D., Klaue, A., Nezat, C. A., *et al.* (2002). Mycorrhizal weathering of apatite as an important calcium source in base-poor forest ecosystems. *Nature*, **417**, 729–31.

Boddey, R. M., Deoliveira, O. C., Urquiaga, S., *et al.* (1995). Biological nitrogen fixation associated with sugar cane and rice contribution and prospects for improvement. *Plant and Soil*, **174**, 195–209.

Bodley, J. H. & Benson, F. C. (1980). Stilt-root walking by an Iriarteoid palm in Peruvian Amazon. *Biotropica*, **12**, 67–71.

Boege, K. & Marquis, R. J. (2005). Facing herbivory as you grow up: the ontogeny of resistance in plants. *Trends in Ecology & Evolution*, **20**, 441–8.

Boeken, B., Ariza, C., Gutterman, Y., & Zaady, E. (2004). Environmental factors affecting dispersal, germination and distribution of *Stipa capensis* in the Negev Desert, Israel. *Ecological Research*, **19**, 533–40.

Boeken, B., Lipchin, C., Gutterman, Y., & Van Rooyen, N. (1998). Annual plant community responses to density of small-scale soil disturbances in the Negev, Israel. *Oecologia*, **114**, 106–17.

Boeken, B. & Orenstein, D. (2001). The effect of plant litter on ecosystem properties in a Mediterranean semi-arid shrubland. *Journal of Vegetation Science*, **12**, 825–32.

Boeken, B. & Shachak, M. (1994). Desert plant communities in human-made patches – implications for management. *Ecological Applications*, **4**, 702–16.

Boeken, B. & Shachak, M. (1998a). The dynamics of abundance and incidence of annual plant species during colonization in a desert. *Ecography*, **21**, 63–73.

Boeken, B. & Shachak, M. (1998b). Colonization by annual plants of an experimentally altered desert landscape: source–sink relationships. *Journal of Ecology*, **86**, 804–14.

Boeken, B. & Shachak, M. (2006). Linking community and ecosystem processes: the role of minor species. *Ecosystems*, **9**, 119–27.

Boeken, B., Shachak, M., Gutterman, Y., & Brand, S. (1995). Patchiness and disturbance: plant community responses to porcupine diggings in the central Negev. *Ecography*, **18**, 410–22.

Boerner, R. E. J. & Brinkman, J. A. (1996). Ten years of tree seedling establishment and mortality in an Ohio deciduous forest complex. *Bulletin of the Torrey Botanical Club*, **123**, 309–17.

Böhning-Gaese, K., Gaese, B. H., & Rabemanantsoa, S. B. (1999). Importance of primary and secondary seed dispersal in the Malagasy tree *Commiphora guillaumini*. *Ecology*, **80**, 821–32.

Bokdam, J. (1977). Seedling morphology of some African Sapotaceae and its taxonomic significance. *Mededelingen Lanbouwhogeschool, Wageningen*, **77**, 1–84.

Bonaccorso, F. J., Glanz, W. E., & Sanford, C. M. (1980). Feeding assemblages of mammals at fruiting *Dipteryx panamensis* (Papilionaceae) trees in Panama: seed predation, dispersal and parasitism. *Revista Biologica Tropical*, **28**, 61–72.

Bond, G. (1983). Taxonomy and distribution of non-legume nitrogen-fixing systems. In *Biological Nitrogen Fixation in Forest Ecosystems: Foundations and Applications*, ed. J. C. Gordon & C. T. Wheeler. The Hague: Martinus Nijhoff, pp. 55–8.

Bond, W. J. (1989). The tortoise and the hare – ecology of angiosperm dominance and gymnosperm persistence. *Biological Journal of the Linnean Society*, **36**, 227–49.

Bond, W. J., Honig, M., & Maze, K. E. (1999). Seed size and seedling emergence: an allometric relationship and some ecological implications. *Oecologia*, **120**, 132–6.

Bond, W. J. & Keeley, J. E. (2005). Fire as global 'herbivore': the ecology and evolution of flammable ecosystems. *Trends in Ecology & Evolution*, **20**, 387–94.

Bond, W. J. & van Wilgen, B. W. (1996). *Fire and Plants*. New York: Chapman & Hall.

Bonello, P., Bruns, T. D., & Gardes, M. (1998). Genetic structure of a natural population of the ectomycorrhizal fungus *Suillus pungens*. *New Phytologist*, **138**, 533–42.

Bonnis, A. & Lepart, J. (1994). Vertical structure of seed banks and the impact of depth of burial on recruitment in two temporary marshes. *Vegetatio*, **112**, 127–39.

Boon, J. D. (2006). *Cephalotus follicularis*, Western Australian pitcher plant, Albany pitcher plant (http://www.aqph26.dsl.pipex.com, March 14, 2007).

Booth, M. G. (2004). Mycorrhizal networks mediate overstorey–understorey competition in a temperate forest. *Ecology Letters*, **7**, 538–46.

Borchers, S. L. & Perry, D. A. (1990). Growth and ectomycorrhiza formation of Douglas-fir seedlings grown in soils collected at different distances from pioneering hardwoods in southwest Oregon clear-cuts. *Canadian Journal of Forest Research*, **20**, 712–15.

Bormann, F. H. & Likens, G. E. (1979). *Pattern and Process in a Forested Ecosystem*. Berlin: Springer-Verlag.

Bornkamm, R., Darius, F., & Prasse, R. (1999). On the life cycle of *Stipagrostis scoparia* hillocks. *Journal of Arid Environments*, **42**, 177–86.

Bosy, J. L. & Reader, R. J. (1995). Mechanisms underlying the suppression of forb seedling emergence by grass (*Poa pratensis*) litter. *Functional Ecology*, **9**, 635–9.

Boucher, D. H. (1983). *Quercus oleoides* (Roble Encino, Oak). In *Costa Rican Natural History*, ed. D. H. Janzen. Chicago: University of Chicago Press, pp. 319–20.

Boulton, A. J. & Brock, M. A. (1999). *Australian Freshwater Ecology: Processes and Management*. Glen Osmond: Gleneagles Publishing.

Bouwmeester, H. J., Matusova, R., Zhongkui, S., & Beale, M. H. (2003). Secondary metabolite signaling in host–parasitic plant interactions. *Current Opinion in Plant Biology*, **6**, 358–64.

Bower, F. O. (1935). *Primitive Land Plants*. London: Macmillan.

Bowers, J. E. & Pierson, E. A. (2001). Implications of seed size for seedling survival in *Carnegiea gigantea* and *Ferocactus wislizeni* (Cactaceae). *Southwestern Naturalist*, **46**, 272–81.

Bowers, J. E., Turner, R. M., & Burgess, T. L. (2004). Temporal and spatial patterns in emergence and early survival of perennial plants in the Sonoran Desert. *Plant Ecology*, **172**, 107–19.

Box, E. O. (1996). Plant functional types and climate at the global scale. *Journal of Vegetation Science*, **7**, 309–20.

Boyce, C. K. (2005). Patterns of segregation and convergence in the evolution of fern and seed plant leaf morphologies. *Paleobiology*, **31**, 117–40.

Boyd, L. (1932). Monocotylous seedlings: morphological studies in the post-seminal development of the embryo. *Transactions and Proceedings of the Botanical Society of Edinburgh*, **31**, 5–224.

Breckle, S. W. (2002). *Walter's Vegetation of the Earth*, 4th edn. Berlin: Springer.

Bremer, K. (1985). Summary of green plant phylogeny and classification. *Cladistics*, **1**, 369–85.

Brewer, J. S. (1999a). Short-term effects of fire and competition on growth and plasticity of the yellow pitcher plant, *Sarracenia alata* (Sarraceniaceae). *American Journal of Botany*, **86**, 1264–71.

Brewer, J. S. (1999b). Effects of fire, competition, and soil disturbances on regeneration of a carnivorous plant (*Drosera capillaris*). *American Midland Naturalist*, **141**, 28–42.

Brewer, J. S. (2001). A demographic analysis of fire-stimulated seedling establishment of *Sarracenia alata* (Sarraceniaceae). *American Journal of Botany*, **88**, 1250–7.

Briede, J. W. & McKell C. M. (1992). Germination of seven perennial arid land species, subject to soil moisture stress. *Journal of Arid Environments*, **23**, 263–70.

Brighigna, L., Fiordi, A. C., & Palandri, M. R. (1990). Structural comparison between free and anchored roots in *Tillandsia* (Bromeliaceae) species. *Caryologia*, **43**, 27–42.

Brittingham, S. & Walker, L. R. (2000). Facilitation of *Yucca brevifolia* recruitment by Mojave Desert shrubs. *Western North American Naturalist*, **60**, 374–83.

Brock, M. A. (1998). Are temporary wetlands resilient? Evidence from seed banks of Australian and South African wetlands. In *Wetlands for the Future: Contributions from INTECOL's V International Wetlands Conference*, ed. J. McComb & J. A. Davis. Adelaide: Gleneagles Publishing.

Brockie, R. (1992). *A Living New Zealand Forest*. Auckland: David Bateman.

Brodribb, T. J., Holbrook, N. M., & Hill, R. S. (2005). Seedling growth in conifers and angiosperms: impacts of contrasting xylem structure. *Australian Journal of Botany*, **53**, 749–55.

Brooks, S. M. & Spencer, T. (1997). Changing soil hydrology due to rain forest logging, an example from Sabah Malaysia. *Journal of Environmental Management*, **49**, 297–310.

Brown, J. & Venable, D. (1986). Evolutionary ecology of seed-bank annuals in temporally varying environments. *American Naturalist*, **127**, 31–47.

Brown, J. S. & Venable, D. L. (1991). Life history evolution of seed-bank annuals in response to seed predation. *Evolutionary Ecology*, **5**, 12–29.

Brown, N. A. C. & van Staden, J. (1997). Smoke as a germination cue: a review. *Plant Growth Regulation*, **22**, 115–24.

Brown, N. D. & Whitmore, T. C. (1992). Do dipterocarp seedlings really partition tropical rain forest gaps? *Philosophical Transactions of the Royal Society of London*, **335**, 369–78.

Brown, R. L. & Fridley, J. D. (2003). Control of plant species diversity and community invasibility by species immigration: seed richness versus seed density. *Oikos*, **102**, 15–24.

Brown, W. H. (1935). *The Plant Kingdom*. Boston: Ginn and Company.

Bruchmann, H. (1898). *Über die Prothallien und die Keimpflanzen mehrer europaischer Lycopodien*. Gotha: self-published.

Bruchmann, H. (1909). Vom Prothallium der grossen Sporen und von der Keimes-Entwicklung einiger *Selaginella*-arten. *Flora*, **99**, 12–51.

Bruchmann, H. (1912). Zur Embryologie der Selaginellaceen. *Flora*, **104**, 180–224.

Brudvig, L. A. & Evans, C. W. (2006). Competitive effects of native and exotic shrubs on *Quercus alba* seedlings. *Northeastern Naturalist*, **13**, 259–68.

Bruelheide, H. (2002). Climatic factors controlling the eastern and altitudinal distribution boundary of *Digitalis purpurea* L. in Germany. *Flora*, **197**, 475–90.

Brundrett, M., Bougher, N., Dell, B., Grove, T., & Malajczuk, N. (1996). *Working with Mycorrhizas in Forestry and Agriculture*. Canberra: ACIAR Monograph.

Bryan, G. S. (1920). Early stages in development of the sporophyte of *Sphagnum subsecundum*. *American Journal of Botany*, **7**, 296–303.

Bryan, G. S. (1952). The cellular proembryo of *Zamia* and its cap cells. *American Journal of Botany*, **39**, 433–43.

Buddenhagen, C. E. & Ogden, J. (2003). Growth and survival of *Dysoxylum spectabile* (Meliaceae) seedlings in canopy gaps. *New Zealand Journal of Botany*, **41**, 179–83.

Budelsky, R. & Galatowitsch, S. (2000). Effects of water regime and competition on the establishment of a native sedge in restored wetlands. *Journal of Applied Ecology*, **37**, 971–85.

Budelsky, R. & Galatowitsch, S. (2004). Establishment of *Carex stricta* Lam. seedlings in experimental wetlands with implications for restoration. *Plant Ecology*, **175**, 91–105.

Budowski, G. (1987). Living fence posts in tropical America, a widespread agroforestry practice. In *Agroforestry, Realities, Possibilities and Potentials*, ed. H. L. Gholz. Dordrecht, Netherlands: Martinus Nijhoff, pp. 169–78.

Buee, M., Rossignol, M., Jauneau, A., Ranjeva, R., & Becard, G. (2000). The presymbiotic growth of arbuscular mycorrhizal fungi is induced by a branching factor partially purified from plant root exudates. *Molecular Plant–Microbe Interactions*, **13**, 693–8.

Buell, M. F. (1935). Seed and seedling of *Acorus calamus*. *Botanical Gazette*, **96**, 758–65.

Bugmann, H. (2001). A review of forest gap models. *Climatic Change*, **51**, 259–305.

Bullock, J. M. (2000). Gaps and seedling colonization. In *Seeds: The Ecology of Regeneration in Plant Communities*, 2nd edn., ed. M. Fenner. Wallingford: CAB International, pp. 375–95.

Bullock, S. H. (1991). Herbivory and the demography of the chaparral shrub *Ceanothus greggii* (Rhamnaceae). *Madroño*, **38**, 63–72.

Bungard, R. A., Press, M. C., & Scholes, J. D. (2000). The influence of nitrogen on rain forest dipterocarp seedlings exposed to a large increase in irradiance. *Plant Cell and Environment*, **23**, 1183–94.

Burger, D. (1972). *Seedlings of Some Tropical Trees and Shrubs Mainly of South East Asia*. Wageningen: Centre for Agricultural Publishing and Documentation.

Burgess, L. M., Hild, A. L., & Shaw, N. L. (2005). Capsule treatments to enhance seedling emergence of *Gaura neomexicana* spp. *coloradensis*. *Restoration Ecology*, **13**, 8–14.

Burleigh, J. G. & Mathews, S. (2007). Assessing among-locus variation in the inference of seed plant phylogeny. *International Journal of Plant Sciences*, **168**, 111–24.

Burrows, C. J. (1993). Vivipary and effects of maternal tissues on germination of some New Zealand seeds. *Canterbury Botanical Society*, **27**, 47–8.

Burrows, C. J. (1995). The germination behaviour of the seeds of the New Zealand species *Aristotelia serrata, Coprosma robusta, Cordyline australis, Myrtus obcordata* and *Schefflera digitata. New Zealand Journal of Botany*, 33, 257–64.

Burrows, C. J. (1996). Germination behaviour of the seeds of seven New Zealand vine species. *New Zealand Journal of Botany*, 34, 93–102.

Burslem, D. F. R. P., Grubb, P. J., & Turner, I. M. (1995). Responses to nutrient addition among shade-tolerant tree seedlings of lowland tropical rain forest in Singapore. *Journal of Ecology*, 83, 113–22.

Burt, A. (1989). Comparative methods using phyllogenetically independent contrasts. *Oxford Surveys in Evolutionary Biology*, 6, 33–53.

Burton, P. J. & Bazzaz, F. A. (1995). Ecophysiological responses of tree seedlings invading different patches of old-field vegetation. *Journal of Ecology*, 83, 99–112.

Burton, P. J. & Mueller-Dombois, D. (1984). Response of *Metrosideros polymorpha* seedlings to experimental canopy opening. *Ecology*, 65, 779–91.

Burt-Smith, G. S., Grime, J. P., & Tilman, D. (2003). Seedling resistance to herbivory as a predictor of relative abundance in a synthesised prairie community. *Oikos*, 101, 345–53.

Buschmann, H., Keller, M., Porret, N., Dietz, H., & Edwards, P. J. (2005). The effect of slug grazing on vegetation development and plant species diversity in an experimental grassland. *Functional Ecology*, 19, 291–8.

Byers, D. L., Platenkamp, G. A. J., & Shaw, R. G. (1997). Variation in seed characters in *Nemophila menziesii* – evidence of a genetic basis for maternal effect. *Evolution*, 51, 1445–56.

Cabin, R. J., Weller, S. G., Lorence, D. H., *et al.* (2002). Effects of light, alien grass, and native species additions on Hawaiian dry forest restoration. *Ecological Applications*, 12, 1595–1610.

Cairney, J. W. G. & Alexander, I. J. (1992). A study of ageing of spruce [*Picea sitchensis* (Bong.) Carr.] ectomycorrhizas. III. Phosphate absorption and transfer in ageing *Picea sitchensis/Tylospora fibrillosa* (Burt.) Donk ectomycorrhizas. *New Phytologist*, 122, 159–64.

Callaway, R. M. (1995). Positive interactions among plants. *Botanical Review*, 61, 306–49.

Callaway, R. M. (1997). Positive interactions in plant communities and the individualistic-continuum concept. *Oecologia*, 112, 143–9.

Callaway, R. M. & Aschehoug., E. T. (2000). Invasive plants versus their new and old neighbors: a mechanism for invasion. *Science*, 290, 521–3.

Callaway, R. M., Brooker, R. W., Choler, P., *et al.* (2002). Positive interactions among alpine plants increase with stress. *Nature*, 417, 844–8.

Callaway, R. M. & Davis, F. W. (1998). Recruitment of *Quercus agrifolia* in central California: the importance of shrub-dominated patches. *Journal of Vegetation Science*, 9, 647–56.

Callaway, R. M., de Lucia, E. H., Moore, D., Nowak, R., & Schlesinger, W. H. (1996). Competition and facilitation: contrasting effects of *Artemisia tridentata* on desert vs. montane pines. *Ecology*, 77, 2130–41.

Callaway, R. M. & Walker, L. R. (1997). Competition and facilitation: a synthetic approach to interactions in plant communities. *Ecology*, 78, 1958–65.

Cameron, G. N. & Spencer, S. R. (1989). Rapid leaf decay and nutrient release in a Chinese tallow forest. *Oecologia*, 80, 222–8.

Campbell, E. J. F. & Newbery, D. M. C. (1993). Ecological relationships between lianas and trees in lowland rain forest in Sabah, East Malaysia. *Journal of Tropical Ecology*, 9, 469–90.

Campbell, E. O. (1928). *The Structure and Development of Mosses and Ferns*, 3rd edn. New York: Macmillan.

Campbell, R. S. (1929). Vegetative succession in the *Prosopis* sand dunes of southern New Mexico. *Ecology*, **10**, 392–8.

Canham, C. D. (1988). Growth and canopy architecture of shade-tolerant trees: response to canopy gaps. *Ecology*, **69**, 786–95.

Canham, C. D., Berkovwitz, A. R., Kelly, V. R., et al. (1996). Biomass allocation and multiple resource limitation in tree seedlings. *Canadian Journal of Forest Research*, **26**, 1521–30.

Canham, C. D., Kobe, R. K., Latty, E. F., & Chazdon, R. L. (1999). Interspecific and intraspecific variation in tree seedling survival: effects of allocation to roots versus carbohydrate reserves. *Oecologia*, **121**, 1–11.

Caprio, A. C. & Lineback, P. (2002). Pre-twentieth century fire history of Sequoia and Kings Canyon National Parks: a review and evaluation of our knowledge. In *Fire in California Ecosystems: Integrating Ecology, Prevention and Management*, ed. M. Morales & T. Morales. Berkeley: Association for Fire Ecology, pp. 180–99.

Carafa, A., Duckett, J. G., & Ligrone, R. (2003). The placenta in *Monoclea forsteri* Hook. and *Treubia lacunosa* (Col.) Prosk: insights into placental evolution in liverworts. *Annals of Botany*, **92**, 299–307.

Cardel, Y., Rico-Gray, V., Garcia-Franco, J. G., & Thien, L. B. (1997). Ecological status of *Beaucarnea gracilis*, an endemic species of the semiarid Tehuacan Valley, Mexico. *Conservation Biology*, **11**, 367–74.

Carlyle, C. N. & Fraser, L. H. (2006). A test of three juvenile plant competitive response strategies. *Journal of Vegetation Science*, **17**, 11–18.

Carnago, C. E. D. & Ferreira, A. W. P. (2005). Genetic control of wheat seedling growth. *Scientia Agricola*, **62**, 325–30.

Carnevale, N. & Montagnini, F. (2002). Facilitating regeneration of secondary forests with the use of mixed and pure plantations of indigenous tree species. *Forest Ecology and Management*, **163**, 217–27.

Carrillo-Garcia, A., Bashan, Y., & Bethlenfalvay, G. J. (2000). Resource-island soils and the survival of the giant cactus, cardon, of Baja California Sur. *Plant and Soil*, **218**, 207–14.

Carrillo-Garcia, A., de la Luz, J. L. L., Bashan, Y., & Bethlenfalvay, G. J. (1999). Nurse plants, mycorrhizae, and plant establishment in a disturbed area of the Sonoran Desert. *Restoration Ecology*, **7**, 321–35.

Casanova, M. & Brock, M. (2004). How do depth, duration, and frequency of flooding influence the establishment of wetland plant communities? *Plant Ecology*, **147**, 237–50.

Casler, M. D. & Undersander, D. J. (2006). Selection for establishment capacity in reed canarygrass. *Crop Science*, **46**, 1277–85.

Casperson, J. P. & Kobe, R. K. (2001). Interspecific variation in sapling mortality in relation to growth and soil moisture. *Oikos*, **92**, 160–8.

Castro, J., Zamora, R., Hodar, J. A., & Gomez, J. M. (2002). Use of shrubs as nurse plants: a new technique for reforestation in Mediterranean mountains. *Restoration Ecology*, **10**, 297–305.

Castro, J., Zamora, R., Hódar, J. A., & Gómez, J. M. (2004a). Seedling establishment of a boreal tree species (*Pinus sylvestris*) at its southernmost distribution limit: consequences of being in a marginal Mediterranean habitat. *Journal of Ecology*, **92**, 266–77.

Castro, J., Zamora, R., Hodar, J. A., Gomez, J. M., & Gomez-Aparico, L. (2004b). Benefits of using shrubs as nurse plants for reforestation in Mediterranean mountains: a 4-year study. *Restoration Ecology*, **12**, 352–8.

Castro-Diez, P., Puyravaud, J. P., & Cornelissen, J. H. C. (2000). Leaf structure and anatomy as related to leaf mass per area variation in seedlings of a wide range of woody plant species and types. *Oecologia*, **124**, 476–86.

Castro-Diez, P., Puyravaud, J. P., Cornelissen, J. H. C., & Villarsalvador, P. (1998). Stem anatomy and relative growth rate in seedlings of a wide range of woody plant species and types. *Oecologia*, **116**, 57–66.

Caswell, H. (1989). *Matrix Population Models*. Sunderland: Sinauer.

Caswell, H. (2001). *Matrix Population Models*, 2nd edn. Sunderland: Sinauer.

Catalán, L., Balzarini, M., Taleisnik, E., Sereno, R., & Karlin, U. (1994). Effects of salinity on germination and seedling growth of *Prosopis flexuosa* (Dc). *Forest Ecology and Management*, **63**, 347–57.

Catovsky, S. & Bazzaz, F. A. (2002). Nitrogen availability influences regeneration of temperate tree species in the understory seedling bank. *Ecological Applications*, **12**, 1056–70.

Catovsky, S., Kobe, R. K., & Bazzaz, F. A. (2002). Nitrogen-induced changes in seedling regeneration and dynamics of mixed conifer–broad-leaved forests. *Ecological Applications*, **12**, 1611–25.

Cavelier, J., Aide, T. M., Santos, C., Eusse, A. M., & Dupuy, J. M. (1998). The savannizations of moist forests in the Sierra Nevada de Santa Marta, Colombia. *Journal of Biogeography*, **25**, 901–12.

Cázares, E. & Smith, J. E. (1996). Occurrence of vesicular-arbuscular mycorrhizae in *Pseudotsuga menziesii* and *Tsuga heterophylla* seedlings grown in Oregon Coast Range soils. *Mycorrhiza*, **6**, 65–7.

Cázares, E. & Trappe, J. M. (1994). Spore dispersal of ectomycorrhizal fungi on a glacier forefront by mammal mycophagy. *Mycologia*, **86**, 507–10.

Cerabolini, B., Ceriani, R. M., Caccianiga, M., De Andreis, R., & Raimondi, B. (2003). Seed size and shape and persistence in soil: a test on Italian flora from alps to Mediterranean coasts. *Seed Science Research*, **13**, 75–85.

Cervera, J. C., Andrade, J. L., Simá, J. L., & Graham, E. A. (2006). Microhabitats, germination, and establishment for *Mammillaria gaumeri* (Cactaceae), a rare species from Yucatan. *International Journal of Plant Science*, **167**, 311–19.

Chacon, P. & Armesto, J. J. (2005). Effect of canopy openness on growth, specific leaf area, and survival of tree seedlings in a temperate rainforest of Chiloe Island, Chile. *New Zealand Journal of Botany*, **43**, 71–81.

Chamberlain, C. J. (1919). *The Living Cycads*. Chicago: University of Chicago Press.

Chamberlain, C. J. (1935). *Gymnosperms: Structure and Evolution*. Chicago: University of Chicago Press.

Chambers, J. C. (1995). Disturbance, life history strategies, and seed fates in alpine herbfield communities. *American Journal of Botany*, **82**, 421–33.

Chambers, J. C., Farleigh, K., Tausch, R. J., *et al.* (1998). Understanding long- and short-term changes in vegetation and geomorphic processes: the key to riparian restoration. In *Proceedings: Rangeland Management and Water Resources*, ed. D. F. Potts. Middleburg: Water Resources Association and Society for Range Management, pp. 101–10.

Chambers, J. C., Vander Wall, S. B., & Schupp, E. W. (1999). Seed and seedling ecology of pinyon and juniper species in the pygmy woodlands of western North America. *Botanical Reviews*, **65**, 1–38.

Chang, S.-M. (2006). Female compensation through the quantity and quality of progeny in a gynodioecious plant, *Geranium maculatum* (Geraniaceae). *American Journal of Botany*, **93**, 263–70.

Chao, W. S., Horvath, D. P., Anderson, J. V., & Foley, M. E. (2005). Potential model weeds to study genomics, ecology, and physiology in the 21st century. *Weed Science*, **53**, 929–37.

Chapin, F. S., III (1980). The mineral nutrition of wild plants. *Annual Review of Ecology and Systematics*, **11**, 233–60.

Chapin, F. S., III (1989). The cost of tundra plant structures: evaluation of concepts and currencies. *American Naturalist*, **133**, 1–19.

Chapin, F. S., III, Autumn, K., & Pugnaire, F. (1993). Evolution of suites of traits in response to environmental stress. *American Naturalist*, **142**, S78–S92.

Chapman, C. A. & Overdonk, D. A. (1998). Forests without primates: primate/plant codependency. *American Journal of Primatology*, **45**, 127–41.

Chapman, V. J. (1976). *Mangrove Vegetation*. Vaduz: J. Cramer.

Charnov, E. L. (1993). *Life History Invariants: Some Explorations of Symmetry in Evolutionary Ecology*. Oxford: Oxford University Press.

Chase, J. M. (2003). Community assembly: when should history matter? *Oecologia*, **136**, 489–98.

Chase, M. W. (2005). Classification of Orchidaceae in the age of DNA data. *Curtis's Botanical Magazine*, **22**, 2–7.

Chauvel, A., Grimaldi, M., & Tessier, D. (1991). Changes in soil pore-space distribution following deforestation and revegetation, an example from the Central Amazon Basin, Brazil. *Forest Ecology and Management*, **38**, 259–71.

Chave, J. (1999). Study of structural, successional and spatial patterns in tropical rain forests using TROLL, a spatially explicit forest model. *Ecological Modelling*, **124**, 233–54.

Chazdon, R. L. (2003). Tropical forest recovery, legacies of human impact and natural disturbance. *Perspectives in Plant Ecology, Evolution and Systematics*, **6**, 51–71.

Chen, J.-G., Ullah, H., Temple, B., *et al.* (2006). RACK1 mediates multiple hormone responsiveness and development processes in *Arabidopsis*. *Journal of Experimental Botany*, **57**, 2697–708.

Chen, Y.-F., Etheridge, N., & Schaller, G. E. (2005). Ethylene signal transduction. *Annals of Botany*, **95**, 901–15.

Cheplick, G. P. (1982). The role of differential allocation to aerial and subterranean propagules in the population dynamics and survival strategies of *Amphicarpum pershii* Kunth. MS thesis, Rutgers, State University of New Jersey, USA.

Cheplick, G. P. (1983). Differences between plants arising from aerial and subterranean seeds in the amphicarpic annual *Cardamine chenopodifolia* (Cruciferae). *Bulletin of the Torrey Botanical Club*, **110**, 442–8.

Cheplick, G. P. (1998). Seed dispersal and seedling establishment in grass populations. In *Population Biology of Grasses*, ed. G. P. Cheplick. Cambridge: Cambridge University Press, pp. 84–105.

Cheplick, G. P. & Quinn, J. A. (1988). Subterranean seed production and population responses to fire in *Amphicarpum purshii* (Gramineae). *Journal of Ecology*, **76**, 263–73.

Cherry, J. A. & Gough, L. (2006). Temporary floating island formation maintains wetland plant species richness: the role of the seed bank. *Aquatic Botany*, **85**, 29–36.

Chiarello, N., Field, C., & Mooney, H. (1987). Midday wilting in a tropical pioneer tree. *Functional Ecology*, **1**, 3–11.

Chick, E. (1903). The seedling of *Torreya Myristica*. *New Phytologist*, **2**, 83–91.

Chittenden, F. J., ed. (1951). *The Royal Horticultural Society Dictionary of Gardening.* Oxford: Clarendon Press.

Chuck, G. & Hake, S. (2005). Regulation of developmental transitions. *Current Opinion in Plant Biology*, **8**, 67–70.

Cione, N. K., Padgett, P. E., & Allen, E. B. (2002). Restoration of a native shrubland impacted by exotic grasses, frequent fire, and nitrogen deposition in southern California. *Restoration Ecology*, **10**, 376–84.

Cipollini, D. (2004). Stretching the limits of plasticity: can a plant defend against both competitors and herbivores? *Ecology*, **85**, 28–37.

Cipollini, D., Enright, S., Traw, M. B., & Bergelson, J. (2004). Salicylic acid inhibits jasmonic acid-induced resistance of *Arabidopsis thaliana* to *Spodoptera exigua*. *Molecular Ecology*, **13**, 1643–53.

Cisse, N. & Ejeta, G. (2003). Genetic variation and relationships among seedling vigor traits in *Sorghum*. *Crop Science*, **43**, 824–8.

Claessen, D., Gilligan, C. A., Lutman, P. J. W., & van den Bosch, F. (2005). Which traits promote persistence of feral GM crops? Part 1: implications of environmental stochasticity. *Oikos*, **110**, 20–9.

Clark, C. J., Poulsen, J. R., Connor, E. F., & Parker, V. T. (2004). Fruiting trees as dispersal foci in a semi-deciduous tropical forest. *Oecologia*, **139**, 66–75.

Clark, D. A. & Clark, D. B. (1984). Spacing dynamics of a tropical rain forest tree: evaluation of the Janzen–Connell model. *American Naturalist*, **124**, 769–88.

Clark, D. B. & Clark, D. A. (1989). The role of physical damage in the seedling mortality regime of a neotropical rain forest. *Oikos*, **55**, 225–30.

Clark, D. B. & Clark, D. A. (1991). The impact of physical damage on canopy tree regeneration in tropical rain forest. *Journal of Ecology*, **79**, 447–57.

Clark, D. B., Palmer, M. W., & Clark, D. A. (1999). Edaphic factors and the landscape-level distributions of tropical rain forest trees. *Ecology*, **80**, 2662–75.

Clark, J. S., Beckage, B., Camill, P., *et al.* (1999). Interpreting recruitment limitation in forests. *American Journal of Botany*, **86**, 1–16.

Clark, S. E. (1997). Organ formation at the vegetative shoot meristem. *Plant Cell*, **9**, 1067–76.

Clauss, M. J. & Venable, D. L. (2000). Seed germination in desert annuals: an empirical test of adaptive bet hedging. *American Naturalist*, **155**, 168–86.

Claveau, Y., Messier, C., & Comeau, P. G. (2005). Interacting influence of light and size on aboveground biomass distribution in sub-boreal conifer saplings with contrasting shade tolerance. *Tree Physiology*, **25**, 373–84.

Clay, K. (1990). Fungal endophytes of grasses. *Annual Review of Ecology and Systematics*, **21**, 275–97.

Clay, K. & Holah, J. (1999). Fungal endophyte symbiosis and plant diversity in successional fields. *Science*, **285**, 1742–4.

Clearwater, M. J., Susilawaty, R., Effendi, R., & van Gardingen, P. R. (1999). Rapid photosynthetic acclimation of *Shorea johorensis* seedlings after logging disturbance in Central Kalimantan. *Oecologia*, **121**, 478–88.

Climent, J. M., Aranda, I., Alonso, J., Pardos, J. A., & Gil, L. (2006). Developmental constraints limit the response of Canary Island pine seedlings to combined shade and drought. *Forest Ecology and Management*, **231**, 164–8.

Clipson, N. J. W., Tomos, A. D., Flowers, T. J., & Jones, R. G. W. (1985). Salt tolerance in the halophyte *Suaeda maritima* L Dum – the maintenance of turgor pressure and water-potential gradients in plants growing at different salinities. *Planta*, **165**, 392–6.

Close, T. J. (1997). Dehydrins: a commonality in the response of plants to dehydration and low temperature. *Physiologia Plantarum*, **100**, 291–6.

Clouse, S. D. & Sasse, J. M. (1998). Brassinosteroids: essential regulators of plant growth and development. *Annual Review of Plant Physiology and Molecular Biology*, **49**, 427–51.

Cochrane, M. A., Alencar, A., Schulze, M. D., et al. (1999). Positive feedbacks in the fire dynamic of closed canopy tropical forests. *Science*, **284**, 1832–5.

Cochrane, M. A. & Laurance, W. F. (2002). Fire as a large-scale edge effect in Amazonian forests. *Journal of Tropical Ecology*, **18**, 311–25.

Cochrane, M. A. & Schulze, M. D. (1999). Fire as a recurrent event in tropical forests of the eastern Amazon: effects on forest structure, biomass, and species composition. *Biotropica*, **31**, 2–16.

Cockayne, L. (1928). *The Vegetation of New Zealand*. In *Die Vegetation der Erde*, ed. A. Engler & O. Drude, XIV, 2nd edn. Liepzig.

Cody, M. L. (1993). Do cholla cacti (*Opuntia* spp., subgenus *cylindropuntia*) use or need nurse plants in the Mojave Desert? *Journal of Arid Environments*, **24**, 139–54.

Cody, M. L. (2000). Slow-motion population dynamics in Mojave Desert perennial plants. *Journal of Vegetation Science*, **11**, 351–8.

Cohen, A. L., Singhakumara, B. M. P., & Ashton, P. M. S. (1995). Releasing rain forest succession: a case study in the *Dicranopteris linearis* fernlands of Sri Lanka. *Restoration Ecology*, **3**, 261–70.

Cohen, D. (1967). Optimizing reproduction in a randomly varying environment when a correlation may exist between the conditions at the time a choice has to be made and the subsequent outcome. *Journal of Theoretical Biology*, **16**, 1–14.

Coley, P. D. (1983). Herbivory and defensive characteristics of tree species in lowland tropical forest. *Ecological Monographs*, **53**, 209–33.

Coley, P. D. & Barone, J. A. (1996). Herbivory and plant defenses in tropical forests. *Annual Review of Ecology and Systematics*, **27**, 305–35.

Coley, P. D., Bryant, J. P., & Chapin, F. S., III. (1985). Resource availability and plant anti-herbivore defense. *Science*, **230**, 895–9.

Collier, M. H., Vankat, J. L., & Hughes, M. R. (2002). Diminished plant richness and abundance below *Lonicera maackii*, an invasive shrub. *American Midland Naturalist*, **147**, 60–71.

Collins, R. J. & Carson, W. P. (2004). The effects of environment and life stage on *Quercus* abundance in the eastern deciduous forest, USA: are sapling densities most responsive to environmental gradients? *Forest Ecology and Management*, **201**, 241–58.

Collins, S. L. & Good, R. E. (1987). The seedling regeneration niche: habitat structure of tree seedlings in an oak-pine forest. *Oikos*, **48**, 89–98.

Colosi, J. C. & McCormick, J. F. (1978). Population structure of *Iva imbricata* in five coastal sand dune habitats. *Bulletin of the Torrey Botanical Club*, **105**, 175–86.

Colpaert, J. V., Vandenkoornhuyse, P., Adriaensen, K., & Vangronsveld, J. (2000). Genetic variation and heavy metal tolerance in the ectomycorrhizal basidiomycete *Suillus luteus*. *New Phytologist*, **147**, 367–79.

Comisky, L., Royo, A. A., & Carson, W. P. (2005). Deer browsing creates rock refugia gardens on large boulders in the Allegheny National Forest, Pennsylvania. *American Midland Naturalist*, **154**, 201–6.

Comita, L. S., Aguilar, S., Pérez, R., Lao, S., & Hubbell, S. P. (2007). Patterns of woody plant species abundance and diversity in the seedling layer of a tropical forest. *Journal of Vegetation Science*, **18**, 163–74.

Cona, A., Rea, G., Angelini, R., Federico, R., & Tavladoraki, P. (2006). Functions of amine oxidases in plant development and defense. *Trends in Plant Science*, **11**, 80–8.

Conard, H. S. (1905). *The Waterlilies: a Monograph of the Genus Nymphaea*. Baltimore: Lord Baltimore Press.

Condit, R., Ashton, P. S., Manokaran, N., *et al.* (1999). Dynamics of the forest communities at Pasoh and Barro Colorado, comparing two 50 ha plots. *Philosophical Transactions of the Royal Society of London*, **354**, 1739–48.

Condit, R., Watts, K., Bohlman, S. A., *et al.* (2000). Quantifying the deciduousness of tropical forest canopies under varying climates. *Journal of Vegetation Science*, **11**, 649–58.

Congdon, R. A. & Herbohn, J. L. (1993). Ecosystem dynamics of disturbed and undisturbed sites in north Queensland wet tropical rain forest. I. Floristic composition, climate and soil chemistry. *Journal of Tropical Ecology*, **9**, 349–63.

Connell, J. H. (1971). On the role of natural enemies in preventing competitive exclusion in some marine animals and rain forest trees. In *Dynamics of Populations*, ed. P. J. Boer & G. R. Gradwell. Wageningen: Centre for Agricultural Publishing and Documentation, pp. 298–310.

Connell, J. H. & Green, P. T. (2000). Seedling dynamics over thirty-two years in a tropical rain forest tree. *Ecology*, **81**, 568–84.

Connell, J. H. & Slatyer, R. O. (1977). Mechanisms of succession in natural communities and their role in community stability and organization. *American Naturalist*, **111**, 1119–44.

Conway, W. C., Smith, L. M., & Bergan, J. F. (2002). Potential allelopathic interference by the exotic Chinese tallow tree (*Sapium sebiferum*). *American Midland Naturalist*, **148**, 43–53.

Cook, C. D. K. (1987). Dispersion in aquatic and amphibious vascular plants. In *Plant Life in Aquatic and Amphibious Habitats*, ed. R. M. M. Crawford. Oxford: Blackwell Scientific Publications, pp. 179–90.

Cook, C. D. K. (1999). The number and kinds of embryo-bearing plants which have become aquatic: a survey. *Perspectives in Plant Ecology, Evolution and Systematics*, **2**, 79–102.

Cook, R. E. (1979). Patterns of juvenile mortality and recruitment in plants. In *Topics in Plant Population Biology*, ed. O. T. Solbrig, S. Jain, G. B. Johnson, & P. H. Raven. New York: Columbia University Press, pp. 206–31.

Cooke, D. A. (1983). The seedling of *Tirthuria* (Hydatellaceae). *Victorian Naturalist*, **100**, 68–9.

Coomes, D. A. & Grubb, P. J. (2003). Colonization, tolerance, competition and seed size variation within functional groups. *Trends in Ecology & Evolution*, **18**, 283–91.

Cooper, D. J., Merritt, D. M., Andersen, D. C., & Chimner, R. A. (1999). Factors controlling the establishment of Fremont cottonwood seedlings on the upper Green River, USA. *Regulated Rivers: Research and Management*, **15**, 419–40.

Cooper, E. J., Alsos, I. G., Hagen, D., *et al.* (2004). Plant recruitment in the High Arctic: seed bank and seedling emergence on Svalbard. *Journal of Vegetation Science*, **15**, 115–24.

Cooper, J. B. & Long, S. R. (1994). Morphogenetic rescue of *Rhizobium meliloti* nodulation mutants by trans-zeatin secretion. *The Plant Cell*, **6**, 215–25.

Corbin, J. D. & D'Antonio, C. M. (2004). Competition between native perennial and exotic annual grasses: implications for an historical invasion. *Ecology*, **85**, 1273–83.

Cordeiro, N. J. & Howe, H. F. (2001). Low recruitment of trees dispersed by animals in African forest fragments. *Conservation Biology*, **15**, 1733–41.

Corlett, R. (1998). Frugivory and seed dispersal by vertebrates in the Oriental (Indomalayan region). *Biological Reviews*, **73**, 413–48.

Cornelissen, J. H. C. (1999). A triangular relationship between leaf size and seed size among woody species: allometry, ontogeny, ecology and taxonomy. *Oecologia*, **118**, 248–55.

Cornelissen, J. H. C., Diez, P. C., & Hunt, R. (1996). Seedling growth, allocation and leaf attributes in a wide range of woody plant species and types. *Journal of Ecology*, **84**, 755–65.

Cornelissen, J. H. C., Werger, M. J. A., Castro-Diez, P., van Rheenen, J. W. A., & Rowland, A. P. (1997). Foliar nutrients in relation to growth, allocation and leaf traits in seedlings of a wide range of woody plant species and types. *Oecologia*, **111**, 460–9.

Cornelissen, J. H. C., Werger, M. J. A., & Zhong, Z. C. (1994). Effects of canopy gaps on the growth of tree seedlings from subtropical broad-leaved evergreen forests of Southern China. *Vegetatio*, **110**, 43–54.

Corner, E. J. H. (1976). *The Seeds of Dicotyledons, Volumes 1 and 2*. Cambridge: Cambridge University Press.

Cornett, M. W., Puettmann, K. J., Frelich, L. E., & Reich, P. B. (2001). Comparing the importance of seedbed and canopy type in the restoration of upland *Thuja occidentalis* forests of northeastern Minnesota. *Restoration Ecology*, **9**, 386–96.

Coruzzi, G. & Zhou, L. (2001). Carbon and nitrogen sensing and signaling in plants: emerging 'matrix effects.' *Current Opinion in Plant Biology*, **4**, 247–53.

Côté, S. D., Rooney, T. P., Tremblay, J.-P., Dussault, C., & Waller, D. M. (2004). Ecological impact of deer overabundance. *Annual Review of Ecology, Evolution and Systematics*, **35**, 113–47.

Cowling, R. M., Kirkwood, D., Midgley, J. J., & Pierce, S. M. (1997). Invasion and persistence of bird-dispersed, subtropical thicket and forest species in fire-prone coastal fynbos. *Journal of Vegetation Science*, **8**, 475–88.

Cox, P. A. (1991). Hydrophilous pollination of a dioecious seagrass, *Thallasodendron ciliatum* (Cymodoceaceae) in Kenya. *Biotropica*, **23**, 159–65.

Cox, P. A. & Knox, R. B. (1988). Pollination postulates and two-dimensional pollination in hydrophilous monocotyledons. *Annals of the Missouri Botanical Garden*, **75**, 811–18.

Crain, C. M. & Bertness M. D. (2005). Community impacts of a tussock-forming sedge: is ecosystem engineering important in physically benign habitats? *Ecology*, **86**, 2695–704.

Craven, D., Braden, D., Ashton, M. S., *et al.* (2007). Between and within-site comparisons of structural and physiological characteristics and foliar nutrient content of 14 tree species at a wet, fertile site and a dry, infertile site in Panama. *Forest Ecology and Management*, **238**, 335–46.

Crawley, M. J. (1997a). Life history and environment. In *Plant Ecology*, ed. M. J. Crawley. Oxford: Blackwell Science, pp. 73–131.

Crawley, M. J. (1997b). *Plant Ecology*. Oxford: Blackwell Science.

Creelman, R. A. & Mullet, J. A. (1997). Biosynthesis and action of jasmonates in plants. *Annual Review of Plant Physiology and Molecular Biology*, **48**, 355–81.

Cribb, P. J., Kell, S. P., Dixon, K. W., & Barrett, R. L. (2003). Orchid conservation: a global perspective. In *Orchid Conservation*, ed. K. W. Dixon, S. P. Kell, R. L. Barrett, & P. J. Cribb. Sabah: Natural History Publications, pp. 1–24.

Croat, T. B. (1983). *Diffenbachia* (Loaterías, Dumb Cane). In *Costa Rican Natural History*, ed. D. H. Janzen. Chicago: University of Chicago Press, pp. 234–6.

Cronk, Q. C. B. & Fuller, J. L. (1995). *Plant Invaders: The Threat to Natural Ecosystems*. London: Chapman & Hall.

Cronquist, A. (1981). *An Integrated System of Classification of Flowering Plants*. New York: Columbia University Press.

Crowder, A. A., Pearson, M. C., Grubb, P. J., & Langlois, P. H. (1990). Biological flora of the British Isles. *Drosera* L. *Journal of Ecology*, **78**, 233–67.

Csapody, V. (1968). *Keimlings-Bestimmungsbuch der Dikotyledonen*. Budapest: Akadémiai Kiadó.

Cullings, K. W., Parker, V. T., Finley, S. K., & Vogler, D. R. (2000). Ectomycorrhizal specificity patterns in a mixed *Pinus contorta* and *Picea engelmannii* forest in Yellowstone National Park. *Applied and Environmental Microbiology*, **66**, 4988–91.

Curran, L. M., Caniago, I., Paoli, G. D., *et al.* (1999). Impact of El Niño and logging on canopy tree recruitment in Borneo. *Science*, **286**, 2184–8.

Curran, L. M. & Webb, C. O. (2000). Experimental tests of the spatiotemporal scale of seed predation in mast-fruiting Dipterocarpaceae. *Ecological Monographs*, **70**, 129–48.

Curtis, J. T. (1943). Germination and seedling development in five species of *Cypripedium* L. *American Journal of Botany*, **30**, 199–206.

Curtis, J. T. (1959). *The Vegetation of Wisconsin*. Madison: University of Wisconsin Press.

Dallimore, J. W. & Jackson, A. B. (1966). *Handbook of Coniferae and Ginkgoaceae*. London: Edward Arnold Publishers, Ltd.

Dalling, J. W. (2005). The fate of seed banks: factors influencing seed survival for light-demanding species in moist tropical forests. In *Seed Fate: Predation, Dispersal and Seedling Establishment*, ed. P.-M. Forget, J. E. Lambert, P. E. Hulme, & S. B. Vander Wall. Wallingford: CAB International, pp. 31–44.

Dalling, J. W. & Burslem, D. F. R. P. (2005). Role of trade-offs in the equalization and differentiation of tropical tree species. In *Biotic Interactions in the Tropics*, ed. D. F. R. P Burslem, M. A. Pinard, & S. E. Hartley. Cambridge: Cambridge University Press, pp. 65–88.

Dalling J. W. & Denslow, J. S. (1998). Changes in soil seed bank composition along a chronosequence of lowland secondary tropical forest, Panama. *Journal of Vegetation Science*, **9**, 669–78.

Dalling, J. W. & Hubbell, S. P. (2002). Seed size, growth rate and gap microsite conditions as determinants of recruitment success for pioneer species. *Journal of Ecology*, **90**, 557–68.

Dalling, J. W., Muller-Landau, H. C., Wright, S. J., & Hubbell, S. P. (2002). Role of dispersal in the recruitment limitation of neotropical pioneer species. *Journal of Ecology*, **90**, 714–27.

Dalling, J. W., Winter, K., Nason, J. D., *et al.* (2001). The unusual life history of *Alseis blackiana*: A shade-persistent pioneer tree? *Ecology*, **82**, 933–45.

D'Antonio, C. M. (1993). Mechanisms controlling invasion of coastal plant communities by the alien succulent *Carpobrotus edulis*. *Ecology*, **74**, 83–95.

D'Antonio, C. M. & Vitousek, P. M. (1992). Biological invasions by exotic grasses, the grass/fire cycle, and global change. *Annual Reviews of Ecology and Systematics*, **23**, 63–87.

Darwin, C. (1859). *On the Origin of Species by Means of Natural Selection or the Preservation of Favored Races in the Struggle for Life*. New York: D. Appleton and Co. (1869).

Davidson, E. A., Reis de Carvalho, C. J., Vieira, I. C. G., *et al.* (2004). Nitrogen and phosphorus limitation of biomass growth in a tropical secondary forest. *Ecological Applications*, **14**, S150–63.

Davies, P. J., ed. (1995). *Plant Hormones: Physiology, Biochemistry, and Molecular Biology*. Dordrecht: Kluwer Academic.

Davis, A. S., Landis, D. A., Nuzzo, V., *et al.* (2006). Demographic models inform selection of biocontrol agents for garlic mustard (*Alliaria petiolata*). *Ecological Applications*, **6**, 2399–410.

Davis, M. A., Grime J. P., & Thompson, K. (2000). Fluctuating resources in plant communities: a general theory of invasibility. *Journal of Ecology*, **88**, 528–34.

Davis, M. A., Wrage, K. J., & Reich, P. B. (1998). Competition between tree seedlings and herbaceous vegetation: support for a theory of resource supply and demand. *Journal of Ecology*, **86**, 652–61.

Daws, M. I., Burslem, D. F. R. P., Crabtree, L. M., *et al.* (2002). Differences in seed germination responses may promote coexistence of four sympatric *Piper* species. *Functional Ecology*, **16**, 258–67.

Daws, M. I., Orr, D., Burslem, D., & Mullins, C. E. (2006). Effect of high temperature on chalazal plug removal and germination in *Apeiba tibourbou* Aubl. *Seed Science and Technology*, **34**, 221–5.

Deacon, J. W. & Flemming, L. V. (1992). Interactions of ectomycorrhizal fungi. In *Mycorrhizal Functioning: an Integrative Plant-fungal Process*, ed. M. F. Allen. New York: Chapman and Hall, pp. 249–300.

Debussche, M., Escarré, J., & Lepart, J. (1982). Ornithochory and plant succession in Mediterranean abandoned orchards. *Vegetatio*, **48**, 255–66.

de Kroon, H., Plaisier, A., van Groenendael, J., & Caswell, H. (1986). Elasticity: the relative contribution of demographic parameters to population growth rate. *Ecology*, **67**, 1427–31.

de Kroon, H., van Groenendael, J. M., & Ehrlén, J. (2000). Elasticities: a review of methods and model limitations. *Ecology*, **81**, 607–18.

Delagrange, S., Messier, C., Lechowicz, M. J., & Dizengremel, P. (2004). Physiological, morphological and allocational plasticity in understory deciduous trees: importance of plant size and light availability. *Tree Physiology*, **24**, 775–84.

de la Providencia, I. E., de Souza, F. A., Fernandez, F., Séjalon-Delmas, N., & Declerck, S. (2005). Arbuscular mycorrhizal fungi reveal distinct patterns of anastomosis formation and hyphal healing mechanisms between different phylogenetic groups. *New Phytologist*, **165**, 261–71.

Del Tredici, P. (1997). Lignotuber development in *Ginkgo biloba*. In *Ginkgo biloba, a Global Treasure: From Biology to Medicine*, ed. T. Hori, R. W. Ridge, W. Tulecke, P. D. Tredici, J. Trémouillaux-Guiller, & H. Tobe. Tokyo: Springer-Verlag, pp. 119–26.

DeLucia, E. H., Sipe, T. W., Herrick, J., & Maherali, H. (1998). Sapling biomass allocation and growth in the understory of a deciduous hardwood forest. *American Journal of Botany*, **85**, 955–63.

Denham, A. J. & Auld, T. D. (2004). Survival and recruitment of seedlings and suckers of trees and shrubs of the Australian arid zone following habitat management and the outbreak of Rabbit Calicivirus Disease (RCD). *Austral Ecology*, **29**, 585–99.

Denk T. & Oh, I.-C. (2006). Phylogeny of Schisandraceae based on morphological data: evidence from modern plants and the fossil record. *Plant Systematics and Evolution*, **256**, 113–45.

de Oliveira Wittmann, A., Piedade, M. T. F., Parolin, P., & Wittmann, F. (2007). Germination in four low-várzea tree species of Central Amazonia. *Aquatic Botany*, **86**, 197–203.

Desbiez, M.-O. & Boyer, N. (1981). Hypocotyl growth and peroxidases of *Bidens pilosus*. *Plant Physiology*, **68**, 41–3.

DeSimone, S. A. & Zedler, P. H. (1999). Shrub seedling recruitment in unburned California coastal sage scrub and adjacent grassland. *Ecology*, **80**, 2018–32.

De Smet, I., Zhang, H., Inzé, D., & Beeckman, T. (2006). A novel role for abscisic acid emerges from underground. *Trends in Plant Science*, **11**, 434–9.

De Souza, F. A. & Declerck, S. (2003). Mycelium development and architecture, and spore production of *Scutellospora reticulata* in monoxenic culture with Ri T-DNA transformed roots. *Mycologia*, **95**, 1004–12.

De Steven, D. (1991). Experiments on mechanisms of tree establishment in old-field succession seedling survival and growth. *Ecology*, **72**, 1076–88.

De Steven, D. (1994). Tropical tree seedling dynamics: recruitment patterns and their population consequences for three canopy species in Panama. *Journal of Tropical Ecology*, **10**, 369–83.

de Vogel, E. F. (1980). *Seedlings of Dicotyledons: Structure, Development, Types, Descriptions of 150 Woody Malesian Taxa*. Wageningen: Centre for Agricultural Publishing and Documentation.

Devoto, A. & Turner, J. G. (2005). Jasmonate-regulated *Arabidopsis* stress signalling network. *Physiologia Plantarum*, **123**, 161–72.

DeWalt, S. J. (2006). Population dynamics and potential for biological control of an exotic invasive shrub in Hawaiian rainforests. *Biological Invasions*, **8**, 1145–58.

DeWitt, T. J., Sih, A., & Wilson, D. S. (1998). Costs and limits of phenotypic plasticity. *Trends in Ecology & Evolution*, **13**, 77–81.

D'Haeze, W., De Rycke, R., Mathis, R., *et al.* (2003). Reactive oxygen species and ethylene play a positive role in lateral root base nodulation of a semiaquatic legume. *Proceedings of the National Academy of Sciences* (USA), **100**, 11789–94.

Diamond, J. M. (1975). Assembly of species communities. In *Ecology and Evolution of Communities*, ed. M. L. Cody & J. M. Diamond. Cambridge: Harvard University Press, pp. 342–444.

Dick, C. W., Etchelecu, G., & Austerlitz, G. (2003). Pollen dispersal of tropical trees (*Dinizia excelsa*, Fabaceae) by native insects and African honeybees in pristine and fragmented Amazonian rainforest. *Molecular Ecology*, **12**, 753–64.

Dickie, I. A., Guza, R. C., Krazewski, S. E., & Reich, P. B. (2004). Shared ectomycorrhizal fungi between an herbaceous perennial (*Helianthemum bicknellii*) and oak (*Quercus*) seedlings. *New Phytologist*, **164**, 375–82.

Dickie, I. A., Koide, R. T., & Steiner, K. C. (2002). Influences of established trees on mycorrhizas, nutrition, and growth of *Quercus rubra* seedlings. *Ecological Monographs*, **72**, 505–21.

Dickie, I. A. & Reich, P. B. (2005). Ectomycorrhizal fungal communities at forest edges. *Journal of Ecology*, **93**, 244–55.

Dickie, I. A., Schnitzer, S. A., Reich, P. B., & Hobbie, E. A. (2005). Spatially disjunct effects of co-occurring competition and facilitation. *Ecology Letters*, **8**, 1191–1200.

Dickie, I. A., Schnitzer, S. A., Reich, P. B., & Hobbie, S. E. (2007). Is oak establishment in old-fields and savanna openings context dependent? *Journal of Ecology*, **95**, 309–20.

Diekmann, M. (2003). Species indicator values as an important tool in applied plant ecology – a review. *Basic and Applied Ecology*, **4**, 493–506.

Dirzo, R. (1984). Herbivory: a phytocentric overview. In *Perspectives in Population Ecology*, ed. R. Dirzo & J. Sarukhan. Sutherland: Sinauer Associates Inc., pp. 141–65.

Dirzo, R. (1985). The role of the grazing animal. In *Studies in Plant Demography*, ed. J. White. London: Academic Press, pp. 343–55.

Dirzo, R. (2001). Plant–mammal interactions, lessons for our understanding of nature and implications for biodiversity conservation. In *Ecology, Achievement and Challenge*, ed. M. C. Press, N. J. Huntly, & S. Levin. Oxford: Blackwell Science, pp. 319–35.

Dirzo, R. & Domínguez, C. A. (1986). Seed shadows, seed predation and the advantages of dispersal. In *Frugivores and Seed Dispersal*, ed. A. Estrada & T. H. Fleming. Dordrecht: Junk, pp. 237–49.

Dirzo, R. & Miranda, A. (1990). Contemporary neotropical defaunation and forest structure, function, and diversity – a sequel. *Conservation Biology*, **4**, 444–7.

Dixon, K. W. (1991). Seeder/clonal concepts in Western Australian orchids. In *Population Ecology of Terrestrial Orchids*, ed. T. C. E. Wells & J. H. Willems. The Hague: SPB Academic Publishing bv, pp. 111–24.

Dixon, K. W., Kell, S. P., Barrett, R. L., & Cribb, P. J. ed. (2003). *Orchid Conservation*. Sabah: Natural History Publications.

Dmowski, K. & Kozakiewicz, M. (1990). Influence of a shrub corridor on movements of passerine birds to a lake littoral zone. *Landscape Ecology*, **4**, 99–108.

Dobbins, D. R. & Kuijt, J. (1974a). Anatomy and fine structure of the mistletoe haustorium (*Phthirusa pyrifolia*). I. Development of the young haustorium. *American Journal of Botany*, **61**, 535–43.

Dobbins, D. R. & Kuijt, J. (1974b). Anatomy and fine structure of the mistletoe haustorium (*Phthirusa pyrifolia*). II. Penetration attempts and formation of the gland. *American Journal of Botany*, **61**, 544–50.

Dodd, A. N., Salathera, N., Hall, A., *et al.* (2005). Plant circadian clocks increase photosynthesis, growth, survival, and competitive advantage. *Science*, **309**, 630–33.

Dolan, R. W. (1984). The effect of seed size and maternal source on individual size in a population of *Ludwigia leptocarpa* Onagraceae. *American Journal of Botany*, **71**, 1302–7.

Domingo, F., Villagarcia, L., Boer, M. M., Alados-Arboledas, L., & Puigdefabregas, J. (2001). Evaluating the long-term water balance of arid zone stream bed vegetation using evapotranspiration modelling and hillslope runoff measurements. *Journal of Hydrology*, **243**, 17–30.

Don, R. (2003). *Handbook for Seedling Evaluation*, 3rd edn. Bassersdorf: International Seed Testing Association.

Dong, Y. J., Ogawa, T., Lin, D. Z., *et al.* (2006). Molecular mapping of quantitative trait loci for zinc toxicity tolerance in rice seedling (*Oryza sativa* L.). *Field Crops Research*, **95**, 420–5.

Dörr, I. (1990). Sieve elements in haustoria of parasitic angiosperms. In *Sieve Elements-Comparative Structure, Induction, and Development*, ed. H.-D. Behnke & R. D. Sjolund. Heidelberg: Springer, pp. 163–70.

dos Santos, U. M., Jr., de Carvalho Gonçalves, J. F., & Feldpausch, T. R. (2006). Growth, leaf nutrient concentration and photosynthetic nutrient use efficiency in tropical tree species planted in degraded areas in central Amazonia. *Forest Ecology and Management*, **226**, 299–309.

Doube, B. M. (1994). Enhanced root nodulation of subterranean clover (*Trifolium subterraneum*) by *Rhizobium leguminosarium* biovar. *trifolii* in the presence of the earthworm *Aporrectodea trapezoides* (Lumbricidae). *Biology and Fertility of Soils*, **18**, 169.

Douglas, I., Spencer, T., Greer, T., *et al.* (1992). The impact of selective commercial logging on stream hydrology, chemistry and sediment loads in the Ulu Segama rain forest, Sabah, Malaysia. *Philosophical Transactions of the Royal Society Series B*, **335**, 397–406.

Dowling, R. M. & McDonald, T. J. (1982). Structure, function and management. In *Mangrove Ecosystems in Australia*, ed. B. F. Clough. Canberra: Australian Institute of Marine Science and Australian National University Press, pp. 79–93.

Downs, R. J. & Hellmers, H. (1975). *Environment and the Experimental Control of Plant Growth*. London: Academic Press.

Downton, W. J. S. (1982). Growth and osmotic relations of the mangrove *Avicennia marina*, as influenced by salinity. *Australian Journal of Plant Physiology*, **9**, 519–28.

Doyle, J. (1963). Proembryogeny in *Pinus* in relation to that of other conifers – a survey. *Proceedings of the Royal Irish Academy*, **62B**, 181–216.

Doyle, J. A. (2006). Seed ferns and the origin of angiosperms. *Journal of the Torrey Botanical Society*, **133**, 169–209.

Doyle, J. A. & Donoghue, M. J. (1986). Seed plant phylogeny and the origin of angiosperms – an experimental cladistic approach. *Botanical Review*, **52**, 321–431.

Doyle, J. A. & Endress, P. K. (2000). Morphological phylogenetic analysis of basal angiosperms: comparison and combination with molecular data. *International Journal of Plant Sciences*, **161**, S121–53.

Doyle, J. A. & Hickey, L. J. (1976). Pollen and leaves from the mid-Cretaceous Potomac Group and their bearing on early angiosperm evolution. In *Origin and Early Evolution of Angiosperms*, ed. C. B. Beck. New York: Columbia University Press, pp. 139–206.

Drake, J. A., Mooney, H. A., diCastri, F., *et al.* (1989). *Biological Invasions: A Global Perspective*. New York: John Wiley & Sons.

Dregne, H. E. (1986). Desertification of arid lands. In *Physics of Desertification*, ed. F. El-Baz & M. H. A. Hassan. Dordrecht: Martinus Nijhoff.

Dressler, R. (1981). *The Orchids, Natural History and Classification*. Cambridge: Harvard University Press.

Dressler, R. (1983). *Phylogeny and Classification of the Orchid Family*. London: Cambridge University Press.

Driscoll, C. T., Driscoll, K. M., Mitchell, M. J., & Raynal, D. J. (2003). Effects of acidic deposition on forest and aquatic ecosystems in New York State. *Environmental Pollution*, **123**, 327–36.

Drumm-Herrel, H. & Mohr, H. (1985). Photosensitivity of seedlings differing in their potential to synthesize anthocyanin. *Physiologia Plantarum*, **64**, 60–5.

Dubrovsky, J. G. (1997a). Determinate primary growth in *Stenocereus gummosus* (Cactaceae). In *The Biology of Root Formation and Development*, ed. A. Altman & Y. Waisel. New York: Plenum Press, pp. 13–20.

Dubrovsky, J. G. (1997b). Determinate primary-root growth in seedlings of Sonoran Desert Cactaceae: its organization, cellular basis, and ecological significance. *Planta*, **203**, 85–92.

Dubrovsky, J. G. & Gómez-Lomelí, L. F. (2003). Water deficit accelerates determinate developmental program of the primary root and does not affect lateral root initiation in a Sonoran Desert cactus (*Pachycereus pringlei*, Cactaceae). *American Journal of Botany*, **90**, 823–31.

Duchok, R., Kent, K., Khumbongmayum, A. D., Paul, A., & Khan, M. L. (2005). Population structure and regeneration status of a medicinal tree *Illicium griffithii* in relation to disturbance gradients in temperate broad-leaved forest of Arunachal Pradesh. *Current Science*, **89**, 673–6.

Ducker, S. C. & Knox, R. B. (1976). Submarine pollination of seagrasses. *Nature (London)*, **263**, 705–6.

Duckett, J. G. & Ligrone, R. (1992). A light and electron microscope study of the fungal endophytes in the sporophyte and gametophyte of *Lycopodium cernuum* L. with observations on the gametophyte–sporophyte junction. *Canadian Journal of Botany*, **70**, 58–72.

Duckett, J. G. & Ligrone, R. (2003). The structure and development of haustorial placentas in leptosporangiate ferns provide a clear-cut distinction between euphyllophytes and lycophytes. *Annals of Botany*, **92**, 513–21.

Duckworth, J. C., Kent, M., & Ramsay, P. M. (2000). Plant functional types: an alternative to taxonomic plant community description in biogeography? *Progress in Physical Geography*, **24**, 515–42.

Dudash, M. R. (1991). Plant size effects on female and male function in hermaphroditic *Sabatia angularis* (Gentianaceae). *Ecology*, **72**, 1004–12.

Duke, J. A. (1965). Keys for the identification of seedlings of some prominent woody species in eight forest types in Puerto Rico. *Annals of the Missouri Botanical Garden*, **52**, 314–50.

Dulmer, K. (2006). Mycorrhizal associations of American chestnut seedlings: a lab and field bioassay. MS thesis, State University of New York, Syracuse, USA.

Dunbar, R. B. (2000). El Niño – clues from corals. *Nature*, **407**, 956–9.

Dunham, S. M., Kretzer, A., & Pfrender, M. E. (2003). Characterization of Pacific golden chanterelle (*Cantharellus formosus*) genet size using co-dominant microsatellite markers. *Molecular Ecology*, **12**, 1607–18.

Dunkerley, D. L. (1997). Banded vegetation: Survival under drought and grazing pressure based on a simple cellular automation model. *Journal of Arid Environments*, **35**, 419–28.

Dunne, J. A. & Parker, V. T. (1999). Seasonal soil water potential patterns and establishment of *Pseudotsuga menziesii* seedlings in chaparral. *Oecologia*, **119**, 36–45.

Durand, E. J. (1908). The development of the sexual organs and sporogonium of *Marchantia polymorpha*. *Bulletin of the Torrey Botanical Club*, **35**, 321–35.

Dyer, A. R. & Rice, K. J. (1997). Intraspecific and diffuse competition: the response of *Nassella pulchra* in a California grassland. *Ecological Applications*, **7**, 484–92.

Dyer, M. I., Turner, C. L., & Seastedt, T. R. (1993). Herbivory and its consequences. *Ecological Applications*, **3**, 10–16.

Eames, A. J. (1961). *Morphology of the Angiosperms*. New York: McGraw Hill Book Company, Inc.

Eapen, D., Barroso, M. L., Ponce, G., Campos, M. E., & Cassab, G. I. (2005). Hydrotropism: root growth responses to water. *Trends in Plant Science*, **10**, 44–50.

Ebbett, R. L. & Ogden, J. (1998). Comparative seedling growth of five endemic New Zealand podocarp species under different light regimes. *New Zealand Journal of Botany*, **36**, 189–201.

Eccles, N. S., Esler, K. J., & Cowling, R. M. (1999). Spatial pattern analysis in Namaqualand desert plant communities: evidence for general positive interactions. *Plant Ecology*, **142**, 71–85.

Edwards, P. J., Kollmann, J., & Fleischmann, K. (2002). Life history evolution in *Lodoicea maldivica* (Arecaceae). *Nordic Journal of Botany*, **22**, 227–37.

Edwards, W., Gadek, P., Webber, E., & Warboys, S. (2001). Idiosyncratic phenomenon of regeneration from cotyledons in the idiot fruit tree, *Idiospermum australiense*. *Austral Ecology*, **26**, 254–8.

Egerton, J. J. G., Banks, J. C. G., Gibson, A., Cunningham, R. B., & Ball, M. C. (2000). Facilitation of seedling establishment: reduction in irradiance enhances winter growth of *Eucalyptus paciflora*. *Ecology*, **81**, 1437–49.

Egerton-Warburton, L. M. & Allen, M. F. (2001). Endo- and ectomycorrhizas in *Quercus agrifolia* Nee. (Fagaceae): patterns of root colonization and effects on seedling growth. *Mycorrhiza*, **11**, 283–90.

Ehrenfeld, J. G. & Schneider, J. P. (1991). *Chamaecyparis thyoides* wetlands and suburbanization: effects on hydrology, water quality and plant community composition. *Journal of Applied Ecology*, **28**, 467–90.

Ehrlén, J. (2002). Assessing the life-time consequences of animal interactions with a perennial herb, *Lathyrus vernus* (Fabaceae). *Perspectives in Plant Ecology, Evolution and Systematics*, **5**, 145–63.

Ehrlén, J. (2003). Fitness components versus total demographic effects: evaluating herbivore impacts on a perennial herb. *American Naturalist*, **162**, 796–810.

Ehrlén, J. & Eriksson, O. (1996). Seedling recruitment in the perennial herb *Lathyrus vernus*. *Flora*, **191**, 377–83.

Ehrlén, J. & Eriksson, O. (2000). Dispersal limitation and patch occupancy in forest herbs. *Ecology*, **81**, 1667–74.

Ehrlén, J. & Lehtilä, K. (2002). How perennial are perennial plants? *Oikos*, **98**, 308–22.

Ehrlén, J., Münzbergova, Z., Diekmann, M., & Eriksson, O. (2006). Long-term assessment of seed limitation in plants: results from an 11-year experiment. *Journal of Ecology*, **94**, 1224–32.

Ehrlén, J., Syrjänen, K., Leimu, R., Garcia, M. B., & Lehtilä, K. (2005). Land use and population growth of *Primula veris*: an experimental demographic approach. *Journal of Applied Ecology*, **42**, 317–26.

Eklund, H., Doyle, J. A., & Herendeen, P. S. (2004). Morphological phylogenetic analysis of living and fossil Chloranthaceae. *International Journal of Plant Sciences*, **165**, 107–51.

El-Bana, M. I., Nijs, I., & Khedr, A. H. A. (2003). The importance of phytogenic mounds (nebkhas) for restoration of arid degraded rangelands in Northern Sinai. *Restoration Ecology*, **11**, 317.

Elberse, W. T. & Breman, H. (1989). Germination and establishment of Sahelian rangeland species. I. Seed properties. *Oecologia*, **80**, 477–84.

Eldred, R. A. & Maun M. A. 1982. A multivariate approach to the problem of decline in vigour in *Ammophila*. *Canadian Journal of Botany*, **60**, 1371–80.

Eldridge, D. J., Zaady, E., & Shachak, M. (2000). Infiltration through three contrasting biological soil crusts in patterned landscapes in the Negev, Israel. *Catena*, **40**, 323–36.

El Harti, A., Saghi, M., Molina, J. A. E., & Teller, G. (2001). Production by the earthworm (*Lumbricus terrestris*) of a rhizogenic substance similar to indolacetic acid. *Canadian Journal of Zoology*, **79**, 1911–20.

Eliason, S. A. & Allen, E. B. (1997). Exotic grass competition in suppressing native shrubland re-establishment. *Restoration Ecology*, **5**, 245–55.

Ellenberg, H., Weber, H. E., Düll, R., *et al.* (1991). Zeigerwerte von Pflanzen in Mitteleuropa. *Scripta Geobotanica*, **18**, 1–248.

Ellis, R. H., Hong, T. D., & Roberts, E. H. (1985). *Handbook of Seed Technology for Genebanks*, Vol. II. *Compendium of Specific Germination Information and Test Recommendations*. Rome: International Board Plant Genetic Resources.

Ellis, S. & Mellor, A. (1995). *Soils and Environment*. New York: Routledge.

Ellison, A. M., Denslow, J. S., Loiselle, B. A., & Brenés M. D. (1993). Seed and seedling ecology of neotropical Melastomataceae. *Ecology*, **74**, 1733–49.

Ellison, A. M. & Gotelli, N. J. (2001). Evolutionary ecology of carnivorous plants. *Trends in Ecology & Evolution*, **16**, 623–9.

Ellison, A. M. & Parker, J. N. (2002). Seed dispersal and seedling establishment of *Sarracenia purpurea* (Sarraceniaceae). *American Journal of Botany*, **89**, 1024–6.

Ellner, S. & Shmida, A. (1981). Why are adaptations for long-range seed dispersal rare in desert plants? *Oecologia*, **51**, 133–44.

Ellner, S. P., Hairston, N. G., Jr., & Babai, D. (1998). Long-term diapause and spreading of risk across the life cycle. *Ergebnisse der Limnologie*, **52**, 297–312.

Elmarsdottir, A., Aradottir, A. L., & Trlica, M. J. (2003). Microsite availability and establishment of native species on degraded and reclaimed sites. *Journal of Applied Ecology*, **40**, 815–23.

Elmqvist, T. & Cox, P. A. (1996). The evolution of vivipary in flowering plants. *Oikos*, **77**, 3–9.

Elton, C. S. (1958). *The Ecology of Invasions by Animals and Plants*. London: Methuen.

Emerson, F. W. (1921). Subterranean organs of bog plants. *Botanical Gazette*, **72**, 359–74.

Emmerson, L. M. (1999). Persistence mechanisms of *Erodiophyllum elderi*. PhD thesis, University of Adelaide, Australia.

Endo, Y. & Ohashi, H. (1998). The features of cotyledon areoles in Leguminosae and their systematic utility. *American Journal of Botany*, **85**, 753–9.

Engelbrecht, B. M. J., Comita, L. S., Condit, R., *et al.* (2007). Drought sensitivity shapes species distribution patterns in tropical forests. *Nature*, **447**, 80–2.

Engelbrecht, B. M. J., Dalling, J. W., Pearson, T. R. H., *et al.* (2006). Short dry spells in the wet season increase mortality of tropical pioneer seedlings. *Oecologia*, **148**, 258–69.

Engelbrecht, B. M. J., Kursar, T. A., & Tyree, M. T. (2005). Drought effects on seedling survival in a tropical moist forest. *Trees*, **19**, 312–21.

Enoch, I. C. (1980). Morphology of germination. In *Recalcitrant Crop Seeds*, ed. H. F. Chin & E. H. Roberts. Kuala Lumpur: Tropical Press SDN. BHD., pp. 6–37.

Enright, N. J. & Watson, A. D. (1992). Population dynamics of the Nikau palm *Rhopalostylis sapida* Wendl. et Drude in a temperate forest remnant near Auckland, New Zealand. *New Zealand Journal of Botany*, **30**, 29–43.

Eppley, S. M. (2001). Gender-specific selection during early life history stages in the dioecious grass *Distichlis spicata*. *Ecology*, **82**, 2022–31.

Eranen, J. K. & Kozlov, M. V. (2006). Physical sheltering and liming improve survival and performance of mountain birch seedlings: a 5-year study in a heavily polluted industrial barren. *Restoration Ecology*, **14**, 77–86.

Eriksson, O. (1989). Seedling recruitment and life histories in clonal plants. *Oikos*, **55**, 231–8.

Eriksson, O. (1993). Dynamics of genets in clonal plants. *Trends in Ecology & Evolution*, **8**, 313–16.

Eriksson, O. (2002). Ontogenetic niche shifts and their implications for recruitment in three clonal *Vaccinium* shrubs: *Vaccinium myrtillus*, *Vaccinium vitis-idaea*, and *Vaccinium oxycoccos*. *Canadian Journal of Botany*, **80**, 635–41.

Eriksson, O. (2005). Game theory provides no explanation for seed size variation in grasslands. *Oecologia*, **114**, 98–105.

Eriksson, O. & Ehrlén, J. (1992). Seed and microsite limitation in plant populations. *Oecologia*, **91**, 360–4.

Eriksson, O., Friis, E.-M., & Crane, P. R. (2000a). Seed size, fruit size, and dispersal systems in angiosperms from the Early Cretaceous to the Late Tertiary. *American Naturalist*, **156**, 47–58.

Eriksson, O., Friis, E.-M., Pedersen, K. R., & Crane, P. R. (2000b). Seed size and dispersal systems of Early Cretaceous angiosperms from Famalicao, Portugal. *International Journal of Plant Sciences*, **161**, 319–29.

Eriksson, O. & Fröborg, H. (1996). Windows of opportunity for recruitment in long-lived clonal plants: experimental studies of seedling recruitment in *Vaccinium* shrubs. *Canadian Journal of Botany*, **74**, 1369–74.

Erskine, P. D., Lamb, D., & Bristow, M. (2006). Tree species diversity and ecosystem function: can tropical multi-species plantations generate greater productivity? *Forest Ecology and Management*, **233**, 205–10.

Esler, K. J. (1999). Plant reproductive ecology. In *The Karoo, Ecological Patterns and Processes*, ed. W. R. J. Dean & S. J. Milton. Cambridge: Cambridge University Press, pp. 123–44.

Esler, K. J. & Cowling, R. M. (1995). The comparison of selected life-history characteristics of Mesembryanthema species occurring on and off Mima-like mounds (*heuweltjies*) in semi arid southern Africa. *Vegetatio*, **116**, 41–50.

Esler, K. J. & Phillips, N. (1994). Experimental effects of water stress on semi-arid succulent Karoo seedlings: implications for field seedling survivorship. *Journal of Arid Environments*, **26**, 325–37.

Espelta, J. M., Riba, M., & Retana, J. (1995). Patterns of seedling recruitment in West-Mediterranean *Quercus ilex* forests influenced by canopy development. *Journal of Vegetation Science*, **6**, 465–72.

Esseling, J. J. & Emons, A. M. (2004). Dissection of Nod factor signalling in legumes: cell biology, mutants and pharmacological approaches. *Journal of Microscopy*, **214**, 104–13.

Evans, D. E. (2004). Aerenchyma formation. *New Phytologist*, **161**, 35–49.

Evenari, M., Shanan, L., & Tadmor, N. (1982). *The Negev: The Challenge of a Desert*. Cambridge: Harvard University Press.

Eviner, V. T. & Chapin, F. S., III (2003). Gopher–plant–fungal interactions affect establishment of an invasive grass. *Ecology*, **84**, 120–8.

Fabian, P., Kohlpainter, M., & Rollenbeck, R. (2005). Biomass burning the Amazon – fertilizer for the mountainous rain forest of Ecuador. *Environmental Science and Pollution Research*, **12**, 290–6.

Facelli, J. M. (1994). Multiple indirect effects of plant litter affect the establishment of woody seedlings in old fields. *Ecology*, **75**, 1727–35.

Facelli, J. M. & Brock, D. J. (2000). Patch dynamics in arid lands: localized effects of *Acacia papyrocarpa* on soils and vegetation of open woodlands of south Australia. *Ecography*, **23**, 479–91.

Facelli, J. M. & Chesson, P. (2008). Cyclic dormancy, temperature and water availability control germination of *Carrichtera annua*, an invasive species in chenopod shrublands. *Austral Ecology*, **33**, 324–8.

Facelli, J. M., Chesson, P., & Barnes, N. (2005). Differences in seed biology of annual plants in arid lands: a key ingredient of the storage effect. *Ecology*, **86**, 2998–3006.

Facelli, J. M. & Facelli, E. (1993). Interactions after death – plant litter controls priority effects in a successional plant community. *Oecologia*, **95**, 277–82.

Facelli, J. M. & Kerrigan, R. (1996). Effects of ash and four types of litter on the establishment of *Eucalyptus obliqua*. *Ecoscience*, **3**, 319–24.

Facelli, J. M. & Ladd, B. (1996). Germination requirements and responses to leaf litter of four species of eucalypt. *Oecologia*, **107**, 441–5.

Facelli, J. M. & Pickett, S. T. A. (1991a). Plant litter: its dynamics and effects on plant community structure. *Botanical Review*, **57**, 1–32.

Facelli, J. M. & Pickett, S. T. A. (1991b). Plant litter: light interception and effects on an old-field plant community. *Ecology*, **72**, 1024–31.

Facelli, J. M., Williams, R., Fricker, S., & Ladd, B. (1999). Establishment and growth of seedlings of *Eucalyptus obliqua*: interactive effects of litter, water and pathogens. *Australian Journal of Ecology*, **24**, 484–94.

Fahn, A. (1979). *Secretory Tissues in Plants*. London: Academic Press.

Fang, W., Taub, D. R., Fox, G. A., *et al.* (2006). Sources of variation in growth, form, and survivial in dwarf and normal-stature pitch pines (*Pinus rigida*, Pinaceae) in long-term transplant experiments. *American Journal of Botany*, **93**, 1125–33.

Fang, X., Wang, X., Li, H., Chen, K., & Wang, G. (2006). Responses of *Caragana korshinskii* to different aboveground shoot removal: combining defense and tolerance strategies. *Annals of Botany*, **98**, 203–11.

FAO (1997). *Drylands Development and Combating Desertification: Bibliographic Study of Experiences in China*. Rome: Food and Agriculture Organization of the United Nations.

Farnsworth, E. (2000). The ecology and physiology of viviparous and recalcitrant seeds. *Annual Review of Ecology and Systematics*, **31**, 107–38.

Farnsworth, E. (2004). Hormones and shifting ecology through plant development. *Ecology*, **85**, 5–15.

Farnsworth, E. J. & Ellison, A. M. (1991). Patterns of herbivory in Belizean mangrove swamps. *Biotropica*, **23**, 555–67.

Farnsworth, E. J. & Ellison, A. M. (1996). Sun–shade adaptability of the red mangrove, *Rhizophora mangle* (Rhizophoraceae): changes through ontogeny at several levels of biological organization. *American Journal of Botany*, **83**, 1131–43.

Farnsworth, E. J. & Farrant, J. M. (1998). Reductions in abscisic acid are linked with viviparous reproduction in mangroves. *American Journal of Botany*, **85**, 760–9.

Farooq, M., Basra, S. M. A., Khalid, M., Tabassum, R., & Mahmood, T. (2006). Nutrient homeostasis, metabolism of reserves, and seedling vigor as affected by seed priming in coarse rice. *Canadian Journal of Botany*, **84**, 1196–202.

Feild, T. S. & Arens, N. C. (2005). Form, function, and environments of the early angiosperms: merging extant phylogeny and ecophysiology with fossils. *New Phytologist*, **166**, 383–408.

Feild, T. S. & Arens, N. C. (2007). The ecophysiology of early angiosperms. *Plant, Cell, & Environment*, **30**, 291–309.

Feild, T. S., Arens, N. C., Doyle, J. A., Dawson, T. E., & Donoghue, M. J. (2004). Dark and disturbed: a new image of early angiosperm ecology. *Paleobiology*, **30**, 82–107.

Feild, T. S., Brodribb T., Jaffre, T., & Holbrook, N. M. (2001). Acclimation of leaf anatomy, photosynthetic light use, and xylem hydraulics to light in *Amborella trichopoda* (Amborellaceae). *International Journal of Plant Sciences*, **162**, 999–1008.

Feild, T. S., Franks, P. J., & Sage, T. L. (2003). Ecophysiological shade adaptation in the basal angiosperm, *Austrobaileya scandens* (Austrobaileyaceae). *International Journal of Plant Sciences*, **164**, 313–24.

Feild, T. S., Sage, T. L., Czerniak, C., & Iles, W. J. D. (2005). Hydathodal leaf teeth of *Chloranthus japonicus* (Chloranthaceae) prevent guttation-induced flooding of the mesophyll. *Plant, Cell, & Environment*, **28**, 1179–90.

Feinbrun-Dothan, N. (1986). *Flora Palestina*, Vol IV. Jerusalem: Israel Academy of Sciences and Humanities.

Fenner, M. (1985). *Seed Ecology*. London: Chapman and Hall.

Fenner, M. (1986). A bioassay to determine the limiting minerals for seeds from nutrient-deprived *Senecio vulgaris* plants. *Journal of Ecology*, **74**, 497–505.

Fenner, M. (1987). Seedlings. *New Phytologist*, **106**, 35–47.

Fenner, M., Hanley, M. E., & Lawrence, R. (1999). Comparison of seedling and adult palatability in annual and perennial plants. *Functional Ecology*, **13**, 546–51.

Fenner, M. & Kitajima, K. (2000). Ecology of seedling regeneration. In *Seeds: the Ecology of Regeneration in Plant Communities*, ed. M. Fenner. Wallingford: CAB International, pp. 331–59.

Fenner, M. & Lee, W. G. (1989). Growth of seedlings of pasture grasses and legumes deprived of single mineral nutrients. *Journal of Applied Ecology*, **26**, 223–32.

Fenner, M. & Thompson, K. (2005). *The Ecology of Seeds*. Cambridge: Cambridge University Press.

Fenster, C. B. (1991). Effect of male pollen donor and female seed parent on allocation of resources to developing seeds and fruit in *Chamaecrista fasciculata* (Leguminosae). *American Journal of Botany*, **78**, 13–23.

Feret, P. P. (1973). Early flowering in *Ailanthus*. *Forest Science*, **19**, 237–9.

Fetcher, N., Strain, B. R., & Oberbauer, S. F. (1983). Effects of light regime on the growth, leaf morphology, and water relations of two species of tropical trees. *Oecologia*, **58**, 314–9.

Fichtner, K. & Schulze, E. D. (1992). The effect of nitrogen nutrition on growth and biomass partitioning of annual plants originating from habitats of different nitrogen availability. *Oecologia*, **92**, 236–41.

Figueroa, J. A. & Lusk, C. H. (2001). Germination requirements and seedling shade tolerance are not correlated in a Chilean temperate rain forest. *New Phytologist*, **152**, 483–9.

Finch-Savage, W. E. & Leubner-Metzger, G. (2006). Seed dormancy and the control of germination. *New Phytologist*, **171**, 501–23.

Fine, P. V. A., Mesones, I., & Coley, P. D. (2004). Herbivores promote habitat specialization by trees in Amazonian forests. *Science*, **305**, 663–5.

Fine, P. V. A., Miller, Z. J., Mesones, I., *et al.* (2006). The growth-defense tradeoff and habitat specialization by plants in Amazonian forests. *Ecology*, **87**, S150–62.

Finkelstein, R. R., Gampala, S. S. L., & Rock, C. D. (2002). Abscisic acid signaling in seeds and seedlings. *Plant Cell*, **14**, S15–45.

Finlay, R. D. & Read, D. J. (1986) The structure and function of the vegetative mycelium of ectomycorrhizal plants. II. The uptake and distribution of phosphorus by mycelial strands interconnecting host plants. *New Phytologist*, **103**, 157–65.

Finzi, A. C., Canham, C. D., & Van Breeman, N. (1998). Canopy tree soil interactions within temperate forests: species effects on pH and cations. *Ecological Applications*, **8**, 447–54.

Fiordi, A. C., Palandra, M. R., Turicchia, S., Tani, G., & Falco, P. D. (2001). Characterization of the seed reserve in *Tillandsia* (Bromeliaceae) and ultrastructural aspects of their use at germination. *Caryologia*, **54**, 1–16.

Firn, R. D., Wagstaff, C., & Digby, J. (2000). The use of mutants to probe models of gravitropism. *Journal of Experimental Botany*, **51**, 1323–40.

Fisher, B. L., Howe, H. F., & Wright, S. J. (1991). Survival and growth of *Virola surinamensis* yearlings, water augmentation in gap understory. *Oecologia*, **86**, 292–7.

Fisher, R. A. (1930). *The Genetical Theory of Natural Selection*. Oxford: Oxford University Press.

Fitter, A. (2005). Darkness visible: reflections on underground ecology. *Journal of Ecology*, **93**, 231–43.

Fitter, A. H., Graves, J. D., Watkins, N. K., Robinson, D., & Scrimgeour, C. (1998). Carbon transfer between plants and its control in networks of arbuscular mycorrhizas. *Functional Ecology*, **12**, 406–12.

Fitzjohn, C., Ternan, J. L., & Williams, A. G. (1998). Soil moisture variability in a semi-arid gully catchment: Implications for runoff and erosion control. *Catena*, **32**, 55–70.

FloraWeb. (2006). *Daten und Informationen zu Wildpflanzen und zur Vegetation Deutschlands*. Bonn: Bundesamt fÿr Naturschutz (http://www.floraweb.de/).

Florence, R. G. (1981). The biology of the eucalypt forest. In *The Biology of Australian Plants*, ed. J. S. Pate & A. J. McComb. Nedlands: University of Western Australia Press, pp. 147–80.

Flores, J., Briones, O., Flores, A., & Sánchez-Colón, S. (2004). Effect of predation and solar exposure on the emergence and survival of desert seedlings of contrasting life-forms. *Journal of Arid Environments*, **58**, 1–18.

Flores, J. & Jurado, E. (2003). Are nurse–protégé interactions more common among plants from arid environments? *Journal of Vegetation Science*, **14**, 911–6.

Folgarait, P. J., Marquis, R. J., Ingvarsson, P., Braker, H. E., & Arguedas, M. (1995). Patterns of attack by insect herbivores and a fungus on saplings in a tropical tree plantation. *Environmental Entomology*, **24**, 1487–94.

Folgarait, P. J. & Sala, O. E. (2002). Granivory rates by rodents, insects, and birds at different microsites in the Patagonian steppe. *Ecography*, **25**, 417–27.

Forbis, T. A., Floyd, S. K., & de Queiroz, A. (2002). The evolution of embryo size in angiosperms and other seed plants: implications for the evolution of seed dormancy. *Evolution*, **56**, 2112–25.

Forget, P.-M. (2004). Post-dispersal predation and scatter-hoarding of *Dipteryx panamensis* seeds by rodents in Panama. *Oecologia*, **94**, 255–61.

Forman, R. T. T. (1995). *Land Mosaics: The Ecology of Landscapes and Regions*. Cambridge: Cambridge University Press.

Forterre, Y., Skotheim, J. M., Dumais, J., & Makhadevan, L. (2005). How the Venus flytrap snaps. *Nature*, **433**, 421–5.

Fountain, D. W. & Outred, H. A. (1990). Seed development in *Phaseolus vulgaris* L. cv Seminole. II. Precocious germination in late maturation. *Plant Physiology*, **93**, 1089–93.

Fountain, D. W. & Outred, H. A. (1991). Germination requirements of New Zealand native seeds: a review. *New Zealand Journal of Botany*, **29**, 311–16.

Fowells, H. A. (1965). *Silvics of Forest Trees of the United States*. Washington: U.S. Department of Agriculture.

Fowells, H. A. & Stark, N. B. (1965). Natural regeneration in relation to environment in the mixed conifer forest type of California. USDA Forest Service, Research Paper PSW-24, Berkeley: Pacific Southwest Forest and Range Experiment Station.

Fowler, N. L. (1986). Microsite requirement for germination and establishment of three grass species. *American Midland Naturalist*, **114**, 131–45.

Fowler, N. L. (1988). What is a safe site? Neighbor, litter, germination date, and patch effects. *Ecology*, **69**, 947–61.

Fox, J. F. (1979). Intermediate disturbance hypothesis. *Science*, **204**, 1344–5.

Francis, R. & Read, D. J. (1994). The contributions of mycorrhizal fungi to the determination of plant community structure. *Plant and Soil*, **159**, 11–25.

Francis, R. & Read, D. J. (1995). Mutualism and antagonism in the mycorrhizal symbiosis, with special reference to impacts on plant community structure. *Canadian Journal of Botany*, **73**, S1301–9.

Franco, A. C. & Nobel, P. S. (1988). Interactions between seedlings of *Agave deserti* and the nurse plant *Hilaria rigida*. *Ecology*, **69**, 1731–40.

Franco, A. C. & Nobel, P. S. (1989). Effect of nurse plants on the microhabitat and growth of cacti. *Journal of Ecology*, **77**, 870–86.

Franco, A. C. & Nobel, P. S. (1990). Influences of root distribution and growth on predicted water uptake and interspecific competition. *Oecologia*, **82**, 151–7.

Franco, M. & Silvertown, J. (2004). Comparative demography of plants based upon elasticities of vital rates. *Ecology*, **85**, 531–8.

Frank, B. (2005). On the nutritional dependence of certain trees on root symbiosis with belowground fungi (an English translation of A. B. Frank's classic paper of 1885). *Mycorrhiza*, **15**, 267–75.

Freckelton, R. P. & Lewis, O. T. (2006). Pathogens, density dependence and the coexistence of tropical trees. *Proceedings of the Royal Society, B*, **273**, 2909–16.

Fredericksen, T. S. & Pariona, W. (2002). Effect of skid disturbance on commercial tree regeneration in logging gaps in a Bolivian tropical forest. *Forest Ecology and Management*, **171**, 223–30.

Frey, W., Hofmann, M., & Hilger, H. H. (1996). The sporophyte–gametophyte junction in *Hymenophyton* and *Symphyogyna* (Metzgeriidae, Hepaticae): structure and phylogenetic implications. *Flora*, **191**, 245–52.

Frey, W., Hofmann, M., & Hilger, H. H. (2001). The gametophyte–sporophyte junction: unequivocal hints for two evolutionary lines of archegoniate land plants. *Flora*, **196**, 431–45.

Friedman, J., Stein, Z., & Rushkin, E. (1981). Drought tolerance of germinating seeds and young seedlings of *Anastatica hierochuntica* L. *Oecologia*, **51**, 400–3.

Friedman, W. E. (2006). Embryological evidence for developmental lability during early angiosperm evolution. *Nature*, **441**, 337–40.

Friml, J., Vieten, A., Sauer, M., *et al.* (2003). Efflux-dependent auxin gradients establish the apical-basal axis of *Arabidopsis*. *Nature*, **426**, 147–53.

Frumin, S. & Friis, E.-M. (1999). Magnoliid reproductive organs from the Cenomanian-Turonian of north-western Kazakhstan: Magnoliaceae and Illiciaceae. *Plant, Systematics and Evolution*, **216**, 265–88.

Fujiyoshi, M., Kagawa, A., Nakatsubo, T., & Masuzawa, T. (2006). Effects of arbuscular mycorrhizal fungi and soil developmental stages on herbaceous plants growing in the early stage of primary succession on Mount Fuji. *Ecological Research*, **21**, 278–84.

Fukuda, H. (2004). Signals that control plant vascular cell differentiation. *Nature Reviews in Molecular Cell Biology*, **5**, 379–91.

Funes, G., Basconcelo, S., Diaz, S., & Cabido, M. (1999). Seed size and shape are good predictors of seed persistence in soil in temperate mountain grasslands of Argentina. *Seed Science Research*, **9**, 341–5.

Gagne, J. M. & Houlé, G. (2001). Facilitation of *Leymus mollis* by *Honckenya peploides* on coastal dunes in subarctic Quebec, Canada. *Canadian Journal of Botany*, **79**, 1327–31.

Galatowitsch, S., Budelsky, R., & Yetka, L. (1999). Revegetation strategies for northern temperate glacial marshes and meadows. In *An International Perspective on Wetland Rehabilitation*, ed. W. Streever. Dordrecht: Kluwer Academic Publishers, pp. 225–41.

Galatowitsch, S. & van der Valk, A. (1996). The vegetation of restored and natural prairie wetlands. *Ecological Applications*, **6**, 102–12.

Galetti, M., Donatti, C. I., Pires, A. C., Guimaraes, P. R., & Jordano, P. (2006). Seed survival and dispersal of an endemic Atlantic forest palm: the combined effects of defaunation and forest fragmentation. *Botanical Journal Linnean Society*, **151**, 141–9.

Gallery, R., Dalling, J. W., & Arnold, A. E. (2007). Diversity, host affinity and distribution of seed-infecting fungi: a case-study with neotropical *Cecropia*. *Ecology*, **88**, 582–8.

Garbary, D. J., Renzaglia, K. S., & Duckett, J. G. (1993). The phylogeny of land plants: a cladistic analysis based on male gametogenesis. *Plant Systematics and Evolution*, **188**, 237–69.

Garcia, D. & Obeso, J. R. (2003). Facilitation by herbivore-mediated nurse plants in a threatened tree, *Taxus baccata*: local effects and landscape level consistency. *Ecography*, **26**, 739–50.

Garcia, D., Obeso, J. R., & Martinez, I. (2005). Spatial concordance between seed rain and seedling establishment in bird-dispersed trees: does scale matter? *Journal of Ecology*, **93**, 693–704.

Garcia, D., Zamora, R., Gomez, J. M., Jordano, P., & Hodar, J. A. (2000). Geographical variation in seed production, predation and abortion in *Juniperus communis* throughout its range in Europe. *Journal of Ecology*, **88**, 436–46.

Garcia-Castaño, J. L., Kollmann, J., & Jordano, P. (2006). Spatial variation of post-dispersal seed removal by rodents in highland microhabitats of Spain and Switzerland. *Seed Science Research*, **16**, 213–22.

García-Fayos, P. & Verdú, M. (1998). Soil seed bank, factors controlling germination and establishment of a Mediterranean shrub: *Pistacia lentiscus* L. *Acta Oecologica*, **19**, 357–66.

Garcia-Guzmán, G., & Benítez-Malvido, J. (2003). Effect of litter on the incidence of leaf-fungal pathogens and herbivory in seedlings of the tropical tree *Nectandra ambigens*. *Journal of Tropical Ecology*, **19**, 171–7.

García-Guzmán, G. & Dirzo, R. (2001). Patterns of leaf-pathogen infection in the understorey of a Mexican rain forest: incidence, spatio-temporal variation, and mechanism of infection. *American Journal of Botany*, **88**, 634–45.

García-Martinez, J. & Gil, J. (2002). Light regulation of gibberellin biosynthesis and mode of action. *Journal of Plant Growth Regulation*, **20**, 354–68.

Gardes, M. & Bruns, T. D. (1996). Community structure of ectomycorrhizal fungi in a *Pinus muricata* forest: above- and below-ground views. *Canadian Journal of Botany*, **74**, 1572–83.

Garwood, N. C. (1996). Functional morphology of tropical tree seedlings. In *The Ecology of Tropical Forest Tree Seedlings*, ed. M. D. Swaine. Paris: UNESCO & The Parthenon Publishing Group, pp. 59–129.

Gascon, C., Williamson, G. B., & da Fonseca, G. A. B. (2000). Receding forest edges and vanishing reserves. *Science*, **288**, 1356–58.

Gay, G., Normand, L., Marmeisse, R., Sotta, B., & Debaud, J. C. (1994). Auxin overproducer mutants of *Hebeioma cylindrosporum* Romagnesi have increased mycorrhizal activity. *New Phytologist*, **128**, 645–57.

Gazzarrini, S. & McCourt, P. (2001). Genetic interactions between ABA, ethylene and sugar signaling pathways. *Current Opinion in Plant Biology*, **4**, 387–91.

Geddes, P. (1893). *Chapters in Modern Botany*. London: John Murray.

Gedroc, J. J., McConnaughay, K. D. M., & Coleman, J. S. (1996). Plasticity in root shoot partitioning: Optimal, ontogenetic, or both? *Functional Ecology*, **10**, 44–50.

Gehring, C., Denich, M., Kanishiro M., & Vlek, P. L. G. (1999). Response of secondary vegetation in eastern Amazonia to relaxed nutrient availability constraints. *Biogeochemistry*, **45**, 223–41.

Gehring, C. A. & Whitham, T. G. (1991). Herbivore-driven mycorrhizal mutualism in insect-susceptible pinyon pine. *Nature*, **35**, 556–7.

Gehring, C. A., Wolf, J. E., & Theimer, J. C. (2002). Terrestrial vertebrates promote arbuscular mycorrhizal fungal diversity and inoculum potential in a rain forest soil. *Ecology Letters*, **5**, 540–8.

Geneve, R. L. (2005). Vigor testing in flower seeds. In *Flower Seeds: Biology and Technology*, ed. M. B. McDonald & F. Y. Kwong. Wallingford: CAB International, pp. 311–32.

Gentry, A. H. (1983). *Macfadyena unguis-cati* (Uña de Gato, Cat-claw Bignone). In *Costa Rican Natural History*, ed. D. H. Janzen. Chicago: University of Chicago Press, pp. 272–3.

George, L. O. & Bazzaz, F. A. (1999). The fern understory as an ecological filter: emergence and establishment of canopy-tree seedlings. *Ecology*, **80**, 833–45.

Geritz, S. A. H. (1995). Evolutionarily stable seed polymorphism and small-scale spatial variation in seedling density. *American Naturalist*, **146**, 685–707.

Geritz, S. A. H., van der Meijden, E., & Metz, J. A. J. (1999). Evolutionary dynamics of seed size and seedling competitive ability. *Theoretical Population Biology*, **55**, 324–43.

Germino, M. J., Smith, W. K., & Resor, A. C. (2002). Conifer seedling distribution and survival in an alpine-treeline ecotone. *Plant Ecology*, **162**, 157–68.

Ghazoul, J. (2005). Implications of plant spatial distribution for pollination and seed production. In *Biotic Interactions in the Tropics*, ed. D. F. R. P. Burslem, M. A. Pinard, & S. E. Hartley. Cambridge: Cambridge University Press, pp. 241–66.

Ghazoul, J., Liston, K. A., & Boyle, T. J. B. (1998). Disturbance-induced density-dependent seed set in *Shorea siamensis* (Dipterocarpaceae), a tropical forest tree. *Journal of Ecology*, **86**, 462–73.

Gibson, G. (2002). Microarrays in ecology and evolution: a preview. *Molecular Ecology*, **11**, 17–24.

Gibson, S. I. (2005). Control of plant development and gene expression by sugar signaling. *Current Opinion in Plant Biology*, **8**, 93–102.

Gifford, E. M. (1983). Concept of apical cells in bryophytes and pteridophytes. *Annual Review of Plant Physiology*, **34**, 419–40.

Gifford, E. M. & Foster, A. S. (1988). *Morphology and Evolution of Vascular Plants*, 3rd edn. New York: Freeman and Company.

Gilbert, B., Wright, S. J., Muller-Landau, H. C., Kitajima, K., & Hernandéz, A. (2006). Life-history trade-offs in tropical trees and lianas. *Ecology*, **87**, 1281–8.

Gilbert, G. S., Foster, R. B., & Hubbell, S. P. (1994). Density and distance-to-adult effects of a canker disease of trees in a moist tropical forest. *Oecologia*, **98**, 100–8.

Gill, D. S., & Marks, P. L. (1991). Tree and shrub seedling colonization of old fields in central New York. *Ecological Monographs*, **61**, 183–205.

Giovannetti, M., Azzonlini, D., & Citernesi, A. S. (1999). Anastomosis formation and nuclear and protoplasmic exchange in arbuscular mycorrhizal fungi. *Applied and Environmental Microbiology*, **65**, 5571–5.

Girlanda, M., Selosse, M.-A., Cafasso, *et al.* (2006). Inefficient photosynthesis in the Mediterranean orchid *Limodorum abortivum* is mirrored by specific association to ectomycorrhizal Russulaceae. *Molecular Ecology*, **15**, 491–504.

Givnish, T. J. (1988). Adaptation to sun and shade: a whole-plant perspective. *Australian Journal of Plant Physiology*, **15**, 63–92.

Givnish, T. J. (1995). Plant stems: biomechanical adaptation for energy capture and influence on species distributions. In *Plant Stems: Physiology and Functional Morphology*, ed. B. L. Gartner. San Diego: Academic Press, pp. 3–41.

Gleadow, R. M., & Ashton, D. H. (1981). Invasion by *Pittosporum undulatum* of the forests of Central Victoria. 1. Invasion patterns and plant morphology. *Australian Journal of Botany*, **29**, 705–20.

Glizenstein, J. S., Platt, W. J., & Streng, D. R. (1995). Effects of fire regime and habitat on tree dynamics in north Florida longleaf pine savannas. *Ecological Monographs*, **65**, 441–76.

Goebel, K. (1928). *Organographie der Pflanzen*. Erster Teil, 3rd edn. Jena: G. Fischer.

Goff, F. G., & Zedler, P. H. (1968). Structural gradient analysis of upland forests in the western Great Lakes area. *Ecological Monographs*, **38**, 65–86.

Goffinet, B. (2000). Origin and phylogenetic relationships of bryophytes. In *Bryophyte Biology*, ed. J. Shaw, & B. Goffinet. Cambridge: Cambridge University Press, pp. 124–49.

Gollotte, A., Tuinen, D., & Atkinson, D. (2004). Diversity of arbuscular mycorrhizal fungi colonising roots of the grass species *Agrostis capillaris*. *Mycorrhiza*, **14**, 111–17.

Golodets, C., & Boeken, B. (2006). Moderate sheep grazing in semiarid shrubland alters small-scale soil surface structure and patch properties. *Catena*, **65**, 285–91.

Gómez, J. M. (2004). Bigger is not always better: conflicting selective pressures on seed size in *Quercus ilex*. *Evolution*, **58**, 71–80.

Gómez-Aparicio, L., Gómez, J. M., & Zamora, R. (2005). Microhabitats shift rank in suitability for seedling establishment depending on habitat type and climate. *Journal of Ecology*, **93**, 1194–202.

Gómez-Aparicio, L., Gómez, J. M., & Zamora R. (2007). Spatiotemporal patterns of seed dispersal in a wind-dispersed Mediterranean tree (*Acer opalus* subsp. *granatense*): implications for regeneration. *Ecography*, **30**, 13–22.

Gómez-Aparicio, L., Gómez, J. M., Zamora, R., & Boettinger, J. L. (2005). Canopy vs. soil effects of shrubs facilitating tree seedlings in Mediterranean montane ecosystems. *Journal of Vegetation Science*, **16**, 191–8.

Gómez-Aparicio, L., Zamora, R., Gómez, J. M., *et al.* (2004). Applying plant facilitation to forest restoration in Mediterranean ecosystems: a meta-analysis of the use of shrubs as nurse plants. *Ecological Applications*, **14**, 1128–38.

Good, R. E., Good, N. F., & Andresen, J. W. (1979). The Pine Barrens plains. In *Pine Barrens: Ecosystem and Landscape*, ed. R. T. T. Forman. New York: Academic Press, pp. 283–95.

Gosz, J. R., Likens, G. E., & Bormann, F. H. (1973). Nutrient release from decomposing leaf and branch litter in the Hubbard Brook forest, New Hampshire. *Ecological Monographs*, **43**, 173–91.

Gotelli, N. J., & Entsminger, G. L. (2006). *EcoSim: Null Models Software for Ecology*, version 7. VT: Acquired Intelligence Inc. and Keasey-Bear. Jericho (http://garyentsminger.com/ecosim.htm, accessed January 2007).

Gould, A. M. A., & Gorchov, D. L. (2000). Effects of the exotic invasive shrub *Lonicera maackii* on the survival and fecundity of three species of native annuals. *American Midland Naturalist*, **144**, 36–50.

Gould, S. J., & Lewontin, R. C. (1979). Spandrels of San Marco and the Panglossian paradigm – a critique of the adaptationist program. *Proceedings of the Royal Society, Series B*, **205**, 581–98.

Goulet, F. (1995). Frost heaving of forest tree seedlings: a review. *New Forests*, **9**, 67–94.

Graham, L. E., & Wilcox, L. W. (2000). *Algae*. Upper Saddle River: Prentice Hall.

Granados J., & Körner, C. (2002). In deep shade, elevated CO_2 increases the vigor of tropical climbing plants. *Global Change Biology*, **8**, 1109–17.

Grant, C. D., Bell, D. T., Koch, J. M., & Loneragan, W. A. (1996). Implications of seedling emergence to site restoration following bauxite mining in Western Australia. *Restoration Ecology*, **4**, 146–54.

Grant, M., & Lamb, C. (2006). Systemic immunity. *Current Opinion in Plant Biology*, **9**, 414–20.

Graves, J. D., Watkins, N. K., Fitter, A. H., Robinson, D., & Scrimgeour, C. (1997). Intraspecific transfer of carbon between plants linked by a common mycorrhizal network. *Plant and Soil*, **192**, 153–9.

Gray, A. N., & Spies, T. A. (1996). Gap size, within-gap position and canopy structure effects on conifer seedling establishment. *Journal of Ecology*, **84**, 635–45.

Greacen, E. L., & Sands, R. (1980). Compaction of forest soils – a review. *Australian Journal of Soil Research*, **18**, 163–89.

Green, D. S. (2005). Adaptive strategies in seedlings of three co-occurring, ecologically distinct northern coniferous tree species across an elevational gradient. *Canadian Journal of Forest Research*, **35**, 910–17.

Green, P. T., & Juniper, P. A. (2004a). Seed mass, seedling herbivory and the reserve effect in tropical rainforest seedlings. *Functional Ecology*, **18**, 539–47.

Green, P. T., & Juniper, P. A. (2004b). Seed-seedling allometry in tropical rain forest trees: seed mass-related patterns of resource allocation and the 'reserve effect.' *Journal of Ecology*, **92**, 397–408.

Greene, D. F., & Johnson, E. A. (1994). Estimating the mean annual seed production of trees. *Ecology*, **75**, 642–7.

Greenlee, J. T., & Callaway, R. M. (1996). Abiotic stress and the relative importance of interference and facilitation in montane bunchgrass communities in western Montana. *American Naturalist*, **148**, 386–96.

Grice, A. C., & Barchia, I. (1992). Does grazing reduce survival of indigenous perennial grasses of the semi-arid woodlands of western New South Wales? *Australian Journal of Ecology*, **17**, 195–205.

Grime, J. P. (1973). Competitive exclusion in herbaceous vegetation. *Nature*, **242**, 344–7.

Grime, J. P. (1977). Evidence for the existence of three primary strategies in plants and its relevance to ecological and evolutionary theory. *American Naturalist*, **111**, 1169–94.

Grime, J. P. (1979). *Plant Strategies and Vegetation Processes*, New York: John Wiley & Sons.

Grime, J. P. (1989). Seed banks in ecological perspective. In *Ecology of Soil Seed Banks*, ed. M. Leck, V. T. Parker, & R. L. Simpson. San Diego: Academic Press, pp. xv–xxii.

Grime, J. P. (2001). *Plant Strategies, Vegetation Processes, and Ecosystem Properties.* Chichester: Wiley.

Grime, J. P. & Curtis, A. V. (1976). The interaction of drought and mineral nutrient stress in calcareous grassland. *Journal of Ecology*, **64**, 975–88.

Grime, J. P., Hodgson, J. G., & Hunt, R. (1988). *Comparative Plant Ecology: a Functional Approach to Common British Species.* London: Unwin Hyman.

Grime, J. P. & Jeffrey, D. W. (1965). Seedling establishment in vertical gradients of sunlight. *Journal of Ecology*, **53**, 621–42.

Grime, J. P., Mackey, J. M. L., Hillier, S. H., & Read, D. J. (1987). Floristic diversity in a model system using experimental microcosms. *Nature*, **328**, 420–2.

Grime, J. P., Mason, G., Curtis, A. V., *et al.* (1981). A comparative study of germination characteristics in a local flora. *Journal of Ecology*, **69**, 1017–59.

Grisez, T. J. (1974). *Prunus.* In *Seeds of Woody Plants in the United States*, tech. coord. C. S. Scopmeyer. Washington: Forest Service, USDA, pp. 658–73.

Groover, A. T. (2005). What genes make a tree a tree? *Trends in Plant Science*, **10**, 210–14.

Gross, K. L. (1984). Effects of seed size and growth form on seedling establishment of six monocarpic perennials. *Journal of Ecology*, **72**, 369–87.

Grotkopp, E., Rejmánek, M., & Rost, T. L. (2002). Toward a causal explanation of plant invasiveness: seedling growth and life-history strategies of 29 pine (*Pinus*) species. *American Naturalist*, **159**, 396–419.

Grubb, P. J. (1977). The maintenance of species–richness in plant communities: the importance of the regeneration niche. *Biological Reviews*, **52**, 107–45.

Grubb, P. J. (1980). Review of Grime (1979) *Plant Strategies and Vegetation Processes.* *New Phytologist*, **86**, 123–4.

Grubb, P. J. (1985). Plant populations and vegetation in relation to habitat, disturbance and competition: problems of generalisation. In *The Population Structure of Vegetation*, ed. J. White. Dordrecht: Junk Publisher, pp. 595–621.

Grubb, P. J. (1987). Global trends in species-richness in terrestrial vegetation: a view from the northern hemisphere. In *Organization of Communities*, ed. J. H. R. Gee & P. S. Giller. Oxford: Blackwell Scientific Publications, pp. 99–118.

Grubb, P. J. (1998). A reassessment of the strategies of plants which cope with shortages of resources. *Perspectives in Plant Ecology, Evolution and Systematics*, **1**, 3–31.

Grubb, P. J. & Coomes, D. A. (1997). Seed mass and nutrient content in nutrient-starved tropical rainforest in Venezuela. *Seed Science Research*, **7**, 269–80.

Grubb, P. J., Coomes, D. A., & Metcalfe, D. J. (2005). Comment on "A brief history of seed size." *Science*, **310**, 783a.

Grubb, P. J., Lee, W. G., Kollmann, J., & Wilson, J. B. (1996). Interaction of irradiance and soil nutrient supply on growth of seedlings of ten European tall-shrub species and *Fagus sylvatica. Journal of Ecology*, **84**, 827–40.

Grubb, P. J. & Metcalfe, D. J. (1996). Adaptation and inertia in the Australia tropical lowland rainforest flora: contradictory trends in intergeneric and intrageneric comparisons of seed size in relation to light demand. *Functional Ecology*, **10**, 512–20.

Guariguata, M. R., Arias-LeClaire, H., & Jones, G. (2002). Tree seed fate in a logged and fragmented forest landscape, northeastern Costa Rica. *Biotropica*, **34**, 405–15.

Guariguata, M. R. & Dupuy, J. M. (1997). Forest regeneration in abandoned logging roads in lowland Costa Rica. *Biotropica*, **29**, 15–28.

Guevara, S., Meave, J., Moreno-Casasola, P., & Laborde, J. (1992). Floristic composition and structure of vegetation under isolated trees in neotropical pastures. *Journal of Vegetation Science*, **3**, 655–64.

Guidot, A., Debaud, J.-C., Effosse, A., & Marmeisse, R. (2003). Below-ground distribution and persistence of an ectomycorrhizal fungus. *New Phytologist*, **161**, 539–47.

Gulmon, S. L. (1992). Patterns of seed germination in Californian serpentine grassland species. *Oecologia*, **89**, 27–31.

Guo, Q., Brown, J. H., Valone, T. J., & Kachman, S. D. (2000). Constraints of seed size on plant distribution and abundance. *Ecology*, **81**, 2149–55.

Guo, Q., Thompson, D. B., Valone, T. J., & Brown, J. H. (1995). The effects of vertebrate granivores and foliovores on plant community structure in the Chihuahuan Desert. *Oikos*, **73**, 251–9.

Guppy, H. B. (1906). *Observations of a Naturalist in the Pacific between 1896 and 1899*, Vol. II, *Plant Dispersal*. London: Macmillan and Co., Ltd.

Guppy, H. B. (1912). *Studies in Seeds and Fruits*. London: Williams & Norgate.

Gustafson, D. J., Gibson, D. J., & Nickrent, D. L. (2005). Using local seeds in prairie restoration. *Native Plants*, **6**, 25–8.

Gutterman, Y. (1972). Delayed seed dispersal and rapid germination as survival mechanisms of the desert plant *Blepharis persica* (Burm.) Kuntze. *Oecologia*, **10**, 145–9.

Gutterman, Y. (1982). Observations on the feeding habit of the Indian crested porcupine (*Hystrix indica*) and the distribution of some hemicryptophytes and geophytes in the Negev Desert Highlands. *Journal of Arid Environments*, **5**, 261–8.

Gutterman, Y. (1990). Do germination mechanisms differ in plants originating in deserts receiving winter or summer rain? *Israel Journal of Botany – Basic and Applied Plant Sciences*, **39**, 355–72.

Gutterman, Y. (1993). *Seed Germination in Desert Plants*. Berlin: Springer-Verlag.

Gutterman, Y. (1994). Strategies of seed dispersal and germination in plants inhabiting deserts. *Botanical Review*, **60**, 373–425.

Gutterman, Y. (1997). Ibex diggings in the Negev Desert highlands of Israel as microhabitats for annual plants: soil salinity, location and digging depth affecting variety and density of plant species. *Journal of Arid Environments*, **37**, 665–81.

Gutterman, Y. (2000). Environmental factors and survival strategies of annual plant species in the Negev Desert, Israel. *Plant Species Biology*, **15**, 113–25.

Gutterman, Y. (2001). *Regeneration of Plants in Arid Ecosystems Resulting from Patch Disturbance*. Dordrecht: Kluwer.

Gutterman, Y. (2002). *Survival Strategies of Annual Desert Plants*. Heidelberg: Springer.

Gutterman, Y. & Ginott, S. (1994). Long-term protected "seed bank" in dry inflorescences of *Asteriscus pygmaeus*: achene dispersal mechanism and germination. *Journal of Arid Environments*, **26**, 149–63.

Gutterman, Y., Golan, T., & Garsani, M. (1990). Porcupine diggings as a unique ecological system in a desert environment. *Oecologia*, **85**, 122–7.

Gutterman, Y. & Shem-Tov, S. (1996). Structure and function of the mucilaginous seed coats of *Plantago coronopus* inhabiting the Negev Desert of Israel. *Israel Journal of Plant Sciences*, **44**, 125–34.

Gutterman, Y. & Shem-Tov, S. (1997a). Mucilaginous seed coat structure of *Carrichtera annua* and *Anastatica hierochuntica* from the Negev Desert highlands of Israel, and its adhesion to the soil crust. *Journal of Arid Environments*, **35**, 695–705.

Gutterman, Y. & Shem-Tov, S. (1997b). The efficiency of the strategy of mucilaginous seeds of some common annuals of the Negev adhering to the soil crust to delay collection by ants. *Israel Journal of Plant Sciences*, **45**, 317–27.

Gworek, J. R., Vander Wall, S. B., & Brussard, P. F. (2007). Changes in biotic interactions and climate determine recruitment of Jeffrey pine along an elevation gradient. *Forest Ecology and Management*, **239**, 57–68.

Haag, R. W. (1983). Emergence of seedlings of aquatic macrophytes from lake sediments. *Canadian Journal of Botany*, **61**, 148–56.

Haas, C. A. (1995). Dispersal and use of corridors by birds in wooded patches on an agricultural landscape. *Conservation Biology*, **9**, 845–54.

Haase, P., Pugnaire, F. I., Clark, S. C., & Incoll, L. D. (1999). Environmental control of canopy dynamics and photosynthetic rate in the evergreen tussock grass *Stipa tenacissima*. *Plant Ecology*, **145**, 327–39.

Haccius, B. (1952). Die Embryoentwicklung bei *Ottelia alismoides* und das Problem des terminalen Monokotylen-Keimblatts. *Planta*, **40**, 443–60.

Haccius, B. (1954). Embryologische und histogenetische studien an "monokotylen dikotylen." I. *Claytonia virginica*. *Osterreichische Botanische Zeitschrift*, **101**, 285–303.

Haccius, B. & Lakshmanan, K. K. (1966). Vergleichende Untersuchung der Entwicklung von Kotyledon und Spross-scheitel bei *Pistia stratiotes* und *Lemna gibba*, ein Beitrag zum Problem der sogenannten terminalen Blattorgane. *Beitrage zur Biologie der Pflanzen*, **42**, 425–43.

Hadley, G. (1970). Non-specificity of symbiotic infection in orchid mycorrhiza. *New Phytologist*, **69**, 1015–23.

Hadley, G. & Williamson, B. (1971). Analysis of the post-infection growth stimulus in orchid mycorrhiza. *New Phytologist*, **70**, 445–55.

Hager, H. A. (2004). Competitive effect versus competitive response of invasive and native wetland plant species. *Oecologia*, **139**, 140–9.

Haines, R. W. & Lye, K. A. (1975). Seedlings of Nymphaeaceae. *Botanical Journal of the Linnean Society*, **70**, 255–65.

Halfhill, M. D., Sutherland, J. P., Hong S. M., *et al.* (2005). Growth, productivity, and competitiveness of introgressed weedy *Brassica rapa* hybrids selected for the presence of Bt cry1Ac and gfp transgenes. *Molecular Ecology*, **14**, 3177–89.

Hallé, F. & Oldeman, R. A. A. (1975). *Essay on the Architecture and Dynamics of Growth of Tropical Trees*. Kuala Lumpur: Penerbit University.

Hallé, F., Oldeman, R. A. A., & Tomlinson, P. B. (1978). *Tropical Trees and Forests*. Berlin: Springer-Verlag.

Halter, M. R., Chanway, C. P., & Harper, G. J. (1993). Growth reduction and root deformation of containerized lodgepole pine saplings 11 years after planting. *Forest Ecology and Management*, **56**, 131–46.

Hamerlynck, E. P., Huxman, T. E., Loik, M. E., & Smith, S. D. (2000). Effects of extreme high temperature, drought and elevated CO_2 on photosynthesis of the Mojave Desert evergreen shrub, *Larrea tridentata*. *Plant Ecology*, **148**, 183–93.

Hammond, D. S. (1995). Post-dispersal seed and seedling mortality of tropical dry forest trees after shifting agriculture, Chiapas, Mexico. *Journal of Tropical Ecology*, **11**, 295–313.

Hammond, D. S. & Brown, V. K. (1998). Disturbance, phenology and life-history characteristics: factors influencing frequency-dependent attach on tropical seeds and seedlings. In *Dynamics of Tropical Communities*, ed. D. M. Newbury, N. Brown, & H. H. T. Prins. Oxford: Blackwell Science, pp. 51–78.

Hanf, M. (1990). *Ackerunkrauter Europas mit ihren Keimlingen und Samen.* München: BLV.

Hanley, M. E. (1998). Seedling herbivory, community composition and plant life history traits. *Perspectives in Plant Ecology, Evolution and Systematics*, **1**, 191–205.

Hanley, M. E. (2004). Seedling herbivory and the influence of plant species richness in seedling neighbourhoods. *Plant Ecology*, **170**, 35–41.

Hanley, M. E. & Fenner, M. (1997). Seedling growth of four fire-following Mediterranean plant species deprived of single mineral nutrients. *Functional Ecology*, **11**, 398–405.

Hanley, M. E., Fenner, M., & Edwards, P. J. (1995). The effect of seedling age on the likelihood of herbivory by the slug *Deroceras reticulatum*. *Functional Ecology*, **9**, 754–9.

Hanley, M. E., Fenner, M., Whibley, H., & Darvill, B. (2004). Early plant growth: identifying the end point of the seedling phase. *New Phytologist*, **163**, 61–6.

Hanley, M. E., Hilhorst, H. W. M., & Karssen, C. M. (2000). Effect of chemical environment on seed germination. In *Seeds: The Ecology of Regeneration in Plant Communities*, 2nd edn., ed. M. Fenner. Wallingford: CAB International, pp. 293–310.

Hanley, M. E. & Lamont, B. B. (2001). Herbivory, serotiny and seedling defense in Western Australian Proteaceae. *Oecologia*, **126**, 409–17.

Hanley, M. E. & Lamont, B. B. (2002). Relationships between physical and chemical attributes of congeneric seedlings: how important is seedling defense? *Functional Ecology*, **16**, 216–22.

Hanley, M. E. & May, O. C. (2006). Cotyledon damage at the seedling stage affects growth and flowering potential in mature plants. *New Phytologist*, **169**, 243–50.

Hanslin, H. M. & Karlsson, P. S. (1996). Nitrogen uptake from prey and substrate as affected by prey capture level and plant reproductive status in four carnivorous plant species. *Oecologia*, **106**, 370–5.

Hanson, T., Brunsfeld, S., & Finegan, B. (2006). Variation in seedling density and seed predation indicators for the emergent tree *Dipteryx panamensis* in continuous and fragmented rain forest. *Biotropica*, **38**, 770–4.

Hara, M. (1985). Analysis of seedling banks of a climax beech forest: ecological importance of seedlings sprouts. *Vegetatio*, **71**, 67–74.

Hardesty, B. D., Hubbell, S. P., & Bermingham, E. (2006). Genetic evidence of frequent long-distance recruitment in a vertebrate-dispersed tree. *Ecology Letters*, **9**, 516–25.

Harley, J. L. & Harley, E. L. (1987). A checklist of mycorrhiza in the British flora. *New Phytologist*, **105**, 1–102.

Harley, J. L. & Smith, S. E. (1983). *Mycorrhizal Symbiosis.* London: Academic Press.

Harms, K. E. & Dalling, J. W. (1997). Damage and herbivory tolerance through resprouting as an advantage of large seed size in tropical trees and lianas. *Journal of Tropical Ecology*, **13**, 617–21.

Harms, K. E., Dalling, J. W., & Aizprua, R. (1997). Regeneration from cotyledons in *Gustavia superba* (Lecythidaceae). *Biotropica*, **29**, 234–7.

Harms, K. E., Wright, S. J., Calderon, O., Hernandez, A., & Herre, E. A. (2000). Pervasive density-dependent recruitment enhances seedling diversity in a tropical forest. *Nature*, **404**, 493–5.

Harper, J. L. (1977). *Population Biology of Plants.* London: Academic Press.

Harper, J. L., Lovell, P. H., & Moore, K. G. (1970). The shapes and sizes of seeds. *Annual Review of Ecology and Systematics*, **1**, 327–56.

Harper, J. L. & Ogden, J. (1970). The reproductive strategy of higher plants. I. The concept of strategy with special reference to *Senecio vulgaris* L. *Journal of Ecology*, **58**, 681–98.

Harper, J. L. & White, J. (1974). The demography of plants. *Annual Review of Ecology and Systematics*, **5**, 419–63.

Harper, J. L., Williams, J. T., & Sagar, G. R. (1965). The behaviour of seeds in soil. 1. The heterogeneity of soil surfaces and its role in determining the establishment of plants from seed. *Journal of Ecology*, **53**, 273–86.

Harris, D. & Davy, A. J. (1987). Seedling growth in *Elymus farctus* after episodes of burial with sand. *Annals of Botany*, **60**, 587–93.

Harrison, C. R. & Arditti, J. (1978). Physiological changes during the germination of *Cattleya aurantiaca* (Orchidaceae). *Botanical Gazette*, **139**, 180–9.

Hartman, K. M. & McCarthy, B. C. (2004). Restoration of a forest understory after the removal of an invasive shrub, Amur honeysuckle (*Lonicera maackii*). *Restoration Ecology*, **12**, 154–65.

Hartnett, D. C., Hetrick, B. A. D., Wilson, G. W. T., & Gibson, D. J. (1993). Mycorrhizal influence on intraspecific and interspecific neighbor interactions among co-occurring prairie grasses. *Journal of Ecology*, **81**, 787–95.

Hartshorn, G. S. (1980). Neotropical forest dynamics. *Biotropica*, **12**, S23–30.

Hartshorn, G. S. (1983). *Pentaclethra macroloba* (Gavilán). In *Costa Rican Natural History*, ed. D. H. Janzen. Chicago. University of Chicago Press, pp. 301–3.

Harvis, G. & Hadley, G. (1967). The relation between host and endophyte in orchid mycorrhiza. *New Phytologist*, **66**, 205–15.

Hastwell, G. T. & Facelli, J. M. (2003). Facilitation in a pulsed environment: Differing effects of facilitation on growth and survivorship of the chenopod shrub *Enchylaena tomentosa*. *Journal of Ecology*, **91**, 941–50.

Hatcher, P. E., Moore, J., Taylor, J. E., Tinney, G. W., & Paul, N. D. (2004). Phytohormones and plant–herbivore–pathogen interactions: integrating the molecular with the ecological. *Ecology*, **85**, 59–69.

Hättenschwiler, S. (2001). Tree seedling growth in natural deep shade: functional traits related to interspecific variation in response to elevated CO_2. *Oecologia*, **129**, 31–42.

Hättenschwiler, S. & Körner, C. (2003). Does elevated CO_2 facilitate naturalization of the non-indigenous *Prunus laurocerasus* in Swiss temperate forests? *Functional Ecology*, **17**, 778–85.

He, X., Critchley, C., Ng, H., & Bledsoe, C. (2005). Nodulated N^2-fixing *Casuarina cunninghamiana* is the sink for net N transfer from non-N^2-fixing *Ecualyptus maculata* via an ectomycorrhizal fungus *Pisolithus* sp. using $^{15}NH_4^+$ or $^{15}NO_3^-$ supplied as ammonium nitrate. *New Phytologist*, **167**, 897–912.

Heggie, L. & Halliday, K. J. (2005). The highs and lows of plant life: temperature and light interactions in development. *International Journal of Developmental Biology*, **49**, 675–87.

Heinemann, K. & Kitzberger, T. (2006). Effects of position, understorey vegetation and coarse woody debris on tree regeneration in two environmentally contrasting forests of north-western Patagonia: a manipulative approach. *Journal of Biogeography*, **33**, 1357–67.

Heisey, R. M. (1990). Allelopathic and herbicidal effects of extracts from Tree of Heaven (*Ailanthus altissima*). *American Journal of Botany*, **77**, 662–70.

Helgason, T., Daniell, T. J., Husband, R., Fitter, A. H., & Young, J. P. Y. (1998). Ploughing up the wood-wide web? *Nature*, **394**, 431.

Helm, D. J., Allen, E. B., & Trappe, J. M. (1996). Mycorrhizal chronosequence near Exit Glacier, Alaska. *Canadian Journal of Botany*, **74**, 1496–506.

Henderson, F. M. (2006). Morphology and anatomy of palm seedlings. *Botanical Review*, **72**, 273–329.

Hendrix, S. D. (1991). Are seedlings from small seeds always inferior to seedlings from large seeds? Effects of seed biomass on seedling growth in *Pastinaca sativa* L. *New Phytologist*, **119**, 299–306.

Hendrix, S. D. & Sun, I. F. (1989). Inter- and intraspecific variation in seed mass in seven species of umbellifers. *New Phylologist*, **112**, 445–52.

Henery, M. L. & Westoby, M. (2001). Seed mass and seed nutrient content as predictors of seed output variation between species. *Oikos*, **92**, 479–90.

Hensen, I. (1999). Reproductive patterns in five semi-desert perennials of south-eastern Spain. *Botanische Jahrbuecher fuer Systematik, Pflanzengeschichte und Pflanzengeographie*, **121**, 491–505.

Hensen, I. & Oberprieler, C. (2005). Effects of population size on genetic diversity and seed production in the rare *Dictamnus albus* (Rutaceae) in central Germany. *Conservation Genetics*, **6**, 63–73.

Hernández, J. C. C., Wolf, J. H. D., Garcia-Franco, J. G., & González-Espinosa, M. (1999). The influence of humidity, nutrients and light on the establishment of the epiphytic bromeliad *Tillandsia guatemalensis* in the highlands of Chiapas, Mexico. *Revista de Biologia Tropical*, **47**, 763–73.

Herrera, C. M. (2000). Flower-to-seedling consequences of different pollination regimes in an insect-pollinated shrub. *Ecology*, **81**, 15–29.

Herrera, C. M. & Jordano, P. (1981). *Prunus mahaleb* and birds, the high-efficiency seed dispersal system of a temperate fruiting tree. *Ecological Monographs*, **51**, 203–18.

Herrera, C. M., Jordano, P., López-Soria, L., & Amat, J. A. (1994). Recruitment of a mast-fruiting, bird-dispersed tree: bridging frugivore activity and seedling establishment. *Ecological Monographs*, **64**, 315–44.

Hetrick. B. A. D., Wilson, G. W. T., & Hartnett, D. C. (1989). Relationship between mycorrhizal dependence and competitive ability of two tallgrass prairie grasses. *Canadian Journal of Botany*, **67**, 2608–15.

Hetrick, B. A. D., Wilson, G. W. T., & Todd, T. C. (1992). Relationship of mycorrhizal symbiosis, rooting strategy and phenology among tall grass prairie forbs. *Canadian Journal of Botany*, **70**, 1521–8.

Hett, J. M. & Loucks, O. L. (1971). Sugar maple (*Acer saccharum* Marsh.) seedling mortality. *Journal of Ecology*, **59**, 507–20.

Hewitt, N. (1998). Seed size and shade-tolerance: a comparative analysis of North American temperate trees. *Oecologia*, **114**, 432–40.

Heydecker, W. (1956). Establishment of seedlings in the field. I. Influence of sowing depth on seedling emergence. *Journal of Horticultural Science*, **31**, 76–87.

Heywood, V. H., ed. (1978). *Flowering Plants of the World*. Oxford: Oxford University Press.

Hickey, L. J. & Doyle, J. A. (1977). Early Cretaceous fossil evidence for angiosperm evolution. *Botanical Review*, **43**, 1–104.

Hierro, J. L., Maron, J. L., & Callaway, R. M. (2005). A biogeographical approach to plant invasions: the importance of studying exotics in their introduced and native range. *Journal of Ecology*, **93**, 5–15.

Hietz, P. & Hietz, U. S. (1995). Intra- and interspecific relations within an epiphyte community in a Mexican humid montane forest. *Selbyana*, **16**, 135–40.

Hilger, H. H., Weigend, M., & Frey, W. (2002). The gametophyte–sporophyte junction in *Isoetes boliviensis* (Isoëtales, Lycopodiophyta). *Phyton*, **42**, 149–57.

Hilhorst, H. W. M., & Karssen, C. M. (2000). Effects of chemical environment on seed germination. In *Seeds: The Ecology of Regeneration in Plant Communities*, 2nd. edn., ed. M. Fenner, Wallingford, CAB International, pp. 293–310.

Hille Ris Lambers, J., Clark, J. S., & Beckage, B. (2002). Density dependent mortality and the latitudinal gradient in species diversity. *Nature*, **417**, 732–5.

Hladik, A. & Miquel, S. (1990). Seedling types and plant establishment in an African rain forest. In *Reproductive Ecology of Tropical Forest Plants*, ed. K. S. Bawa & M. Hadley. Carnforth: Parthenon, pp. 261–82.

Hobbie, E. A. (2006). Carbon allocation to ectomycorrhizal fungi correlates with belowground allocation in culture studies. *Ecology*, **87**, 563–9.

Hobbie, J. E. & Hobbie, E. A. (2006). ^{15}N in symbiotic fungi and plants estimates nitrogen and carbon flux rates in arctic tundra. *Ecology*, **87**, 816–22.

Hobbs, R. J. & Huenneke, L. F. (1992). Disturbance, diversity and invasion: implications for conservation. *Conservation Biology*, **6**, 324–37.

Hobbs, R. J. & Mooney, H. A. (1998). Broadening the extinction debate: population deletions and additions in California and western Australia. *Conservation Biology*, **12**, 271–83.

Hoekstra, T. W., Allen, T. F. H., & Flather, L. H. (1991) Implicit scaling in ecological research. *BioScience*, **41**, 148–54.

Hoffmann, W. A. (2000). Post-establishment seedling success in the Brazilian cerrado: a comparison of savanna and forest species. *Biotropica*, **32**, 62–9.

Hoffmann, W. A., Orthen, B., & Franco, A. C. (2004). Constraints to seedling success of savanna and forest trees across the savanna-forest boundary. *Oecologia*, **140**, 252–60.

Hofmeister, W. (1862). *On the Germination, Development, and Fructification of the Higher Cryptogamia and on the Fructification of the Coniferae*. London: Ray Society.

Hogarth, P. J. (1999). *The Biology of Mangroves*. Oxford: Oxford University Press.

Holdaway, R. J. & Sparrow, A. D. (2006). Assembly rules operating along a primary riverbed grassland successional sequence. *Journal of Ecology*, **94**, 1092–102.

Holdridge, L. R., Grenke, W. C., Hatheway, W. H., Liang, T., & Tosi, J. A. (1971). *Forest Environments in Tropical Life Zones*. Oxford: Pergamon Press.

Holdsworth, A. R. & Uhl, C. (1997). Fire in the eastern Amazonian logged rain forest and the potential for fire reduction. *Ecological Applications*, **7**, 713–25.

Holl, K. D. (2003). Tropical moist forest. In *Handbook of Ecological Restoration*, ed. M. Perrow & A. Davy. Cambridge: Cambridge University Press, pp. 539–58.

Holl, K. D., Loik, M. E., Lin, E. H. V., & Samuels, I. A. (2000). Tropical montane forest restoration in Costa Rica: Overcoming barriers to dispersal and establishment. *Restoration Ecology*, **8**, 339–49.

Holla, T. A. & Knowles, P. (1988). Age structure analysis of a virgin white pine, *Pinus strobus*, population. *Canadian Field-Naturalist*, **102**, 221–6.

Holmes, P. M. & Newton, R. J. (2004). Patterns of seed persistence in South African fynbos. *Plant Ecology*, **172**, 143–58.

Holmes, P. M. & Richardson, D. M. (1999). Protocols for restoration based on recruitment dynamics, community structure, and ecosystem function: perspectives from South African fynbos. *Restoration Ecology*, **7**, 215–30.

Holmgren, M. (2000). Combined effects of shade and drought on tulip poplar seedlings: trade-off in tolerance or facilitation? *Oikos*, **90**, 67–78.

Holmgren, M., Scheffer, M., & Huston, M. A. (1997). The interplay of facilitation and competition in plant communities. *Ecology*, **78**, 1966–75.

Holzapfel, C. & Mahall, B. E. (1999). Bidirectional facilitation and interference between shrubs and annuals in the Mojave Desert. *Ecology*, **80**, 1747–61.

Honnay, O., Verheyen, K., Butaye, J., Jacquemyn, H., Bossuyt, B., & Hermy, M. (2002). Possible effects of habitat fragmentation and climate change on the range of forest plant species. *Ecology Letters*, **5**, 525–30.

Hooper, E., Condit R., & Legendre, P. (2002). Responses of 20 native species to reforestation strategies for abandoned farmland in Panama. *Ecological Applications*, **12**, 1626–41.

Hori, H., Lim, B.-L., & Osawa, S. (1985). Evolution of green plants as deduced from 5S rRNA sequences. *Proceedings of the National Academy of Sciences (USA)*, **82**, 820–3.

Horsley, S. B., Stout, S. L., & de Calesta, D. S. (2003). White-tailed deer impact on the vegetation dynamics of a northern hardwood forest. *Ecological Applications*, **13**, 98–118.

Horton, J. L. & Hart, S. C. (1998). Hydraulic lift: A potentially important ecosystem process. *Trends in Ecology & Evolution*, **13**, 232–5.

Horton, T. R. & Bruns, T. D. (1998). Multiple-host fungi are the most frequent and abundant ectomycorrhizal types in a mixed stand of Douglas fir (*Pseudostuga menziesii*) and bishop pine (*Pinus muricata*). *New Phytologist*, **139**, 331–9.

Horton, T. R. & Bruns, T. D. (2001). The molecular revolution in ectomycorrhizal ecology: peeking into the black-box. *Molecular Ecology*, **10**, 1855–71.

Horton, T. R., Bruns, T. D., & Parker, V. T. (1999). Ectomycorrhizal fungi associated with *Arctostaphylos* contribute to *Pseudotsuga menziesii* establishment. *Canadian Journal of Botany*, **77**, 93–102.

Horton, T. R., Cázares, E., & Bruns, T. D. (1998). Ectomycorrhizal, vesicular-arbuscular and dark septate fungal colonization of bishop pine (*Pinus muricata*) seedlings in the first 5 months of growth after wildfire. *Mycorrhiza*, **8**, 11–18.

Horton, T. R., Molina, R., & Hood, K. (2005). Douglas-fir ectomycorrhizae in 40- and 400-year-old stands: mycobiont availability to late successional western hemlock. *Mycorrhiza*, **15**, 393–403.

Horvitz, C. C. & Schemske, D. W. (1995). Spatiotemporal variation in demographic transitions of a tropical understory herb: projection matrix analysis. *Ecological Monographs*, **65**, 155–92.

Horvitz, C. C., Schemske, D. W., & Caswell, H. (1997). The relative "importance" of life-history stages to population growth: prospective and retrospective analyses. In *Structured Population Models in Marine, Terrestrial and Freshwater Systems*, ed. S. Tuljapurkar & H. Caswell. New York: Chapman and Hall, pp. 247–72.

Hoshizaki, K., Suzuki, W., & Sasaki, S. (1997). Impacts of secondary seed dispersal and herbivory on seedling survival in *Aesculus turbinata*. *Journal of Vegetation Science*, **8**, 735–42.

Houlé, G. (1991). Regenerative traits of tree species in a deciduous forest of northeastern North America. *Holoartic Ecology*, **14**, 142–51.

Houlé, G. (1992). Spatial relationship between seed and seedling abundance and mortality in a deciduous forest of north-eastern North America. *Journal of Ecology*, **80**, 99–108.

Houlé, G. (1994). Spatiotemporal patterns in the components of regeneration of four sympatric tree species – *Acer rubrum, A. saccharum, Betula alleghaniensis* and *Fagus grandifolia*. *Journal of Ecology*, **82**, 39–53.

Houter, N. C. & Pons, T. L. (2005). Gap size effects on photoinhibition in understorey saplings in tropical rainforest. *Plant Ecology*, **179**, 43–51.

Howard, T. G. & Goldberg, D. E. (2001). Competitive response hierarchies for germination, growth, and survival, and their influence on abundance. *Ecology*, **82**, 979–90.

Howe, H. F., Brown, J. S., & Zorn-Arnold, B. (2002). A rodent plague on prairie diversity. *Ecology Letters*, **5**, 30–6.

Howe, H. F. & Smallwood, J. (1982). Ecology of seed dispersal. *Annual Review of Ecology and Systematics*, **13**, 201–28.

Howlett, B. E. & Davidson, D. W. (2003). Effects of seed availability, site conditions, and herbivory on pioneer recruitment after logging in Sabah, Malaysia. *Forest Ecology and Management*, **184**, 369–83.

Huang, Z. & Gutterman, Y. (1998). *Artemisia monosperma* achene germination in sand: effects of sand depth, sand/water content, cyanobacterial sand crust and temperature. *Journal of Arid Environments*, **38**, 27–43.

Huang, Z. & Gutterman, Y. (2004). Seedling desiccation tolerance of *Leymus racemous* (Poaceae) (wild rye), a perennial sand-dune grass inhabiting the Junggar Basin of Xinjiang, China. *Seed Science Research*, **14**, 233–9.

Huante, P., Rincón, E., & Acosta, I. (1995). Nutrient availability and growth rate of 34 woody species from a tropical deciduous forest in Mexico. *Functional Ecology*, **9**, 849–58.

Hubbell, S. P. (2001). *The Unified Neutral Theory of Biodiversity and Biogeography*. Princeton: Princeton University Press.

Hubbell, S. P. (2004). Two decades of research on the BCI forest dynamics plot: where have we been and where are we going. In *Tropical Forest Diversity and Dynamism: Findings from a Large-Scale Plot Network*, ed. E. C. Loso & E. G. Leigh, Jr. Chicago: University of Chicago Press, pp. 8–30.

Hubbell, S. P. (2005). Neutral theory in community ecology and the hypothesis of functional equivalence. *Functional Ecology*, **19**, 166–72.

Hubbell, S. P. & Foster, R. B. (1992). Short-term dynamics of a neotropical forest: why ecological research matters to tropical conservation and management. *Oikos*, **63**, 48–61.

Hubbell, S. P., Foster, R. B., O'Brien, S. T., *et al.* (1999). Light gaps, recruitment limitation, and tree diversity in a neotropical forest. *Science*, **283**, 554–7.

Huber-Sanwald, E. & Pike, D. A. (2005). Establishing native grasses in a big sagebrush-dominated site: an intermediate restoration step. *Restoration Ecology*, **13**, 292–301.

Hulme, P. E. (1994). Seedling herbivory in grassland: relative importance of vertebrate and invertebrate herbivores. *Journal of Ecology*, **82**, 873–80.

Hulme, P. E. (1996). Herbivory, plant regeneration, and species coexistence. *Journal of Ecology*, **84**, 609–15.

Hulme, P. E. & Kollmann, J. (2005). Seed predator guilds, spatial variation in post-dispersal seed predation and potential effects on plant demography – a temperate perspective. In *Seed Fate: Predation, Dispersal and Seedling Establishment*, ed. P.-M. Forget, J. E. Lambert, P. E. Hulme, & S. B. Vander Wall. Wallingford: CAB International, pp. 9–30.

Hultine, K. R., Scott, R. L., Cable, W. L., Goodrich, D. C., & Williams, D. G. (2004). Hydraulic redistribution by a dominant, warm-desert phreatophyte: seasonal patterns and response to precipitation pulses. *Functional Ecology*, **18**, 530–8.

Humbert, L., Gagnon, D., Kneeshaw, D. D., & Messier, C. (2007). A shade tolerance index for common understory species of northeastern North America. *Ecological Indicators*, **7**, 195–207.

Hupy, J. P. (2004). Influence of vegetation cover and crust type on wind-blown sediment in a semi-arid climate. *Journal of Arid Environments*, **58**, 167–79.

Hurtt, G. C. & Pacala, S. W. (1995). The consequences of recruitment limitation: reconciling chance, history and competitive differences between plants. *Journal of Theoretical Biology*, **176**, 1–12.

Husband, R., Herre, E. A., & Young, J. P. W. (2002). Temporal variation in the arbuscular mycorrhizal communities colonising seedlings in a tropical forest. *FEMS Microbiology Ecology*, **42**, 131–6.

Huxman, T. E., Wilcox, B. P., Breshears, D. D., *et al.* (2005). Ecohydrological implications of woody plant encroachment. *Ecology*, **86**, 308–19.

Hyatt, L. A. & Araki, S. (2006). Comparative population dynamics of an invading species in its native and novel ranges. *Biological Invasions*, **8**, 261–75.

Hyatt, L. A., Rosenberg, M. S., Howard, T. G., *et al.* (2003). The distance dependence prediction of the Janzen-Connell hypothesis: a meta-analysis. *Oikos*, **103**, 590–602.

Ibarra-Manríquez, G., Ramos, M. M., & Oyama, K. (2001). Seedling functional types in a lowland rain forest in Mexico. *American Journal of Botany*, **88**, 1801–12.

Ichie, T., Ninomiya, I., & Ogino, K. (2001). Utilization of seed reserves during germination and early seedling growth by *Dryobalanops lanceolata* (Dipterocarpaceae). *Journal of Tropical Ecology*, **17**, 371–8.

Ickes, K. (2001). Hyper-abundance of native wild pigs (*Sus scrofa*) in a lowland dipterocarp rain forest of Peninsular Malaysia. *Biotropica*, **33**, 682–90.

Inderjit & Callaway, R. M. (2003). Experimental designs for the study of allelopathy. *Plant and Soil*, **256**, 1–11.

Inderjit & Weston, L. A. (2003). Root exudates: an overview. In *Root Ecology*, ed. H. de Kroon & E. J. W. Visser. Berlin: Springer-Verlag, pp. 235–55.

Inghe, O. & Tamm, C. O. (1985). Survival and flowering of perennial herbs. IV. The behaviour of *Hepatica nobilis* and *Sanicula europaea* on permanent plots during 1943–81. *Oikos*, **45**, 400–20.

IPCC. (2007). Impacts, adaptation and vulnerability. *Working Group II contribution to the Intergovernmental Panel on Climate Change Fourth Assessment Report. Summary for policymakers*, Geneva: Intergovernmental Panel on Climate Change.

Isaac, F. M. (1969). Floral structure and germination in *Cymodocea ciliata*. *Phytomorphology*, **19**, 44–51.

Ishida, T. A., Nara, K., & Hogetsu, T. (2006). Host effects on ectomycorrhizal fungal communities: insight from eight host species in mixed conifer-broadleaf forests. *New Phytologist*, **174**, 430–40.

Izhaki, I., Walton, P. B., & Safriel, U. N. (1991). Seed shadows generated by frugivorous birds in an eastern Mediterranean scrub. *Journal of Ecology*, **79**, 575–90.

Jackson, M. W., Stinchcombe, J. R., Korves, T. M., & Schmitt, J. (2004). Costs and benefits of cold tolerance in transgenic *Arabidopsis thaliana*. *Molecular Ecology*, **13**, 3609–15.

Jacquemyn, H., Brys, R., & Hermy, M. (2001). Within and between plant variation in seed number, seed mass and germinability of *Primula elatior*: Effect of population size. *Plant Biology*, **3**, 561–8.

Jaffe, K., Michelangeli, F., Gonzalez, J. M., Miras, B., & Ruiz, M. C. (1992). Carnivory in pitcher plants of the genus *Heliamphora* (Sarraceniaceae). *New Phytologist*, **122**, 733–44.

Jakobsen, I. (1999). Transport of phosphorus and carbon in arbuscular mycorrhizas. In *Mycorrhiza. Structure, Function, Molecular Biology and Biotechnology*, ed. A. Varma & B. Hock. Berlin: Springer-Verlag, pp. 305–32.

Jakobsen, I., Abbott, L. K., & Robson, A. D. (1992). External hyphae of vesicular-arbuscular mycorrhizal fungi associated with *Trifolium subterraneum* L. I. Spread of hyphae and phosphorus inflow into roots. *New Phytologist*, **120**, 371–80.

Jakobsson, A. & Eriksson, O. (2000). A comparative study of seed number, seed size, seedling size and recruitment in grassland plants. *Oikos*, **88**, 493–502.

Jakobsson, A., Bruun, H. H., & Eriksson, O. (2006). Local seed rain and seed bank in species-rich grassland: effects of plant abundance and seed size. *Canadian Journal Botany*, **84**, 1870–81.

Jambois, A., Dauphin, A., Kawano, T., *et al.* (2005). Competitive antagonism between IAA and indole alkaloid hypaphorine must contribute to regulate ontogenesis. *Physiologia Plantarum*, **123**, 120–9.

James, C. D., Landsberg, J., & Morton, S. R. (1999). Provision of watering points in the Australian arid zone: a review of effects on biota. *Journal of Arid Environments*, **41**, 87–121.

James, T. Y., Kauff, F., Schoch, C., *et al.* (2006). Reconstructing the early evolution of fungi using a six-gene phylogeny. *Nature*, **433**, 818–22.

Janos, D. P. (1980a). Mycorrhizae influence tropical succession. *Biotropica*, **12**, 56–95.

Janos, D. P. (1980b). Vesicular-arbuscular mycorrhizae affect lowland tropical rain forest plant growth. *Ecology*, **61**, 151–62.

Janos, D. P., Sahley, C. T., & Emmons, L. H. (1995). Rodent dispersal of vesicular-arbuscular mycorrhizal fungi in Amazonian Peru. *Ecology*, **76**, 1852–8.

Janzen, D. H. (1970). Herbivores and the number of tree species in tropical forests. *American Naturalist*, **104**, 501–28.

Janzen, D. H. (1983). *Mora megistosperma* (Alcornoque, Mora). In *Costa Rican Natural History*, ed. D. H. Janzen. Chicago: University of Chicago Press, pp. 280–2.

Ji, X.-B. & Ye, N.-G. (2003). The seedling types of dicots and their evolutionary relationships. *Acta Phytotaxonomica Sinica*, **41**, 447–64 (in Chinese).

Jiménez, J. A. (1994). *Los Manglares del Pacífico de Centroamérica*. Heredia, Costa Rica: EFUNA.

Johansen, D. A. (1950). *Plant Embryology*. Waltham: Chronica Botanica.

John, R., Dalling, J. W., Harms, K. E., *et al.* (2007). Soil nutrients influence spatial distributions of tropical tree species. *Proceedings of the National Academy of Sciences* (USA), **104**, 864–9.

Johns, A. D. (1992). Species conservation in managed tropical forests. In *Tropical Deforestation and Species Extinctions*, ed. T. C. Whitmore & J. A. Sayer. London: Chapman and Hall, pp. 59–77.

Johns, A. D. (1997). *Timber Production and Biodiversity Conservation in Tropical Rain Forests*. Oxford: Oxford University Press.

Johnsen, T. N., Jr., & Alexander, R. A. (1974). *Juniperus* L. Juniper. In *Seeds of Woody Plants in the United States*, Agriculture Handbook No. 450, tech. coord. C. S. Schopmeyer. Washington, D.C.: Forest Service, U.S. Department of Agriculture, pp. 460–9.

Johnson, D., Leake, J. R., & Read, D. J. (2001). Novel in-growth core system enables functional studies of grassland mycorrhizal mycelial networks. *New Phytologist*, **152**, 555–62.

Johnson, D. A. (1986). Seed and seedling relations of crested wheatgrass: a review. In *Crested Wheatgrass: Its Values, Problems, and Myths*. ed. K. Johnson, Logan: Utah State University, pp. 65–90.

Johnson, J. D., Tognetti, R., Michelozzi, M., *et al.* (1997). Ecophysiological responses of *Fagus sylvatica* seedlings to changing light conditions. 2. The interaction of light environment and soil fertility on seedling physiology. *Physiologia Plantarum*, **101**, 124–34.

Johnson, N. C. (1993). Can fertilization of soil select less mutualistic mycorrhizae. *Ecological Applications*, **3**, 749–57.

Johnson, N. C., Graham, J. H., & Smith, F. A. (1997). Functioning of mycorrhizal associations along the mutualism-parasitism continuum. *New Phytologist*, **135**, 575–85.

Jones, M. D., Durall, D. M., & Tinker, P. B. (1991). Fluxes of carbon and phosphorus between symbionts in willow ectomycorrhizas and their changes with time. *New Phytologist*, **119**, 99–106.

Jordan, P. W. & Nobel, P. S. (1979). Infrequent establishment of seedlings of *Agave desertii* (Agavaceae) in the northwestern Sonoran Desert. *American Journal of Botany*, **66**, 1079–84.

Jordano, P. (1992). Fruits and frugivory. In *Seeds: the Ecology of Regeneration in Natural Plant Communities*, ed. M. Fenner. Wallingford: CAB International, pp. 105–50.

Jordano, P. & Godoy, J. A. (2002). Frugivore-generated seed shadows: a landscape view of demographic and genetic effects. In *Seed Dispersal and Frugivory: Ecology, Evolution and Conservation*, ed. D. J. Levey, W. R. Silva, & M. Galetti. Wallingford: CAB International, pp. 305–21.

Jordano, P. & Schupp, E. W. (2000). Seed disperser effectiveness: the quantity component and patterns of seed rain for *Prunus mahaleb*. *Ecological Monographs*, **70**, 591–615.

Joshi, G. V., Pimlaskar, M., & Bhosale, L. J. (1972). Physiological studies in germination of mangroves. *Botanica Marina*, **5**, 91–5.

Jouve, L., Gaspar, T., Kevers, C., Greppin, H., & Degi Agosti, R. (1999). Involvement of indole-3-acetic acid in the circadian growth of the first internode of *Arabidopsis*. *Planta*, **209**, 136–42.

Jubinsky, G. & Anderson, L. C. (1996). The invasive potential of Chinese tallow-tree (*Sapium sebiferum* Roxb.) in the Southeast. *Castanea*, **61**, 226–31.

Jumpponen, A. (2003). Soil fungal community assembly in a primary successional glacier forefront ecosystem as inferred from rDNA sequence analysis. *New Phytologist*, **158**, 569–78.

Juniper, B. E. & Jeffree, C. E. (1983). *Plant Surfaces*. London: Edward Arnold.

Juniper, B. E., Robins, R. J., & Joel, D. M. (1989). *The Carnivorous Plants*. London: Academic Press.

Jurado, E. & Flores, J. (2005). Is seed dormancy under environmental control or bound to plant traits? *Journal of Vegetation Science*, **16**, 559–64.

Jurado, E. & Westoby, M. (1992). Seedling growth in relation to seed size among species of arid Australia. *Journal of Ecology*, **80**, 407–16.

Jurado, E., Westoby, M., & Nelson, D. (1991). Diaspore weight, dispersal, growth form and perenniality of central Australian plants. *Journal of Ecology*, **79**, 811–30.

Jurik, T. W., Wang, S. C., & van der Valk, A. G. (1994). Effects of sediment load on seedling emergence from wetland seed banks. *Wetlands*, **14**, 159–65.

Jusoff, K. & Majid, N. M. (1992). An analysis of soil disturbance from a logging operation in a hill forest of peninsular Malaysia. *Forest Ecology and Management*, **47**, 323–33.

Kabeya, D. & Sakai, S. (2003). The role of roots and cotyledons as storage organs in early stages of establishment in *Quercus crispula*: a quantitative analysis of the nonstructural carbohydrate in cotyledons and roots. *Annals of Botany*, **92**, 537–45.

Kadmon, R. (1993). Population dynamic consequences of habitat heterogeneity: an experimental study. *Ecology*, **74**, 816–25.

Kang, H., Jaschek, G., & Bawa, K. S. (1992). Variation in seed and seedling traits in *Pithecellobium pedicellare*, a tropical rain-forest tree. *Oecologia*, **91**, 239–44.

Karlsson, P. S., Nordell, K. O., Eirefelt, S., & Svensson, A. (1987). Trapping efficiency of three carnivorous *Pinguicula* species. *Oecologia*, **73**, 518–21.

Karnieli, A., Kidron, G. J., Glaesser, C., & Ben-Dor, E. (1999). Spectral characteristics of cyanobacteria soil crust in semiarid environments. *Remote Sensing of Environment*, **69**, 67–75.

Karrenberg, S., Blaser, S., Kollmann, J., Speck, T., & Edwards, P. J. (2003). Root anchorage of saplings and cuttings of woody pioneer species in a riparian environment. *Functional Ecology*, **17**, 170–7.

Karrenberg, S., Edwards, P. J., & Kollmann, J. (2002). The life history of Salicaceae living in the active zone of flood plains. *Freshwater Biology*, **47**, 733–48.

Karrenberg, S. & Suter, M. (2003). Phenotypic trade-offs in the sexual reproduction of Salicaceae from flood plains. *American Journal of Botany*, **90**, 749–54.

Kato, M. & Akiyama, H. (2005). Interpolation hypothesis for origin of the vegetative sporophyte of land plants. *Taxon*, **54**, 443–50.

Katoh, K., Takeuchi, K., Jiang, D., Nan, Y., & Kou, Z. (1998). Vegetation restoration by seasonal exclosure in the Kerqin Sandy Land, Inner Mongolia. *Plant Ecology*, **139**, 133–44.

Keddy, P. A., Fraser L. H., & Wisheu, I. C. (1998). A comparative approach to examine competitive response of 48 wetland plant species. *Journal of Vegetation Science*, **9**, 777–86.

Keddy, P. A. & Reznicek, A. A. (1982). The role of seed banks in the persistence of Ontario's coastal plain flora. *American Journal of Botany*, **69**, 13–22.

Keddy, P. A, Twolan-Strutt, L., & Wisheu, I. C. (1994). Competitive effect and response rankings in 20 wetland plants: are they consistent across three environments? *Journal of Ecology*, **82**, 635–43.

Keeley, J. E. (1991). Seed germination and life history syndromes in the California Chaparral. *Botanical Review*, **57**, 81–116.

Keeley, J. E. (1992). Recruitment of seedlings and vegetative sprouts in unburned chaparral. *Ecology*, **73**, 1194–208.

Keeley, J. E. (1998). Coupling demography, physiology and evolution in chaparral shrubs. In *Landscape Degradation and Biodiversity in Mediterranean-Type Ecosystems*, ed. P. W. Rundel, G. Montenegro, & F. M. Jaksic. New York: Springer, pp. 257–64.

Keeley, J. E. (2000). Chaparral. In *North American Terrestrial Vegetation*, 2nd edn., ed. M. G. Barbour & W. D. Billings. Cambridge: Cambridge University Press, pp. 203–53.

Keeley, J. E. (2002). Fire management of California shrubland landscapes. *Environmental Management*, **29**, 395–408.

Keeley, J. E. & Bond, W. J. (1999). Mast flowering and semelparity in bamboos: the bamboo fire cycle hypothesis. *American Naturalist*, **154**, 383–91.

Keeley, J. E. & Bond, W. J. (2001). On incorporating fire into our thinking about natural ecosystems: a response to Saha and Howe. *American Naturalist*, **158**, 664–70.

Keeley, J. E. & Fotheringham, C. J. (2006). Wildfire management on a human dominated landscape: California chaparral wildfires. In *Wildlife – A Century of Failed Forest Policy*, ed. G. Wuerther. Covelo: Island Press, pp. 69–75.

Keeley, J. E., Fotheringham, C. J., & Baer-Keeley, M. (2006). Demographic patterns of postfire regeneration in mediterranean-climate shrublands of California. *Ecological Monographs*, **76**, 235–55.

Keeley, J. E. & Rundel, P. H. (2005). Fire and the Miocene expansion of C_4 grasslands. *Ecology Letters*, **8**, 683–90.

Keeley, J. E. & Stephenson, N. L. (2000). Restoring natural fire regimes in the Sierra Nevada in an era of global change. In *Wilderness Science in a Time of Change Conference*, ed. D. N. Cole, S. F. McCool, & J. O'Loughlin, RMRS-P-15, Vol. 5. Missoula: USDA Forest Service, Rocky Mountain Research Station, pp. 255–65.

Keenan, R., Lamb, D., Woldring, O., Irvine, T., & Jensen, R. (1997). Restoration of plant biodiversity beneath tropical tree plantations in northern Australia. *Forest Ecology and Management*, **99**, 117–31.

Keller, M. & Kollmann, J. (1999). Effects of seed provenance on germination of herbs for agricultural compensation sites. *Agriculture, Ecosystems and Environment*, **72**, 87–99.

Keller, M., Kollmann, J., & Edwards, P. J. (1999). Palatability of weeds from different European origins to the slugs *Deroceras reticulatum* Muller and *Arion lusitanicus* Mabille. *Acta Oecologica*, **20**, 109–18.

Keller, M., Kollmann, J., & Edwards, P. J. (2000). Genetic introgression from distant provenances reduces fitness in local weed populations. *Journal of Applied Ecology*, **37**, 647–59.

Kelly, D. & Sork, V. L. (2002). Mast seeding in perennial plants: why, how, where? *Annual Review of Ecology and Systematics*, **33**, 427–47.

Kennedy, D. N. & Swaine, M. D. (1992). Germination and growth of colonizing species in artificial gaps of different sizes in dipterocarp rain forest. *Philosophical Transactions of Royal Society London, Series B*, **335**, 357–66.

Kennedy, P. G., Hausmann, N. J., Wenk, E. H., & Dawson, T. E. (2004). The importance of seed reserves for seedling performance: an integrated approach using morphological, physiological, and stable isotope techniques. *Oecologia*, **141**, 547–54.

Kennedy, P. G., Izzo, A. D., & Bruns, T. D. (2003). There is high potential for the formation of common mycorrhizal networks between understorey and canopy trees in a mixed evergreen forest. *Journal of Ecology*, **91**, 1071–80.

Kennedy, P. G. & Sousa, W. P. (2006). Forest encroachment into a Californian grassland: examining the simultaneous effects of facilitation and competition on tree seedling recruitment. *Oecologia*, **148**, 464–74.

Kepinski, S. (2006). Integrating hormone signaling and patterning mechanisms in plant development. *Current Opinion in Plant Biology*, **9**, 28–34.

Kerley, G. I. H. (1991). Seed removal by rodents, birds and ants in the semi-arid Karoo, South Africa. *Journal of Arid Environments*, **20**, 63–9.

Kern, R. A. (1996). A comparative field study of growth and survival of Sierran conifer seedlings. PhD dissertation, Duke University, USA.

Kéry, M., Gregg, K. B., & Schaub, M. (2005). Demographic estimation methods for plants with unobservable life-states. *Oikos*, **108**, 307–20.

Kery, M., Matthies, D., & Spillmann, H.-H. (2000). Reduced fecundity and offspring performance in small populations of the declining grassland plants *Primula veris* and *Gentiana lutea*. *Journal of Ecology*, **88**, 17–30.

Kettenring, K. (2006). Seed ecology of wetland *Carex* spp. – implications for restoration. PhD dissertation, University of Minnesota-Twin Cities, USA.

Khan, R. (1943). Contributions to the morphology of *Ephedra foliata* Boss. II. Fertilization and embryogeny. *Proceedings of the National Academy of Sciences* (India), **13**, 357–75.

Khurana, E. & Singh, J. S. (2001a). Ecology of tree seed and seedlings: implications for tropical forest conservation and restoration. *Current Science*, **80**, 748–57.

Khurana, E. & Singh, J. S. (2001b). Ecology of seed and seedling growth for conservation and restoration of tropical dry forest: a review. *Environmental Conservation*, **28**, 39–52.

Khurana, E. & Singh, J. S. (2006). Impact of life-history traits on response of seedlings of five tree species of tropical dry forest to shade. *Journal of Tropical Ecology*, **22**, 653–61.

Kidd, F. (1914). The controlling influence of carbon dioxide in the maturation, dormancy and germination of seeds. *Proceedings of the Royal Society of London (Series B)*, **87**, 609–25.

Kidron, G. J., Yaalon, D. H., & Vonshak, A. (1999). Two causes for runoff initiation on microbiotic crusts: Hydrophobicity and pore clogging. *Soil Science*, **164**, 18–27.

Kidson, R. & Westoby, M. (2000). Seed mass and seedling dimensions in relation to seedling establishment. *Oecologia*, **125**, 11–17.

Kiers, E. T., Lovelock, C, E., Krueger, E. L., & Herre, E. A. (2000). Differential effects of tropical arbuscular mycorrhizal fungal inocula on root colonization and tree seedling growth: implications for tropical forest diversity. *Ecology Letters*, **3**, 106–13.

Kiers, E. T. & van der Heijden, M. G. A. (2006). Mutualistic stability in the arbuscular mycorrhizal symbiosis: Exploring hypotheses of evolutionary cooperation. *Ecology*, **87**, 1627–37.

King, D. A. (2003). Allocation of above-ground growth is related to light in temperate deciduous saplings. *Functional Ecology*, **17**, 482–8.

Kirby, K. R., Laurance, W. F., Albernaz, A. K., *et al.* (2006). The future of deforestation in the Brazilian Amazon. *Futures*, **38**, 432–53.

Kistner, C. & Parniske, M. (2002). Evolution of signal transduction in intracellular symbiosis. *Trends in Plant Science*, **7**, 511–18.

Kitajima, K. (1992a). The importance of cotyledon functional morphology and patterns of seed reserve utilization for the physiological ecology of neotropcial tree seedlings. PhD thesis, University of Illinois, USA.

Kitajima, K. (1992b). Relationship between photosynthesis and thickness of cotyledons for tropical tree species. *Functional Ecology*, **6**, 582–9.

Kitajima, K. (1994). Relative importance of photosynthetic traits and allocation patterns as correlates of seedling shade tolerance of 13 tropical trees. *Oecologia*, **98**, 419–28.

Kitajima, K. (1996a). Cotyledon functional morphology, patterns of seed reserve utilisation and regeneration niches of tropical tree seedlings. In *The Ecology of Tropical Forest Tree Seedlings*, ed. M. D. Swaine. New York: Pantheon, pp. 193–210.

Kitajima, K. (1996b). Ecophysiology of tropical tree seedlings. In *Tropical Forest Plant Ecophysiology*, ed. S. S. Mulkey, R. L. Chazdon, & A. P. Smith. New York: Chapman and Hall, pp. 559–96.

Kitajima, K. (2002). Do shade-tolerant tropical tree seedlings depend longer on seed reserves? Functional growth analysis of three Bignoniaceae species. *Functional Ecology*, **16**, 433–44.

Kitajima, K. (2007). Seed and seedling ecology. In *Functional Plant Ecology*, 2nd edn., ed. F. I. Pugnaire & V. Valladores. New York: Marcel Deckker, pp. 549–79.

Kitajima, K. & Bolker, B. M. (2003). Testing performance rank reversals among coexisting species: crossover point irradiance analysis by Sack & Grubb (2001) and alternatives. *Functional Ecology*, **17**, 276–81.

Kitajima, K. & Fenner, M. (2000). Ecology of seed regeneration. In *The Ecology of Seed Regeneration in Plant Communities*, ed. M. Fenner. Wallingford: CAB International, pp. 331–59.

Kitajima, K., Fox, A. M., Sato, T., & Nagamatsu, D. (2006). Cultivar selection prior to introduction may increase invasiveness: evidence from *Ardisia crenata*. *Biological Invasions*, **8**, 1471–82.

Kitajima, K. & Hogan, K. P. (2003). Increases of chlorophyll a/b ratios during acclimation of tropical woody seedlings to nitrogen limitation and high light. *Plant Cell and Environment*, **26**, 857–65.

Kiviniemi, K. (2001). Evolution of recruitment features in plants: a comparative study of species in the Rosaceae. *Oikos*, **94**, 250–62.

Kling, J. (1996). Could transgenic supercrops one day breed superweeds? *Science*, **274**, 180–1.

Klironomos, J. N. (2002). Feedback with soil biota contributes to plant rarity and invasiveness in communities. *Nature*, **417**, 67–70.

Klironomos, J. N. (2003). Variation in plant response to native and exotic arbuscular mycorrhizal fungi. *Ecology*, **84**, 2292–301.

Kneeshaw, D. D., Kobe, R. K., Coates, K. D., & Messier, C. (2006). Sapling size influences shade tolerance ranking among southern boreal tree species. *Journal of Ecology*, **94**, 471–80.

Knevel, I. C, Bekker, R. M., Bakker, J. P., & Kleyer, M. (2003). Life-history traits of the northwest European flora: The LEDA database. *Journal of Vegetation Science*, **14**, 611–14.

Knight H., Zarka, D. G., Okamoto, H., Thomashow, M. E., & Knight, M. R. (2004). Abscisic acid induces CBF gene transcription and subsequent induction of cold-regulated genes via the CRT promoter element. *Plant Physiology*, **135**, 1710–17.

Knoll, A. H. (1986). Patterns of change in plant communities through geological time. In *Community Ecology*, ed. J. Diamond & T. J. Case. New York: Harper & Row, pp. 126–41.

Kobe, R. K. (1997). Carbohydrate allocation to storage as a basis of interspecific variation in sapling survivorship and growth. *Oikos*, **80**, 226–33.

Kobe, R. K. (1999). Light gradient partitioning among tropical tree species through differential seedling mortality and growth. *Ecology*, **80**, 187–201.

Koch, J. M., Richardson, J., & Lamont, B. B. (2004). Grazing by kangaroo limits the establishment of grass trees *Xanthorroea gracilis* and *X. priesii* in restored bauxite mines in eucalypt forest of southwestern Australia. *Restoration Ecology*, **12**, 297–305.

Koenig, W. D. & Knops, J. M. H. (2002). The behavioral ecology of masting in oaks. In *Oak Forest Ecosystems. Ecology and Management for Wildlife*, ed.

W. J. McShea & W. M. Healy. Baltimore: Johns Hopkins University Press, pp. 129–48.

Kollmann, J. (1995). Regeneration window for fleshy-fruited plants during scrub development on abandoned grassland. *Ecoscience*, **2**, 213–22.

Kollmann, J. (2000). Dispersal of fleshy-fruited species: a matter of spatial scale? *Perspectives in Plant Ecology, Evolution and Systematics*, **3**, 29–51.

Kollmann, J. & Bañuelos, M. J. (2004). Latitudinal trends in growth and phenology of the invasive alien plant *Impatiens glandulifera* (Balsaminaceae). *Diversity and Distributions*, **10**, 377–85.

Kollmann, J., Frederiksen, L., Vestergaard, P., & Bruun, H. H. (2006). Limiting factors for emergence and establishment of the invasive non-native *Rosa rugosa* in a coastal dune system. *Biological Invasions*, **9**, 31–42.

Kollmann, J. & Grubb, P. J. (1999). Recruitment of fleshy-fruited species under different shrub species: control by under-canopy environment. *Ecological Research*, **14**, 63–74.

Kollmann, J. & Reiner, S. A. (1996). Light demands of shrub seedlings and their establishment within scrublands. *Flora*, **191**, 191–200.

Kollmann, J. & Schill, H.-P. (1996). Spatial patterns of dispersal, seed predation and germination during colonization of abandoned grassland by *Quercus petraea* and *Corylus avellana*. *Vegetatio*, **125**, 193–205.

Kollmann, J. & Schneider, B. (1997). Effects of landscape structure on seed dispersal of fleshy-fruited species along forest edges. *Bulletin of the Geobotanical Institute ETH*, **63**, 77–86.

Kollmann, J. & Schneider, B. (1999). Landscape structure and diversity of fleshy-fruited species at forest edges. *Plant Ecology*, **144**, 37–48.

Končalová, H. (1990). Anatomical adaptations to waterlogging in roots of wetland graminoids: limitations and drawback. *Aquatic Botany*, **38**, 127–34.

Kondo, K., Segawa, M., & Nehira, K. (1978). Anatomical studies on seeds and seedlings of some *Utricularia* (Lentibulariaceae). *Brittonia*, **30**, 89–95.

Koop, A. L. & Horvitz, C. C. (2005). Projection matrix analysis of the demography of an invasive, nonnative shrub (*Ardisia elliptica*). *Ecology*, **86**, 2661–72.

Körner, C. (2003). *Alpine Plant Life: Functional Plant Ecology of High Mountain Ecosystems*. Berlin: Springer.

Kosovsky, A. (1994). Generation of runoff in first order drainage basins in a semi-arid region, Lahav hills, Negev, Israel. MSc thesis, The Hebrew University of Jerusalem, Israel.

Kost, C. & Heil, M. (2006). Herbivore-induced plant volatiles induce an indirect defense in neighboring plants. *Journal of Ecology*, **94**, 619–28.

Kostel-Hughes, F., Young, T. P., & Wehr, J. D. (2005). Effects of leaf litter depth on the emergence and seedling growth of deciduous forest tree species in relation to seed size. *Journal of the Torrey Botanical Society*, **132**, 50–61.

Kotler, B. P., Ayal, Y., & Subach, A. (1994). Effects of predatory risk and resource renewal on the timing of foraging activity in a gerbil community. *Oecologia*, **100**, 391–6.

Kotler, B. P. & Brown, J. S. (1999). Mechanisms of coexistence of optimal foragers as determinants of local abundances and distributions of desert granivores. *Journal of Mammalogy*, **80**, 361–74.

Kozlowski, T. T. & Pallardy, S. G. (2002). Acclimation and adaptive responses of woody plants to environmental stresses. *Botanical Review*, **68**, 270–334.

Krannitz, P. G., Aarssen, L. W., & Dow, J. M. (1991). The effect of genetically based differences in seed size on seedling survival in *Arabidopsis thaliana* (Brassicaceae). *American Journal of Botany*, **78**, 446–50.

Krause, A., Ramakumar, A., Bartels, D., *et al.* (2006). Complete genome of the mutualistic, N 2-fixing grass endophyte *Azoarcus* sp. strain BH72. *Nature Biotechnology*, **24**, 1385–91.

Krause, G. H., Gallé, A., Virgo, A., *et al.* (2006). High-light stress does not impair biomass accumulation of sun-acclimated tropical tree seedlings (*Calophyllum longifolium* Willd. and *Tectona grandis* L. f.). *Plant Biology*, **8**, 31–41.

Krause, G. H., Koroleva, O. Y., Dalling, J. W., & Winter, K. (2001). Acclimation of tropical tree seedlings to excessive light in simulated treefall gaps. *Plant Cell and Environment*, **24**, 1345–52.

Kretzer, A., Dunham, S., Molina, R., & Spatafora, J. W. (2004). Microsatellite markers reveal the below ground distribution of genets in two species of *Rhizopogon* forming tuberculate ectomycorrhizas on Douglas fir. *New Phytologist*, **161**, 313–20.

Kretzer, A., Li, Y., Szaro, T. M., & Bruns, T. D. (1996). Internal transcibed spacer sequences from 38 recognized species of *Suillus* sensu lato: phylogenetic and taxonomic implications. *Mycologia*, **88**, 776–85.

Kretzer, A. M., Dunham, S., Molina, R., & Spatafora, J. W. (2005). Patterns of vegetative growth and gene flow in *Rhizopogon vinicolor* and *R. vesiculosus* (Boletales, Basidiomycota). *Molecular Ecology*, **14**, 2259–68.

Krishna, P. (2003). Brassinosteroid-mediated stress responses. *Journal of Plant Growth Regulation*, **22**, 289–97.

Krueger, L. M. & Peterson, C. J. (2006). Effects of white-tailed deer on *Tsuga canadensis* regeneration: evidence of microsites as refugia from browsing. *American Midland Naturalist*, **156**, 353–62.

Kubo, K., Iwama, K., Yanagisawa, A., *et al.* (2006). Genotypic variation of the ability of roots to penetrate hard soil layers among Japanese wheat cultivars. *Plant Production Science*, **9**, 47–55.

Kubota, M., McGonigle, T. P., & Hyakumachi, M. (2001). *Clethra barbinervis*, a member of the order Ericales, forms arbuscular mycorrhizae. *Canadian Journal of Botany*, **79**, 300–6.

Kuijt, J. (1969). *The Biology of Parasitic Flowering Plants*. Berkeley: University of California Press.

Kuijt, J. (1982). Seedling morphology and its systematic significance in Loranthaceae of the New World, with supplementary comments on Eremolepidaceae. *Botanische Jahrbucher*, **103**, 305–42.

Kummer, A. P. (1951). *Weed Seedlings*. Chicago: The University of Chicago Press.

Kuo, J. & Kirkman, H. (1990). Anatomy of the viviparous seagrass seedlings of *Amphibolus* and *Thallasodendron* and their nutrient supply. *Botanica Marina*, **33**, 117–26.

Kursar, T. A. & Coley, P. D. (1991). Nitrogen content and expansion rate of young leaves of rain forest species: implications for herbivory. *Biotropica*, **23**, 140–50.

Kurtz, H. D., Jr., & Netoff, D. I. (2001). Stabilization of friable sandstone surfaces in a desiccating, wind-abraded environment of south-central Utah by rock surface microorganisms. *Journal of Arid Environments*, **48**, 89–100.

Kutiel, P. (1998). Possible role of biogenic crusts in plant succession on the Sharon Sand Dunes, Israel. *Israel Journal of Plant Sciences*, **46**, 279–86.

Kwit, C., Schwartz, M. W., Platt, W. J., & Geaghan, J. P. (1998). The distribution of tree species in steepheads of the Apalachicola River Bluffs, Florida. *Journal of the Torrey Botanical Society*, **125**, 309–18.

Kytoviita, M. M., Vestberg, M., & Tuom, J. (2003). A test of mutual aid in common mycorrhizal networks: established vegetation negates benefit in seedlings. *Ecology*, **84**, 898–906.

Ladd, B. M. & Facelli, J. M. (2005). Effects of competition, resource availability and invertebrates on tree seedling establishment. *Journal of Ecology*, 93, 968–77.

Lal, R. (1987). *Tropical Ecology and Physical Edaphology*, London: Wiley.

Lamattina, L., García-Mata, C., Graziano, M., & Pagnussat, G. (2003). Nitric oxide: the versatility of an extensive signal molecule. *Annual Review of Plant Biology*, 54, 109–36.

Lamb, D., Erskine, P. D., & Parrotta, J. A. (2005). Restoration of degraded tropical forest landscapes. *Science*, 310, 1628–32.

Lamb, E. G. & Cahill, J. F. (2006). Consequences of differing competitive abilities between juvenile and adult plants. *Oikos*, 112, 502–12.

Lambers, H., Chapin, F. S., III, & Pons, T. J. (1998). *Plant Physiological Ecology*. New York: Springer.

Lambers, J. H. R. & Clark, J. S. (2005). The benefits of seed banking for red maple (*Acer rubrum*): maximizing seedling recruitment. *Canadian Journal of Forest Research*, 35, 806–13.

Lambers, J. H. R., Clark, J. S., & Beckage, B. (2002). Density-dependent mortality and the latitudinal gradient in species diversity. *Nature*, 417, 732–5.

Lambrecht-McDowell, S. C. & Radosevich, S. R. (2005). Population demographics and trade-offs to reproduction of an invasive and noninvasive species of *Rubus*. *Biological Invasions*, 7, 281–95.

Lambrinos, J. G. (2006). Spatially variable propagule pressure and herbivory influence invasion of chaparral shrubland by an exotic grass. *Oecologia*, 147, 327–34.

Lamont, B. (1983). Germination of mistletoes. In *The Biology of Mistletoes*, ed. M. Calder & P. Bernhardt. Sydney: Academic Press, pp. 129–43.

Lamont, B. & Perry, M. (1977). The effects of light, osmotic potential and atmospheric gases on germination of the mistletoe, *Amyema preisii*. *Annals of Botany*, 41, 203–9.

Lamont, B. B., Witkowski, E. T. F., & Enright, N. J. (1993). Post-fire litter microsites: safe for seeds, unsafe for seedlings. *Ecology*, 74, 501–12.

La Motte, C. (1933). Morphology of the megagametophyte and the embryo sporophyte of *Isoetes lithophila*. *American Journal of Botany*, 20, 217–33.

La Motte, C. (1937). Morphology and orientation of the embryo of *Isoetes*. *Annals of Botany* (n.s.), 1, 695–716.

Land, W. J. G. (1907). Fertilization and embryology in *Ephedra trifurca*. *Botanical Gazette*, 44, 273–92.

Lange, R. T. & Purdie, R. (1976). Western myall (*Acacia sowdenii*), its survival prospects and management needs. *Australian Rangeland Journal*, 1, 64–9.

Langenheim, J. H., Osmond, C. B., Brooks, A., & Ferrar, P. J. (1984). Photosynthetic responses to light in seedlings of selected Amazonian and Australian rainforest tree species. *Oecologia*, 63, 215–24.

Lapeyrie, F. F. & Chilvers, G. A. (1985). An endomycorrhiza–ectomycorrhiza succession associated with enhanced growth of *Eucalyptus domosa* seedlings planted in a calcereous soil. *New Phytologist*, 100, 93–104.

Laroche, J. (1968). Contributions à l'étude de l'*Equisetum arvense* L. II. Etude embryologique. Caractères morphologiques, histologiques et anatomiques de la première pousse transitorie. *Review Cytolologie et de Biologie Végétales*, 31, 155–216.

Laube, S. & Zotz, G. (2003). Which abiotic factors limit vegetative growth in a vascular epiphyte? *Functional Ecology*, 17, 598–604.

Lauenroth, W. K. & Gill, R. (2003). Turnover of root systems. In *Root Ecology*, ed. H. de Kroon & E. J. W. Visser. Berlin: Springer-Verlag, pp. 61–89.

Laurance W. F. (1991). Edge effects in tropical forest fragments: application of a model for the design of nature reserves. *Biological Conservation*, **57**, 205–19.

Laurance, W. F. (1998). A crisis in the making: responses of Amazonian forests to land use and climate change. *Trends in Ecology & Evolution*, **13**, 411–5.

Laurance, W. F. (2003). Slow burn: the insidious effects of surface fires on tropical forests. *Trends in Ecology & Evolution*, **18**, 209–12.

Laurance, W. F. (2005). The alteration of biotic interactions in fragmented tropical forests. In *Biotic Interactions in the Tropics*, ed. D. F. R. P Burslem, M. A. Pinard, & S. E. Hartley. Cambridge: Cambridge University Press, pp. 442–58.

Laurance W. F., Lovejoy, T. E., Vasconcelos H. L., *et al.* (2002). Ecosystem decay of Amazonian forest fragments: a 22-year investigation. *Conservation Biology*, **16**, 605–18.

Laurance W. F., Nascimento, H. E. M, Laurance, S. G., *et al.* (2006). Rain forest fragmentation and the proliferation of successional trees. *Ecology*, **87**, 469–82.

Laurance, W. F., Pérez-Salicrup, D., Delamônica, P., *et al.* (2001). Rain forest fragmentation and the structure of liana communities. *Ecology*, **82**, 105–16.

Lavorel, S., McIntyre, S., Landsberg, J., & Forbes, T. D. A. (1997). Plant functional classifications: from general groups to specific groups based on response to disturbance. *Trends in Ecology & Evolution*, **12**, 474–8.

Lawrence, J. G., Colwell, A., & Sexton, O. J. (1991). The ecological impact of allelopathy in *Ailanthus altissima* (Simaroubaceae). *American Journal of Botany*, **78**, 948–58.

Lazaro, A., Traveset, A., & Castillo, A. (2006). Spatial concordance at a regional scale in the regeneration process of a circum-Mediterranean relict (*Buxus balearica*): connecting seed dispersal to seedling establishment. *Ecography*, **29**, 683–96.

Leake, J. R. (1994). The biology of myco-heterotrophic ('saprophytic') plants. *New Phytologist*, **127**, 171–216.

Leake, J. R., McKendrick, S. L., Bidartondo, B. I., & Read, D. J. (2004). Symbiotic germination and development of the myco-heterotroph *Monotropa hypopitys* in nature and its requirement for locally distributed *Tricholoma* spp. *New Phytologist*, **163**, 405–23.

Leck, M. A. (1996). Germination of macrophytes from a Delaware River wetland. *Bulletin of the Torrey Botanical Club*, **123**, 48–67.

Leck, M. A. & Brock, M. A. (2000). Ecological and evolutionary trends in wetlands: evidence from seeds and seed banks in New South Wales, Australia and New Jersey, USA. *Plant Species Biology*, **15**, 97–112. [Corrigendum. (2001). **16**, 183–4.]

Leck, M. A. & Schütz, W. (2005). Regeneration of Cyperaceae, with particular reference to seed ecology and seed banks. *Perspectives in Plant Ecology, Evolution and Systematics*, **7**, 95–133.

Leck, M. A. & Simpson, R. L. (1993). Seeds and seedlings of the Hamilton Marshes, a Delaware River tidal freshwater wetland. *Proceedings of the Academy of Natural Sciences of Philadelphia*, **144**, 267–81.

Leck, M. A. & Simpson, R. L. (1994). Tidal freshwater wetland zonation: seed and seedling dynamics. *Aquatic Botany*, **47**, 61–75.

Leck, M. A. & Simpson, R. L. (1995). Ten year seed bank and vegetation dynamics of a tidal freshwater wetland. *American Journal of Botany*, **82**, 1547–57.

Leck, M. A., Parker, V. T., & Simpson, R. L., ed. (1989a). *Ecology of Soil Seed Banks*. San Diego: Academic Press.

Leck, M. A., Simpson, R. L., & Parker, V. T. (1989b). The seed bank of a freshwater tidal wetland and its relationship to vegetation dynamics. In *Proceedings Symposium on Freshwater Wetlands and Wildlife*, ed. R. Sharitz & J. W. Gibbons. Oak Ridge: USDOE Office of Scientific and Technical Information, pp. 198–205.

Lee, D. (2007). *Nature's Palette: The Science of Plant Color*. Chicago: The University of Chicago Press.

Lee, D. W., Baskaran, K., Mansor, M., Mohamad, H., & Yap, S. K. (1996). Irradiance and spectral quality affect Asian tropical rain forest tree seedling development. *Ecology*, **77**, 568–80.

Lee, D. W., Oberbauer, S. F., Krishnapilay, B., *et al.* (1997). Effects of irradiance and spectral quality on seedling development of two southeast Asian *Hopea* species. *Oecologia*, **110**, 1–9.

Lee, J. A. & Harmer, R. (1980). Vivipary, a reproductive strategy in response to environmental stress? *Oikos*, **35**, 254–65.

Lee, S. S., Alexander, I. J., Moura-Costa, P., & Yap, S. W. (1996). Mycorrhizal infection of dipterocarp seedlings in logged and undisturbed forest. In *Proceedings of Fifth Round-Table Conference on Dipterocarps*, ed. S. Appanah & K. C. Khoo. Kuala Lumpur: Forest Research Institute of Malaysia, pp. 157–64.

Lee, W. G. & Fenner, M. (1989). Mineral nutrient allocation in seeds and shoots of twelve *Chionochloa* species in relation to soil fertility. *Journal of Ecology*, **77**, 704–16.

Legard, N. J. (1979). First-year loses of *Pinus mugo* seed and seedlings on an exposed high country subsoil. *New Zealand Journal of Forestry*, **24**, 90–100.

Lehmann–Baerts, M. (1967). Etudes sur les Gnétales. XII. Ovule, gamétophyte femelle et embryogenèse chez *Ephedra distachya*. *Cellule*, **67**, 51–87.

Le Houérou, H. N. (1992). The role of saltbushes *Atriplex* spp. in arid land rehabilitation in the mediterranean basin. A review. *Agroforestry Systems*, **18**, 107–48.

Lehtilä, K., Syrjänen, K., Leimu, R., Garcia, M. B., & Ehrlén, J. (2006). Land use and population growth of *Primula veris*: an experimental demographic approach. *Conservation Biology*, **20**, 833–43.

Leibold, M. A., Holyoak, M., Mouquet, N., *et al.* (2004). The metacommunity concept: a framework for multi-scale ecology. *Ecology Letters*, **7**, 601–13.

Leigh, E. G., Dividar, P., Dick, C. W., *et al.* (2004). Why do some tropical forests have so many species of trees? *Biotropica*, **33**, 447–73.

Leigh, E. G., Jr., Wright, S. J., Herre, E. A., & Putz, F. E. (1993). The decline of tree diversity on newly isolated tropical islands: a test of a null hypothesis and the implications. *Evolutionary Ecology*, **7**, 76–102.

Leimu, R. (2004). Variation in the mating system of *Vincetoxicum hirundinaria* (Asclepiadaceae) in peripherial island populations. *Annals of Botany*, **93**, 107–13.

Leishman, M. R. (2001). Does the seed size/number trade-off model determine plant community structure? An assessment of the model mechanisms and their generality. *Oikos*, **93**, 294–302.

Leishman, M. R. & Westoby, M. (1994a). Hypotheses on seed size tests using the semiarid flora of Western New South Wales, Australia. *American Naturalist*, **143**, 890–906.

Leishman, M. R. & Westoby, M. (1994b). The role of large seed size in shaded conditions – experimental-evidence. *Functional Ecology*, **8**, 205–14.

Leishman, M. R. & Westoby, M. (1998). Seed size and shape are not related to persistence in soil in Australia in the same way as in Britain. *Functional Ecology*, **12**, 480–5.

Leishman, M. R., Westoby, M., & Jurado, E. (1995). Correlates of seed size variation: a comparison among five temperate floras. *Journal of Ecology*, **83**, 517–30.

Leishman, M. R., Wright, I. J., Moles, A. T., & Westoby, M. (2000). The evolutionary ecology of seed size. In *Seeds: The Ecology of Regeneration in Plant Communities*, 2nd edn., ed. M. Fenner. Wallingford: CAB International, pp. 31–57.

Leite, A. M. C. & Rankin, J. M. (1981). Ecologia de sementes de *Pithecelobium racemosum* Ducke. *Acta Amazonica*, **11**, 309–18.

Le Maitre, D. C. & Midgely, J. J. (1992). Plant reproductive ecology. In *The Ecology of Fynbos: Nutrients, Fire and Diversity*, ed. R. Cowling. Cape Town: Oxford University Press, pp. 135–74.

Lerat, S., Rachel, R., Catford, J. G., *et al.* (2002). ^{14}C transfer between the spring ephemeral *Erythronium americanum* and sugar maple saplings via arbuscular mycorrhizal fungi in natural stands. *Oecologia*, **132**, 181–7.

Leroux, G., Barabé, D., & Vieth, J. (1997). Morphogenesis of the protocorm of *Cypripedium acaule* (Orchidaceae). *Plant Systematics and Evolution*, **205**, 53–72.

Lesica, P. & Antibus, R. K. (1990). The occurrence of mycorrhizae in vascular epiphytes of two Costa Rican rain forests. *Biotropica*, **22**, 250–8.

Leuchtmann, A. (1992). Systematics, distribution and host specificity of grass endophytes. *Natural Toxins*, **1**, 150–62.

Leverenz, J. W. (1996). Shade-shoot structure, photosynthetic performance in the field, and photosynthetic capacity of evergreen conifers. *Tree Physiology*, **16**, 109–14.

Levey, D. J. & Byrne M. M. (1993). Complex ant-plant interactions: rain forest ants as secondary dispersers and post-dispersal seed predators. *Ecology*, **74**, 1802–12.

Levin, D. A. (1974). The oil content of seeds: an ecological perspective. *American Naturalist*, **108**, 193–206.

Levin, S. (1992). The problem of pattern and scale in ecology. *Ecology*, **73**, 1943–67.

Levine, J. M. & Murrell, D. J. (2003). The community-level consequences of seed dispersal patterns. *Annual Review of Ecology and Systematics*, **34**, 549–74.

Lewis, S. L. & Tanner, E. V. J. (2000). Effects of above- and below-ground competition on the growth and survival of rain forest tree seedlings. *Ecology*, **81**, 2525–38.

Leyser, O. & Day, S. (2003). *Mechanisms in Plant Development*. Oxford: Blackwell Science, Ltd.

Li, F. R., Zhang, H., Zhao, L. Y., Shirato, Y., & Wang, X. Z. (2003). Pedoecological effects of a sand-fixing poplar (*Populus simonii* Carr.) forest in a desertified sandy land of Inner Mongolia, China. *Plant and Soil*, **256**, 431–42.

Li, X.-L., George, E., & Marschner, H. (1991). Extension of the phosphorus depletion zone in VA-mycorrhizal white clover in a calcareous soil. *Plant and Soil*, **136**, 41–8.

Li, X. R., Wang, X. P., Li, T., & Zhang, J. G. (2002). Microbiotic soil crust and its effect on vegetation and habitat on artificially stabilized desert dunes in Tengger Desert, North China. *Biology and Fertility of Soils*, **35**, 147–54.

Li, X. R., Xiao, H. L., Zhang, J. G., & Wang, X. P. (2004). Long-term ecosystem effects of sand-binding vegetation in the Tengger Desert, northern China. *Restoration Ecology*, **12**, 376–90.

Lian, C. L., Narimatsu, M., Nara, K., & Hogetsu, T. (2006). *Tricholoma matsutake* in a natural *Pinus densiflora* forest: correspondence between above- and below-ground genets, association with multiple host trees and alteration of existing ectomycorrhizal communities. *New Phytologist*, **171**, 825–36.

Lienert, J. & Fischer, M. (2004). Experimental inbreeding reduces seed production and germination independent of fragmentation of populations of *Swertia perennis*. *Basic and Applied Ecology*, **5**, 43–52.

Ligrone, R., Duckett, J. G., & Renzaglia, K. S. (1993). The gametophyte-sporophyte junction in land plants. *Advances in Botanical Research*, **19**, 231–317.

Lin, C. (2002). Blue light receptors and signal transduction. *The Plant Cell*, **14**, S207–25.

Lindquist, E. S. & Carroll, C. R. (2004). Differential seed and seedling predation by crabs: impacts on tropical coastal forest composition. *Oecologia*, **141**, 661–71.

Linhart, Y. B. (1988). Intrapopulational differentiation in annual plants. 3. The contrasting effects of intraspecific and interspecific competition. *Evolution*, **42**, 1042–64.

Linhart, Y. B. & Baker, I. (1973). Intrapopulation differentiation of physiological response to flooding in a population of *Veronica peregrina* L. *Nature*, **242**, 275–7.

Lipow, S. R. & Wyatt, R. (2000). Towards an understanding of the mixed breeding system of swamp milkweed (*Asclepias incarnata*). *Journal of the Torrey Botanical Society*, **127**, 193–9.

Liptay, A. & Geier, T. (1983). Mechanism of emergence of tomato (*Lycopersicon esculentum* L.) seedlings through surface-crusted or compressed soil. *Annals of Botany*, **51**, 409–12.

Liu, Z. M., Thompson, K., Spencer, R. E., & Reider, R. J. (2000). A comparative study of morphological responses of seedling roots to drying soil in 20 species from different habitats. *Acta Botanica Sinica*, **42**, 628–35.

Lloret, F., Peñuelas, J., & Estiarte, M. (2005). Effects of vegetation canopy and climate on seedling establishment in Mediterranean shrubland. *Journal of Vegetation Science*, **16**, 67–76.

Lloyd, F. E. (1976). *The Carnivorous Plants*. New York: Dover Publications Inc.

Loach, K. (1967). Shade tolerance in tree seedlings. I. Leaf photosynthesis and respiration in plants raised under artificial shade. *New Phytologist*, **66**, 607–21.

Loffler, C., Czygan, F. C., & Proksch, P. (1999). Role of indole-3-acetic acid in the interaction of the phanerogamic parasite *Cuscuta* and host plants. *Plant Biology*, **1**, 613–7.

Loha, A., Tigabu, M., Teketey, D., Lundkvist, K., & Fries, A. (2006). Provenance variation in seed morphometric traits, germination, and seedling growth of *Cordia africana* Lam. *New Forests*, **32**, 71–86.

Lohne, C., Brosch, T., & Wiersema, J. H. (2007). Phylogenetic analysis of Nymphaeales using fast-evolving and noncoding chloroplast markers. *Botanical Journal of the Linnean Society*, **154**, 141–63.

Lokesha, R., Hegde, S. G., Shaanker, R. U., & Ganeshaiah, K. N. (1992). Dispersal mode as a selective force in shaping the chemical composition of seeds. *American Naturalist*, **140**, 520–5.

Long, J. A., Ohio, C., Smith, Z. R., & Meyerowitz, E. M. (2006) *TOPLESS* regulates apical embryonic fate in *Arabidopsis*. *Science*, **312**, 1520–3.

Lonsdale, W. M. & Abrecht, D. G. (1989). Seedling mortality in *Mimosa pigra*, an invasive tropical shrub. *Journal of Ecology*, **77**, 371–85.

López-Bucio, J., Cruz-Ramírez, A., & Herrera-Estrella, L. (2003). The role of nutrient availability in regulating root architecture. *Current Opinion in Plant Biology*, **6**, 280–7.

López De Buen, L. & Ornelas, J. F. (2002). Host compatibility of the cloud forest mistletoe *Psittacanthus schiedeanus* (Loranthaceae) in central Veracruz, Mexico. *American Journal of Botany*, **89**, 95–102.

López-Molina, L., Mongrand, S., & Chua, N.-H. (2001). A postgermination developmental arrest checkpoint is mediated by abscisic acid and requires the ABI5 transcription factor in *Arabidopsis*. *Proceedings of the National Academy of Sciences* (USA), **98**, 4782–7.

Lortie, C. J. & Aarssen, L. W. (1996). The specialization hypothesis for phenotypic plasticity in plants. *International Journal of Plant Sciences*, **157**, 484–7.

Louda, S. M. (1989). Predation in the dynamics of seed regeneration. In *Ecology of Soil Seed Banks*, ed. M. A. Leck, V. T. Parker, & R. L. Simpson. San Diego: Academic Press, pp. 25–51.

Lovelock, C. Kyllo, D., Popp, M., *et al.* (1997). Symbiotic vesicular-arbuscular mycorrhizae influence maximum rates of photosynthesis in tropical tree seedlings grown under elevated CO_2. *Australian Journal of Plant Physiology*, **24**, 185–94.

Lubbock, J. (1892). *A Contribution to Our Knowledge of Seedlings*. Vols. 1 & 2 (Kegan Paul, Trench, Trubner & Co.) Reprinted 1978. New York: Allanheld, Osmun / Universe Books.

Ludwig, J. A., Wilcox, B. P., Breshears, D. D., Tongway, D. J., & Imeson, A. C. (2005). Vegetation patches and runoff-erosion as interacting ecohydrological processes in semiarid landscapes. *Ecology*, **86**, 288–97.

Lugo, A. E. (1997). The apparent paradox of reestablishing species richness on degraded lands with tree monocultures. *Forest Ecology and Management*, **99**, 9–19.

Lugo, A. E. & Snedaker, S. C. (1974). The ecology of mangroves. *Annual Review of Ecology and Systematics*, **5**, 39–64.

Luken, J. O. (2005a). Habitats of *Dionaea muscipula* (Venus' Fly Trap), Droseraceae, associated with Carolina Bays. *Southeastern Naturalist*, **4**, 573–84.

Luken, J. O. (2005b). *Dionaea muscipula* (Venus flytrap) establishment, release, and response of associated species in mowed patches on the rims of Carolina Bays. *Restoration Ecology*, **13**, 678–84.

Lumba, S. & McCourt, P. (2005). Preventing leaf identity theft with hormones. *Current Opinion in Plant Biology*, **8**, 501–5.

Lundgren, M. R. & Sultan, S. E. (2005). Seedling expression of cross-generational plasticity depends on reproductive architecture. *American Journal of Botany*, **92**, 377–81.

Lunt, P. H. & Hedger, J. N. (2003). Effects of organic enrichment of mine spoil on growth and nutrient uptake in oak seedlings inoculated with selected ectomycorrhizal fungi. *Restoration Ecology*, **11**, 125–30.

Lupia, R., Lidgard, S., & Crane, P. R. (1999). Comparing palynological abundance and diversity: implications for biotic replacement during the Cretaceous angiosperm radiation. *Paleobiology*, **25**, 305–40.

Lusk, C. H. (2004) Leaf area and growth of juvenile temperate evergreens in low light: species of contrasting shade tolerance change rank during ontogeny. *Functional Ecology*, **18**, 820–8.

Lusk, C. H., Contreras, O., & Figueroa, J. (1997). Growth, biomass allocation and plant nitrogen concentration in Chilean temperate rainforest tree seedlings: effects of nutrient availability. *Oecologia*, **109**, 49–58.

Lusk, C. H. & Del Pozo, A. (2002). Survival and growth of seedlings of 12 Chilean rainforest trees in two light environments: gas exchange and biomass distribution correlates. *Austral Ecology*, **27**, 173–82.

Lusk, C. H., Falster, D. S., Perez-Millaqueo, M., & Saldana, A. (2006). Ontogenetic variation in light interception, self-shading and biomass distribution of seedlings of the conifer *Araucaria araucana* (Molina) K. Koch. *Revista Chilena De Historia Natural*, **79**, 321–8.

Lusk, C. H. & Kelly, C. K. (2003). Interspecific variation in seed size and safe sites in a temperate rain forest. *New Phytologist*, **158**, 535–41.

Lusk, C. H. & Piper, F. I. (2007). Seedling size influences relationships of shade tolerance with carbohydrate-storage patterns in a temperate rainforest. *Functional Ecology*, **21**, 78–86.

Lüttge, U. (1983). Ecophysiology of carnivorous plants. In *Encyclopedia of Plant Physiology*, Vol. 12C, ed. O. L. Lange, P. S. Nobel, C. B. Osmond, & H. Ziegler. Berlin: Springer-Verlag, pp. 489–517.

Lwanga, J. S. (2003). Forest succession in Kibale National Park, Uganda: implications for forest restoration and management. *African Journal of Ecology*, **41**, 9–22.

Mabberley, D. J. (1997). *The Plant Book*. Cambridge: Cambridge University Press.

MacArthur, R. H. & Levins, R. (1964). Competition, habitat selection and character displacement. *Proceedings of the National Academy of Sciences* (USA), **51**, 2581–93.

Mack, A. L. (1997). Spatial distribution, fruit production and seed removal of a rare, dioecious canopy tree species (*Aglaia* aff. *flavida* Merr. et Perr.) in Papua New Guinea. *Journal of Tropical Ecology*, **13**, 305–16.

Mack, R. N. (1996). Predicting the identity and fate of plant invasives: emergent and emerging approaches. *Biological Conservation*, **78**, 107–21.

Mack, R. N. & Pyke, D. A. (1984). The demography of *Bromus tectorum*: the role of microclimate, grazing and disease. *Journal of Ecology*, **72**, 731–48.

Madison, M. (1977). Vascular epiphytes: their systematic occurrence and salient features. *Selbyana*, **2**, 1–13.

Maguire, D. A. & Forman, R. T. T. (1983). Herb cover effects on tree seedling patterns in a mature hemlock-hardwood forest. *Ecology*, **64**, 1367–80.

Maheshwari, P. (1950). *An Introduction to the Embryology of Angiosperms*. New York: McGraw-Hill.

Main, R. R. (1981). Plants as animal food. In *The Biology of Australian Plants*, ed. J. S. Pate & A. J. McComb. Nedlands: University of Western Australia Press, pp. 342–60.

Maiti, R. K., Ramaiah, K. V., Bisen, S. S., & Chidley, V. L. (1984). Comparative study of the haustorial development of *Striga asiatica* on Sorhum cultivars. *Annals of Botany*, **54**, 447–58.

Makana, J. R. & Thomas, S. C. (2004). Dispersal limits natural recruitment of African mahoganies. *Oikos*, **106**, 67–72.

Malm, T. (2006). Reproduction and recruitment of the seagrass *Halophila stipulacea*. *Aquatic Botany*, **85**, 345–9.

Malmer, A. & Grip, H. (1990). Soil disturbance and loss of infiltratability caused by mechanized and manual extractions of tropical rainforest in Sabah, Malaysia. *Forest Ecology and Management*, **38**, 1–12.

Mangan, J. M., Overpeck, J. T., Webb, R. S., Wessman, C., & Goetz, A. F. H. (2004). Response of Nebraska Sand Hills natural vegetation to drought, fire, grazing, and plant functional type shifts as simulated by the century model. *Climatic Change*, **63**, 49–90.

Mangan, S. A., & Adler, G. H. (1999). Consumption of arbuscular mycorrhizal fungi by spiny rats (*Proechimys semispinosus*) in eight isolated populations. *Journal of Tropical Ecology*, **15**, 779–90.

Mantegazza, R., Möller, M., Harrison, C. J., *et al.* (2007). Anisocotyly and meristem initiation in an unorthodox plant, *Streptocarpus rexii* (Gesneriaceae). *Planta*, **225**, 653–63.

Mapes, G., Rothwell, G. W., & Haworth, M. T. (1989). Evolution of seed dormancy. *Nature*, **337**, 645–6.

Maranon, T. & Grubb, P. J. (1993). Physiological basis and ecological significance of the seed size and relative growth rate relationship in Mediterranean annuals. *Functional Ecology*, **7**, 591–9.

Marcar, N. E., Crawford, D. F., Saunders, A., *et al.* (2002). Genetic variation among and within provenances and families of *Eucalyptus grandis* W. Hill and *E. globulus* Labill. subsp. *globulus* seedlings in response to salinity and waterlogging. *Forest Ecology and Management*, **162**, 231–49.

Mares, M. A. & Rosenzweig, M. L. (1978). Granivory in North and South American deserts: rodents, birds, and ants. *Ecology*, **59**, 235–41.

Marks, P. L. & Gardescu, S. (1998). A case study of sugar maple (*Acer saccharum*) as a forest seedling bank species. *Journal of the Torrey Botanical Society*, **125**, 287–96.

Marlette, G. M. & Anderson, J. E. (1986). Seed banks and propagule dispersal in crested-wheatgrass stands. *Journal of Applied Ecology*, **23**, 161–75.

Maron, J. L., Vilà, M., Bommarco, R., Elmendorf, S., & Beardsley, P. (2004). Rapid evolution of an invasive plant. *Ecological Monographs*, **74**, 261–80.

Marr, D. L. & Marshall, M. L. (2006). The role of fungal pathogens in flower size and seed mass variation in three species of *Hydrophyllum* (Hydrophyllaceae). *American Journal of Botany*, **93**, 389–98.

Marrero, J. (1942). A seed storage study of Maga. *Caribbean Forester*, **3**, 173–84.

Marshall, J. D. & Ehleringer, J. R. (1990). Are xylem-tapping mistletoes partially heterotrophic? *Oecologia*, **84**, 244–8.

Marthews, T. R. (2007). Modelling regeneration in tropical forests. PhD dissertation. Aberdeen University, UK.

Martin, A. C. (1946). The comparative internal morphology of seeds. *American Midland Naturalist*, **36**, 513–660.

Martin, L. M. & Wilsey, B. J. (2006). Assessing grassland restoration success: relative roles of seed additions and native ungulate activities. *Journal of Applied Ecology*, **43**, 1098–109.

Martin, P. H. & Marks, P. L. (2006). Intact forests provide only weak resistance to a shade-tolerant invasive Norway maple (*Acer platanoides* L.). *Journal of Ecology*, **94**, 1070–9.

Martin, T. J. & Ogden, J. (2002). The seed ecology of *Ascarina lucida*: a rare New Zealand tree adapted to disturbance. *New Zealand Journal of Botany*, **40**, 397–404.

Martin, T. J. & Ogden, J. (2005). Experimental studies on the drought, waterlogging, and frost tolerance of *Ascarina lucida* Hook. f (Chloranthaceae) seedlings. *New Zealand Journal of Ecology*, **29**, 53–9.

Martínez, I., García, D., & Obeso, J. R. (2007). Allometric allocation in fruit and seed packaging conditions the conflict among selective pressures on seed size. *Evolutionary Ecology*, **21**, 517–33.

Marx, D. H. (1969). The influence of ectotrophic ectomycorrhizal fungi on the resistance of pine roots to pathogenic infections. I. Antagonism of mycorrhizal fungi to pathogenic fungi and soil bacteria. *Phytopathology*, **59**, 153–63.

Marx, D. H. (1973). Mycorrhizae and feeder root diseases. In *Ectomycorrhizae: Their Ecology and Physiology*, ed. G. C. Marks & T. T. Kozlowski. New York: Academic Press, pp. 351–82.

Marx, L. M. & Walters, M. B. (2006). Effects of nitrogen supply and wood species on *Tsuga canadensis* and *Betula alleghaniensis* seedling growth on decaying wood. *Canadian Journal of Forest Research*, **36**, 2873–84.

Massey, F. P., Massey, K., Press, M. C., & Hartley, S. E. (2006). Neighborhood composition determines growth, architecture and herbivory in tropical rain forest tree seedlings. *Journal of Ecology*, **94**, 646–55.

Masuhara, G. & Katsuya, K. (1994). *In situ* and *in vitro* specificity between *Rhizoctonia* spp. and *Spiranthes sinensis* (Persoon.) Ames. var. *amoena* (M. Bieberstein) Hara (Orchidaceae). *New Phytologist*, **127**, 711–18.

Mathews, S. & Donoghue, M. J. (1999). The root of angiosperm phylogeny inferred from duplicate phytochrome genes. *Science*, **286**, 947–50.

Matsubayashi, Y. & Sakagami, Y. (2006). Peptide hormones in plants. *Annual Review of Plant Biology*, **57**, 649–74.

Maun, M. A. (1985). Population biology of *Ammophila breviligulata* and *Calamovilfa longifolia* on Lake Huron sand dunes. 1. Habitat, growth form, reproduction, and establishment. *Canadian Journal of Botany*, **63**, 113–24.

Maun, M. A. (1989). Population biology of *Ammophila breviligulata* and *Calamovilfa longifolia* on Lake Huron sand dunes. 3. Dynamic changes in plant community structure. *Canadian Journal of Botany*, **67**, 1267–70.

Maun, M. A. (1994). Adaptations enhancing survival and establishment of seedlings on coastal dune systems. *Vegetatio*, **111**, 59–70.

Maun, M. A. (1998). Adaptations of plants to burial in coastal sand dunes. *Canadian Journal of Botany*, **76**, 713–38.

Maun, M. A., Elberling, H., & D'Ulisse, A. (1996). The effects of burial by sand on survival and growth of Pitcher's thistle (*Circium pitcheri*) along Lake Huron. *Journal of Coastal Conservation*, **2**, 3–12.

Maun, M. A. & Lapierre, J. (1986). Effects of burial by sand on seed germination and seedling emergence of four dune species. *American Journal of Botany*, **73**, 450–5.

Maxwell, G. S. (1995). Ecogeographic variation in *Kandelia candel* from Brunei, Hong Kong and Thailand. *Hydrobiologia*, **295**, 59–65.

Mayaux, P., Holmgren, P., Achard, F., *et al.* (2005). Tropical forest cover change in the 1990s and the options for future monitoring. *Philosophical Transactions of the Royal Society Series B*, **360**, 373–84.

Mayer, A. M. & Poljakoff-Mayber, A. (1963). *The Germination of Seeds*. Oxford: Pergamon Press, Ltd.

Mazon, G., Kidron, G. J., Vonshak, A., & Abeliovich, A. (1996). The role of cyanobacterial exopolysaccharides in structuring desert microbial crusts. *FEMS Microbiology Ecology*, **21**, 121–30.

McCarthy, B. C. & Facelli, J. M. (1990). Microdisturbances in oldfields and forests: implications for woody seedling establishment. *Oikos*, **58**, 55–60.

McClure, F. (1966). *The Bamboos – a Fresh Perspective*. Cambridge: Harvard University Press.

McComb, A. J., Cambridge, M. L., Kirkman, H., & Kuo, J. (1981). The biology of Australian seagrasses. In *The Biology of Australian Plants*, ed. J. S. Pate & A. J. McComb. Nedlands: University of Western Australia Press, pp. 258–93.

McCormick, J. F. (1995). A review of the population dynamics of selected tree species in the Luquillo experimental forest, Puerto Rico. In *Tropical Forests: Management and Ecology*, ed. A. E. Lugo & C. Lowe. New York: Springer-Verlag, pp. 224–57.

McCormick, M. K., Whigham, D. F., & O'Neill, J. (2004). Mycorrhizal diversity in photosynthetic terrestrial orchids. *New Phytologist*, **163**, 425–38.

McCormick, M. K., Whigham, D. F., Sloan, D., O'Malley, K., & Hopkinson, B. (2006). Orchid-fungal fidelity: a marriage meant to last? *Ecology*, **87**, 903–11.

McCourt, P. (1999). Genetic analysis of hormone signaling. *Annual Review of Plant Physiology and Plant Molecular Biology*, **50**, 219–43.

McCourt, P., Lumba, S., Tsuchiya, Y., & Gazzarrini, S. (2005). Crosstalk and abscisic acid: the roles of terpenoid hormones in coordinating development. *Physiologia Plantarum*, **123**, 147–52.

McEvoy, P. B. & Coombs, E. M. (1999). A parsimonious approach to biological control of plant invaders. *Ecological Applications*, **9**, 387–401.

McGraw, J. B., Gottschalk, K. W., Vavrek, M. C., & Chester, A. L. (1990). Interactive effects of resource availabilities and defoliation on photosynthesis, growth, and mortality of red oak seedlings. *Tree Physiology*, **7**, 247–54.

McKay, J. K., Christian, C. E., Harrison, S., & Rice, K. J. (2005). "How local is local?" – Review of practical and conceptual issues in the genetics of restoration. *Restoration Ecology*, **13**, 432–40.

McKendrick, S. L., Leake, J. R., & Read, D. J. (2000a). Symbiotic germination and development of myco-heterotrophic plants in nature: transfer of carbon from ectomycorrhizal *Salix repens* and *Betula pendula* to the orchid *Corallorhiza trifida* through shared hyphal connections. *New Phytologist*, **145**, 539–48.

McKendrick, S. L., Leake, J. R., Taylor, D. L., & Read, D. J. (2000b). Symbiotic germination and development of myco-heterotrophic plants in nature: ontogeny of *Corallorhiza trifida* and characterization of its mycorrhizal fungi. *New Phytologist*, **145**, 523–37.

McKendrick, S. L., Leake, J. R., Taylor, D. L., & Read, D. J. (2002). Symbiotic germination and development of the myco-heterotrophic orchid *Neottia nidus-avis* in nature and its requirement for locally distributed *Sebacina* spp. *New Phytologist*, **154**, 233–47.

McLeod, K. W. & McPherson, J. K. (1973). Factors limiting the distribution of *Salix nigra*. *Bulletin of the Torrey Botanical Club*, **100**, 102–10.

McMillan, C. (1974). Salt tolerance of mangroves and submerged aquatic plants. In *Ecology of Halophytes*, ed. R. J. Reinold & W. H. Queen. New York: Academic Press, pp. 379–90.

McPherson, K. & Williams, K. (1998). The role of carbohydrate reserves in the growth, resilience, and persistence of cabbage palm seedlings (*Sabal palmetto*). *Oecologia*, **117**, 460–8.

Meeuse, A. D. J. 1963. A possible case of interdependence between a mammal and a higher plant. *Archives Neerlandaises de Zoologie*, **13**, 314–18.

Mehra P. N. & Handoo, O. N. (1953). Morphology of *Anthoceros erectus* and *Anthoceros himalayensis*, and the phylogeny of the Anthocerotales. *Botanical Gazette*, **114**, 371–82.

Meiners, S. J. & Handel, S. N. (2000). Additive and nonadditive effects of herbivory and competition on tree seedling mortality, growth, and allocation. *American Journal of Botany*, **87**, 1821–6.

Meiners, S. J., Pickett, S. T. A., & Cadenasso, M. L. (2001). Beyond biodiversity: individualistic controls of invasion in a self-assembled community. *Ecology Letters*, **7**, 121–6.

Meir, A. & Tsoar, H. (1996). International borders and range ecology: the case of Bedouin transborder grazing. *Human Ecology*, **24**, 39–64.

Mendez, M. (1997). Sources of variation in seed mass in *Arum italicum*. *International Journal of Plant Sciences*, **158**, 298–305.

Merryweather, J. & Fitter, A. (1996). Phosphorus nutrition of an obligately mycorrhizal plant treated with the fungicide benomyl in the field. *New Phytologist*, **132**, 307–11.

Mesquita, R. C. G., Ickes, K., Ganade, G., & Williamson, G. B. (2001) Alternative successional pathways in the Amazon Basin. *Journal of Ecology*, **89**, 528–37.

Messier, C., Doucet, R., Ruel, J.-C., Clavewa, Y., Kelly, C., & Lechowicz, M. J. (1999). Functional ecology of advance regeneration in relation to light in boreal forests. *Canadian Journal of Forest Research*, **29**, 812–23.

Metcalf, C. J. E., Rees, M., Alexander, J. M., & Rose, K. (2006). Growth-survival trade-offs and allometries in rosette-forming perennials. *Functional Ecology*, **20**, 217–25.

Metcalfe, D. J. (2005). Hedera helix L. *Journal of Ecology*, **93**, 632–48.

Metcalfe, D. J. & Grubb, P. J. (1997). The responses to shade of seedlings of very small-seeded tree and shrub species from tropical rain forest in Singapore. *Functional Ecology*, **11**, 215–21.

Metcalfe, D. J., Grubb, P. J., & Turner, I. M. (1998). The ecology of very small-seeded shade-tolerant trees and shrubs in lowland rain forest in Singapore. *Plant Ecology*, **134**, 131–49.

Meyer, S. E. & Carlson, S. L. (2001). Achene mass variation in *Ericameria nauseosus* (Asteraceae) in relation to dispersal ability and seedling fitness. *Functional Ecology*, **15**, 274–81.

Miao, S. L., Borer, R. E., & Sklar, F. H. (1997). Sawgrass seedling responses to transplanting and nutrient additions. *Restoration Ecology*, **5**, 162–8.

Michaels, H. J., Benner, B., Hartgerinck, A. P., *et al.* (1988). Seed size variation: magnitude, distribution, and ecological correlates. *Evolutionary Ecology*, **2**, 157–66.

Mihulka, S., Pyšek, P., Martínková, J., & Jarošík, V (2006). Invasiveness of *Oenothera* congeners alien to Europe: jack of all trades, master of invasion? *Perspectives in Plant Ecology, Evolution & Systematics*, **8**, 83–96.

Mikola, P. (1970). Mycorrhizal inoculation in afforestation. *International Review of Forest Research*, **3**, 123–96.

Milberg, P. & Lamont, B. B. (1997). Seed/cotyledon size and nutrient content play a major role in early performance of species on nutrient-poor soils. *New Phytologist*, **137**, 665–72.

Milcu, A., Schumacher, J., & Scheu, S. (2006). Earthworms (*Lumbricus terrestris*) affect plant seedling recruitment and microhabitat heterogeneity. *Functional Ecology*, **20**, 261–8.

Miller, G. & Mittler, R. (2006). Could heat shock transcription factors function as hydrogen peroxide sensors in plants? *Annals of Botany*, **98**, 279–88.

Miller, J. H. & Miller, K. V. (2005). *Food Plants of the Southeast and Their Wildlife Uses*. Athens: University of Georgia.

Miller, R. D. (1972). Freezing and heaving of saturated and unsaturated soils. *Highway Research Record*, **393**, 1–11.

Milton, S. J. & Dean, W. R. J. (2000). Disturbance, drought and dynamics of desert dune grassland, South Africa. *Plant Ecology*, **150**, 37–51.

Milton, S. J., Dean, W. R. J., & Klotz, S. (1997). Thicket formation in abandoned fruit orchards: processes and implications for the conservation of semi-dry grasslands in Central Germany. *Biodiversity and Conservation*, **6**, 275–90.

Milton, S. J., Yeaton, R. I., Dean, W. R. J., & Vlok, J. H. H. (1997). Succulent karoo. In *Vegetation of Southern Africa*, ed. R. Cowling, D. Richardson, & S. Pierce. Cambridge: Cambridge University Press. pp. 131–66.

Miquel, S. (1987). Morphologie fonctionnelle de plantules d'espèces forestières du Gabon. *Bulletin du Musèum National d'Histoire Naturelle, 4c série, section B, Andansonia*, **9**, 101–21.

Miriti, M. N. (2006). Ontogenic shift from facilitation to competition in a desert shrub. *Journal of Ecology*, **94**, 973–9.

Mischler, B. D., Lewis, L. A., Buchheim, M. A., *et al.* (1994). Phylogenetic relationships of the "green algae" and "bryophytes." *Annals of the Missouri Botanical Garden*, **81**, 451–83.

Mitchell-Olds, T. (2001). *Arabidopsis thaliana* and its wild relatives: a model system for ecology and evolution. *Trends in Ecology & Evolution*, **16**, 693–700.

Miyanishi, K. & Kellman, M. (1986). The role of root nutrient reserves in regrowth of two savanna shrubs. *Canadian Journal of Botany*, **64**, 1244–8.

Modjo, H. S. & Hendrix, J. W. (1986). The mycorrhizal fungus *Glomus macrocarpum* as a cause of tobacco stunt disease. *Phytopathology*, **76**, 688–91.

Moegenburg, S. M. (1996). Sabal palmetto seed size – causes of variation, choices of predators, and consequences for seedlings. *Oecologia*, **106**, 539–43.

Mogie, M., Latham, J. R., & Warman, E. A. (1990). Genotype-independent aspects of seed ecology in *Taraxacum*. *Oikos*, **59**, 175–82.

Molau, U. (1992). On the occurrence of sexual reproduction in *Saxifraga ceruna* and *S. foliolosa* (Saxifragaceae). *Nordic Journal of Botany*, **12**, 197–203.

Moles, A. T., Ackerly, D. D., Tweddle, J. C., *et al.* (2007). Global patterns in seed size. *Global Ecology and Biogeography*, **16**, 106–16.

Moles, A. T., Ackerly, D. D., Webb, C. O., *et al.* (2005a). Factors that shape seed mass evolution. *Proceedings of the National Academy of Sciences* (USA), **102**, 10540–4.

Moles, A. T., Ackerly, D. D., Webb, C. O., *et al.* (2005b). A brief history of seed size. *Science*, **307**, 576–80.

Moles, A. T., Ackerly, D. D., Webb, C. O., *et al.* (2005c). Response to comment on "A brief history of seed size." *Science*, **310**, 783.

Moles, A. T., Falster, D. S., Leishman, M. R., & Westoby, M. (2004). Small-seeded species produce more seeds per square metre of canopy per year, but not per individual per lifetime. *Journal of Ecology*, **92**, 384–96.

Moles, A. T., Hodson, D. W., & Webb, C. J. (2000). Do seed size and shape predict persistence in soil in New Zealand? *Oikos*, **89**, 541–5.

Moles, A. T., Warton, D. I., Stevens, R. D., & Westoby, M. (2004). Does a latitudinal gradient in seedling survival favor larger seeds in the tropics? *Ecology Letters*, **7**, 911–14.

Moles, A. T., Warton, D. I., & Westoby, M. (2003). Do small-seeded species have higher survival through seed predation than large-seeded species? *Ecology*, **84**, 3148–61.

Moles, A. T. & Westoby, M. (2002). Seed addition experiments are more likely to increase recruitment in larger-seeded species. *Oikos*, **99**, 241–8.

Moles, A. T. & Westoby, M. (2003). Latitude, seed predation and seed mass. *Journal of Biogeography*, **30**, 105–28.

Moles, A. T. & Westoby, M. (2004a). Seedling survival and seed size: a synthesis of the literature. *Journal of Ecology*, **92**, 372–83.

Moles, A. T. & Westoby, M. (2004b). Seedling establishment after fire in Kuring-gai Chase National Park, Sydney, Australia. *Austral Ecology*, **29**, 383–90.

Moles, A. T. & Westoby, M. (2004c). What do seedlings die from, and what are the implications for evolution of seed size? *Oikos*, **106**, 193–9.

Moles, A. T. & Westoby, M. (2006a). Seed size and plant strategy across the whole life cycle. *Oikos*, **113**, 91–105.

Moles, A. T. & Westoby, M. (2006b). Seedling survival and seed size: a synthesis of the literature. *Journal of Ecology*, **92**, 372–83.

Molina, R., Massicotte, H., & Trappe, J. M. (1992). Specificity phenomena in mycorrhizal symbioses: community-ecological consequences and practical implications. In *Mycorrhizal Functioning: An Integrative Plant-fungal Process*, ed. M. F. Allen. New York: Chapman and Hall, pp. 357–423.

Molina, R. & Trappe, J. M. (1982). Lack of mycorrhizal specificity by the ericaceous hosts *Arbutus menziesii* and *Arctostaphylos uva-ursi*. *New Phytologist*, **90**, 485–509.

Molina, R. & Trappe, J. M. (1994). Biology of the ectomycorrhizal genus, *Rhizopogon*. I. Host associations, host-specificity and pure culture syntheses. *New Phytologist*, **126**, 653–75.

Molina, R., Trappe, J. M., Grubisha, L. C., & Spatafora, J. W. (1999). *Rhizopogon*. In *Ectomycorrhizal Fungi: Key Genera in Profile*, ed. J. W. G. Cairney & S. M. Chambers. Berlin: Springer-Verlag, pp. 129–61.

Molino, J.-F. & Sabatier, D. (2001). Tree diversity in a tropical rain forest: a validation of the intermediate disturbance hypothesis. *Science*, **294**, 1702–4.

Montagnini, F., Gonzalez, E., Porras, C., & Rheingans, R. (1995). Mixed and pure forest plantations in the humid neotropics: a comparison of early growth, pest damage and establishment costs. *Commonwealth Forestry Review*, **74**, 306–14.

Montaña, C. (1992). The colonization of bare areas in two-phase mosaics of an arid ecosystem. *Journal of Ecology*, **80**, 315–27.

Mooney, H. A. & Drake, J. A. (1989). Biological invasions: a SCOPE program overview. In *Biological Invasions: A Global Perspective*, ed. J. A. Drake, H. A. Mooney, F. diCastri, *et al.*, New York: John Wiley & Sons, pp. 491–507.

Mooney, H. A., Gulmon, L., & Johnson, N. D. (1983). Physiological constraints on plant chemical defenses. In *Plant Resistance to Insects*, ed. P. A. Hedin. Washington: American Chemical Society, pp. 21–36.

Mopper, S. & Agrawal, A. A. (2004). Phytohormonal ecology. *Ecology*, **85**, 3–4.

Mopper, S., Wang, Y., Criner, C., & Hasenstein, K. (2004). *Iris hexagona* hormonal responses to salinity stress, leaf miner herbivory, and phenology. *Ecology*, **85**, 38–47.

Morris, A. B., Bell, C. D., Clayton, J. W., *et al.* (2007). Phylogeny and divergence time estimation in *Illicium* with implications for New World biogeography. *Systematic Botany*, **32**, 236–49.

Morse, D. H. & Schmitt, J. (1985). Propagule size, dispersal ability, and seedling performance in *Asclepias syriaca*. *Oecologia*, **67**, 372–9.

Mott, J. J. (1972). Germination studies on some annual species from arid region of Western Australia. *Journal of Ecology*, **60**, 293–304.

Mousseau, T. A. & Fox, C. W., ed. (1998). *Maternal Effects as Adaptations*. New York: Oxford University Press.

Moyes, A. B., Witter, M. S., & Gamon, J. A. (2005). Restoration of native perennials in a California annual grassland after prescribed spring burning and solarization. *Restoration Ecology*, **13**, 659–66.

Mrkvicka, A. C. (1990). Neue Bobachtungen zu Samenkeimung und Entwicklung von *Liparis loeselii* (L.) Rich. *Mitteilungsblatt des Arbeitskreis heimische Orchideen Baden-Wurttembert*, **22**, 172–80.

Muenscher, W. C. (1944). *Aquatic Plants of the United States*. Ithaca: Comstock Publishing Co., Inc.

Mukhammedov, G., Durikov, M., & Nechaeva, N. T. (1999). The technology of desert pasture improvement. In *Desert Problems and Desertification in Central Asia*, ed. A. G. Babaev. Berlin: Springer, pp. 101–14.

Mulder, L., Hogg, B., Bersoult, A., & Cullimore, J. V. (2005). Integration of signalling pathways in the establishment of the legume-rhizobia symbiosis. *Physiologia Plantarum*, **123**, 207–18.

Muller, F. M. (1978). *Seedlings of the North-western European Lowland: A Flora of Seedlings*. Wageningen: Dr. W. Junk B. V. Publishers.

Muller-Landau, H. C. (2004). Interspecific and inter-site variation in wood specific gravity of tropical trees. *Biotropica*, **36**, 20–32.

Muller-Landau, H. C., Wright, S. J., Calderón, O., Hubbell, S. P., & Foster, R. (2002). Assessing recruitment limitation: concepts, methods and case-studies from a tropical forest. In *Seed Dispersal and Frugivory: Ecology, Evolution and Conservation*, ed. D. J. Levey, W. R. Silva & M. Galetti. Wallingford: CAB International, pp. 35–53.

Münzbergová, Z. & Herben, T. (2005). Seed, dispersal, microsite, habitat and recruitment limitation: identification of terms and concepts in studies of limitations. *Oecologia*, **145**, 1–8.

Murawski, D. A., Gunatilleke, I. A. U. N., & Bawa, K. S. (1994). The effects of selective logging on inbreeding in *Shorea megistophylla* (Dipterocarpaceae) from Sri Lanka. *Conservation Biology*, **8**, 997–1002.

Musselman, L. J. & Dickison, W. C. (1975). The structure and development of the haustorium in parasitic Scrophulariaceae. *Botanical Journal of the Linnean Society*, **70**, 183–212.

Musselman, L. J. & Press, M. C. (1995). Introduction to parasitic plants. In *Parasitic Plants*, ed. M. C. Press & J. D. Graves. London: Chapman and Hall, pp. 1–13.

Mutch, L. S. & Parsons, D. J. (1998). Mixed conifer forest mortality and establishment before and after prescribed fire in Sequoia National Park, California. *Forest Science*, **44**, 341–55.

Myers, J. A. & Kitajima, K. (2007). Carbohydrate storage enhances seedling shade and stress tolerance in a neotropical forest. *Journal of Ecology*, **95**, 383–95.

Myster, R. W. (1993). Tree invasion and establishment in old fields at Hutcheson Memorial Forest. *Botanical Review*, **59**, 251–72.

Nadarajah, P. & Nawawi, A. (1993). Mycorrhizal status of epiphytes in Malaysian oil palm plantations. *Mycorrhiza*, **4**, 21–5.

Nagalingum, N., Drinnan, A. N., McLoughlin, S., & Lupia, R. (2002). Patterns of fern diversification in the Cretaceous of Australia. *Review of Palaeobotany and Palynology*, **119**, 69–92.

Napier, R., M., David, K. M., & Perrot-Rechenmann, C. (2002). A short history of auxin-binding proteins. *Plant Molecular Biology*, **49**, 339–48.

Nara, K. (2006a). Ectomycorrhizal networks and seedling establishment during early primary succession. *New Phytologist*, **169**, 169–78.

Nara, K. (2006b). Pioneer dwarf willow may facilitate tree succession by providing late colonizers with compatible ectomycorrhizal fungi in a primary successional volcanic dessert. *New Phytologist*, **171**, 187–98.

Nara, K. & Hogetsu, T. (2004). Ectomycorrhizal fungi on established shrubs facilitate subsequent seedling establishment of successional plant species. *Ecology*, **85**, 1700–7.

Nathan, R. & Muller-Landau, H. C. (2000). Spatial patterns of seed dispersal, their determinants and consequences for recruitment. *Trends in Ecology & Evolution*, **15**, 278–85.

Nechaeva, N. T. (1985). *Improvement of Desert Ranges in Soviet Central Asia*. Chur: Harwood Academic Publishers.

Neill, S. J., Horgan, R., & Rees, A. G. (1987). Seed development and vivipary in *Zea mays* L. *Planta*, **171**, 358–64.

Nemhauser, J. L., Hong, F., & Chory, J. (2006). Different plant hormones regulate similar processes through largely nonoverlapping transcriptional responses. *Cell*, **126**, 467–75.

Nepstad, D. C., Uhl, C., & Serrão, E. A. S. (1991). Recuperation of a degraded Amazonian landscape: forest recovery and agricultural restoration. *Ambio*, **20**, 248–55.

Nepstad, D. C., Verissimo, A., Alencar, A., *et al.* (1999). Large-scale impoverishment of Amazonian forests by logging and fire. *Nature*, **398**, 505–8.

Newberry, D. M., Alexander, I. J., & Rother, J. A. (2000). Does proximity to conspecific adults influence the establishment of ectomycorrhizal trees in rain forest. *New Phytologist*, **147**, 401–9.

Newell, E. A., McDonald, E. P., Strain, B. R., & Denslow, J. S. (1993). Photosynthetic responses of *Miconia* species to canopy openings in a lowland tropical rain forest. *Oecologia*, **94**, 49–56.

Newman, E. I. (1988). Mycorrhizal links between plants: their functioning and ecological significance. *Advances in Ecological Research*, **18**, 243–70.

Newsham, K. K., Fitter, A. H., & Merryweather, J. W. (1995). Multifunctionality and biodiversity in arbuscular mycorrhizas. *Tree*, **10**, 407–11.

Newton, A. C., Baker, P., Ramnarine, S., Mesen, J. F., & Leakey, R. R. B. (1993). The mahogany shoot borer: prospects for control. *Forest Ecology and Management*, **57**, 301–28.

Ng, F. S. P. (1978). Strategies of establishment in Malayan forest trees. In *Tropical Trees as Living Systems*, ed. P. B. Tomlinson & M. H. Zimmermann. Cambridge: Cambridge University Press, pp. 129–62.

Ng, F. S. P. (1992). *Manual of Forest Fruits, Seeds and Seedlings*. Kuala Lumpur: Forest Research Institute of Malaysia.

Nichols, J. D., Wagner, M. R., Agyeman, V. K., Bosu, P., & Cobbinah, J. R. (1998). Influence of artificial gaps in tropical forest on the survival, growth and *Phytolyma lata* attack on *Milicia excelsa*. *Forest Ecology and Management*, **110**, 353–62.

Nickrent, D. L. (2003). Parasitic Plant Connection. Southern Illinois University at Carbondale (http://www.parasiticplants.siu.edu/ParPlantNumbers.pdf; November 6, 2006).

Nickrent, D. L., Duff, R. J., Colwell, A. E., *et al.* (1998). Molecular phylogenetic and evolutionary studies of parasitic plants. In *Molecular Systematics of Plants. II. DNA Sequencing*, ed. D. E. Soltis, P. S. Soltis, & J. J. Doyle. Boston: Kluwer Academic, pp. 211–41.

Nicotra, A. B., Babicka, N., & Westoby, M. (2002). Seedling root anatomy and morphology: an examination of ecological differentiation with rainfall using phylogenetically independent contrasts. *Oecologia*, **130**, 136–45.

Nicotra, A. B., Chazdon, R. L., & Iriate, S. V. B. (1999). Spatial heterogeneity of light and woody seedling regeneration in tropical wet forests. *Ecology*, **80**: 1908–26.

Niering, W. A., Whittaker, R. H., & Lowe, C. H. (1963). The saguaro: a population in relation to its environment. *Science*, **142**, 15–23.

Nieuwland, J. A. (1916). Habits of waterlily seedlings. *American Midland Naturalist*, **4**, 291–7.

Nievola, C. C. & Mercier, H. (1996). The importance of leaf and root systems in nitrate assimilation in *Vriesea fosteriana*. *Bromélia*, **3**, 14–17.

Niinemets, Ü. (2006). The controversy over traits conferring shade-tolerance in trees: ontogenetic changes revisited. *Journal of Ecology*, **94**, 464–70.

Niinemets, Ü., Bilger, W., Kull, O., & Tenhunen, J. D. (1998). Acclimation to high irradiance in temperate deciduous trees in the field: changes in xanthophyll cycle pool size and in photosynthetic capacity along a canopy light gradient. *Plant Cell and Environment*, **21**, 1205–18.

Niinemets, Ü. & Kull, K. (1994). Leaf weight per area and leaf size of 85 Estonian woody species in relation to shade tolerance and light availability. *Forest Ecology and Management*, **70**, 1–10.

Niklas, K. J. (1983). The influence of Paleozoic ovule and cupule morphologies on wind pollination. *Evolution*, **37**, 968–86.

Niklas, K. J. (1997). *The Evolutionary Biology of Plants*. Chicago: University of Chicago Press.

Niklas, K. J. & Paolillo, D. J., Jr. (1990). Biomechanical and morphometric differences in *Triticum aestivum* seedlings differing in *Rht* gene-dosage. *Annals of Botany*, **65**, 365–77.

Nilsen, E. T. & Orcutt, D. M. (1996). *Physiology of Plants Under Stress*. New York: John Wiley & Sons, Inc.

Nishiyama, T. & Kato, M. (1999). Molecular phylogenetic analysis among bryophytes and tracheophytes based on combined data of plastid coded genes and the 18S rRNA gene. *Molecular Biology and Evolution*, **16**, 1027–36.

Nobel, P. S. (1984). Extreme temperatures and thermal tolerances for seedlings of desert succulents. *Oecologia*, **62**, 310–17.

Nobel, P. S. (1989). Temperature, water availability, and nutrient levels at various soil depths consequences for shallow-rooted desert succulents, including nurse plant effects. *American Journal of Botany*, **76**, 1486–92.

Nobel, P. S. & Bobich, E. G. (2002). Plant frequency, stem and root characteristics, and CO_2 uptake for *Opuntia acanthocarpa*: elevational correlates in the northwestern Sonoran Desert. *Oecologia*, **130**, 165–72.

Nobel, P. S. & Linton, M. J. (1997). Frequencies, microclimate and root properties for three codominant perennials in the northwestern Sonoran Desert on north- vs. south-facing slopes. *Annals of Botany*, **80**, 731–9.

Nobel, P. S. & Zutta, B. R. (2005). Morphology, ecophysiology, and seedling establishment for *Fouquieria splendens* in the northwestern Sonoran Desert. *Journal of Arid Environments*, **62**, 251–65.

North, G. B. & Nobel, P. S. (1997). Drought-induced changes in soil contact and hydraulic conductivity for roots of *Opuntia ficus-indica* with and without rhizosheaths. *Plant and Soil*, **191**, 249–58.

North, M., Hurteau, M., Fiegener, R., & Barbour, M. (2005). Influence of fire and El Niño on tree recruitment varies by species in Sierran mixed conifer. *Forest Science*, **51**, 186–97.

Norton, D. A. & Carpenter, M. A. (1998). Mistletoes as parasites: host specificity and speciation. *Trends in Ecology & Evolution*, **13**, 101–5.

Noy-Meir, I. (1974). Desert ecosystems: higher trophic levels. *Annual Review of Ecology and Systematics*, **5**, 195–214.

Noy-Meir, I. (1993). Compensating growth of grazed plants and its relevance to the use of rangelands. *Ecological Applications*, **3**, 32–4.

Nozue, K. & Maloof, J. N. (2006). Diurnal regulation of plant growth. *Plant, Cell and Environment*, **29**, 396–408.

Nussbaum, R., Anderson, J., & Spencer, T. (1995a). Effects of selective logging on soil characteristics and growth of planted dipterocarp seedlings in Sabah. In *Ecology, Conservation and Management of Southeast Asian Rainforests*, ed. R. B. Primack & T. E. Lovejoy. New Haven: Yale University Press, pp. 105–15.

Nussbaum, R., Anderson, J., & Spencer, T. (1995b). Factors limiting the growth of indigenous tree seedlings planted on degraded rainforest soils in Sabah, Malaysia. *Forest Ecology and Management*, **74**, 149–59.

Obeso, J. R. (1993). Seed mass variation in the perennial herb *Asphodelus albus*: sources of variation and position effect. *Oecologia*, **93**, 571–5.

Odling-Smee, F. J., Laland, K. N., & Feldman, M. W. (2003). *Niche Construction. The Neglected Process in Evolution*. Princeton: Princeton University Press.

Offer, Z. Y., Zaady, E., & Shachak, M. (1998). Aeolian particle input to the soil surface at the northern limit of the Negev Desert. *Arid Soil Research and Rehabilitation*, **12**, 55–62.

Oh, I.-C., Denk, T., & Friis, E. M. (2003). Evolution of *Illicium* (Illiciaceae): mapping morphological characters on the molecular tree. *Plant Systematics and Evolution*, **240**, 175–209.

Ohara, M., Tomimatsu, H., Takada, T., & Kawano, S. (2006). Importance of life history studies for conservation of fragmented population: a case study of the understory herb, *Trillium camschatcense*. *Plant Species Biology*, **21**, 1–12.

Ohlson, M., Okland, R. H., Nordbakken, J. F., & Dahlberg, B. (2001). Fatal interactions between Scots pine and *Sphagnum* mosses in bog ecosystems. *Oikos*, **94**, 425–32.

Okonkwo, S. N. C. & Nwoke, F. I. O. (1978). Initiation, development and structure of the primary haustorium in *Striga gesnerioides* (Scrophulariacae). *Annals of Botany*, **42**, 455–63.

Olff, H., Vera, F. W. M., Bokdam, J., *et al.* (1999). Shifting mosaics in grazed woodlands driven by the alternation of plant facilitation and competition. *Plant Biology*, **1**, 127–37.

Olmstead, R. G., Depamphilis, C. W., Wolfe, A. D., *et al.* (2001). Disintegration of the Scrophulariaceae. *American Journal of Botany*, **88**, 348–61.

Olson, B. E. & Wallander R. T. (2002). Effects of invasive forb litter on seed germination, seedling growth and survival. *Basic and Applied Ecology*, **3**, 309–17.

Olsson, P. A., Jakobsen, I., & Wallander, H. (2002). Foraging and resource allocation strategies of mycorrhizal fungi in a patchy environment, In *Mycorrhizal Ecology*, ed. M. G. A. van der Heijden & I. R. Sanders. Berlin: Springer-Verlag, pp. 93–115.

Olszyk, D., Wise, C. VanEss, E., & Tingey, D. (1998). Elevated temperature but not elevated CO_2 affects long-term patterns of stem diameter and height of Douglas-fir seedlings. *Canadian Journal of Forest Research*, **28**, 1046–54.

O'Neill, R. V. (1989). Perspectives in hierarchy and scale. In *Perspectives in Ecological Theory*, ed. J. Roughgarden, R. M. May, & S. A. Levin. Princeton: Princeton University Press, pp. 140–56.

Onguene, N. A. & Kuyper, T. W. (2002). Importance of the ectomycorrhizal network for seedling survival and ectomycorrhizal formation in rain forests of south Cameroon. *Mycorrhiza*, **12**, 13–17.

Opik, M., Moora, M., Liira, J., & Zobel, M. (2006). Composition of root-colonizing arbuscular mycorrhizal fungal communities in different ecosystems around the globe. *Journal of Ecology*, **94**, 778–90.

Oren, Y. (2000). Patchiness, disturbances, and flows of matter and organisms in an arid landscape: a multiscale experimental approach. PhD thesis, Ben-Gurion University of the Negev, Israel.

Orlovsky, L., Dourikov, M., & Babaev, A. (2004). Temporal dynamics and productivity of biogenic soil crusts in the central Karakum desert, Turkmenistan. *Journal of Arid Environments*, **56**, 579–601.

Orr, G. L., Haidar, M. A., & Orr, D. A. (1996). Smallseed dodder (*Cuscuta planiflora*) phototropism toward far-red when in white light. *Weed Science*, **44**, 233–40.

Osborne, D. J. & Berjak, P. (1997). The making of mangroves: the remarkable pioneering role played by seeds of *Avicennia marina*. *Endeavour*, **21**, 143–7.

Osem, Y., Perevolotsky, A., & Kigel, J. (2002). Grazing effect on diversity of annual plant communities in a semi-arid rangeland: interactions with small-scale spatial and temporal variation in primary productivity. *Journal of Ecology*, **90**, 936–46.

Osmond, C. B. (1981). Photo-respiration and photoinhibition some implications for the energetics of photosynthesis. *Biochimica and Biophysica Acta*, **639**, 77–98.

Osumi, K. & Sakurai, S. (2002). The unstable fate of seedlings of the small-seeded pioneer tree species, *Betula maximowicziana*. *Forest Ecology and Management*, **160**, 85–95.

Osunkoya, O. O., Ash, J. E., Hopkins, M. S., & Graham, A. W. (1994). Influence of seed size and seedling ecological attributes on shade-tolerance of rain-forest tree species in northern Queensland. *Journal of Ecology*, **82**, 149–63.

Osunkoya, O. O., Othman, F. E., & Kahar, R. S. (2005). Growth and competition between seedlings of an invasive plantation tree, *Acacia mangium*, and those of a native Borneo heath-forest species, *Melastoma beccarianum*. *Ecological Research*, **20**, 205–14.

Outred, H. A. (1973). Studies on the respiration of mangrove seedlings. PhD thesis, University of Auckland, NZ.

Ozinga, W. A., VanAndel, J., & McDonnell-Alexander, M. P. (1997). Nutritional soil heterogeneity and mycorrhiza as determinants of plant species diversity. *Acta Botanica Neerlandica*, **46**, 237–54.

Pacala, S. W., Canham, C. D., Saponara, J., Silander, J. A., & Kobe, R. K. (1996). Forest models defined by field measurements: estimation, error analysis and dynamics. *Ecological Monographs*, **66**, 1–43.

Packer, A. & Clay, K. (2000). Soil pathogens and spatial patterns of seedling mortality in a temperate tree. *Nature*, **404**, 278–81.

Packer, A. & Clay, K, (2003). Soil pathogens and *Prunus serotina* seedling and sapling growth near conspecific trees. *Journal of Ecology*, **84**, 108–19.

Packer, A. & Clay, K. (2004). Development of negative feedback during successive growth cycles of black cherry. *Proceedings of the Royal Society Series B*, **271**, 317–24.

Padilla, F. M. & Pugnaire, F. I. (2006). The role of nurse plants in the restoration of degraded environments. *Frontiers in Ecology and the Environment*, **4**, 196–202.

Pake, C. E. & Venable, D. L. (1995). Is coexistence of Sonoran Desert annuals mediated by temporal variability in reproductive success? *Ecology*, **76**, 246–61.

Pake, C. E. & Venable, D. L. (1996). Seed banks in desert annuals: implications for persistence and coexistence in variable environments. *Ecology*, **77**, 1427–35.

Palmer, J. D., Soltis, D. E., & Chase, M. W. (2004). The plant tree of life: an overview and some points of view. *American Journal of Botany*, **91**, 1437–45.

Pannier, F. & Rodriguez, M. D. P. (1967). The ß-inhibitor complex and its relation to vivipary in *Rhizophora mangle* L. *Internationale Revue der Gesamten Hydrobiologie*, **52**, 783–92.

Parciak, W. (2002). Seed size, number, and habitat of a fleshy-fruited plant: Consequences for seedling establishment. *Ecology*, **83**, 794–808.

Pareliussen, I., Olsson, E. G. A., & Armbruster, W. S. (2006). Factors limiting the survival of native tree seedlings used in conservation efforts at the edges of forest fragments in upland Madagascar. *Restoration Ecology*, **14**, 196–203.

Parihar, N. S. (1962). *An Introduction to Embryophyta*. Vol. 1. *Bryophyta*. Allahabad: Central Book Depot.

Parker, I. M., Simberloff, D., Lonsdale, W. M., *et al.* (1999). Impact: Toward a framework for understanding the ecological effects of invasives. *Biological Invasions*, **1**, 3–19.

Parker, I. M. (2000). Invasion dynamics of *Cytisus scoparius*: a matrix model approach. *Ecological Applications*, **10**, 726–43.

Parker, M. A., Malek, W., & Parker, I. M. (2006). Growth of an invasive legume is symbiont limited in newly occupied habitats. *Diversity and Distributions*, **12**, 563–71.

Parker, V. T. (1990). Problems encountered while mimicking nature in vegetation management: an example from fire-prone vegetation. In *Ecosystem Management: Rare Species and Significant Habitats*, ed. R. Mitchell, C. Sheviak & D. Leopold. Proceedings of the 15th Annual Natural Areas Conference, *New York State Museum Bulletin*, **471**, 231–4.

Parker, V. T. & Leck, M. A. (1985). Relationship of seed banks to plant distribution patterns in a freshwater tidal wetland. *American Journal of Botany*, **72**, 161–74.

Parker, V. T., Simpson, R. L., & Leck, M. A. (1989). Pattern and process in the dynamics of seed banks. In *Ecology of Soil Seed Banks*, ed. M. A. Leck, V. T. Parker, & R. L. Simpson. San Diego: Academic Press, pp. 367–84.

Parrish, J. A. D. & Bazzaz, F. A. (1985). Ontogenetic niche shifts in old-field annuals. *Ecology*, **66**, 1296–302.

Parrotta, J. A. (1992). The role of plantation forests in rehabilitating degraded tropical ecosystems. *Agriculture, Ecosystems, and Environment*, **41**, 115–33.

Parson, W. F. J., Ehrenfeld, J. G., & Handel, S. N. (1998). Vertical growth and mycorrhizal infection of woody plant roots as potential limits to the restoration of woodlands on landfills. *Restoration Ecology*, **6**, 280–9.

Parsons, D. J. & DeBenedetti, S. H. (1979). Impact of fire suppression on a mixed-conifer forest. *Forest Ecology and Management*, **2**, 21–33.

Pate, J. S. (1989). Australian micro stilt plants. *TREE*, **4**, 45–9.

Pate, J. S. (1995). Mineral relationships of parasites and their hosts. In *Parasitic Plants*, ed. M. C. Press & J. D. Graves. London: Chapman and Hall, pp. 80–102.

Pate, J. S. & Dixon, K. W. (1981). Plants with fleshy underground storage organs – a Western Australian survey. In *The Biology of Australian Plants*, ed. J. S. Pate & A. J. McComb. Nedlands: University of Western Australia Press, pp. 181–215.

Pate, J. S. & Dixon, K. W. (1982). *Tuberous, Cormous and Bulbous Plants: Biology of an Adaptive Strategy in Western Australia*. Perth: University of Western Australia Press.

Pate, J. S., Rasins, E., Rullo, J., & Kuo, J. (1985). Seed nutrient reserves of Proteaceae with special reference to protein bodies and their inclusions. *Annals of Botany*, **57**, 747–70.

Pate, J. S., True, K. C., & Kuo, J. (1991). Xylem transport and storage of amino acids by S. W. Australian mistletoes and their hosts. *Journal of Experimental Botany*, **42**, 441–51.

Paz, H. (2003). Root/shoot allocation and root architecture in seedlings: variation among forest sites, microhabitats, and ecological groups. *Biotropica*, **35**, 318–32.

Paz, H., Mazer, S. J., & Martinez-Ramos, M. (2005). Comparative ecology of seed mass in *Psychotria* (Rubiaceae): within- and between-species effects of seed mass on early performance. *Functional Ecology*, **19**, 707–18.

Pearson, A. K., Pearson, O. P., & Gomez, I. A. (1994). Biology of the bamboo *Chusquea culeou* (Poaceae: Bambusoideae) in southern Argentina. *Vegetatio*, **111**, 93–126.

Pearson, T. R. H., Burslem, D., Mullins, C. E., & Dalling, J. W. (2002). Germination ecology of neotropical pioneers: interacting effects of environmental conditions and seed size. *Ecology*, **83**, 2798–807.

Pearson, T. R. H., Burslem, D., Goeriz, R. E., & Dalling, J. W. (2003a). Regeneration niche partitioning in neotropical pioneers: effects of gap size, seasonal drought and herbivory on growth and survival. *Oecologia*, **137**, 456–65.

Pearson, T. R. H., Burslem, D. F. R. P., Goeriz, R. E., & Dalling, J. W. (2003b). Interactions of gap size and herbivory on establishment, growth and survival of three species of neotropical pioneer trees. *Journal of Ecology*, **91**, 785–96.

Peco, B., Traba, J., Levassor, C., Sanchez, A. M., & Azcarate, F. M. (2003). Seed size, shape and persistence in dry Mediterranean grass and scrublands. *Seed Science Research*, **13**, 87–95.

Pemadasa, M. A. & Lovell, P. H. (1975). Factors controlling germination of some dune annuals. *Journal of Ecology*, **63**, 41–59.

Perala, D. A. & Alm, A. A. (1990). Reproductive ecology of birch: a review. *Forest Ecology and Management*, **32**, 1–38.

Peres, C. A. (2000). Effects of subsistence hunting on vertebrate community structure in Amazonian forests. *Conservation Biology*, **14**, 240–53.

Peres, C. A. (2001). Synergistic effects of subsistence hunting and habitat fragmentation on Amazonian forest vertebrates. *Conservation Biology*, **15**, 1490–505.

Perez-Salicrup, D. R. & Barker, M. G. (2000). Effect of liana cutting on water potential and growth of adult *Senna multijuga* (Caesalpinioideae) trees in a Bolivian tropical forest. *Oecologia*, 124, 469–75.

Perry, D. A., Amaranthus, M. P., Borcher, J. G., Borcher, S. L., & Brainerd, R. E. (1989). Bootstrapping in ecosystems. *BioScience*, **39**, 230–7.

Perry, L. G., Galatowitsch, S. M., & Rosen, C. J. (2004). Competitive control of invasive vegetation: a native wetland sedge suppresses *Phalaris arundinacea* in carbon-enriched soil. *Journal of Applied Ecology*, **41**, 151–62.

Peters, H. A. (2003). Neighbour-regulated mortality: the influence of positive and negative density dependence on tree populations in species-rich tropical forests. *Ecology Letters*, **6**, 757–65.

Peters, V. S., MacDonald, E., & Dale, M. R. T. (2005). The interaction between masting and fire is key to white spruce regeneration. *Ecology*, **86**, 1744–50.

Peterson, C. J. & Facelli J. M. (1992). Contrasting germination and seedling growth of *Betula alleghaniensis* and *Rhus typhina* subjected to various amounts and types of plant litter. *American Journal of Botany*, **79**, 1209–16.

Peterson, C. J. & Haines, B. L. (2000a). Patterns and potential facilitation of woody plant colonization by rotting logs in premontane Costa Rican pastures. *Restoration Ecology*, **8**, 361–9.

Peterson, C. J. & Haines, B. L. (2000b). Early successional patterns and potential facilitation of woody plant colonization by rotting logs in premontane Costa Rican pastures. *Restoration Ecology*, **8**, 361–9.

Petersen, J. E., Cornwell, J. C., & Kemp, W. M. (1999). Implicit scaling in the design of experimental aquatic ecosystems. *Oikos*, **85**, 3–18.

Peterson, L. R., Massicotte, H. B., & Melville, L. H. (2004). *Mycorrhizas: Anatomy and Cell Biology*. Ottawa: NRC Research Press.

Peterson, R. L., Uetake, Y., & Armstrong, L. N. (1998). Fungal symbioses with orchid protocorms. *Symbiosis*, **25**, 29–55.

Philbrick, C. T. & Novelo, R. A. (2004). Monograph of *Podostemum* (Podostemaceae). *Systematic Botany Monographs*, **70**, 1–106.

Philipp, M. (1992). Reproductive biology of *Geranium sessifolium*. III. Population ecology of two populations and three leaf colour morphs. *New Zealand Journal of Botany*, **30**, 151–61.

Phillips, H. R. (1985). *Growing and Propagating Wild Flowers*. Chapel Hill: The University of North Carolina Press.

Phillips, O. L., Núñez, P., Monteagudo, L. A., *et al.* (2003). Habitat association among Amazonian tree species, a landscape-scale approach. *Journal of Ecology*, **91**, 757–75.

Phillips, O. L., Vásquez, M. R., Arroyo, L., *et al.* (2002). Increasing dominance of large lianas in Amazonian forests. *Nature*, **418**, 770–4.

Phillips, T. L. (1979). Reproduction of heterosporous arborescent lycopods in the Mississippian-Pennsylvanian of Euramerica. *Review of Palaeobotany and Palynology*, **27**, 239–89.

Phillips, T. L., Avcin, M. J., & Schopf, J. M. (1975). Gametophytes and young sporophyte development in *Lepidocarpon* (abstract). Corvallis, Oregon: *Botanical Society of America*, p. 23.

Phoenix, G. K. & Press, M. C. (2005). Linking physiological traits to impacts on community structure and function: the role of root hemiparasitic Orobanchaceae (ex-Scrophulariaceae). *Journal of Ecology*, **93**, 67–78.

Pianka, E. R. (1970). On r- and K-selection. *American Naturalist*, **104**, 492–597.

Pickett, S. T. A., Collins, S. L., & Armesto, J. J. (1987). A hierarchical consideration of causes and mechanisms of succession. *Vegetatio*, **69**, 109–14.

Pierik, R., Tholen, D., Poorter, H, Visser, E. J. W., & Voesenek, L. A. C. J. (2006). The Janus face of ethylene: growth inhibition and stimulation. *Trends in Plant Science*, **11**, 176–83.

Pigliucci, M. (1998). Developmental phenotypic plasticity: where internal programming meets the external environment. *Current Opinion in Plant Biology*, **1**, 87–91.

Pimentel, D., Lach, L., Zuniga, R., & Morrison, D. (2000). Environmental and economic costs of nonindigenous species in the United States. *BioScience*, **50**, 53–64.

Pinard, M. A., Barker, M. G., & Tay, J. (2000). Soil disturbance and post-logging forest recovery on bulldozer paths in Sabah, Malaysia. *Forest Ecology and Management*, **130**, 213–25.

Pinard, M. A., Howlett, B., & Davidson, D. (1996). Site conditions limit pioneer tree recruitment after logging of Dipterocarp forests in Sabah Malaysia. *Biotropica*, **28**, 2–12.

Piper, F. I., Cavieres, L. A., Reyes-Diaz, M., & Corcuera, L. J. (2006). Carbon sink limitation and frost tolerance control performance of the tree *Kageneckia angustifolia* D. Don (Rosaceae) at the treeline in central Chile. *Plant Ecology*, **185**, 29–39.

Pizo, M. A., Von Allmen, C., & Morellato, P. C. (2006). Seed size variation in the palm *Euterpe edulis* and the effects of seed predators on germination and seedling survival. *Acta Oecologica*, **29**, 311–15.

Platt, W. J. & Strong, D. R. (1989). Special feature – treefall gaps and forest dynamics – gaps in forest ecology. *Ecology*, **70**, 535–76.

Platt, W. J. & Weis, I. M. (1977). Resource partitioning and competition within a guild of fugitive prairie plants. *American Naturalist*, **111**, 479–513.

Plattner, I. & Hall, I. R. (1995). Parasitism of non-host plants by the mycorrhizal fungus *Tuber melanosporum*. *Mycological Research*, **99**, 1367–70.

Pluess, A. R. & Stöcklin, J. (2004). Population genetic diversity of the clonal plant *Geum reptans* (Rosaceae) in the Swiss Alps. *American Journal of Botany*, **91**, 2013–21.

Pohlman, C. L., Nicotra, A. B., & Murray, B. R. (2005). Geographic range size, seedling ecophysiology and phenotypic plasticity in Australian *Acacia* species. *Journal of Biogeography*, **32**, 341–51.

Poljakoff-Mayber, A., Somers, G. F., Werker, E., & Gallagher, J. L. (1994). Seeds of *Kosteletzkya virginica* (Malvaceae) – their structure, germination, and salt tolerance. 2. Germination and salt tolerance. *American Journal of Botany*, **81**, 54–9.

Poore, M. E. D. (1964). Integration in the plant community. *Journal of Animal Ecology*, **33**, 213–26.

Poorter, H. & Garnier, E. (1999). Ecological significance of inherent variation in relative growth rate and its components. In *Handbook of Functional Plant Ecology*, ed. F. I. Pugnaire & F. Valladares. New York: Marcel Dekker, Inc., pp. 81–120.

Poorter, H. & Garnier, E. (2007). Ecological significance of inherent variation in relative growth rate. In *Functional Plant Ecology*, 2nd edn., F. I. Pugnaire & F. Valladares. New York: CRC Press, pp. 67–100.

Poorter, H. & Remkes, C. (1990). Leaf area ratio and net assimilation rate of 24 wild species differing in relative growth rate. *Oecologia*, **83**, 553–9.

Poorter, L. (1999). Growth responses of 15 rain-forest tree species to a light gradient: the relative importance of morphological and physiological traits. *Functional Ecology*, **13**, 396–410.

Poorter, L. (2007). Are species adapted to their regeneration niche, adult niche, or both? *American Naturalist*, **169**, 433–42.

Poorter, L. & Bongers, F. (2006). Leaf traits are good predictors of plant performance across 53 rain forest species. *Ecology*, **87**, 1733–43.

Poorter, L. & Kitajima, K. (2007). Carbohydrate storage and light requirements of tropical moist and dry forest tree species. *Ecology*, **88**, 1000–11.

Poorter, L. & Rose, S. (2005). Light-dependent changes in the relationship between seed mass and seedling traits: a meta-analysis for rain forest tree species. *Oecologia*, **142**, 378–87.

Poorter, L., Wright, S. J., Paz, H., *et al.* (in press). Are functional traits good predictors of demographic rates? Evidence from five Neotropical forests. *Ecology*.

Popma, J., Bongers, F., & Werger, M. J. A. (1992). Gap-dependence and leaf characteristics of trees in a tropical lowland rain-forest in Mexico. *Oikos*, **63**, 207–14.

Portsmuth, A. & Niinemets, U. (2007). Structural and physiological plasticity in response to light and nutrients in five temperate deciduous woody species of contrasting shade tolerance. *Functional Ecology*, **21**, 61–77.

Posada, J. M., Aide, T. M., & Cavelier, J. (2000). Cattle and weedy shrubs as restoration tools of tropical montane rain forest. *Restoration Ecology*, **8**, 370–9.

Prasse, R. & Bornkamm, R. (2000). Effect of microbiotic soil surface crusts on emergence of vascular plants. *Plant Ecology*, **150**, 65–75.

Prentice, I. C., Sykes, M. T., & Cramer, W. (1993). A simulation model for the transient effects of climate change on forest landscapes. *Ecological Modelling*, **65**, 51–70.

Press, M. C. (1995). Carbon and nitrogen relations. In *Parasitic Plants*, ed. M. C. Press & J. D. Graves. London: Chapman and Hall, pp. 103–40.

Press, M. C. & Graves, J. D., ed. (1995). *Parasitic Plants*. London: Chapman and Hall.

Press, M. C. & Phoenix, G. K. (2005). Impacts of parasitic plants on natural communities. *New Phytologist*, **166**, 737–51.

Pridgeon, A. (2003). Modern species concepts and practical consideration for conservation of Orchidaceae. In *Orchid Conservation*, ed. K. W. Dixon, S. P. Kell, R. L. Barrett, & P. J. Cribb. Sabah: Natural History Publications, pp. 43–53.

Pryer, K. M., Schneider, H., Smith, A. R., *et al.* (2001). Horsetails and ferns are a monophyletic group and the closest living relatives to seed plants. *Nature*, **409**, 618–22.

Pryer, K. M., Schuettpelz, E., Wolf, P. G., *et al.* (2004). Phylogeny and evolution of ferns (monilophytes) with a focus on the early leptosporangiate divergences. *American Journal of Botany*, **91**, 1582–98.

Pujol, B., Muhlen, G., Garwood, N., *et al.* (2005). Evolution under domestication: contrasting functional morphology of seedlings in domesticated cassava and its closest wild relative. *New Phytologist*, **166**, 305–18.

Putz, F. E. (1984). The natural history of lianas on Barro Colorado Island. *Ecology*, **65**, 1713–24.

Pywell, R. F., Bullock, J. M., Hopkins, A., *et al.* (2002). Restoration of a species-rich grassland on arable land: assigning the limiting processes using a multi-species experiment. *Journal of Applied Ecology*, **39**, 294–309.

Pywell, R. F., Bullock, J. M., Roy, D. B., *et al.* (2003). Plant traits as predictors of performance in ecological restoration. *Journal of Applied Ecology*, **40**, 65–77.

Pywell, R. F., Bullock, J. M., Walker, K. J., *et al.* (2004). Facilitating grassland diversification using the hemiparasitic plant *Rhinanthus minor*. *Journal of Applied Ecology*, **41**, 880–7.

Quested, H. & Eriksson, O. (2006). Litter species composition influences the performance of seedlings of grassland herbs. *Functional Ecology*, **20**, 522–32.

Rabatin, S. C., Stinner, B. R., & Paoletti, M. G. (1993). Vesicular-arbuscular mycorrhizal fungi, particularly *Glomus tenue*, in Venezuelan bromeliad epiphytes. *Mycorrhiza*, **4**, 17–20.

Rabinowitz, D. (1978a). Dispersal properties of mangrove propagules. *Biotropica*, **10**, 47–57.

Rabinowitz, D. (1978b). Early growth of mangrove seedlings in Panama and an hypothesis concerning the relationship of dispersal and germination. *Journal of Biogeography*, **5**, 113–33.

Rafferty, C., Lamont, B. B., & Hanley, M. E. (2005). Selective feeding by Kangaroos (*Macropus fuliginosus*) on seedlings of *Hakea* species: effects of chemical and physical defenses. *Plant Ecology*, **177**, 201–8.

Raghavan, V. (1986). *Embryogenesis in Angiosperms: A Developmental and Experimental Study*. Cambridge: Cambridge University Press.

Rai, A. N., Soderback, E., & Bergman, B. (2000). Cyanobacterium-plant symbioses. *New Phytologist*, **147**, 449–81.

Rains, K. C., Nadkarni, N. M., & Bledsoe, C. S. (2003). Epiphytic and terrestrial mycorrhizas in a lower montane Costa Rican cloud forest. *Mycorrhiza*, **13**, 257–64.

Raju, M. V. S. (1975). Experimental studies on leafy spurge (*Euphorbia esula* L.). I. Ontogeny and distribution of buds and shoots on the hypocotyl. *Botanical Gazette*, **136**, 254–61.

Ranwell, D. S. (1972). *Ecology of Salt Marshes and Sand Dunes*. London: Chapman & Hall.

Rao, M., Terborgh, J., & Nuñez, P. (2001). Increased herbivory in forest isolates: Implications for plant community structure and composition. *Conservation Biology*, **15**, 624–33.

Rashotte, A. M., Chae, H. S., Maxwell, B. B., & Kieber, J. J. (2005). The interaction of cytokinin with other signals. *Physiologia Plantarum*, **123**, 184–94.

Rasmussen, H. N. (1994). The roles of fungi in orchid life history. In *Proceedings of the 14th World Orchid Conference*, ed. A. Pridgeon. London: HMSO, pp. 130–7.

Rasmussen, H. N. (1995). *Terrestrial Orchids from Seed to Mycotrophic Plant*. London: Cambridge University Press.

Rasmussen, H. N. & Whigham, D. F. (1993). Seed ecology of dust seeds *in situ*: a new study technique and its application in terrestrial orchids. *American Journal of Botany*, **80**, 1374–8.

Rasmussen, H. N. & Whigham, D. F. (1998a). The underground phase: a special challenge in studies of terrestrial orchid populations. *Botanical Journal of the Linnean Society*, **126**, 49–64.

Rasmussen, H. N. & Whigham, D. F. (1998b). Importance of woody debris in seed germination of *Tipularia discolor* (Orchidaceae). *American Journal of Botany*, **85**, 829–34.

Raubeson, L. A. & Jansen, R. K. (1992). Chloroplast DNA evidence on their ancient evolutionary split in vascular land plants. *Science*, **255**, 1697–9.

Rauh, R. A. & Basile, D. V. (2000). Induction of phyletic phenocopies in *Streptocarpus* (Gesneriaceae) by three antagonists of hydroxyproline-protein synthesis. In *Cell and Developmental Biology of Arabinogalactan Proteins*, ed. E. A. Nothnagel, A. Bacic, & A. E. Clarke. New York: Kluwer Academic/Plenum Publishers, pp. 191–203.

Raunkiær, C. (1934). *The Life Forms of Plants and Statistical Plant Geography*. Oxford: Clarendon Press.

Raven, J. A. (1983). Phytophages of xylem and phloem: a comparison between animal and plant sap-feeders. *Advances in Ecological Research*, **13**, 135–235.

Read, D. J. (1991). Mycorrhizas in ecosystems. *Experientia*, **47**, 376–91.

Read, D. J. & Birch, C. P. D. (1988). The effects and implications of disturbance of mycorrhizal mycelial systems. *Proceedings of the Royal Society of Edinburgh*, **94B**, 13–24.

Read, D. J., Francis, R., & Finlay, R. D. (1985). Mycorrhizal mycelia and nutrient cycling in plant communities. In *Ecological Interactions in Soil*, ed. A. H. Fitter, D. Atkinson, D. J. Read, & M. B. Usher. Oxford: Blackwell Science Publishers, pp. 193–217.

Reader, R. J., Jalili, A., Grime, J. P., Spencer, R. E., & Matthews, N. (1993). A comparative study of plasticity in seedling rooting depth in drying soil. *Journal of Ecology*, **81**, 543–50.

Rebetzke, G. J., Richards, R. A., Sirault, X. R. R., & Morrison, A. D. (2004). Genetic analysis of coleoptile length and diameter in wheat. *Australian Journal of Agricultural Research*, **55**, 733–43.

Reddell, P., Spain, A. V., & Hopkins, M. (1997). Dispersal of spores of mycorrhizal fungi in scats of native mammals in tropical forests of northeastern Australia. *Biotropica*, **29**, 184–92.

Reddy, A. S., Komariah, M., & Reddy, S. M. (1980). Cellulase activity in haustoria of *Cassytha filiformis* L. *Current Science*, **49**, 670–1.

Reddy, A. S., Komariah, M., & Reddy, S. M. (1981). Production of pectin enzymes in *Cassytha filiformis* L. *Current Science*, **50**, 283.

Redecker, D., Kodner, R., & Graham, L. E. (2000). Glomalean fungi from the Ordovician. *Science*, **289**, 1920–1.

Redecker, D., Szaro, T. M., Bowman, R. J., & Bruns, T. D. (2001). Small genets of *Lactarius xanthogalactus*, *Russula cemoricolor* and *Amanita francheti* in late-stage ectomycorrhizal successions. *Molecular Ecology*, **10**, 1025–34.

Rees, M. & Westoby, M. (1997). Game theoretical evolution of seed mass in multi-species ecological models. *Oikos*, **78**, 116–26.

Reich, P. B. (2000). Do tall trees scale physiological heights? *Trends in Ecology & Evolution*, **15**, 41–2.

Reich, P. B., Buschena, C., Tjoelker, M. G., *et al.* (2003). Variation in growth rate and ecophysiology among 34 grassland and savanna species under contrasting N supply: a test of functional group differences. *New Phytologist*, **157**, 617–31.

Reich, P. B., Tjoelker, M. G., Walters, M. B., Vanderklein, D. W., & Bushena, C. (1998). Close association of RGR, leaf and root morphology, seed mass and shade tolerance in seedlings of nine boreal tree species grown in high and low light. *Functional Ecology*, **12**, 327–38.

Reich, P. B., Wright, I. J., Cavender-Bares, J., *et al.* (2003). The evolution of plant functional variation: traits, spectra, and strategies. *International Journal of Plant Sciences*, **164**, S143–64.

Reichman, O. J. (1984). Spatial and temporal variation of seed distributions in Sonoran Desert soils. *Journal of Biogeography*, **11**, 26–30.

Reid, N., Stafford Smith, M., & Yan, Z. (1995). Ecology and population biology of mistletoes. In *Forest Canopies*, ed. M. D. Lowman & N. M. Nadkarni. San Diego: Academic Press, pp. 285–310.

Reilly, A. (1978). *Park's Success with Seeds*. Greenwood: Geo. W. Park Seed Co., Inc.

Reinhardt, C. (2004). Effectively controlling *Phalaris arundinacea* L. in wet meadow restorations and subsequent native species establishment. PhD dissertation, University of Minnesota-Twin Cities, USA.

Reinhart, K. O., Greene, E., & Callaway, R. M. (2005a). Effects of *Acer platanoides* invasion on understory plant communities and tree regeneration in the northern Rocky Mountains. *Ecography*, **28**, 573–82.

Reinhart, K. O., Royo, A. A., van der Putten, W. H., & Clay, K. (2005b). Soil feedback and pathogen activity associated with *Prunus serotina* throughout its native geographic range. *Journal of Ecology*, **93**, 890–8.

Reisman-Berman, O. (2004). Mechanisms controlling spatiotemporal patterns of shrubland patchiness: the case study of *Sarcopoterium spinosum* (L.) Spach. PhD thesis, Ben-Gurion University of the Negev, Israel.

Renker C., Zobel, M., Öpik, M., *et al.* (2004). Structure, dynamics, and restoration of plant communities: do arbuscular mycorrhizae matter? In *Assembly Rules and Restoration Ecology*, ed. V. Temperton, R. Hobbs, T. Nuttle, & S. Halle. London: Island Press, pp. 189–229.

Renzaglia, K. S., Duff, R. J., Nickrent, D. L., & Garbary, D. (2000). Vegetative and reproductive innovations of early land plants: implications for a unified phylogeny. *Transactions of the Royal Society, London*, **355**, 769–93.

Restrepo, C., Sargent, S., Levey, D. J., & Watson, D. M. (2002). The role of vertebrates in the diversification of New World mistletoes. In *Seed Dispersal and*

Frugivory: Ecology, Evolution and Conservation, ed. D. J. Levey, W. R. Silva, & M. Galetti. Wallingford: CAB International, pp. 83–98.

Restrepo, C. & Vargas, A. (1999). Seeds and seedlings of two neotropical montane understory shrubs respond differently to anthropogenic edges and treefall gaps. *Oecologia*, **119**, 419–26.

Retallack, G. J. & Dilcher, D. L. (1988). Reconstructions of selected seed ferns. *Annals of the Missouri Botanical Garden*, **75**, 1010–57.

Rey, P. J. & Alcantara, J. M. (2000). Recruitment dynamics of a fleshy-fruited plant (*Olea europaea*): connecting patterns of seed dispersal to seedling establishment. *Journal of Ecology*, **88**, 622–33.

Rhoades, C. C., Eckert, G. E., & Coleman, D. C. (1998). Effect of pasture trees on soil nitrogen and organic matter: implications for tropical montane forest restoration. *Restoration Ecology*, **6**, 262–70.

Rice, B. A. (2006). *Growing Carnivorous Plants*. Portland: Timber Press.

Rice, K. J. & Emery, N. C. (2003). Managing microevolution: restoration in the face of global change. *Frontiers in Ecology and the Environment*, **9**, 469–78.

Richardson, K. A., Peterson, R. L., & Currah, R. S. (1992). Seed reserves and early symbiotic protocorm development of *Platanthera hyperborea* (Orchidaceae). *Canadian Journal of Botany*, **70**, 291–300.

Riches, C. R. & Parker, C. (1995). Parasitic plants as weeds. In *Parasitic Plants*, ed. M. C. Press & J. D. Graves. London: Chapman and Hall, pp. 226–55.

Ridley, H. N. (1930). *The Dispersal of Plants Throughout the World*. Ashford: L. Reeve & Co., Ltd.

Riopel, J. L. & Timko, M. P. (1995). Haustorial initiation and differentiation. In *Parasitic Plants*, ed. M. C. Press & J. D. Graves. London: Chapman and Hall, pp. 39–79.

Riviere, T., Natarajan, K., & Dreyfus, B. (2006). Spatial distribution of ectomycorrhizal Basidiomycete *Russula* subsect. Foetentinae populations in a primary dipterocarp rainforest. *Mycorrhiza*, **16**, 143–8.

Roberts, M. L. & Haynes, R. R. (1983). Ballistic seed dispersal in *Illicium* (Illiciaceae). *Plant Systematics and Evolution*, **143**, 227–32.

Roberts, P. R. & Oosting, H. J. (1958). Responses of venus fly trap (*Dionaea muscipula*) to factors involved in its endemism. *Ecological Monographs*, **28**, 193–218.

Robichaux, R. H., Rundel, P. W., Stemmermann, L., *et al.* (1984). Tissue water deficits and plant growth in wet tropical environments. In *Plants of the Wet Tropics*, ed. E. H. Medina, H. C. Mooney, & C. Vasquez-Yanez. The Hague: W. Junk, pp. 99–112.

Robinson, D., Hodge, A., & Fitter, A. (2003). Constraints on the form and function of root systems. In *Root Ecology*, ed. H. de Kroon & E. J. W. Visser. Berlin: Springer-Verlag, pp. 1–31.

Robinson, J. G., Redford, K. H., & Bennett, E. L. (1999). Wildlife harvest in logged tropical forests. *Science*, **284**, 595–6.

Roche, S., Koche, J., & Dixon, K. W. (1997). Smoke-enhanced seed germination for mine rehabilitation in the southwest of Western Australia. *Restoration Ecology*, **5**, 191–203.

Rochefort, L., Quinty, F., Campeau, S., Johnson, K., & Malterer, T. (2003). North American approach to the restoration of *Sphagnum* dominated peatlands. *Wetlands Ecology and Management*, **11**, 3–20.

Roels, B., Donders, S., Werger, M. J. A., & Dong, M. (2001). Relation of wind-induced sand displacement to plant biomass and plant sand-binding capacity. *Acta Botanica Sinica*, **43**, 979–82.

Rogers, D. L. & Montalvo, A. M. (2004). Genetically appropriate choices for plant materials to maintain biological diversity. Report to the USDA Forest Service, Rocky Mountain Region, Lakewood, CO (http://www.fs.fed.us/r2/publications/botany/plantgenetics.pdf).

Rogers, W. E. & Siemann, E. (2004). Invasive ecotypes tolerate herbivory more effectively than native ecotypes of the Chinese tallow tree *Sapium sebiferum*. *Journal of Applied Ecology*, **41**, 561–70.

Rohde, K. (1992). Latitudinal gradients in species diversity: the search for the primary cause. *Oikos*, **65**, 514–27.

Rokich, D. P., Dixon, K. W., Sivasithamparam, K., & Meney, K. A. (2002). Smoke, mulch, and seed broadcasting effects on woodland restoration in Western Australia. *Restoration Ecology*, **10**, 185–94.

Roldán A. I. & Simonetti, J. A. (2001). Plant–mammal interactions in tropical Bolivian forests with different hunting pressures. *Conservation Biology*, **15**, 617–23.

Rolland, F., Moore, B., & Sheen, J. (2002). Sugar sensing and signaling in plants. *The Plant Cell*, **14**, S185–205.

Romme, W. H., Turner, M. G., Tuskan, G. A., & Reed, R. A. (2005). Establishment, persistence, and growth of aspen (*Populus tremuloides*) seedlings in Yellowstone National Park. *Ecology*, **86**, 404–18.

Rooney, T. P., McCormick, R. J., Solheim, S. L., & Waller, D. M. (2000). Regional variation in recruitment of hemlock seedlings and saplings in the upper Great Lakes, USA. *Ecological Applications*, **10**, 1119–32.

Rooney, T. P., Solheim, S. L., & Waller, D. M. (2002). Factors affecting the regeneration of northern white cedar in lowland forests of the upper Great Lakes region, USA. *Forest Ecology and Management*, **163**, 119–30.

Roques, K. G., O'Connor, T. G., & Watkinson, A. R. (2001). Dynamics of shrub encroachment in an African savanna: Relative influences of fire, herbivory, rainfall and density dependence. *Journal of Applied Ecology*, **38**, 268–80.

Rothwell, G. W. & Nixon, K. C. (2006). How does the inclusion of fossil data change our conclusions about the phylogenetic history of euphyllophytes? *International Journal of Plant Science*, **167**, 737–49.

Rothwell, G. W., Scheckler, S. E., & Gillespie, W. H. (1989). *Elkinsia* gen. Nov., a late Devonian gymnosperm with cupulate ovules. *Botanical Gazette*, **150**, 170–89.

Room, P. M. (1971). Some physiological aspects of the relationship between cocoa, *Theobroma cacao*, and the mistletoe *Tapinanthus bangwensis* (Engl. and K. Krause). *Annals of Botany*, **35**, 169–74.

Roundy, B. A. & Call, C. A. (1988). Revegetation of arid and semiarid rangelands. In *Vegetation Science Applications for Rangeland Analysis and Management*, ed. P. Tueller. Dordrecht: Kluwer Academic Publishers, pp. 607–35.

Roxburgh, L. & Nicolson, S. W. (2005). Patterns of host use in two African mistletoes: the importance of mistletoe–host compatibility and avian disperser behaviour. *Functional Ecology*, **19**, 865–73.

Roy, J. (1990). In search of the characteristics of plant invaders. In *Biological Invasions in Europe and the Mediterranean*, ed. F. Castri, A. J. Hansen, & M. Debussche. Dordrecht: Kluwer, pp. 335–52.

Royo, A. A. & Carson, W. P. (2006). On the formation of dense understory layers in forests worldwide: consequences and implications for forest dynamics, biodiversity, and succession. *Canadian Journal of Forest Research*, **36**, 1345–62.

Rudgers, J. A., Mattingly, W. B., & Koslow, J. M. (2005). Mutualistic fungus promotes plant invasion into diverse communities. *Oecologia*, **144**, 462–71.

Rudolf, P. O. (1974). *Taxus*. In *Seeds of Woody Plants of the United States*, Agriculture Handbook No. 450, tech. coord. C. S. Scopmeyer. Washington, D.C.: Forest Service, U. S. Department of Agriculture, pp. 799–802.

Runyon, J. B., Mescher, M. C., & De Moraes, C. M. (2006). Volatile chemical cues guide host location and host selection by parasitic plants. *Science*, **313**, 1964–8.

Russo, S. E., Portnoy, S., & Augspurger, C. K. (2006). Incorporating animal behavior into seed dispersal models: implications for seed shadows. *Ecology*, **87**, 3160–74.

Ryan, C. A., Pearce, G., Scheer, J., & Moura, D. S. (2002). Polypeptide hormones. *The Plant Cell*, **14**, S251–64.

Ryser, P. (1993). Influences of neighbouring plants on seedling establishment in limestone grassland. *Journal of Vegetation Science*, **4**, 195–202.

Ryser, P. & Eek, L. (2000). Consequences of phenotypic plasticity vs. interspecific differences in leaf and root traits for acquisition of aboveground and belowground resources. *American Journal of Botany*, **87**, 402–11.

Saarela, J. M., Rai, H. S., Doyle, J. A., *et al.* (2007). A new branch emerges near the root of angiosperm phylogeny. *Nature*, **446**, 312–15.

Sacchi, D. F. & Price, P. W. (1992). The relative roles of abiotic and biotic factors in seedling demography of arroyo willow (*Salix lasiolepis*: Salicaceae). *American Journal of Botany*, **79**, 395–405.

Sack, L. & Grubb, P. J. (2001). Why do species of woody seedlings change rank in relative growth rate between low and high irradiance? *Functional Ecology*, **15**, 145–54.

Sack, L. & Grubb, P. J. (2002). The combined impacts of deep shade and drought on the growth and biomass of shade-tolerant woody seedlings. *Oecologia*, **131**, 175–85.

Sack, L. & Grubb, P. J. (2003). Crossovers in seedling relative growth rates between low and high irradiance: analyses and ecological potential (reply to Kitajima & Bolker 2003). *Functional Ecology*, **17**, 281–7.

Sadebeck, R. (1902). Equisetales. In *Die Naturlichen Pflanzen Famalien*, Bd. I, ed. A. Engler & K. Prantl. Leipzig: W. Engelmann Verlag, pp. 520–58.

Saenz-Romero, C. & Guries, R. P. (2002). Landscape genetic structure of *Pinus banksiana*: seedling traits. *Silvae Genetica*, **51**, 26–35.

Saha, S. & Howe, H. F. (2001). The bamboo fire cycle hypothesis: a comment. *American Naturalist*, **158**, 659–63.

Saikkonen, K., Faeth, S. H., Helander, M., & Sullivan, T. J. (1998). Fungal endophytes: a continuum of interactions with host plants. *Annual Review of Ecology and Systematics*, **29**, 319–43.

Sakai, A. & Larcher, W. (1987). *Frost Survival of Plants: Responses and Adaptation to Freezing Stress*. Berlin: Springer-Verlag.

Sakai, S. & Sakai, A. (2005). Nature of size–number trade-off: test of the terminal-stream-limitation model for seed production of *Cardiocrium cordatum*. *Oikos*, **108**, 105–14.

Sakakibara, H., Takei, K., & Hirose, N. (2006). Interactions between nitrogen and cytokinin in the regulation of metabolism and development. *Trends in Plant Science*, **11**, 440–8.

Sala, O. E., Chapin, F. S., Armesto, J. J., *et al.* (2000). Biodiversity – global biodiversity scenarios for the year 2100. *Science*, **287**, 1770–4.

Salmon, B. (2001). *Carnivorous Plants of New Zealand*. Auckland: Ecosphere Publications.

Salmon, J. T. (1991). *Native New Zealand Flowering Plants*. Auckland: Reed Books.

Sanders, D., Pelloux, J., Brownlee, C., & Harper, J. F. (2002). Calcium at the crossroads of signaling. *The Plant Cell*, **14**, S401-17.

Sandquist, D. R., Schuster, W. S. F., Donovan, L. A., Phillips, S. L., & Ehleringer, J. R. (1993). Differences in carbon isotope discrimination between seedlings and adults of southwestern desert perennial plants. *Southwestern Naturalist*, **38**, 212-7.

Sangster, T. A. & Queitsch, C. (2005). The HSP90 chaperone complex, an emerging force in plant development and phenotypic plasticity. *Current Opinion in Plant Biology*, **8**, 86–92.

Sansen, U. & Koedam, N. (1996). Use of sod cutting for restoration of wet heathlands: revegetation and establishment of typical species in relation to soil conditions. *Journal of Vegetation Science*, **7**, 483-6.

Saunders, R. M. K. (1998). Monograph of *Kadsura* (Schisandraceae). *Systematic Botany Monographs*, **54**, 1–106.

Saunders, R. M. K. (2000). Monograph of *Schisandra* (Schisandraceae). *Systematic Botany Monographs*, **58**, 1–146.

Saverimuttu, T. & Westoby, M. (1996a). Seedling longevity under deep shade in relation to seed size. *Journal of Ecology*, **84**, 681-9.

Saverimuttu, T. & Westoby, M. (1996b). Components of variation in seedling potential relative growth rate – phylogenetically independent contrasts. *Oecologia*, **105**, 281-5.

Sbrana, C., Nuti, M. P., & Giovannetti, M. (2007). Self-anastomosing ability and vegetative incompatibility of *Tuber borchii* isolates. *Mycorrhiza*, **17**, 667-75.

Schaal, B. A. (1980). Reproductive capacity and seed size in *Lupinus texensis*. *American Journal of Botany*, **67**, 703-9.

Schaefer, M. (1992). *Worterbucher der Biologie. Okologie.* Jena: Gustav Fischer.

Schardl, C. L., Leuchtmann, A., & Spiering, M. J. (2004). Symbioses of grasses with seed borne fungal endophytes. *Annual Review of Plant Biology*, **55**, 315–40.

Schimel, J. P. & Bennett, J. (2004). Nitrogen mineralization: challenges of a changing paradigm. *Ecology*, **85**, 591–602.

Schimpf, D. J. (1977). Seed weight of *Amaranthus retroflexus* in relation to moisture and length of the flowering season. *Ecology*, **58**, 450-3.

Schlesinger, W. H., Abrahams, A. D., Parsons, A. J., & Wainwright, J. (1999). Nutrient losses in runoff from grassland and shrubland habitats in Southern New Mexico: I. rainfall simulation experiments. *Biogeochemistry*, **45**, 21-34.

Schlesinger, W. H., Reynolds, J. F., Cunningham, G. L., *et al.* (1990). Biological feedbacks in global desertification. *Science*, **247**, 1043-8.

Schlising, R. A. (1969). Seedling morphology in *Marah* (Cucurbitaceae) related to the Californian Mediterranean climate. *American Journal of Botany*, **56**, 552-60.

Schmidt, G. & Zotz, G. (2002). Inherently slow growth in two Caribbean epiphytic species: a demographic approach. *Journal of Vegetation Science*, **13**, 527-34.

Schmitt, J., Stinchrcombe, J. R., Heschel, M. S., & Huber, H. (2003). The adaptive evolution of plasticity: phytochrome-mediated shade avoidance responses. *Integrative and Comparative Biology*, **43**, 459–69.

Schnarf, K. (1929). *Embryologie der Angiospermen.* In *Handbuch der Pflanzenanatomie* (II. Abteilung 2. Teil: *Archegoniaten.* Band X/2), ed. K. Linsbauer. Berlin: Gebrüder Borntraeger, pp. 1–689.

Schneider, D. C. (1994). *Quantitative Ecology: Spatial and Temporal Scaling.* San Diego: Academic Press.

Schnell, D. E. (1976). *Carnivorous Plants of the United States and Canada*. Winston-Salem: John F. Blair, Publisher.

Schnitzer, S. A., Dalling, J. W., & Carson, W. P. (2000). The impact of lianas on tree regeneration in tropical forest canopy gaps: evidence for an alternative pathway of gap-phase regeneration. *Journal of Ecology*, **88**, 655–66.

Schnitzer, S. A., Kuzee, M., & Bongers, F. (2005). Disentangling above- and below-ground competition between lianas and trees in a tropical forest. *Journal of Ecology*, **93**, 1115–25.

Schopmeyer, C. S., tech. coord. (1974). *Seeds of Woody Plants in the United States*. Agriculture Handbook No. 450. Washington: Forest Service, U.S. Department of Agriculture.

Schreeg, L. A., Kobe, R. K., & Walters, M. B. (2005). Tree seedling growth, survival and morphology in response to landscape-level variation in soil resource availability in northern Michigan. *Canadian Journal of Forest Research*, **35**, 263–73.

Schüßler, A., Schwarzott, D., & Walker, C. (2001). A new fungal phylum, the Glomeromycota: phylogeny and evolution. *Mycological Research*, **105**, 1413–21.

Schultze, M. & Kondorosi, A. (1998). Regulation of symbiotic root nodule development. *Annual Review of Genetics*, **32**, 33–57.

Schulze, E.-D., Lange, O. L., Ziegler, H., & Gebauer, G. (1991). Carbon and nitrogen isotope ratios of mistletoes growing on nitrogen and non-nitrogen fixing hosts and on CAM plants in the Namib Desert confirm partial heterotrophy. *Oecologia*, **88**, 457–62.

Schulze, W., Schulze, E.-D., Pate, J. S., & Gillison, A. N. (1997). The nitrogen supply from soils and insects during growth of the pitcher plants *Nepenthes mirabilis*, *Cephalotus follicularis* and *Darlingtonia californica*. *Oecologia*, **112**, 464–71.

Schulze, W., Schulze, E.-D., Schulze, I., & Oren, R. (2001). Quantification of insect nitrogen utilization by the venus flytrap *Dionaea muscipula* catching prey with highly variable isotope signatures. *Journal of Experimental Botany*, **52**, 1041–9.

Schupp, E. W. (1988). Seed and early seedling predation in the forest understory and in treefall gaps. *Oikos*, **51**, 71–8.

Schupp, E. W. (1995). Seed-seedling conflicts, habitat choice, and patterns of plant recruitment. *American Journal of Botany*, **82**, 399–409.

Sculthorpe, C. D. (1967). *The Biology of Aquatic Vascular Plants*. London: Edward Arnold, Ltd. Reprinted 1985, Königstein: Koeltz Scientific Books.

Schaal, B. A. (1980). Reproductive capacity and seed size in *Lupinus texensis*. *American Journal of Botany*, **67**, 703–9.

Schatz, G. E. (1996). Malagasy/Indo-australo-malesian phytogeographic connections. In *Biogéographie de Madagascar*, ed. W. R. Lourenço. Paris: ORSTOM, pp. 73–84.

Scheublin, T. R., Ridgway, K. P., Young, J. P. W., & van der Heijden, M. G. A. (2004). Nonlegumes, legumes, and root nodules harbor different arbuscular mycorrhizal fungal communities. *Applied and Environmental Microbiology*, **70**, 6240–6.

Scheublin, T. R. & van der Heijden, M. G. A. (2006). Arbuscular mycorrhizal fungi colonize root nodules of several legume species. *New Phytologist*, **172**, 732–8.

Scheublin, T. R., van Logtestijn R., & van der Heijden, M. G A (2007). Presence and identity of arbuscular mycorrhizal fungi influence competitive interactions between plant species. *Journal of Ecology*, **95**, 631–8.

Schimper, A. F. W. (1898). *Pflanzengeographie auf physiologischer Grundlage.* Jena: Gustav Fischer.

Schlichting, C. D. (1986). The evolution of phenotypic plasticity in plants. *Annual Review of Ecology and Systematics,* **17,** 667–93.

Schmitt, J., Stinchcombe, J. R., Heschel, M. S., & Huber, H. (2003). The adaptive evolution of plasticity: phytochrome-mediated shade avoidance responses. *Integrative and Comparative Biology,* **43,** 459–69.

Schneider, E. L. (1978). Morphological studies of the Nymphaeaceae. IX. Seed of *Barclaya longifolia* Wall. *Botanical Gazette,* **139,** 223–30.

Schneider, E. L. & Ford, E. G. (1978). Morphological studies of the Nymphaeaceae. X. The seed of *Ondinea purpurea* Den Hartog. *Bulletin of the Torrey Botanical Club,* **105,** 192–200.

Schulze, D. M., Walker, J. L., & Spira, T. P. (2002). Germination and seed bank studies of *Macbridea alba* (Lamiaceae), a federally threatened plant. *Castanea,* **67,** 280–9.

Seabloom E. W., Harpole, W. S., Reichman, O. J., & Tilman, D. (2003). Invasion, competitive dominance, and resource use by exotic and native California grassland species. *Proceedings of the National Academy of Sciences* (USA), **100,** 13384–9.

Sehgal, A., Mohan Ram, H. Y., & Bhatt, J. R. (1993). In vitro germination, growth, morphogenesis and flowering of an aquatic angiosperm, *Polypleurum stylosum* (Podostemaceae). *Aquatic Botany,* **45,** 269–83.

Selosse, M.-A., Richard, R., He, X., & Simard, S. W. (2006). Mycorrhizal networks: des liaisons dangereuses. *Trends in Ecology & Evolution,* **21,** 621–8.

Shachak, M. & Lovett, G. M. (1998). Atmospheric deposition to a desert ecosystem and its implications for management. *Ecological Applications,* **8,** 455–63.

Shachak, M., Brand, S., & Gutterman, Y. (1991). Porcupine disturbances and vegetation pattern along a resource gradient in a desert. *Oecologia,* **88,** 141–7.

Shachak, M., Sachs, M., & Moshe, I. (1998). Ecosystem management of desertified shrublands in Israel. *Ecosystems,* **1,** 475–83.

Shannon, E. L. (1953). The production of root hairs by aquatic plants. *American Midland Naturalist,* **50,** 474–9.

Sharifi, M. R. & Rundel, P. W. (1993). The effect of vapour pressure deficit on carbon isotope discrimination in the desert shrub *Larrea tridentata* (Creosote Bush). *Journal of Experimental Botany,* **44,** 481–7.

Shaw, J. & Renzaglia, K. (2004). Phylogeny and diversification of bryophytes. *American Journal of Botany,* **91,** 1557–81.

Shaw, J. D., Hovenden, M. J., & Bergstrom, D. M. (2005). The impact of introduced ship rats (*Rattus rattus*) on seedling recruitment and distribution of a subantarctic megaherb (*Pleurophyllum hookeri*). *Austral Ecology,* **30,** 118–25.

Shefferson, R. P., Kull, T., & Tali, K. (2005). Adult whole-plant dormancy induced by stress in long-lived orchids. *Ecology,* **86,** 3099–104.

Shefferson, R. P., Sandercock, B. K., Proper, J., & Beissinger, S. R. (2001). Estimating dormancy and survival of a rare herbaceous perennial using mark-recapture models. *Ecology,* **82,** 145–56.

Shem-Tov, S., Zaady, E., Groffman, P. M., & Gutterman, Y. (1999). Soil carbon content along a rainfall gradient and inhibition of germination: a potential mechanism for regulating distribution of *Plantago coronopus. Soil Biology and Biochemistry,* **31,** 1209–17.

Shem-Tov, S., Zaady, E., & Gutterman, Y. (2002). Germination of *Carrichtera annua* (Brassicaceae) seeds on soil samples collected along a rainfall gradient in the Negev Desert of Israel. *Israel Journal of Plant Sciences,* **50,** 113–18.

Shen, Y.-Y., Wang, X.-F., Wu, F.-Q., *et al.* (2006). The Mg-chelatase H subunit is an abscisic acid receptor. *Nature*, **443**, 823–6.

Sher, A. A. & Hyatt, L. A. (1999). The disturbed resource-flux invasion matrix: a new framework for patterns of plant invasion. *Biological Invasions*, **1**, 107–14.

Sher, A. A., Marshall, D. L., & Gilbert, S. A. (2000). Competition between native *Populus deltoides* and invasive *Tamarix ramosissima* and the implications of reestablishing flooding disturbance. *Conservation Biology*, **14**, 1744–54.

Sher, A. A., Marshall, D. L., & Taylor, J. (2002). Spatial partitioning within southwestern floodplains: patterns of establishment of native *Populus* and *Salix* in the presence of invasive, non-native *Tamarix*. *Ecological Applications*, **12**, 760–72.

Shimada, T., Genma, T., Furuya, S., & Kondo, Y. (1982). Frost heaving injury in alfalfa. *Journal of the Japanese Society of Grassland Science*, **28**, 147–53.

Shipley, B. & Almeida-Cortez, J. (2003). Interspecific consistency and intraspecific variability of specific leaf area with respect to irradiance and nutrient availability. *Ecoscience*, **10**, 74–9.

Shipley, B. & Dion, J. (1992). The allometry of seed production in herbaceous angiosperms. *American Naturalist*, **139**, 467–83.

Shipley, B., Keddy, P. A., Moore, D. R. J., & Lemky, K. (1989). Regeneration and establishment strategies of emergent macrophytes. *Journal of Ecology*, **77**, 1093–110.

Shipley, B. & Peters, R. H. (1990). The allometry of seed weight and seedling relative growth rate. *Functional Ecology*, **4**, 523–9.

Shirato, Y., Zhang, T. H., Ohkuro, T., Fujiwara, H., & Taniyama, I. (2005). Changes in topographical features and soil properties after exclosure combined with sand-fixing measures in Horqin Sandy Land, Northern China. *Soil Science and Plant Nutrition*, **51**, 61–8.

Shmida, A. & Ellner, S. (1984). Coexistence of plant species with similar niches. *Plant Ecology*, **58**, 29–55.

Shreve, F. (1906). The development and anatomy of *Sarracenia purpurea*. *Botanical Gazette*, **42**, 107–26.

Shugart, H. H. & West, D. C. (1980). Forest succession models. *BioScience*, **30**, 308–13.

Sibly, R. & Calow, P. (1985). Classification of habitats by selection pressures: a synthesis of life-cycle and r-K theory. In *Behavioural Ecology. Ecological Consequences of Adaptive Behaviour*, ed. M. R. Sibly & R. H. Smith. Oxford: Blackwell Scientific Publications, pp. 75–90.

Sidhu, S. S. & Cavers, P. B. (1977). Maturity-dormancy relationships in attached and detached seeds of *Medicago lupulina* L. (Black medick). *Botanical Gazette*, **138**, 174–82.

Siegert, F., Ruecker, G., Hinrichs, A., & Hoffmann, A. A. (2001). Increased damage from fires in logged forests during droughts caused by El Niño. *Nature*, **414**, 437–40.

Siemann, E. & Rogers, W. E. (1993). Changes in light and nitrogen availability under pioneer trees may indirectly facilitate tree invasions of grasslands. *Journal of Ecology*, **91**, 923–31.

Silvertown, J. (1989). The paradox of seed size and adaptation. *Trends in Ecology and Evolution*, **4**, 24–6.

Silvertown, J. & Bullock, J. M. (2003). Do seedlings in gaps interact? A field test of assumptions in ESS seed size models. *Oikos*, **101**, 499–504.

Silvertown, J. & Charlesworth, D. (2001). *Introduction to Plant Population Biology*, 4th edn. Oxford: Blackwell.

Silvertown, J., Franco, M., & Menges, E. (1996). Interpretation of elasticity matrices as an aid to the management of plant populations for conservation. *Conservation Biology*, **10**, 591–7.

Silvertown, J., Franco, M., Pisanty, I., & Mendoza, A. (1993). Comparative plant demography – relative importance of life-cycle components to the finite rate of increase in woody and herbaceous perennials. *Journal of Ecology*, **81**, 465–76.

Silvertown, J. & Wilson, J. B. (1994). Community structure in a desert perennial community. *Ecology*, **75**, 409–17.

Silvertown, J. W. (1981). Seed size, lifespan and germination date as co-adapted features of plant life history. *American Naturalist*, **118**, 860–4.

Simard, S. W. & Durall, D. M. (2004). Mycorrhizal networks: a review of their extent, function, and importance. *Canadian Journal of Botany*, **82**, 1140–65.

Simard, S. W., Durall, D., & Jones, M. (2002). Carbon and nutrient fluxes within and between mycorrhizal plants. In *Mycorrhizal Ecology*, ed. M. G. A. van der Heijden & I. R. Sanders. Berlin: Springer-Verlag, pp. 33–74.

Simard, S. W., Perry, D. A., Jones, M. D., Myrold, D. D., Durall, D. M., & Molina, R. (1997). Net transfer of carbon between ectomycorrhizal tree species in the field. *Nature*, **388**, 579–82.

Simberloff, D., Relva, M. A., & Nuñez, M. (2002). Gringos en al bosque: introduced tree invasion in a native *Nothofagus/Austrocedrus* forest. *Biological Invasions*, **4**, 35–53.

Simmers, S. (2006). Recovery of semi-arid grassland on recontoured and revegetated oil access roads. MS thesis, University of Minnesota-Twin Cities, USA.

Simmons, M. T. (2005). Bullying the bullies: the selective control of an exotic, invasive annual (*Rapistrum rugosum*) by oversowing with a competitive native species (*Gaillardia pulchella*). *Restoration Ecology*, **13**, 609–15.

Simon, L., Bousquet, J., Levesque, R. C., & LaLonde, M. (1993). Origin and diversification of endomycorrhizal fungi and coincidence with vascular land plants. *Nature*, **363**, 67–9.

Simons, P. (1992). *The Action Plant*. Oxford: Blackwell.

Simpson, M. J. A. (1979). Lack of dormancy in seeds of New Zealand plants. *Canterbury Botanical Society*, **13**, 36–7.

Simpson, R. L., Leck, M. A., & Parker, V. T. (1985). The comparative ecology of *Impatiens capensis* Meerb. (Balsaminaceae) in central New Jersey. *Bulletin of the Torrey Botanical Club*, **112**, 295–311.

Simpson, R. L., Leck, M. A., & Parker, V. T. (1989). Seed banks: central concepts and methodological issues. In *Ecology of Soil Seed Banks*, ed. M. A. Leck, V. T. Parker, & R. L. Simpson. San Diego: Academic Press, pp. 3–8.

Singh, G., Bala, N., Rathod, T. R., & Chouhan, S. (2003). Effect of adult neighbours on regeneration and performance of surface vegetation for control of sand drift in Indian desert. *Environmental Conservation*, **30**, 353–63.

Singh, G. & Rathod, T. R. (2002). Plant growth, biomass production and soil water dynamics in a shifting dune of Indian desert. *Forest Ecology and Management*, **171**, 309–20.

Sizer, N. & Tanner E. V. J. (1999). Responses of woody plant seedlings to edge formation in a lowland tropical rainforest, Amazonia. *Biological Conservation*, **91**, 135–42.

Skene, M. (1959). *The Biology of Flowering Plants*, 8th Impr. London: Sidgwick & Jackson, Ltd.

Skenel, K. R. (2001). Cluster roots: model experimental tools for key biological problems. *Journal of Experimental Botany*, **52**, 479–85.

Slik, J. W. F. & Eichhorn, K. A. O. (2003) Fire survival of lowland tropical rain forest trees in relation to stem diameter and topographic position. *Oecologia*, **137**, 446–55.

Slocum, M. G. (2000). Logs and fern patches as recruitment sites in a tropical pasture. *Restoration Ecology*, **8**, 408–13.

Slocum, M. G. (2001). How tree species differ as recruitment foci in a tropical pasture. *Ecology*, **82**, 2547–59.

Slocum, M. G., Aide, T. M., Zimmerman, J. K., & Navarro, L. (2006). A strategy for restoration of montane forest in anthropogenic fern thickets in the Dominican Republic. *Restoration Ecology*, **14**, 526–36.

Smit, C., den Ouden, J., & Müller-Schärer, H. (2006). Unpalatable plants facilitate tree sapling survival in wooded pastures. *Journal of Applied Ecology*, **43**, 305–12.

Smith, C. C. & Fretwell, S. D. (1974). The optimal balance between size and number of offspring. *American Naturalist*, **108**, 499–506.

Smith, C. E., Dudley, M. W. D., & Lynn, G. (1990). Vegetative/parasitic transition: control and plasticity in *Striga* development. *Plant Physiology*, **93**, 208–15.

Smith, C. M. (1931). Development of *Dionaea muscipula*. II. Germination of seed and development of seedling to maturity. *Botanical Gazette*, **91**, 377–94.

Smith, G. F., Nicholas, N. S., & Zedaker, S. M. (1997). Succession dynamics in a maritime forest following hurricane Hugo and fuel reduction burns. *Forest Ecology and Management*, **95**, 275–83.

Smith, G. H. (1955). *Cryptogamic Botany*. Vol. II. *Bryophytes and Pteridophytes*. New York: McGraw-Hill.

Smith, J. E., Johnson, K. A., & Cázares, E. (1998). Vesicular mycorrhizal colonization of seedlings of Pinaceae and Betulaceae after spore inoculation with *Glomus intraradices*. *Mycorrhiza*, **7**, 279–85.

Smith, S. E. & Read, D. J. (1997). *Mycorrhizal Symbiosis*, 2nd edn. London: Academic Press.

Smith, S. E., Riley, E., Tiss, J. L., & Fendenheim, D. M. (2000). Geographical variation in predictive seedling emergence in a perennial desert grass. *Journal of Ecology*, **88**, 139–49.

Smith, S. M. & Snedaker, S. C. (1995). Salinity responses in two populations of viviparous *Rhizophora mangle* L. seedlings. *Biotropica*, **27**, 435–40.

Smits A. J. M., Vanavesaath, P. H., & Vanderveld, E. G. (1990). Germination requirements and seed banks of some nymphaeid macrophytes – *Nymphaea alba*, *Nuphar lutea*, and *Nymphoides peltata*. *Freshwater Biology*, **24**, 315–26.

Smoot, E. L. & Taylor, T. N. (1986). Evidence of simple polyembryony in Permian seeds from Antarctica. *American Journal of Botany*, **73**, 1079–81.

Snyman, H. A. (1999). Quantification of the soil–water balance under different veld condition classes in a semi-arid climate. *African Journal of Range and Forage Science*, **16**, 108–17.

Sorenson, F. C. & Miles, R. S. (1978). Cone and seed weight relationships in Douglas-fir from Western and Central Oregon. *Ecology*, **59**, 641–4.

Soriano, A., Sala, O. E., & Perelman, S. B. (1994). Patch structure and dynamics in a Patagonian arid steppe. *Plant Ecology*, **111**, 127–35.

Souèges, R. (1919). Les premières divisions de l'oeuf et les différenciations du suspenseur chez le *Capsella bursa-pastoralis* Moench. *Annales des Sciences Naturelles*, 10th Série, *Botanique*, **1**, 1–28.

Specht, A. & Harvey-Jones, J. (2000). Improving water delivery to the roots of recently transplanted seedling trees: the use of hydrogels to reduce leaf loss and hasten root establishment. *Forest Research*, **1**, 117–23.

Spiering, M. J., Moon, C. D., Wilkinson, H. H., & Schardl, C. L. (2005). Gene clusters for insecticidal loline alkaloids in the grass-endophytic fungus *Neotyphodium uncinatum*. *Genetics*, **169**, 1403–14.

Sporne, K. R. (1967). *The Morphology of Gymnosperms*. London: Hutchinson University Library.

Sprent, J. I. (2001). *Nodulation in Legumes*. Kew: Royal Botanical Gardens.

Sprent, J. I. & Parsons, R. (2000). Nitrogen fixation in legume and non-legume trees. *Field Crops Research*, **65**, 183–96.

Staff, I. A. & Waterhouse, J. T. (1981). The biology of arborescent monocotyledons, with special reference to Australian species. In *The Biology of Australian Plants*, ed. J. S. Pate & A. J. McComb. Nedlands: University of Western Australia Press, pp. 216–57.

Stamp, N. E. (1984). Self-burial behaviour of *Erodium cicutarium* seeds. *The Journal of Ecology*, **72**, 611–20.

Stanton, M. L. (1984). Seed variation in wild radish *Raphanus raphanistrum*: effect of seed size on components of seedling and adult fitness. *Ecology*, **65**, 1105–12.

Stanton, M. L. (1985). Seed size and emergence time within a stand of wild radish *Raphanus raphanistrum*: the establishment of a fitness hierarchy. *Oecologia*, **67**, 524–31.

Stebbins, G. L. (1965). The probable growth habit of the earliest flowering plants. *Annals of the Missouri Botanical Garden*, **52**, 457–68.

Stebbins, G. L. (1971). Adaptive radiation of reproductive characteristics in angiosperms. II. Seeds and seedlings. *Annual Review of Ecology and Systematics*, **2**, 237–60.

Stebbins, G. L. (1974). *Flowering Plants: Evolution Above the Species Level*. Cambridge: Harvard University Press.

Steele, M. A. & Koprowski, J. L. (2001). *North American Tree Squirrels*. Washington: Smithsonian Institution Press.

Stein, W. E. (1993). Modeling the evolution of the stelar architecture in vascular plants. *International Journal of Plant Sciences*, **154**, 229–63.

Steinger, T., Korner, C., & Schmid, B. (1996). Long-term persistence in a changing climate: DNA analysis suggests very old ages of clones of alpine *Carex curvula*. *Oecologia*, **105**, 94–9.

Stendell, E. R., Horton, T. R., & Bruns, T. D. (1999). Early effects of prescribed fire on the structure of the ectomycorrhizal fungus community in a Sierra Nevada ponderosa pine forest. *Mycological Research*, **103**, 1353–9.

Stepanova, A. N. & Alonso, J. M. (2005). Ethylene signalling and response pathway: a unique signalling cascade with a multitude of inputs and outputs. *Physiologia Plantarum*, **123**, 195–206.

Stephens, E. L. (1912). The structure and development of the haustorium of *Striga lutea*. *Annals of Botany*, **50**, 1–7.

Stephenson, M. & Mari, J. (2005). Laboratory germination testing of flower seed. In *Flower Seeds Biology and Technology*, ed. M. B. McDonald & F. Y. Kwong. Wallingford: CAB International, pp. 263–97.

Stephenson, N. L. (1988). Climatic control of vegetation distribution: The role of the water-balance with examples from North America and Sequoia National Park, California. PhD dissertation, Cornell University, USA.

Stephenson, N. L. (1990). Climatic control of vegetation distribution: the role of the water balance. *American Naturalist*, **135**, 649–70.

Stephenson, N. L. (1998). Actual evapotranspiration and deficit: biologically meaningful correlates of vegetation distribution across spatial scales. *Journal of Biogeography*, **25**, 855–70.

Stephenson, N. L. & van Mantgem, P. J. (2005). Forest turnover rates follow global and regional patterns of productivity. *Ecology Letters*, **8**, 524–31.

Sterck, F. J., van Gelder, H. A., & Poorter, L. (2006). Mechanical branch constraints contribute to life-history variation across tree species in a Bolivian forest. *Journal of Ecology*, **94**, 1192–200.

Stevens, G. C. (1987). Lianas as structural parasites, the *Bursera simaruba* example. *Ecology*, **68**, 77–81.

Stewart, R. B., Wheaton, E., & Spittlehouse, D. L. (1998). Climate change: implications for the Boreal forest. In *Emerging Air Issues for the 21st Century: The Need for Multidisciplinary Management, Proceedings of a Specialty Conference*, ed. A. H. Legge & L. L. Jones. Pittsburgh: Air and Waste Management Association, pp. 86–101.

Stewart, W. N. & Rothwell, G. W. (1993). *Paleobotany and the Evolution of Plants*. Cambridge: Cambridge University Press.

Stinchcombe, J. R., Weinig, C., Ungerer, M., et al. (2004). A latitudinal cline in flowering time in *Arabidopsis thaliana* modulated by the flowering time gene FRIGIDA. *Proceedings of the National Academy of Sciences* (USA), **101**, 4712–17.

Stinson, K. A., Campbell, S. A., Powell, J. R., et al. (2006). Invasive plant suppresses the growth of native tree seedlings by disrupting belowground mutualisms. *Public Library of Science Biology*, **4**, 727–31.

St. John, T. V., Coleman, D. C., & Reid, C. P. P. (1983). Growth and spatial distribution of nutrient absorbing organs: selective placement of soil heterogeneity. *Plant and Soil*, **71**, 487–93.

Stock, W. D., Pate, J. S., & Delfs, J. (1990). Influence of seed size and quality on seedling development under low nutrient conditions in five Australian and South African members of the Proteaceae. *Journal of Ecology*, **78**, 1005–20.

Stockey, R. A. & Rothwell, G. W. (2003). Anatomically preserved *Williamsonia* (Williamsoniaceae): evidence for Bennettitalean reproduction in the Late Cretaceous of Western North America. *International Journal of Plant Sciences*, **164**, 251–62.

Stocklin, J. & Favre, P. (1994). Effects of plant size and morphological constraints on variation in reproductive components in two related species of *Epilobium*. *Journal of Ecology*, **82**, 735–46.

Stohlgren, T. J., Bachand, R. R., Onami, Y., & Binkley, D. (1998). Species–environment relationships and vegetation patterns: effects of spatial scale and tree life-stage. *Plant Ecology*, **135**, 215–28.

Stone, D. E. 1968. Cytological and morphological notes on the southeastern endemic *Schisandra glabra* (Schisandraceae). *Journal of the Elisha Mitchell Scientific Society*, **84**, 351–6.

Stopf, O. (1904). On the fruit of *Melocanna bambusoides*. *Transactions of the Linnean Society, 2nd Ser.*, **6**, 401–25.

Stoutamire, W. P. (1974). Terrestrial orchid seedlings. In *The Orchids. Scientific Studies*, ed. C. L. Withner. New York: Wiley, pp. 101–28.

Strauss-Debenedetti, S. & Bazzaz, F. A. (1991). Plasticity and acclimation to light in tropical Moraceae of different successional positions. *Oecologia*, **87**, 377–87.

Stromberg, J. C. & Patten, D. T. (1990). Seed production and seedling establishment of a southwest riparian tree Arizona Walnut *Juglans major*. *Great Basin Naturalist*, **50**, 47–56.

Stubblefield, S. P. & Rothwell, G. W. (1981). Embryology and reproductive biology of *Bothrodendrostrobus mundus* (Lycopsida). *American Journal of Botany*, **68**, 625–34.

Stubblefield, S. P., Taylor, T. N., & Trappe, J. M. (1987). Fossil mycorrhizae: a case for symbiosis. *Science*, **59**, 236–7.

Stuckenbrock, E. H. & Rosendahl, S. (2005). Distribution of dominant arbuscular mycorrhizal fungi among five plant species in undisturbed vegetation of a coastal grassland. *Mycorrhiza*, **15**, 497–503.

Su, Y. Z., Zhang, T. H., Li, Y. L., & Wang, F. (2005). Changes in soil properties after establishment of *Artemisia halodendron* and *Caragana microphylla* on shifting sand dunes in semiarid Horqin sandy land, Northern China. *Environmental Management*, **36**, 272–81.

Su, Y. Z., Zhao, H., Zhang, T., & Li, Y. L. (2004). Characteristics of plant community and soil properties in the plantation chronosequence of *Caragana microphylla* in Horqin sandy land. *Zhiwu Shengtai Xuebao*, **28**, 93–100.

Suding, K. N. & Goldberg, D. E. (1999). Variation in the effects of vegetation and litter on recruitment across productivity gradients. *Journal of Ecology*, **87**, 436–49.

Sun, G., Ji, Q., Dilcher, D. L., *et al.* (2002). Archaefructaceae, a new basal angiosperm family. *Science*, **296**: 899–904.

Sun, M., Wong, K. C., & Lee, J. S. Y. (1998). Reproductive biology and population genetic structure of *Kandelia candel* (Rhizophoraceae), a viviparous mangrove species. *American Journal of Botany*, **85**, 1631–7.

Sutherland, S. (2004). What makes a weed a weed: life history traits of native and exotic plants in the USA. *Oecologia*, **141**, 24–39.

Svenning, J.-C. & Wright, S. J. (2005). Seed limitation in a Panamanian forest. *Journal of Ecology*, **93**, 853–62.

Swaine, M. D. & Hall, J. B. (1988). The mosaic theory of forest regeneration and the determination of forest composition in Ghana. *Journal of Tropical Ecology*, **4**, 253–69.

Swaine, M. D. & Whitmore, T. C. (1988). On the definition of ecological species groups in tropical rain forests. *Vegetatio*, **75**, 81–6.

Swanborough, P. & Westoby, M. (1996). Seedling relative growth rate and its components in relation to seed size: phylogenetically independent contrasts. *Functional Ecology*, **10**, 176–84.

Sydes, C. & Grime, J. P. (1981). Effect of tree leaf litter on herbaceous vegetation in the deciduous woodlands. I. Field investigations. *Journal of Ecology*, **69**, 237–48.

Symonides, E. (1979). The structure and population dynamics of psammophytes on inland dunes. III. Populations of compact psammophyte communities. *Ekologia Polska*, **27**, 235–57.

Symonides, E. (1988). Population dynamics of annual plants. In *Plant Population Ecology*, ed. A. J. Davy, M. J. Hutchings, & A. R. Watkinson. Oxford: Blackwell, pp. 221–48.

Tali, K. (2002). Dynamics of *Orchis ustulata* populations in Estonia. In *Trends and Fluctuations and Underlying Mechanisms in Terrestrial Orchid Populations*, ed. P. Kindlmann, J. H. Willems, & D. F. Whigham. Leiden: Backhuys Publishers, pp. 33–42.

Tanaka, A., Tapper, B. A., Popay, A., Parker, E. J., & Scott, B. (2005). A symbiosis expressed non-ribosomal peptide synthetase from a mutualistic fungal endophyte of perennial ryegrass confers protection to the symbiotum from insect herbivory. *Molecular Microbiology*, **57**, 1036–150.

Taylor, C. M., Davis, H. G., Civille, J. C., Grevstad, F. S., & Hastings, A. (2004). Consequences of an allee effect in the invasion of a pacific estuary by *Spartina alterniflora*. *Ecology*, **85**, 3254–66.

Taylor, D. L. & Bruns, T. D. (1997). Independent, specialized invasions of ecto-mycorrhizal mutualism by two nonphotosynthetic orchids. *Proceedings of the National Academy of Sciences* (USA), **94**, 4510–15.

Taylor, D. L. & Bruns, T. D. (1999a). Community structure of ectomycorrhizal fungi in a *Pinus muricata* forest: minimal overlap between the mature forest and resistant propagule communities. *Molecular Ecology*, **8**, 1837–50.

Taylor, D. L. & Bruns, T. D. (1999b). Population, habitat and genetic correlates of mycorrhizal specialization in the 'cheating' orchids *Corallorhiza maculata* and *C. mertensiana*. *Molecular Ecology*, **8**, 1719–32.

Taylor, D. L., Bruns, T. D., Leake, J. R., & Read, D. J. (2002). Mycorrhizal specificity and function in myco-heterotrophic plants, In *Mycorrhizal Ecology*, ed. M. G. A. van der Heijden & I. Sanders. Berlin: Springer, pp. 375–413.

Taylor, D. W. & Hickey, L. J. (1992). Phylogenetic evidence for the herbaceous origin of angiosperms. *Plant Systematics and Evolution*, **180**, 137–56.

Taylor, D. W. & Hickey, L. J. (1996). Evidence for and implications of an herba-ceous origin for angiosperms. In *Flowering Plant Origin, Evolution, and Phylo-geny*, ed. D. W. Taylor & L. J. Hickey. New York: Chapman & Hall, pp. 232–66.

Taylor, K. M. & Aarssen, L. W. (1989). Neighbor effects in mast year seedlings of *Acer saccharum*. *American Journal of Botany*, **76**, 546–54.

Taylor, T. N., Remy, W., Hass, H., & Kerp, H. (1995). Fossil arbuscular mycor-rhizae from the early Devonian. *Mycologia*, **87**, 560–73.

Taylor, T. N. & Taylor, E. L. (1993). *The Biology and Evolution of Fossil Plants*. Englewood Cliffs: Prentice Hall.

Teale, W. D., Paponova, I. A., Ditengou, F., & Palme, K. (2005). Auxin and the developing root of *Arabidopsis thaliana*. *Physiologia Plantarum*, **123**, 130–8.

Tecco, P. A., Gurvich, D. E., Diaz, S., Perez-Harguindeguy, N. P., & Cabido, M. (2006). Positive interactions between invasive plants: the influence of *Pyra-cantha angustifolia* on the recruitment of native and exotic woody species. *Austral Ecology*, **31**, 293–300.

Telewski, F. W. (2006). A unified hypothesis of mechanoperception in plants. *American Journal of Botany*, **93**, 1466–76.

Telewski, F. W. & Zeevart, J. A. D. (2002). The 120-year period for Dr. Beal's seed viability experiment. *American Journal of Botany*, **89**, 1285–8.

Terada, K., Sun, G., & Nishida, T. (2005). 3D Models of two species of *Archae-fructus*, one of the earliest angiosperms, reconstructed taking account of their ecological strategies. *Memoir of the Fukui Prefectural Dinosaur Museum*, **4**, 35–44.

Terborgh J. (1988). The big things that run the world – a sequel to E. O. Wilson. *Conservation Biology*, **2**, 402–3.

Terborgh J., Feeley, K., Silman, M., Nuñez, P., & Balukjian, B. (2006). Vegetation dynamics of predator-free land-bridge islands. *Journal of Ecology*, **94**, 253–63.

Tewksbury, J. L. & G. P. Nabhan. (2001). Seed dispersal: directed deterrence by capsaicin in chilies. *Nature*, **412**, 403–4.

Thoday, D. (1951). The haustorial system of *Viscum album*. *Journal of Experimental Botany*, **2**, 1–19.

Thomas, L. K., Jr. (1980). The Impact of Three Exotic Plant Species on a Potomac Island. National Park Service Scientific Monograph Series No. 13. Washington: U.S. Department of the Interior.

Thompson, J. N. (1994). *The Coevolutionary Process*. Chicago: University of Chicago Press.

Thompson, K. (1984). Why biennials are not as few as they ought to be. *American Naturalist*, **123**, 854–61.

Thompson, K., Bakker, J. P., Bekker, R. M., & Hodgson, J. G. (1998). Ecological correlates of seed persistence in soil in the north-west European flora. *Journal of Ecology*, **86**, 163–9.

Thompson, K., Band, S. R., & Hodgson, J. G. (1993). Seed size and shape predict persistence in the soil. *Functional Ecology*, **7**, 236–41.

Thompson, K., Ceriani, R. M., Bakker, J. P., & Bekker, R. M. (2003). Are seed dormancy and persistence in the soil related? *Seed Science Research*, **13**, 97–100.

Thompson, K. & Grime, J. P. (1979). Seasonal variation in the seed banks of herbaceous species in ten contrasting habitats. *Journal of Ecology*, **67**, 893–921.

Thompson, K., Grime, J. P., & Mason, G. (1977). Seed germination in response to diurnal fluctuations of temperature. *Nature*, **267**, 147–9.

Thompson, K., Hodgson, J. G., Grime, J. P., & Burke, M. J. W. (2001). Plant traits and temporal scale: evidence from a 5-year invasion experiment using native species. *Journal of Ecology*, **89**, 1054–60.

Thompson, K., Jalili, A., Hodgson, J. G., *et al.* (2001). Seed size, shape and persistence in the soil in an Iranian flora. *Seed Science Research*, **11**, 345–55.

Thomsen, M. A., D'Antonio, C. M., Suttle, K., & Sousa, W. (2006). Ecological resistance, seed density and their interactions determine patterns of invasion in a California coastal grassland. *Ecology Letters*, **9**, 160–70.

Thoreau, H. D. (1993). *Faith in a Seed: the Dispersion of Seeds and Other Late Natural History Writings*, ed. B. P. Dean. Washington: Island Press.

Thorén, L. M. & Karlsson, P. S. (1998). Effects of supplementary feeding on growth and reproduction of three carnivorous plant species in a subarctic environment. *Journal of Ecology*, **86**, 501–10.

Tielbörger, K. & Kadmon, R. (2000). Indirect effects in a desert plant community: is competition among annuals more intense under shrub canopies? *Plant Ecology*, **150**, 53–63.

Tiffney, B. (1986). Evolution of seed dispersal syndromes according to the fossil record. In *Seed Dispersal*, ed. D. R. Murray. Orlando: Academic Press, pp. 273–305.

Tiffney, B. H. (2004). Vertebrate dispersal of seed plants through time. *Annual Review of Ecology, Evolution, and Systematics*, **35**, 1–29.

Tillich, H.-J. (1990). The seedlings of Nymphaeaceae – monocotylar or dicotylar? *Flora*, **184**, 169–76.

Tillich, H.-J. (1995). Seedlings and systematics in monocotyledons. In *Monocotyledons: Systematics and Evolution*, ed. P. J. Rudall, P. J. Cribb, D. F. Cuttler, & C. J. Humphries. Kew: Royal Botanic Gardens, pp. 303–52.

Tillich, H.-J. (2000). Ancestral and derived character states in seedlings of monocotyledons. In *Monocots: Systematics and Evolution*, ed. K. I. Wilson & D. A. Morrison. Melbourne: CSIRO, pp. 212–29.

Tilman, D. (1988). *Plant Strategies and the Dynamics and Structure of Plant Communities*. Princeton: Princeton University Press.

Tilman, D. (1994). Competition and biodiversity in spatially structured habitats. *Ecology*, **75**, 2–16.

Tilman, D. (1997). Community invasibility, recruitment limitation, and grassland biodiversity. *Ecology*, **78**, 81–92.

Tilman, D., Fargione, J., Wolff, B., *et al.* (2001). Forecasting agriculturally driven global environmental change. *Science*, **292**, 281–4.

Titus, J. E. & Hoover, D. T. (1991). Toward predicting reproductive success in submersed freshwater angiosperms. *Aquatic Botany*, **41**, 111–36.

Titus, J. H. & del Moral, R. (1998). The role of mycorrhizal fungi and microsites in primary succession on Mount St. Helens. *American Journal of Botany*, **85**, 370–5.

Tobe, K., Zhang, L., & Omasa, K. (2006). Seed germination and seedling emergence of three *Artemisia* species (Asteraceae) inhabiting desert sand dunes in China. *Seed Science Research*, **16**, 61–9.

Tobin, M. F., Lopez, O. R., & Kursar, T. A. (1999). Responses of tropical understory plants to a severe drought, tolerance and avoidance of water stress. *Biotropica*, **31**, 570–8.

Toh, I., Gillespie, M., & Lamb, D. (1999). The role of isolated trees in facilitating tree seedling recruitment at a degraded sub-tropical rainforest site. *Restoration Ecology*, **7**, 288–97.

Tomilov, A. A., Tomilova, N. B., Abdallah, I., & Yoder, J. I. (2005). Localized hormone fluxes and early haustorium development in the hemiparasitic plant *Triphysaria versicolor*. *Plant Physiology*, **138**, 1469–80.

Tomlinson, P. B. (1971). The shoot apex and its dichotomous branching in the Nypa palm. *Annals of Botany*, **35**, 865–79.

Tomlinson, P. B. (1986). *The Botany of Mangroves*. Cambridge: Cambridge University Press.

Tomlinson, P. B. (1990). *The Structural Biology of Palms*. Oxford: Clarendon Press Oxford.

Tomlinson, P. B. & Esler, A. E. (1973). Establishment growth in woody monocotyledons native to New Zealand. *New Zealand Journal of Botany*, **11**, 627–44.

Tongway, D. J. & Ludwig, J. A. (1996). Rehabilitation of semiarid landscapes in Australia. I. Restoring productive soil patches. *Restoration Ecology*, **4**, 388–97.

Tooke, F., Ordidge, M., Chiurugwi, T., & Battey, N. (2005). Mechanisms and function of flower and inflorescence reversion. *Journal of Experimental Botany*, **56**, 2587–99.

Topa, M. A. & McLeod, K. W. (1986). Responses of *Pinus clausa*, *Pinus serotina* and *Pinus taeda* seedlings to anaerobic solution culture. I. Changes in growth and root morphology. *Physiologia Plantarum*, **68**, 523–31.

Toth, R. & Kuijt, J. (1977). Cytochemical localization of acid phosphatase in endophyte cells of the semiparasitic angiosperm *Commandra umbellata*. *Canadian Journal of Botany*, **55**, 470–5.

Todzia, C. A. (1988). Chloranthaceae. *Hedyosmum*. *Flora Neotropica Monograph*, **48**, 1–139.

Trappe, J. M. (1977). Selection of fungi for ectomycorrhizal inoculation in nurseries. *Annual Review of Phytopathology*, **15**, 203–22.

Travis, S. E. & Hester, M. W. (2005). A space-for-time substitution reveals the long-term decline in genotypic diversity of a widespread salt marsh plant, *Spartina alterniflora*, over a span of 1500 years. *Journal of Ecology*, **93**, 417–30.

Trenberth, K. E. & Hoar, T. J. (1996). The 1990–1995 El Niño-southern oscillation event: longest on record. *Geophysical Research Letters*, **23**, 57–60.

Tripathi, R. S. & Khan, M. L. (1990). Effects of seed weight and microsite characteristics on germination and seedling fitness in two species of *Quercus* in a subtropical wet hill forest. *Oikos*, **57**, 289–96.

Trudell, S. A., Rygiewicz, P. T., & Edmonds, R. L. (2003). Nitrogen and carbon stable isotope abundances support the myco-heterotrophic nature and host-specificity of certain achlorophyllous plants. *New Phytologist*, **160**, 391–401.

Tsoar, H. & Karnieli, A. (1996). What determines the spectral reflectance of the Negev-Sinai sand dunes. *International Journal of Remote Sensing*, **17**, 513–25.

Tucic, B., Pemac, D., & Ducic J. (2005). Life history responses to irradiance at the early seedling stage of *Picea omorika* (Pancic) Purkynhe: adaptiveness and evolutionary limits. *Acta Oecologica*, **27**, 185–95.

Tungate, K. D., Burton, M. G., Susko, D. J., Sermons, S. M., & Rufty, T. W. (2006). Altered weed reproduction and maternal effects under low-nitrogen fertility. *Weed Science*, **54**, 847–53.

Tuomisto, H., Poulsen, A. D., Moran, R. C., *et al.* (2003). Linking floristic patterns with soil heterogeneity and satellite imagery in Ecuadorian Amazonia. *Ecological Applications*, **13**, 352–71.

Turnbull, L. A., Coomes, D., Hector, A., & Rees, M. (2004). Seed mass and the competition/colonization trade-off: competitive interactions and spatial patterns in a guild of annual plants. *Journal of Ecology*, **92**, 97–109.

Turnbull, L. A., Crawley, M. J., & Rees, M. (2000). Are plant populations seed-limited? A review of seed sowing experiments. *Oikos*, **88**, 225–38.

Turnbull, L. A., Rees, M., & Crawley, M. (1999). Seed mass and the competition/colonization trade-off: a sowing experiment. *Journal of Ecology*, **87**, 899–912.

Turner, D. P. & Franz, E. H. (1985). Size class structure and tree dispersion patterns in old-growth cedar-hemlock forests of the northern Rocky Mountains (USA). *Oecologia*, **68**, 52–6.

Turner, I. M. (2001). *The Ecology of Trees in the Tropical Rain Forest*. Cambridge: Cambridge University Press.

Turner, R. M., Alcorn, S. M., Olin, G., & Booth, J. A. (1966). The influence of shade, soil, and water on saguaro seedling establishment. *Botanical Gazette*, **127**, 95–102.

Turner, S. R., Pearce, B., Rokich, D. P., *et al.* (2006). Influence of polymer seed coatings, soil raking, and time of sowing seedling performance in post-mining restoration. *Restoration Ecology*, **14**, 267–77.

Tyler, G. & Strom, L. (1995). Differing organic-acid exudation pattern explains calcifuge and acidifuge behavior in plants. *Annals of Botany*, **75**, 75–8.

Uchiyama, Y. (1981). Studies on the germination of saltbushes. 1. The relationship between temperature and germination of *Atriplex nummularia* Lindl. *Japanese Journal of Tropical Agriculture*, **25**, 62–7.

Uhl, C. & Kauffman, J. B. (1990). Deforestation, fire susceptibility, and potential tree responses to fire in the eastern Amazon. *Ecology*, **71**, 437–49.

Ungar, I. A. (1996). Effect of salinity on seed germination, growth, and ion accumulation of *Atriplex patula* (Chenopodiaceae). *American Journal of Botany*, **83**, 604–7.

Urbanska, K. M. (1997). Safe sites – interface of plant population ecology and restoration ecology. In *Restoration Ecology and Sustainable Development*, ed. K. M. Urbanska, N. Webb, & P. Edwards. Cambridge: Cambridge University Press, pp. 81–110.

Urbanska, K. M. & Chambers, J. C. (2002). High elevation ecosystems. In *Handbook of Ecological Restoration*, vol. 2., ed. M. Perrow & A. Davy. Cambridge University Press, pp. 376–400.

Uva, R. H., Neal, J. C., & DiTomaso, J. M. (1997). *Weeds of the Northeast*. Ithaca: Cornell University Press.

Vacher, C., Weis, A. E., Hermann, D., *et al.* (2004). Impact of ecological factors on the initial invasion of *Bt* transgenes into wild populations of birdseed rape (*Brassica rapa*). *Theoretical & Applied Genetics*, **109**, 806–14.

Valiente-Banuet, A. & Ezcurra, E. (1991). Shade as a cause of the association between the cactus *Neobuxbaumia tetetzo* and the nurse plant *Mimosa luisana* in the Tehuacan Valley, Mexico. *Journal of Ecology*, **79**, 961–71.

Valiente-Banuet, A., Bolongaro-Crevenna, A., Briones, O., *et al.* (1991a). Spatial relationships between cacti and nurse shrubs in a semi-arid environment in central Mexico. *Journal of Vegetation Science*, **2**, 15–20.

Valiente-Banuet, A., Vite, F., & Zavala-Hurtado, J. A. (1991b). Interaction between the cactus *Neobuxbaumia tetetzo* and the nurse shrub *Mimosa luisana*. *Journal of Vegetation Science*, **2**, 11–14.

Valladares, F., Balaguer, L., Martinez-Ferri, E., Perez-Corona, E., & Manrique, E. (2002). Plasticity, instability and canalization: is the phenotypic variation in seedlings of sclerophyll oaks consistent with the environmental unpredictability of Mediterranean ecosystems? *New Phytologist*, **156**, 457–67.

Valladares, F., Sanchez-Gomez, D., & Zavala, M. A. (2006). Quantitative estimation of phenotypic plasticity: bridging the gap between the evolutionary concept and its ecological applications. *Journal of Ecology*, **94**, 1103–16.

Valladares, F., Villar-Salvador, P., Domínguez, S., *et al.* (2002). Enhancing the early performance of the leguminous shrub *Retama sphaerocarpa* (L.) Boiss.: fertilisation versus *Rhizobium* inoculation. *Plant and Soil*, **240**, 253–62.

Valladares, F., Wright, S. J., Lasso, E., Kitajima, K., & Pearcy, R. W. (2000). Plastic phenotypic response to light of 16 congeneric shrubs from a Panamanian rainforest. *Ecology*, **81**, 1925–36.

Van Auken, O. W. & Bush, J. K. (1990). Influence of light levels soil nutrients and competition on seedling growth of *Baccharis neglecta* Asteraceae. *Bulletin of the Torrey Botanical Club*, **117**, 438–44.

Vandenbussche, F. & Van Der Straeten, D. (2004). Shaping the shoot: a circuitry that integrates multiple signals. *Trends in Plant Science*, **9**, 499–506.

Vandenbussche, F., Verbelen, J.-P., & Van Der Straeten, D. (2005). Of light and length: regulation of hyopocotyl growth in *Arabidopsis*. *BioEssays*, **27**, 275–84.

Vandenkoornhuyse P., Husband, R., Daniell, T. J., *et al.* (2002). Arbuscular mycorrhizal community composition associated with two plant species in a grassland ecosystem. *Molecular Ecology*, **11**, 1555–64.

van der Heijden, M. G. A. (2002). Arbuscular mycorrhizal fungi as a determinant of plant diversity: in search of underlying mechanisms and general principles. In *Mycorrhizal Ecology*, ed. M. G. A. van der Heijden & I. Sanders. Berlin: Springer, pp. 243–65.

van der Heijden, M. G. A. (2004). Arbuscular mycorrhizal fungi as support systems for seedling establishment in grassland. *Ecology Letters*, **7**, 293–303.

van der Heijden, M. G. A., Bakker, R., Verwaal, J., *et al.* (2006a). Symbiotic bacteria as a determinant of plant community structure and plant productivity in dune grassland. *FEMS Microbiology Ecology*, **56**, 178–87.

van der Heijden, M. G. A., Klironomos, J. N., Ursic, M., *et al.* (1998). Mycorrhizal fungal diversity determines plant biodiversity, ecosystem variability and productivity. *Nature*, **396**, 69–72.

van der Heijden, M. G. A. & Scheublin, T. R. (2007). Functional traits in mycorrhizal ecology: their use for predicting the impact of arbuscular mycorrhizal fungal communities on plant growth and ecosystem functioning. *New Phytologist*, **174**, 244–50.

van der Heijden, M. G. A., Streitwolf-Engel, R., Riedl, R., *et al.* (2006b). The mycorrhizal contribution to plant productivity, plant diversity, plant nutrition and soil structure in experimental grassland. *New Phytologist*, **172**, 739–52.

van der Pijl, L. (1982). *Principles of Dispersal in Higher Plants*. Berlin: Springer-Verlag.

Van der Plank, J. E. (1978). *Genetic and Molecular Basis of Plant Pathogenesis*. New York: Springer-Verlag.

van der Valk, A. G. (1978). The role of seed banks in the vegetation dynamics of prairie glacial marshes. *Ecology*, **59**, 322–35.

van der Valk, A. G. (1981). Succession in wetlands: a Gleasonian approach. *Ecology*, **62**, 688–96.

van der Valk, A. G., Bremholm, T. L., & Gordon, E. (1999). The restoration of sedge meadows: seed viability, seed germination requirements, and seedling growth of *Carex* species. *Wetlands*, **19**, 756–64.

Vander Wall, S. B. (1994). Seed fate pathways of antelope bitterbrush: dispersal by seed-caching yellow pine chipmunks. *Ecology*, **75**, 1911–26.

van Gelder, H. A., Poorter, L., & Sterck, F. J. (2006). Wood mechanics, allometry, and life-history variation in a tropical rain forest tree community. *New Phytologist*, **171**, 367–78.

van Mantgem, P. J., Stephenson, N. L., & Keeley, J. E. (2006). Forest reproduction along a climatic gradient in the Sierra Nevada, California. *Forest Ecology and Management*, **225**, 391–9.

van Mantgem, P. J., Stephenson, N. L., Keifer, M. B., & Keeley, J. E. (2004). Effects of an introduced pathogen and fire exclusion on the demography of sugar pine. *Ecological Applications*, **14**, 1590–602.

Van Rheenen, H. M. P. J. B., Boot, R. G. A., Werger, M. J. A., & Ulloa Ulloa, M. (2004). Regeneration of timber trees in a logged tropical forest in North Bolivia. *Forest Ecology and Management*, **200**, 39–48.

Van Splunder, I., Coops, H., Voesenek, L. A. C. J., & Blom, C. W. P. M. (1995). Establishment of alluvial forest species in floodplains: The role of dispersal timing, germination characteristics and water level fluctuations. *Acta Botanica Neerlandica*, **44**, 269–78.

van Staden, J., Brown, N., Pager, A., & Johnson, T. (2000). Smoke as a germination cue. *Plant Species Biology*, **15**, 167–78.

Vaughton, G. & Ramsey, M. (1997). Seed mass variation in the shrub *Banksia spinulosa* (Proteaceae) – resource constraints and pollen source effects. *International Journal of Plant Sciences*, **158**, 424–31.

Vaughton, G. & Ramsey, M. (2001). Relationships between seed mass, seed nutrients, and seedling growth in *Banksia cunninghamii* (Proteaceae). *International Journal of Plant Sciences*, **162**, 599–606.

Vazquez-Yanes, C., Orozco-Segovia, A., Rincon, E., et al. 1990. Light beneath the litter in a tropical forest: effect on seed germination. *Ecology*, **71**, 1952–8.

Veblen, T. T. (1986). Regeneration dynamics. In *Plant Succession. Theory and Prediction*, ed. D. C. Glenn-Lewin, R. K. Peet, & T. T. Veblen. New York: Chapman & Hall, pp. 152–87.

Veenendaal, E. M., Swaine, M. D., Agyeman, V. K., et al. (1995). Differences in plant and soil water relations in and around a forest gap in West Africa during the dry season may influence seedling establishment and survival. *Journal of Ecology*, **83**, 83–90.

Veenendaal, E. M., Swaine, M. D., Lecha, R. T., et al. (1996). Responses of West African forest tree seedlings to irradiance and soil fertility. *Functional Ecology*, **10**, 501–11.

Veevers-Carter, W. (1991). *Riches of the Rain Forest: An Introduction to the Trees and Fruits of the Indonesian and Malaysian Rain Forests*. Oxford: Oxford University Press.

Venable, D. L. (1992). Size-number trade-offs and the variation of seed size with plant resource status. *American Naturalist*, **140**, 287–304.

Venable, D. L. & Brown, J. S. (1988). The selective interactions of dispersal, dormancy and seed size as adaptations for reducing risks in variable environments. *American Naturalist*, **131**, 360–84.

Venable, D. L., Dyreson, E., & Morales, E. (1995). Population dynamic consequences and evolution of seed traits of *Heterosperma pinnatum* (Asteraceae). *American Journal of Botany*, **82**, 410–20.

Venable, D. L., Pake, C. E., & Caprio, A. C. (1993). Diversity and coexistence of Sonoran desert winter annuals. *Plant Species Biology*, **8**, 207–16.

Veneklaas, E. J. & den Ouden, F. (2005). Dynamics of non-structural carbohydrates in two *Ficus* species after transfer to deep shade. *Environmental and Experimental Botany*, **54**, 148–54.

Venning, J. (1988). *Growing Trees for Farms, Parks and Roadsides. A Revegetation Manual.* Melbourne: Lothian.

Verburg, R., Maas, J., & During, H. J. (2000). Clonal diversity in differently-aged populations of the pseudo-annual clonal plant *Circaea lutetiana* L. *Plant Biology*, **2**, 646–52.

Verdcourt, B. (1986). Chloranthaceae. *Flora Malesiana*, **10**, 123–44.

Verhoeven, K. J. F., Biere, A., Nevo, E., & van Damme, J. M. M. (2004). Can a genetic correlation with seed mass constrain adaptive evolution of seedling desiccation tolerance in wild barley? *International Journal of Plant Sciences*, **165**, 281–8.

Verrecchia, E., Yair, A., Kidron, G. J., & Verrecchia, K. (1995). Physical properties of the psammophile cryptogamic crust and their consequences to the water regime of sandy soils, north-western Negev Desert, Israel. *Journal of Arid Environments*, **29**, 427–37.

Vert, G., Nemhauser, J. L., Geldner, N., Hong, F., & Chory, J. (2005). Molecular mechanisms of steroid hormone signaling in plants. *Annual Review of Cell and Developmental Biology*, **21**, 177–201.

Viana, V. M., Tabanez, A. A., & Batista, J. (1997). Dynamics and restoration of forest fragments in the Brazilian Atlantic moist forest. In *Tropical Forest Remnants, Ecology, Management, and Conservation of Fragmented Communities*, ed. W. F. Laurance & R. O. Bierregaard. Chicago: The University of Chicago Press, pp. 351–65.

Villagra, P. E. & Cavagnaro J. B. (2005). Effects of salinity on the establishment and early growth of *Prosopis argentina* and *Prosopis alpataco* seedlings in two contrasting soils: implications for their ecological success. *Austral Ecology*, **30**, 325–35.

Vitousek, P. M. & Walker, L. R. (1989). Biological invasions by *Myrica faya* in Hawaii: plant demography, nitrogen fixation, and ecosystem effects. *Ecological Monographs*, **59**, 247–65.

Vladesco, M. A. (1935). Rescherches morphologiques et expérimentales sur l'embryogénie et l'organogénie des fougères leptosporangiées. *Revue Générale du Botanique*, **47**, 513–28; 564–88.

Voesenek, L. A. C. J. & Blom, C. W. P. M. (1996). Plants and hormones: an ecophysiological view on timing and plasticity. *Journal of Ecology*, **84**, 111–19.

Voesenek, L. A. C. J., Colmer, T. D., Pierik, R., Millenaar, F. F., & Peeters, A. J. M. (2006). How plants cope with complete submergence. *New Phytologist*, **170**, 213–26.

Voesenek, L. A. C. J., Rijnders, J. H. G. M., Peeters, A. J. M., van de Steeg, H. M., & de Kroon, H. (2004). Plant hormones regulate fast shoot elongation under water: from genes to communities. *Ecology*, **85**, 16–27.

Voets, L., de la Providencia, I. E., & Declerck, S. (2006). Glomeraceae and Gigasporaceae differ in their ability to form hyphal networks. *New Phytologist*, **172**, 185–8.

Void, K. M. (1962). Vivipary in bamboo, *Melocanna bambosoides* Trin. *Journal Bombay Natural History Society*, **59**, 696–7.

Volaire, F. & Norton, M. (2006). Summer dormancy in perennial temperate grasses. *Annals of Botany*, **98**, 927–33.

von Teichman, I. & van Wyk, A. E. (1991). Structural aspects and trends in the evolution of recalcitrant seeds in dicotyledons. *Seed Science Research*, **4**, 225–39.

von Wettberg, E. J. & Schmitt, J. (2005). Physiological mechanism of population differentiation in shade-avoidance responses between woodland and clearing genotypes of *Impatiens capensis*. *American Journal of Botany*, **92**, 868–74.

Vourc'h, G., Vila, B., Gillon, D., Escarré, J., & Guibal, F. (2002). Disentangling the causes of damage variation by deer browsing on young *Thuja plicata*. *Oikos*, **98**, 271–83.

Wagner, W. H., Jr. (1952). The fern genus *Diellia*. *University of California Publications in Botany*, **26**, 1–212.

Waisel, Y. (1972). *Biology of Halophytes*. New York: Academic Press.

Walbot, V. (1978). Control mechanisms for plant embryogeny. In *Dormancy and Developmental Arrest*, ed. M. E. Cutter. New York: Academic Press, pp. 113–66.

Walker, L. R. & del Moral, R. (2003). *Primary Succession and Ecosystem Rehabilitation*. Cambridge: Cambridge University Press.

Walker, L. R., Thompson, D. B., & Landau, F. H. (2001). Experimental manipulations of fertile islands and nurse plant effects in the Mojave Desert, USA. *Western North American Naturalist*, **61**, 25–35.

Waller, D. M. (1984). Differences in fitness between seedlings derived from cleistogamous and chasmogamous flowers in *Impatiens capensis*. *Evolution*, **38**, 427–40.

Walters, J. R., Bell, T. L., & Read, S. (2005). Intra-specific variation in carbohydrate reserves and sprouting ability in *Eucalyptus* seedlings. *Australian Journal of Botany*, **53**, 195–203.

Walters, M. B., Kruger, E. L., & Reich, P. B. (1993a). Growth, biomass distribution and CO_2 exchange of northern hardwood seedlings in high and low light: relationships with successional status and shade tolerance. *Oecologia*, **96**, 7–16.

Walters, M. B., Kruger, E. L., & Reich, P. B. (1993b). Relative growth rate in relation to physiological and morphological traits for northern hardwood tree seedlings – species, light environment and ontogenetic considerations. *Oecologia*, **96**, 219–31.

Walters, M. B. & Reich, P. B. (1996). Are shade tolerance, survival, and growth linked? Low light and nitrogen effects on hardwood seedlings. *Ecology*, **77**, 841–53.

Walters, M. B. & Reich, P. B. (2000). Seed size, nitrogen supply, and growth rate affect tree seedling survival in deep shade. *Ecology*, **81**, 1887–901.

Wang, B. & Qiu, Y.-L. (2006) Phylogenetic distribution and evolution of mycorrhizas in land plants. *Mycorrhiza*, **16**, 299–363.

Wang, J. H., Machado, C., Panaccione, D. G., Tsai, H. F., & Schardl, C. L. (2004). The determinant step in ergot alkaloid biosynthesis by an endophyte of perennial ryegrass. *Fungal Genetics and Biology*, **41**, 189–98.

Wang, X. (2004). Lipid signaling. *Current Opinion in Plant Biology*, **7**, 329–36.

Wang, Y.-H. & Augspurger, C. (2006). Comparison of seedling recruitment under arborescent palms in two Neotropical forests. *Oecologia*, **147**, 533–45.

Wang, Z. M. & Lechowitz. M. J. (1998). Effect of sowing date on the germination and establishment of black spruce and jack pine under simulated field conditions. *Ecoscience*, **5**, 95–9.

Ward, D. (2005). Do we understand the causes of bush encroachment in African savannas? *African Journal of Range and Forage Science*, **22**, 101–5.

Wardle, J. A. (1984). *The New Zealand Beeches: Ecology, Utilisation and Management*. Christchurch: NZ Forest Service, The Caxton Press.

Webb, C. O., Gilbert, G. S., & Donoghue, M. J. (2006). Phylodiversity-dependent seedling mortality, size structure, and disease in a bornean rain forest. *Ecology*, **87**, S123–31.

Webb, C. O. & Peart, D. R. (1999). Seedling density dependence promotes coexistence of Bornean rain forest trees. *Ecology*, **80**, 2006–17.

Webb, L. J. (1958). Cyclones as an ecological factor in tropical lowland rain forest in Northern Queensland. *Australian Journal of Botany*, **6**, 220–8.

Weber, A., Karst, J., Gilbert, B., & Kimmins, J. P. (2005). *Thuja plicata* exclusion in ectomycorrhiza-dominated forests: testing the role of inoculum potential of arbuscular mycorrhizal fungi. *Oecologia*, **143**, 148–56.

Weiher, E. & Keddy, P. A. (1995). The assembly of experimental wetland plant communities. *Oikos*, **73**, 323–35.

Weiher, E. & Keddy, P., ed. (1999). *Ecological Assembly Rules. Perspectives, Advances, Retreats*. Cambridge: Cambridge University Press.

Weinbaum, B. S., Allen, M. F., & Allen, E. B. (1996). Survival of arbuscular mycorrhizal fungi following reciprocal transplanting across the Great Basin, USA. *Ecological Applications*, **6**, 1365–72.

Weinig, C. (2000). Differing selection in alternative competitive environments: shade-avoidance responses and germination timing. *Evolution*, **54**, 124–36.

Welch, B. L. (1997). Seeded versus containerized big sagebrush plants for seed-increase gardens. *Journal of Range Management*, **50**, 611–14.

Weller, S. G. (1985). Establishment of *Lithospermum caroliniense* on sand dunes – the role of nutlet mass. *Ecology*, **66**, 1893–901.

Wells, A. G. (1982). Mangrove vegetation of northern Australia. In *Mangrove Ecosystems in Australia: Structure, Function and Management*, ed. B. F. Clough. Canberra: Australian National University Press, pp. 57–78.

Weltzin, J. F. & McPherson, G. R. (1999). Facilitation of conspecific seedling recruitment and shifts in temperate savanna ecotones. *Ecological Monographs*, **69**, 513–34.

Wenny, D. G. (2001). Advantages of seed dispersal: a reevaluation of directed dispersal. *Evolutionary Ecology Research*, **3**, 51–74.

Went, F. W. (1942). The dependence of certain annual plants on shrubs in a Southern California desert. *Bulletin of the Torrey Botanical Club*, **69**, 100–14.

Went, F. W. (1948). Ecology of desert plants. I. Observations on germination in the Joshua Tree National Monument, California. *Ecology*, **29**, 242–53.

Went, F. W. (1949). Ecology of desert plants. II. The effect of rain and temperature on germination and growth. *Ecology*, **30**, 1–13.

Werner, C., Correia, O., & Beyschlag, W. (1999). Two different strategies of Mediterranean macchia plants to avoid photoinhibitory damage by excessive radiation levels during summer drought. *Acta Oecologica-International Journal of Ecology*, **20**, 15–23.

Wernert, S. J., ed. (1982). *North American Wildlife*. Pleasantville: Reader's Digest Association, Inc.

Wesche, K., Ronnenberg, K., & Hensen, I. (2005). Lack of sexual reproduction within mountain steppe populations of the clonal shrub *Juniperus sabina* L. in semi-arid southern Mongolia. *Journal of Arid Environments*, **63**, 390–405.

West, N. E. (1986). Desert ecosystems: desertification or xerification? *Nature*, **321**, 562–3.

Westbrook, T. (1999). Episodic events in the regeneration of *Myoporum platicarpum* spp. *platicarpum* in south east Australia. In *Proceedings of the VI International Rangelands Congress*, ed. D. Eldridge & D. Freudenberger. Townsville: VI International Rangeland Conference, Inc., pp. 212–14.

Westbury, D. B. (2004). Biological flora of the British Isles: *Rhinanthus minor* L. *Journal of Ecology*, **93**, 906–27.

Westoby, M. (1998). A leaf-height-seed (LHS) plant ecology strategy scheme. *Plant and Soil*, **199**, 213–27.

Westoby, M., Falster, D. S., Moles, A. T., Vesk, P. A., & Wright, I. J. (2002). Plant ecological strategies: some leading dimensions of variation between species. *Annual Review of Ecology and Systematics*, **33**, 125–59.

Westoby, M., Jurado, E., & Leishman, M. (1992). Comparative evolutionary ecology of seed size. *Trends in Ecology & Evolution*, **7**, 368–72.

Westoby, M. & Wright, I. J. (2006). Land-plant ecology on the basis of functional traits. *Trends in Ecology & Evolution*, **21**, 261–8.

Wheelwright, N. T. (2004). Fruit size in a tropical tree species: variation, preference by birds and heritability. *Plant Ecology*, 107–8, 163–74.

Wheland R. J. (1995). *The Ecology of Fire*. Cambridge: Cambridge University Press.

Wherry, E. T. (1948). *Wild Flower Guide, Northeastern and Midland United States*. New York: Doubleday & Company, Inc.

Whigham, D. (1984). The influence of vines on the growth of *Liquidambar styraciflua* L. (sweetgum). *Canadian Journal of Forest Research*, **14**, 37–9.

Whigham, D. F., Dickinson, M. B., & Brokow, N. V. L. (1999). Background canopy gap and catastrophic wind disturbances in tropical forests. In *Ecosystems of Disturbed Ground, Ecosystems of the World 16*, ed. L. R. Walker. Amsterdam: Elsevier, pp. 223–52.

Whigham, D. F., O'Neill, J. P., Rasmussen, H. N., Caldwell, B. A., & McCormick, M. K. (2006). Seed longevity in terrestrial orchids – potential for persistent *in situ* seed banks. *Biological Conservation*, **129**, 24–30.

Whitehead, M. R. & Brown, C. A. (1940). The seeds of the spider lily, *Hymenocallis occidentalis*. *American Journal of Botany*, **27**, 199–303.

Whitehouse, M. E. A., Shochat, E., Shachak, M., & Lubin, Y. (2002). The influence of scale and patchiness on spider diversity in a semi-arid environment. *Ecography*, **25**, 395–404.

Whitford, W. G. & Kay, F. R. (1999). Bio perturbation by mammals in deserts: a review. *Journal of Arid Environments*, **41**, 203–30.

Whitman, A. A., Brokaw, N. V. L., & Hagan J. M. (1997). Forest damage caused by selection logging of mahogany (*Swietenia macrophylla*) in northern Belize. *Forest Ecology and Management*, **92**, 87–96.

Whitmore, M. D. & Swaine, T. C. (1988). On the definition of ecological species groups in tropical rain forests. *Vegetatio*, **75**, 81–6.

Whitmore, W. C. (1996). A review of some aspects of tropical rain forest seedling ecology with suggestions for further enquiry. In *The Ecology of Tropical Forest Tree Seedlings*, ed. M. D. Swaine. Paris: UNESCO & The Parthenon Publishing Group, pp. 3–39.

Whittaker, R. H. (1970). *Communities and Ecosystems*. London: Collier-Macmillan.

Wiens, J. A., Addicott, J. F., Case, T. J., & Diamond, J. (1986). Overview: the importance of spatial and temporal scale in ecological investigations. In *Community Ecology*, ed. J. Diamond & T. J. Case. New York: Harper & Row, pp. 145–53.

Wiersema, J. H. (1987). A monograph of *Nymphaea* subgenus *Hydrocallis* (Nymphaeaceae). *Systematic Botany Monographs*, **16**, 1–112.

Wilby, A. & Shachak, M. (2004). Shrubs, granivores and annual plant community stability in an arid ecosystem. *Oikos*, **106**, 209–16.

Willems, J. H. (1982). Establishment and development of a population of *Orchis simia* Lamk. in The Netherlands, 1972 to 1981. *New Phytologist*, **91**, 757–65.

Willems, J. H. (2002). A founder population of *Orchis simia* in The Netherlands: a 30-year struggle for survival. In *Trends and Fluctuations and Underlying Mechanisms in Terrestrial Orchid Populations*, ed. P. Kindlmann, J. H. Willems & D. F. Whigham. Leiden: Backhuys Publishers, pp. 23–32.

Williams, C. E., Lipscomb, M. V., Johnson, W. C., & Nilsen, E. T. (1990). Influence of leaf litter and soil moisture regime on early establishment of *Pinus pungens*. *American Midland Naturalist*, **124**, 142–52.

Williams, J. H., & Friedman, W. E. (2002). Identification of diploid endosperm in an early angiosperm lineage. *Nature*, **415**, 522–6.

Williams-Linera, G. (1990). Origin and early development of forest edge vegetation in Panama. *Biotropica*, **22**, 235–41.

Willig, M. R., Kaufman, D. M., & Stevens, R. D. (2003). Latitudinal gradients of biodiversity: pattern, process, scale and synthesis. *Annual Review of Ecology and Systematics*, **34**, 273–309.

Willis, A. J., Memmott, J., & Forrester, R. I. (2000). Is there evidence for the post-invasion evolution of increased size among invasive plant species? *Ecology Letters*, **3**, 275–83.

Willis, S. G. & Hulme, P. E. (2004). Environmental severity and variation in the reproductive traits of *Impatiens glandulifera*. *Functional Ecology*, **18**, 887–98.

Wilson, J. B. (1999). Assembly rules in plant communities. In *Ecological Assembly Rules. Perspectives, Advances, Retreats*, ed. E. Weiher & P. Keddy. Cambridge: Cambridge University Press, pp. 130–64.

Willson, M. F., Irvine, A. K., & Walsh, N. G. (1989). Vertebrate dispersal syndromes in some Australian and New Zealand plant communities, with geographic comparisons. *Biotropica*, **21**, 133–47.

Wing, S. L. & Boucher, L. D. (1998). Ecological aspects of the Cretaceous flowering plant radiation. *Annual Review of Earth and Planetary Sciences*, **26**, 379–421.

Winn, A. A. (1988). Ecological and evolutionary consequences of seed size in *Prunella vulgaris*. *Ecology*, **69**, 1537–44.

Winn, A. A. (1991). Proximate and ultimate sources of within-individual variation in seed mass in *Prunella vulgaris* Lamiaceae. *American Journal of Botany*, **78**, 838–44.

Winn, A. A. & Werner, P. A. (1987). Regulation of seed yield within and among populations of *Prunella vulgaris*. *Ecology*, **68**, 1224–33.

Wirtz, K. W. (2003). Adaptive significance of C partitioning and regulation of specific leaf area in *Betula pendula*. *Tree Physiology*, **23**, 181–90.

Wolf, L. L., Hainsworth, F. R., & Mercier, T. B. R. (1986). Seed size variation and pollinator uncertainty in *Ipomopsis aggregata* Polemoniaceae. *Journal of Ecology*, **74**, 361–72.

Wolfe, L. N. (1995). The genetics and ecology of seed size variation in a biennial plant, *Hydrophyllum appendiculatum* (Hydrophyllaceae). *Oecologia*, **101**, 343–52.

Wood, D. M. & Morris, W. F. (1990). Ecological constraints to seedling establishment on the Pumice Plains, Mount St. Helens, Washington. *American Journal of Botany*, **77**, 1411–18.

Woodhouse, J. M. & Johnson, M. S. (1991). The effect of gel-forming polymers on seed germination and establishment. *Journal of Arid Environments*, **20**, 375–80.

Wright, H. A. & Bailey, A. W. (1982). *Fire Ecology. United States and Canada*. New York: John Wiley & Sons.

Wright, I. J., Ackerly, D. D., Bongers, F., *et al.* (2007). Relationships among ecologically important dimensions of plant trait variation in seven Neotropical forests. *Annals of Botany*, **99**, 1003–15.

Wright, I. J. & Cannon, K. (2001). Relationships between leaf lifespan and structural defences in a low-nutrient, sclerophyll flora. *Functional Ecology*, **15**, 351–9.

Wright, I. J., Clifford, H. T., Kidson, R., *et al.* (2000). A survey of seed and seedling characters in 1744 Australian dicotyledon species: cross-species trait correlations and correlated trait-shifts within evolutionary lineages. *Biological Journal of the Linnean Society*, **69**, 521–47.

Wright, I. J., Reich, P. B., & Westoby, M. (2001). Strategy shifts in leaf physiology, structure and nutrient content between species of high and low rainfall, and high and low nutrient habitats. *Functional Ecology*, **15**, 423–34.

Wright, I. J., Reich, P. B., Westoby, M., *et al.* (2004). The world-wide leaf economics spectrum. *Nature*, **428**, 821–7.

Wright, I. J. & Westoby, M. (1999). Differences in seedling growth behaviour among species: trait correlations across species, and trait shifts along nutrient compared to rainfall gradients. *Journal of Ecology*, **87**, 85–97.

Wright, I. J. & Westoby, M. (2001). Understanding seedling growth relationships through specific leaf area and leaf nitrogen concentration: generalisations across growth forms and growth irradiance. *Oecologia*, **127**, 21–9.

Wright, I. J., Westoby, M., & Reich, P. B. (2002). Convergence towards higher leaf mass per area in dry and nutrient-poor habitats has different consequences for leaf life span. *Journal of Ecology*, **90**, 534–43.

Wright, S. J. (2002). Plant diversity in tropical forests: a review of mechanisms of species coexistence. *Oecologia*, **130**, 1–14.

Wright, S. J. (2003). The myriad consequences of hunting for vertebrates and plants in tropical forests. *Perspectives in Plant Ecology, Evolution and Systematics*, **6**, 73–86.

Wright, S. J. & Calderón, O. (2006). Seasonal, El Niño and longer term changes in flower and seed production in a moist tropical forest. *Ecology Letters*, **9**, 35–44.

Wright, S. J. Calderón, O., Hernandéz, A., & Paton, S. (2004). Are lianas increasing in importance in tropical forests? A 17-year record from Panama. *Ecology*, **85**, 484–9.

Wright, S. J. & Duber, H. C. (2001). Poachers and forest fragmentation alter seed dispersal, seed survival, and seedling recruitment in the palm *Attalea butyraceae*, with implications for tropical tree diversity. *Biotropica*, **33**, 583–95.

Wright, S. J., Zeballos, H., Dominguez, I., *et al.* (2000). Poachers alter mammal abundance, seed dispersal, and seed predation in a neotropical forest. *Conservation Biology*, **14**, 227–39.

Wu, B., Nara, K., & Hogetsu, T. (2001). Can ^{14}C-labelled photosynthetic products move between *Pinus densiflora* seedlings linked by ectomycorrhizal mycelia? *New Phytologist*, **149**, 137–46.

Wu, G. & Poethig, R. S. (2006). Temporal regulation of shoot development in *Arabidopsis thaliana* by *miR156* and its target *SPL3*. *Development*, **133**, 3539–47.

Wu, H., Pratley, J., Lemerle, D., & Haig, T. (2000). Evaluation of seedling allelopathy in 453 wheat (*Triticum aestivum*) accessions against annual ryegrass (*Lolium rigidum*) by the equal-compartment-agar method. *Australian Journal of Agricultural Research*, **51**, 937–44.

Wulff, R. D. (1986a). Seed size variation in *Desmodium paniculatum*. I. Factors affecting seed size. *Journal of Ecology*, **74**, 87–98.

Wulff, R. D. (1986b). Seed size variation in *Desmodium paniculatum*. II. Effects on seedling growth and physiological performance. *Journal of Ecology*, **74**, 99–114.

Xiong, L., Schumaker, K. S., & Zhu, J.-K. (2002). Cell signaling during cold, drought, and salt stress. *The Plant Cell*, **14**, S165–83.

Xiong, S. & Nilsson, C. (1999). The effects of plant litter on vegetation: a meta-analysis. *Journal of Ecology*, **87**, 984–94.

Xiong, S. J., Johansson, M. E., Hughes, F. M. R., *et al.* (2003). Interactive effects of soil moisture, vegetation canopy, plant litter and seed addition on plant diversity in a wetland community. *Journal of Ecology*, **91**, 976–86.

Yair, A. & Danin, A. (1980). Spatial variations in vegetation as related to the soil moisture regime over an arid limestone hillside, Northern Negev, Israel. *Oecologia*, **47**, 83–8.

Yair, A. & Shachak, M. (1982). A case study of energy, water and soil flow chains in an arid ecosystem. *Oecologia*, **54**, 389–97.

Yair, A., Sharon, D., & Lavee, H. (1980). Trends in runoff and erosion processes over an arid limestone hillside, northern Negev, Israel. *Hydrological Sciences Bulletin*, **25**, 243–55.

Yakovlev, M. S. & Yoffe, M. D. (1957). On the peculiar features in the embryogeny of *Paeonia*. *Phytomorphology*, **7**, 74–82.

Yan, Z. (1993). Resistance to haustorial development of two mistletoes, *Amyema preissii* (Miq.) Tieghem and *Lysiana exocarpi* (Behr.) Tieghem ssp. *exocarpi* (Loranthaceae), on host and non-host species. *International Journal of Plant Science*, **154**, 386–94.

Yeaton, R. I. & Esler, K. J. (1990). The dynamics of a succulent Karoo vegetation. *Vegetatio*, **88**, 103–13.

Yetka, L. A. & Galatowitsch, S. M. (1998). Factors affecting revegetation of *Carex lacustris* Willd. and *Carex stricta* Lam. from rhizomes. *Restoration Ecology*, **7**, 86–97.

Yoder, C. K. & Nowak, R. S. (1999). Hydraulic lift among native plant species in the Mojave Desert. *Plant and Soil*, **215**, 93–102.

Yoder, J. A., Zettler, L. W., & Stewart, S. L. (2000). Water requirements of terrestrial and epiphytic orchid seeds and seedlings, and evidence for water uptake by means of mycotrophy. *Plant Science*, **156**, 145–50.

Yoder, J. I. (2001). Host-plant recognition by parasitic Scrophulariaceae. *Current Opinion in Plant Biology*, **4**, 359–65.

Young, C. A., Bryant, M. K., Christensen, M. J., *et al.* (2005). Molecular cloning and genetic analysis of a symbiosis-expressed gene cluster for lolitrem biosynthesis from a mutualistic endophyte of perennial ryegrass. *Molecular Genetics and Genomics*, **274**, 13–29.

Young, J. A. & Young, C. G. (1992). *Seeds of Woody Plants in North America*. Portland: Dioscorides Press.

Young, J. P., Dickinson, T. A., & Dengler, N. G. (1995). A morphometric analysis of heterophyllous leaf development in *Ranunculus flabellaris*. *International Journal of Plant Sciences*, **156**, 5–11.

Young, T. P. & Evans, R. Y. (2000). Container stock versus direct seeding for woody species in restoration sites. *Proceedings International Plant Propagators' Society*, **50**, 577–82.

Young, T. P., Petersen, D. A., & Clary, J. J. (2005). The ecology of restoration: historical links, emerging issues and unexplored realms. *Ecology Letters*, **8**, 662–73.

Yuan, K. & Wysock-Diller, J. (2006). Phytohormone signalling pathways interact with sugars during seed germination and seedling development. *Journal of Experimental Botany*, **57**, 3359–67.

Yurkonis, K. A., Meiners, S. J., & Wachholder, B. E. (2005). Invasion impacts diversity through altered community dynamics. *Journal of Ecology*, **93**, 1053–61.

Zaady, E., Groffman, P. M., & Shachak, M. (1996a). Release and consumption of nitrogen by snail feces in Negev Desert soils. *Biology and Fertility of Soils*, **23**, 399–404.

Zaady, E., Groffman, P. M., & Shachak, M. (1996b). Litter as a regulator of N and C dynamics in macrophytic patches in Negev desert soils. *Soil Biology and Biochemistry*, **28**, 39–46.

Zaady, E., Groffman, P. M., & Shachak, M. (1998). Nitrogen fixation in macro- and microphytic patches in the Negev desert. *Soil Biology and Biochemistry*, **30**, 449–54.

Zaady, E., Gutterman, Y., & Boeken, B. (1997). The germination of mucilaginous seeds of *Plantago coronopus*, *Reboudia pinnata* and *Carrichtera annua* on cyanobacterial soil crust from the Negev Desert. *Plant and Soil*, **190**, 247–52.

Zaady, E., Levacov, R., & Shachak, M. (2004). Application of the herbicide, simazine, and its effect on soil surface parameters and vegetation in a patchy desert landscape. *Arid Land Research and Management*, **18**, 397–410.

Zaady, E. & Shachak, M. (1994). Microphytic soil crust and ecosystem leakage in the Negev desert. *American Journal of Botany*, **81**, 109.

Zaady, E., Yonatan, R., Shachak, M., & Perevolotsky, A. (2001). The effects of grazing on abiotic and biotic parameters in a semiarid ecosystem: a case study from the Northern Negev Desert, Israel. *Arid Land Research and Management*, **15**, 245–61.

Zahawi, R. A. (2005). Establishment and growth of living fence species: an overlooked tool for the restoration of degraded areas in the tropics. *Restoration Ecology*, **13**, 92–102.

Zahawi, R. A. & Augspurger, C. K. (2006). Tropical forest restoration: tree islands as recruitment foci in degraded lands in Honduras. *Ecological Applications*, **16**, 464–78.

Zanis, M. J., Soltis, D. E., Soltis, P. S., Mathews, S., & Donoghue, M. J. (2002). The root of the angiosperms revisited. *Proceedings of the National Academy of Sciences (USA)*, **99**, 6848–53.

Zanne, A. E., Chapman, C. A., & Kitajima, K. (2005). Evolutionary and ecological correlates of early seedling morphology in East African trees and shrubs. *American Journal of Botany*, **92**, 972–8.

Zas, R., Sampedro, L., Prada, E., & Fernandez-Lopez, J. (2005). Genetic variation of *Pinus pinaster* Ait. seedlings in susceptibility to the pine weevil *Hylobius abietis* L. *Annals of Forest Science*, **62**, 681–8.

Zhang, F., Chen, G., Huang, Q., et al. (2005a). Genetic basis of barley caryopsis dormancy and seedling desiccation tolerance at the germination stage. *Theoretical and Applied Genetics*, **110**, 445–53.

Zhang, J. & Maun, M. A. (1991). Establishment and growth of *Panicum virgatum* L. seedlings on a Lake Erie sand dune. *Bulletin of the Torrey Botanical Club*, **118**, 141–53.

Zhang, J. H. & Maun, M. A. (1990). Seed size variation and its effects on seedling growth in *Agropyron psammophilum*. *Botanical Gazette*, **151**, 106–13.

Zhang, J. H. & Maun, M. A. (1993). Components of seed mass and their relationships to seedling size in *Calamovilfa longifolia*. *Canadian Journal of Botany*, **71**, 551–7.

Zhang, J. H. & Maun, M. A. (1994). Potential for seed bank formation in 7 Great Lakes sand dune species. *American Journal of Botany*, **81**, 387–94.

Zhang, Z. H., Su, L., Li, W., Chen, W., & Zhu, Y. G. (2005b). A major QTL conferring cold tolerance at the early seedling stage using recombinant inbred lines of rice (*Oryza sativa* L.). *Plant Science*, **168**, 527–34.

Zhou, S., Wang, Y., & Qian, J. (1997). *Aerial Seeding on the Moving Dunes of the Maowusu Desert for Vegetation Growing*. Ottawa: International Development Research Centre.

Zhou, Z., Miwa, M., & Hogetsu, T. (1999). Analysis of genetic structure of a *Suillus grevillei* population in a *Larix kaempferi* stand by polymorphism of inter-simple sequence repeat (ISSR). *New Phytologist*, **144**, 55–63.

Zimmerman, J. K. & Olmsted, I. C. (1992). Host tree utilization by vascular epiphytes in a seasonally inundated forest (Tintal) in Mexico. *Biotropica*, **24**, 402–7.

Zohary, M. (1937). Die verbreitungsökologischen Verhältnisse der Pflanzen Palästinas. I. Die antitelechorischen Erscheinungen. *Beihheft zum Botanischen Zentralblatt A*, **56**, 1–55.

Zotz, G. (1998). Demography of the epiphytic orchid, *Dimerandra emarginata*. *Journal of Tropical Ecology*, **14**, 725–41.

Zotz, G., Cueni, N., & Körner, C. (2006). In situ growth stimulation of a temperate zone liana (*Hedera helix*) in elevated CO_2. *Functional Ecology*, **20**, 763–9.

Index

ABA, see abscisic acid
Abies, 36, 119, 265
  A. alba, 152
  A. amabilis, 202
  A. concolor, 257–9, 268
  A. lasiocarpa, 383
Abrotenella linearis var. *apiculata*, 53
abscisic acid (ABA), 35, 153, 155, 156,
  159, 160–4, 167, 168, 169
Abutilon theophrastii, 7
Acacia, 46, 187, 212
  A. mangium, 298
  A. oraria, 26
  A. papyrocarpa, as patch formers,
    67, 319
  A. tortillis, 328
  A. verticillata, 36
Acanthaceae, 318
Acaulaspora, and facilitating
  restoration, 366
acclimation, 186
Acer, 7, 39, 198, 201, 204, 206
  A. platanoides, 297
  A. pseudoplatanus, 42
  A. rubrum, 201, 257, 305
  A. saccharinum, 257
  A. saccharum, 151, 256, 257, 258,
    263, 305, 383
Aceraceae, 7, 39, 42, 151, 198, 257,
  297, 305, 383
Acetobacter, 211
achlorophyllous plants, 82, 84, 100,
  192, 193, 196, 198
Acoraceae, 21
Acorus, 20, 21, 137, 144
Acremonium, 211
Actaea spicata, 246
Actinidia deliciosa, 151
Actinidiaceae, 151
Actinostachys, 108
Actinostrobus pyramidalis, 42
adaptations, patterns of adaptations
  in deserts, 309
Adenanthera, 46
Adenostoma fasciculatum, 200, 262, 271
adult size, and seed production, 228
advanced regeneration, 256
Aegialitis, 51
  A. annulata, 152
Aegiceras, 51

A. corniculatum, 69, 151
Aegilops triuncialis, 304
Aegle marmelos, 26
aerenchyma, 37, 162
Aesculus hippocastanum, 151
Aetanthus, 92, 94
Agathis dammara, 49
Agavaceae, 19, 316, 320
Agave, 316
Aglaodorum griffithii, 55
Agrimonia
  A. eupatoria, 277
  A. procera, 277
Agropyron
  A. cristatum, 302
  A. desertorum, and restoration, 356,
    366
  A. smithii, 366
Ailanthus altissima, 297, 304
Aizoaceae, 301, 309, 310, 318
Aizoon hispanicum, 318
Albizia, 46
  A. lophantha, 49
Albuca fastigiata, 11
Alchornea latifolia, 350
Aldrovanda, 88
  A. vesiculosa, 50, 89
Aleurites moluccana, 45
Alisma plantago-aquatica, 20, 46
Alismataceae, 18, 20, 36, 38, 46, 47,
  48, 127
allelopathy, 19, 64, 209, 297, 302, 304
Alliaria petiolata, 209, 299, 305
Allium
  A. ascolonium, 35
  A. cepa, 151
  A. porrum, 151
allocation patterns, 24, 233, 297
  trade-offs, 179, 180, 217
  to roots, 66
Alnus, 44, 45, 196
Aloe, 31, 50
  A. arborescens, 151
alpine zone
  and facilitating restoration, 365
  and restoration, 360
Alseis blackiana, 185
Alstonia, 49
  A. angustiloba, 50
Alstromeriaceae, 11, 127

AM, see arbuscular mycorrhizae
Amaranthaceae, 44, 384
Amaranthus
  A. cannabinus, 383, 384
  A. hypochondriacus, 44
Amaryllidaceae, 53
Amborella, 135, 141, 143, 146
  A. trichopoda, 142, 143
Ambrosia trifida, 5, 7, 383, 384
Ammophila
  A. arenaria, 73
  A. breviligulata, 77
Amorphophallus albus, 151
amphibious plants, 37
Amphibolis antarctica, 38, 43, 55
Amphicarpum purshii, 29
amphicarpy, 29
Amsinckia intermedia, 42
Amyema preisii, 54
Anacardiaceae, 6, 32, 49, 51, 151,
  271
Anacharis occidentalis, 48
Anastatica hierochuntica, 66, 318
anchorage, 85
Andropogon virginicus, 152
Androsace rotundifolia, 45
angiosperm, 37, 49, 62, 81, 91, 103–5,
  111, 120, 121, 125, 126, 129,
  130–49, 192, 193, 196, 204, 385,
  386
angiosperm phylogeny, 135
Angraecum maculatum, 44
Annonaceae, 49, 132
anoxia, 144, 146, 162
Anthericaceae, 28
Anthoceros, 108, 112, 123
  A. erectus, 110, 112
Anthocerotaceae, 110
anthropogenic disturbances
  nitrate deposition, 345
  types of disturbance in forests, 333
Antirrhinum majus, 152
  and phytohormones, 169
Anthurium scandens, 21
Apiaceae, 32, 42, 47, 151, 284, 286
apical hook, 157
Apocynaceae, 49, 50, 151
Aponogeton distachyon, 47
Aponogetonaceae, 47
Aquifoliaceae, 244, 301

Arabidopsis, 10, 124, 153, 155, 157, 169
  A. thaliana, 46, 50, 151, 304
  and phytohormones, 168
  mutants and phytohormone function, 168
Arabis hirsuta, 208
Araceae, 11, 20, 21, 28, 32, 47, 52, 55, 88, 127, 384
Arachis hypogaea, 25, 151
Araliaceae, 29, 53, 347
Araucaria, 32
  A. araucana, 47
Araucariaceae, 32, 42, 47, 49
arbuscular mycorrhizae (AM) (*See* mycorrhizae), 193, 195
Arbutoideae, 193
Archaefructus, 136
archegonium, 103
Archontophoenix cunninghamiana, 44
arctic zone, and restoration, 360
Arctostaphylos, 204
  A. glandulosa, 200
Ardisia
  A. crenulata, 48
  A. elliptica, 299
  A. escallonioides, 151
  A. polycephala, 49
Arecaceae (*See* Palmae), 151
arid zone, and restoration, 360
Arion
  A. fasciatus, 242
  A. lusitanicus, 286
Arisaema, 20
  A. triphyllum, 32, 47
Artemisia
  A. annua, 151
  A. californica, 271
  A. halodendron, 319, 328
  A. monosperma, 319
  A. tridentata, 269, 357, 366
Artocarpus, 49
  A. altilis, 52
  A. incise, 52
Ascarina, 146
  A. lucida, 141, 145
Asclepiadaceae, 27, 42, 43, 233
Asclepias
  A. incarnata, 233
  A. tuberosa, 27
Ascomycota, 193, 211
Asparagus madagascarensis, 47
Asphodelaceae, 9, 151

Asphodelus
  A. lusitanicus, 9
  A. tenuifolius, 9
assembly rules, 255, 269
Asteraceae, and genetic structure, 253
Asteraceae, 4, 5, 6, 7, 19, 28, 29, 30, 31, 36, 39, 43, 45, 47, 48, 49, 50, 52, 53, 66, 77, 151, 250, 269, 271, 272, 284, 286, 302, 317, 318, 319, 328, 357, 366, 384
Asteriscus pygmaeus, 66
Astrocaryum standleyanum, 339
Atkinsonia, 92
Atriplex, 69
  A. halimus, 69
  A. hortensis, 36, 43
  A. nummularia, 70
  A. patula, 70
Atropa belladonna, 152
Attractylis serratuloides, 319, 322, 327
Augea capensis, 318
Austrobaileya, 141, 146
  A. scandens, 136, 140
Austrobaileyaceae, 135
autotrophic mistletoes, 95
auxin IAA, 153, 156, 157, 158, 159, 160, 161, 163, 164, 167
  and mycorrhizae, 165
  and nodulation, 165
Avena sativa, 151
Avicennia, 34, 35, 48, 49, 51
  A. marina, 68, 69
  A. maritima, 70
Avicenniaceae, 34, 35, 48, 49, 51, 68
Azoarcus, 211

Baldellia ranunculoides, 38
Balsaminaceae, 6, 30, 44, 49, 303, 384
Banksia, 363, 368
  B. australis, 48
  B. spinulosa, 233
Barclaya longifolia, 146
Barringtonia, 42
  B. racemosa, 51
Basidiomycota, 193
Bennettitales, 141
Berberidaceae, 47
Bertholletia excelsa, 42
Beta vulgaris, 151
bet-hedging, 309
Betula, 205, 297, 362
  B. alleghaniensis, 201, 202

  B. ermanii, 203, 204, 207
  B. lenta, 201, 202
  B. maximowicziana, 72
  B. papyrifera, 383
  B. pendula, 151, 198
  B. pubescens, 362
Betulaceae, 44, 72, 151, 193, 196, 198, 202, 204, 297, 362, 383
Bidens
  B. dioica, 28
  B. laevis, 7, 29, 383, 384
  B. pilosus, 28
Bignoniaceae, 30, 174
biological crusts, 313, 317, 324, 360
  and seedling recruitment, 326
biome, 279, 287
Blepharis, 318
Bomaria edulis, 11
Bombacaceae, 23, 45
Boraginaceae, 32, 42, 151, 318
  and genetic structure, 253
Botrychium, 108, 115, 124
  B. dissectum, 115
  B. virginianum, 111, 115, 116
Bouteloua gracilis, 366
Bowiea volubilis, 42, 47
Brasenia, 146
Brassica
  B. balearica, 36, 44
  B. campestris, 151
  B. napus, 151
  B. nigra, 42
  B. oleracea, 151
  B. rapa, 151, 302
Brassicaceae, 10, 26, 27, 29, 36, 43, 44, 46, 49, 50, 66, 120, 121, 151, 153, 190, 204, 208, 209, 284, 299, 302, 304, 317, 318
brassinosteroids, 155, 156, 157, 167, 377
Bromeliaceae, 11, 48, 79
bromeliads, 79, 85
Bromus
  B. inermis, 151
  B. tectorum, 364
Bruguiera, 4, 51, 151
bryophytes, 103–6, 108, 109, 110, 112, 113, 123–6
Bupleurum falcatum, 151
burial, 73
Burseraceae, 32, 49
Butomaceae, 55
Byttneria aculeata, 26

Cabombaceae, 135
Camptotheca acuminata, 151
Cactaceae, 39, 42, 43, 48, 67, 151, 316
Caesalpinaceae, 23, 200
Caladenia arenicola, 84
Calla, 21
Callichlamys latifolia, 174
Calligonum, 328
Callistemon rigidus, 48
Callitrichaceae, 49
Callitriche, 49
    C. stagnalis, 49
Calocedrus decurrens, 257, 268
Calotropis gigantea, 43
Calycanthaceae, 29, 42, 43
Calystegia macrostegia, 271
Campanula rotundifolia, 151
Campanulaceae, 32, 43, 151
Camptotheca, 151
Cananga odorata, 49
Canavalia ensiformis, 151
Canistrum lindenii, 85, 89
Cannabaceae, 30, 47
Cannabis sativa, 30
Caprifoliaceae, 278, 298
Capsella bursa-pastoris, 120, 121
Capsicum annuum, 152, 155, 315
Caragana microphylla, 328
Carapa guianensis, 52
carbohydrate reserves, 182, 184,
    185
carbon balance, 188
Carboniferous, 106, 114, 120, 126
Cardamine
    C. bulbifera, 36
    C. chenopodifolia, 29
Carex, 18, 365, 383, 384
    C. curvula, 72
    C. flacca, 209
    C. lacustris, 358, 359
    C. stricta, 39, 358, 359
Carica papaya, 51, 151
Caricaceae, 51, 151
Carnegiea gigantea, 67
carnivorous plants, 79, 88
carotenoids, 160
Carpinus, 45
Carpobrotus edulis, 301
Carrichtera annua, 66, 317, 318
Carya tomentosa, 7, 39
Caryophyllaceae, 52, 151, 244, 284,
    286, 317
    and genetic structure, 253
Cassipourea elliptica, 151

Cassytha, 97
Castanea, 204
    C. dentata, 202, 204
Catasetum, 80
Catharanthus roseus, 151
Caytonia, 141
Cecropia, 343
    C. insignis, 343
    C. schrebiana, 347
Cecropiaceae, 343
Celastraceae, 53
Celtidaceae, 342
Centaurea
    C. clementei, 45
    C. cyanus, 286
    C. diffusa, 302, 304
    C. maculosa, 304
Cephaelis ipecacuanha, 151
Cephalotaceae, 46, 88
Cephalotaxaceae, 42
Cephalotus follicularis, 46, 88, 89
Ceratobasidium, 193
Ceratonia siliqua, 50
Ceratozamia, 118
Cereus emoryi, 48
Ceriops, 51, 151
Chaerophyllum procumbens, 32
Chamaecrista fasciculata, 233
Chamaecyparis
    C. lawsoniana, 26
    C. thyoides, 39
Chamaecytisus proliferus, 151
chaperone proteins, 153, 155,
    167
charophycean algae, 123, 124
chasmogamy, 29
Chenopodiaceae, 4, 36, 43, 49, 55, 67,
    69, 151, 208, 317, 328, 366
Chenopodium
    C. album, 208
    C. ambrosioides, 49
    C. rubrum, 4
Chimonanthus fragrans, 43
Chloraea membranaceae, 27
Chloranthaceae, 135, 137, 142, 143,
    145
Chloranthus, 138, 141, 146
    C. erectus, 143
    C. japonicus, 144
Chlorophytum, 54
    C. arundinaceum, 151
Chrysophyllum, 37
Cicer arietinum, 151
circadian rhythm, in seedlings, 155

Circium
    C. acaule, 39
    C. vulgare, 250
Citrus
    C. aurantifolia, 23, 30, 152
    C. aurantium, 42, 49
    C. reticulata, 152
    C. sinensis, 152
Cladium jamaicense, 358
Clavicipitaceae, 210
Claytonia virginica, 120
cleistogamy, 29
Clidemia hirta, 299
climate change, 377
    and dryland systems, 329
    and forest lianas, 336
    and plant distribution, 265
clonal offspring, 239, 252
cloud forests, and facilitating
    restoration, 365
Clusiaceae, 88, 347
coastal scrub, and invasives
    inhibiting restoration, 362
Cobaea scandens, 47
Cocos nucifera, 53
Coffea arabica, 42
Colchicum autumnale, 27
cold tolerance, 70
Coleochaete, 124
coleoptile, 20, 122
coleorhiza, 20, 122
Coleus, 46
collar, 20
Combretaceae, 42, 49, 51
community dynamics, 12, 131, 310
competition, 244, 298, 310, 311, 342,
    346, 358, 364
    and seed size, 236
    among seedlings, 237
    and structuring communities, 259
competition/colonization models,
    236, 237, 247, 298
Connaraceae, 51
Connarus grandis, 51
Convallaria majalis, 246
Convallariaceae, 20, 246
Convolvulaceae, 4, 42, 43, 45, 46, 47,
    53, 92, 97, 151, 271, 384
Convolvulus sepium, 43, 47
Coprosma robusta, 52
Corallorhiza, 80, 82, 84
    C. odontorhiza, 81, 83
Cordia africana, and genetic structure,
    253

Cordyline australis, 19
Coriandrum sativum, 151
Cornaceae, 52, 151, 278, 298
Cornus, 298
    C. sanguinea, 278
Corokia macrocarpa, 52
Cortaderia jubata, 300
cotyledon, 18, 20
cotyledonary axis
    adjacent, 20
    remote, 20
cotyledons, 8, 24–5, 126
    burial, 25
    food storing, 25
    function, 127
    functional morphologies, 173
    haustorial, 25
    photoperiod sensitive, 4
    photosynthetic, 25
    storage, 132
    thickness, 231
    types, 10, 24, 25
Crassula quadrifida, 43, 44, 49
Crassulaceae, 43, 44, 49, 317
Cretaceous, 130, 131, 136, 138, 139,
    146
Crinum
    C. capense syn *bulbispermum*, 53
    C. longifolium, 54
Croton, 52
cryptochromes, 158
Cryptocoryne, 55
cryptocotylar, 23, 94, 96
Cryptomeria, 26
    C. japonica, 42
cryptovivipary, see also vivipary, 25,
    34, 169
cryptocotyl, 94, 96
Cucumis
    C. humofructus, and directed
        dispersal by aardvarks, 315
    C. melo, 151
    C. sativus, 151
Cucurbita, 36
    C. pepo, 151
    C. texana, 151
Cucurbitaceae, 28, 36, 43, 44, 45, 53,
    151, 315
Cupressaceae, 26, 33, 39, 42, 46, 72,
    111, 198, 257, 269, 319, 328, 341,
    363
Cuscuta, 92, 97, 151
    C. europaea, 45
    C. gronovii, 92, 383, 384

Cyclamen, 42
    C. persicum, 47, 50
Cymodocea ciliata, 55
Cymodoceaceae, 43, 55
Cyperaceae, 10, 18, 44, 48, 72, 209,
    358, 384
Cyperus, 18
Cypripedium reginae, 84
Cyrophonectria parasitica, 204
Cytisus scoparius, 212, 299
cytokinins, 153, 157, 158, 161, 164
    and nodulation, 165

Dactylis
    D. glomerata, 28
    D. polygama, 28
Dactylorhiza lapponica, 83
Dalechampia capensis, 48
damp, dark and disturbed hypothesis
    (3-D), 134, 144, 145
Darlingtonia, 50, 88
Dasypogonaceae, 27
Dasyprocta punctata, 339
Daucus carota, 151, 286
defenses, 6
DELLA, 153, 162
Delphinium
    D. elatum, 42
    D. staphysagria, 42
Dendrobium moschatum, 151
Deroceras reticulatum, 286
desert patch formation, and a model
    of dynamics, 320
deserts, and facilitating restoration,
    365
development, niche shifts during,
    358
Devonian, 106, 107, 114, 120, 126
Dianthus caryophyllus, 151
Diapensiaceae, 43
Dichopogon strictus, 28
dicot (See dicotyledon), 8–11, 23, 26,
    30, 42–4, 111, 120–2, 127, 128,
    129, 135, 137, 138, 139, 145, 149,
    157, 283, 379, 384–6
dicotyledon (See dicot), 23, 283
Dicraeia stylosa, 47
Dicranopteris
    D. linaeris, and seedling inhibition,
        347
    D. pectinata, and seedling
        inhibition, 347
Dieffenbachia longispatha, 52
Digitalis purpurea, 72

Dimerandra emarginata, 84
Dinochola, 53
dioecy, 30
Dionaea, 88
    D. muscipula, 46, 49, 50, 88, 89, 90,
        91
Dioscorea batatas, 47
Dioscoreaceae, 47
Diospyros embryopteris, 45
diplobiontic life cycle, 103, 106
Dipterix panamensis, 348
Dipterocarpaceae, 37, 46, 51, 151,
    193, 341
dispersal, 8, 11, 36, 37, 86, 90, 94, 97,
    140, 256, 285, 352, 363
    and forest fragmentation, 348
    and perches in restoration, 347
    and seed size, 229
    by birds, 281
    directed, 315
    fleshy-fruits, 285
    low-cost in deserts, 309
    passive transport, 318
    rare long-distance, 90
    secondary, 315
    and spatial patterns, 374
    topochory, 318
    variation within communities, 236
dispersal limitation, 245, 247, 374
Distichlis spicata, and sex-biased
    survival, 252
disturbance, 332, 368
    effects on soils, 337
    effects on soils and growth rates,
        345
    regimes, 378
disturbance-adaptation, 132, 147
disturbed xeric shrub hypothesis,
    133, 134
DMBQ, 96
dodder, 97
Dolichos lablab, 151
dormancy, 5, 8, 141, 155, 160, 162,
    168, 170, 256, 310, 332, 357,
    375
    and biomes, 287
    and persistence in seed banks, 230
    and restoration, 368
    in seedlings, 31, 381
    morphological, 141
    seedling, 85
Downingia pulchella, 32, 43
Dracaena, 20, 52
Dracaenaceae, 20, 127

Drosera, 88, 89, 90, 91
  D. anglica, 89
  D. binata, 44, 46, 49
  D. capensis, 49
  D. capillaris, 90
  D. intermedia, 89, 90
  D. pygmaea, 36
  D. rotundifolia, 36, 89, 91
Droseraceae, 36, 44, 46, 49, 50, 88
Drosophyllum lusitanicum, 49
drought, 65, 156, 162, 243, 252, 263
Dryobalanops, 46
  D. camphora, 51
Dulichium arundinaceum, 383, 384
Durio zibethinus, 23, 45
dust seeds, 81
Dyckia floribunda, 48
Dysoxylum, 37

Ebenaceae, 45
Echinocactus viridescens, 48
Echinocystis lobata, 45
ecosystem engineers, 311, 314, 321
ecosystem functions, 307
ectomycorrhizae (EM), also see
    mycorrhizae, 165, 192, 193, 363
  size of genets, 194
Eichhornia crassipes, 28, 48
elasticity, 249, 250, 299, 300, 308
Elatinaceae, 48
Elatine, 48
Elkinsia polymorpha, 120
Elymus smithii, 366
Elyna myosuroides, 20
Embelia
  E. ribes, 48
  E. viridiflora, 31
Embothrium coccineum, 33
embryo
  development, 112, 120, 122
  endoscopic, 108, 111, 114, 115, 123, 124
  exoscopic, 108, 111, 115, 123
  foot, 108, 109, 124
  haustoria, 109, 111, 118, 125
  polarity, 104, 123
  suspensors, 103, 125
  transfer cells, 124
embryo:seed ratio (E:S), 141
emergence, definition, 276
Emex spinosa, 29
Encelia californica, 271
Enchylaena tomentosa, 67
Encyclia tampensis, 83

Endertia spectabilis, 46
Endogone, 193
endophytic fungi and bacteria, 209
Enhalus acaroides, 55
environmental stress, 38, 57, 59, 86
  and phytohormones, 156
Ephedra, 115, 118
  E. trifurca, 117, 119
epicotyl, dormancy, 32
epicotyl, 9, 85, 122, 143
epigeal germination, 8, 18, 23, 143
Epidendrum, 80
Epifagus virginiana, 97
epiphytes, 79, 87
Equisetum, 108, 115
  E. arvense, 111, 115, 116
Eremolepidaceae, 92, 93
Ericaceae, 88, 98, 193, 204, 240, 291
  and mycorrhizae, 192
Eriocaulaceae, 20
Eriogonum fasciculatum, 271
Eriophyllum confertiflorum, 272
Erodiophyllum elderi, 66
Erodium, 318, 319
Erythronium americanum, 198
Eschweilera tenuifolia, 33
establishment, definition, 276
establishment growth, 22
ethylene, 153, 154, 157, 162, 163, 164
  and nodulation, 165
  and triple response, 157
etiolation, 10, 155, 157, 158
Eucalyptus, 205, 226
  E. albens, 301
  E. globulus, 49, 151, 253
  E. marginata, 359
  E. obliqua, 64, 253
  E. pauciflora, 72
  E. saligna, 151
  E. tereticornis, 49
Eupatorium perfoliatum, 28
Euphorbia
  E. esula, 45
  E. splendens, 49
Euphorbiaceae, 45, 48, 49, 52, 128, 151, 297, 350
Euphrasia disperma, 53
euphyllophytes, 106
Euryops multifidus, 318
Eusideroxylon zwageri, 26
Evolution of increased competitive
    ability EICA, 302

Fabaceae, see Leguminosae
facilitation, 67, 244, 280, 298, 310
  and mycorrhizae, 199
  and restoration, 365, 366
facultative seeding, and fire, 270
Fagaceae, 25, 43, 45, 46, 52, 72, 151,
    187, 193, 204, 233, 243, 244, 257,
    281, 298, 357, 367, 383
Fagopyrum esculentum, 53
Fagus
  F. grandifolia, 43, 201, 202, 383
  F. sylvatica, 43, 151
ferns, 115
Festuca ovina, 208
Ficus, 226
Filago, 317
fires, 268, 338, 368
flooding tolerance, and restoration,
    162, 167, 358
flowering, precocious, 4
Foeniculum vulgare, 42
forests, and restoration, 357, 363, 364
Forsythia suspensa, 36, 43, 44
fragmentation, 340, 352
  in forests, 335
Frankeniaceae, 49
Fraxinus americana, 305
Freycinetia banksii, 19
frost heaving, 74
frost resistance, 33, 161
frugivory, and anthropogenic
    disturbance, 340
fruit, dimorphic, 29
Fucus, 124
Fumariaceae, 44
Funaria, 112, 123
  F. hygrometrica, 110, 112
functional traits, 132, 178–80
functional types, 20, 23, 24, 274
  and life forms, 290
fungal specificity, 99

Gaiadendron, 92, 94
Gaillardia pulchella, 302
Galearis, 80
Galium
  G. angustifolium, 271
  G. aparine, 303
  G. nuttallii, 272
  G. saccharatum, 44
Galphimia glauca, 151
game-theory of coexistence, 236
gametophyte, 103, 109, 111
gamocotyl, 94

gaps, 61, 180, 184, 186, 245, 256, 257, 260, 269, 281, 282
  dependency, 276
  detection, 8
  and seed limitation, 247
  and seed size, 248
Garcinia, 42
Gaura coccinia, 46
Gazella, soil disturbance and facilitation, 319
genetic structure
  differentiation among communities, 281
  in seedling populations, 251
genetically modified organisms GM, 302
Gentiana lutea, 39
Gentianaceae, 39, 98, 233
Geraniaceae, 29, 39, 151, 277, 318
Geranium
  G. columbinum, 277
  G. dissectum, 277
  G. maculatum, 39
  G. sessifolium, 29
germination, 3, 230
  cryptocotylar, 231
  definition, 276
  epigeal, 127, 128, 252, 290
  and fungi, 81
  hypogeal, 127, 128, 231, 252, 290
  timing, 56, 317
Gesnera macrantha, 47
Gesneriaceae, 30, 42, 47
gibberellins GA, 153, 157, 158
Ginkgo, 111, 118
  G. biloba, 27, 31, 50, 103, 115, 117, 126
Ginkgoaceae, 27, 50, 117
gland, mistletoe haustoria, 93
Gleicheniaceae, 347
Gliricidia sepium, 347
Glomeromycota, 190, 192, 193
Glomus
  changes during restoration, 366
  G. intraradices, 196
Glycine max, 151
Gnetales, 117, 118
Gnetum, 118
Gonatopus, 21
Goodyera, 80
  G. pubescens, 81, 83, 84
Gossypium hirsutum, 151
grasslands, and restoration, 356, 360, 364

grazing, and landscape degradation, 323–5
grazing optimization hypothesis, 323
Grevillea, 152
Grossulariaceae, 271
growth curves, 173, 174, 176
Guara neomexicana, seedling performance and differences in seed source, 357
Gunnera, 211
Gunneraceae, 211
Gustavia superba, 29
Gymnarrhena micrantha, 29
Gymnogramme, 108, 116
gymnosperm, 8, 9, 11, 24, 26, 36, 62, 105, 111, 115, 117, 120, 125–7, 129, 131, 132, 136, 139–41, 147, 149, 193, 196, 204, 385
Gymnosporangium juniperi-virginianae, 33

habitat, 279, 282
  specialization, 342
Halophila stipulacea, 39
Haloragidaceae, 5
Haloxylon
  H. aphylla, 319
  H. persicum, 319, 328
Harpagophytum
  H. procumbens, 151
  H. zeyheri, 151
haustoria, 10, 11, 21, 92, 93, 96, 97, 99, 164, 165
Hazardia squarrosa, 271
heat shock proteins, 161
Hedeoma pulegioides, 49
Hedera helix, 29, 53
Hedyosmum, 146
  H. cuatrezacanum, 139
  H. goudotianum, 142, 143, 145
  H. maximum, 142, 143
  H. peruvianum, 139
  H. translucidum, 142, 143
Hedysarum laeve, 319, 328
Helianthus annuus, 50, 127, 151
Helminthostachys, 111, 116
hemiparasitism, 92, 95, 97, 291
Hepatica nobilis, 240
herbivory, 3, 6, 8, 19, 26, 33, 37, 59, 60, 65, 74–6, 138, 154, 161, 162, 167, 169, 172, 180–2, 184, 188, 205, 211, 237, 241–5, 252, 253, 262, 267, 268, 276, 280, 282, 286–8, 291, 301, 309, 311, 329,

340, 345, 348, 365, 367, 368, 373, 376, 379, 387, 389
  and chemical defense, 242, 291
  and restoration, 367
Heterosperma pinnatum, 30
heterosporous, 104
heterotrophic, 81, 95
Hibiscus
  H. diversifolius, 53
  H. moscheutos, 42
  H. trionum, 42
Hippocastanaceae, 151
Hodgsonia macrocarpa, 43, 44
Holcus lanatus, 301
holoparasitic, 92, 95, 97
Honckenya peploides, 244
Hopea
  H. nervosa, 341
  H. odorata, 151
Hordeum
  H. spontaneum, and genetic structure, 253
  H. vulgare, 151
hornworts, 106
host specificity, 97, 98
host suitability, 94
Hottonia palustris, 38
Humulus japonicus, 47
hunting, 339, 340
Hyacinthaceae, 11, 42, 47
Hyacinthus orientalis, 151
Hydatella filamentosa, 136
Hydatellaceae, 135, 136, 141, 143, 144, 146
hydathode, 25, 143
Hydrocharitaceae, 4, 39, 44, 46, 48, 55, 121, 151
hydrophilic polymers, and restoration, 362
Hydrophyllaceae, 32, 48, 233
Hydrophyllum, 32, 233
Hymenocallis occidentalis, 54
Hymenoscyphus ericae, 193
Hypecoum procumbens, 36, 44
hyperphyll, 20
hyphal networks, 189, 191, 197, 207
  and carbon dynamics, 198
  and seedling establishment, 199
hypocotyl, 9, 18, 20, 43–5, 47, 85, 92, 122, 142, 157, 158, 160
  hypogeal germination, 9, 121, 122, 127, 128, 144
  and light, 186
Hypoxidaceae, 11

Hypoxis hygrometrica, 11
Hystrix indica, soil disturbance and
    facilitation, 319

IAA, see auxin
Ibex, soil disturbance and
    facilitation, 319
Idiospermum australiense, 29, 42
Ifloga, 317
    I. spicata, 317
Ilex aquifolium, 244, 301
Illiciaceae, 135
Illicium, 137, 142
    I. floridanum, 144, 145, 146
    I. griffithi, 148
    I. parviflorum, 142
    I. verum, 148
Impatiens
    I. capensis, 6, 7, 30, 44, 383, 384
    I. fruticosa, 49
    I. pallida, 303
    I. parviflora, 44
imprinting soils, and restoration,
    360, 361
Inga, 34, 51
    I. fagifolia, 350
intermediate disturbance hypothesis,
    323
invasion, 244
invasive species, 212, 244, 277, 278,
    280, 281, 295, 296–306, 362, 369,
    370
Ipomoea, 7
    I. batatus, 151
    I. dissecta, 47
    I. glaberrima, 53
    I. leptophylla, 45
    I. peltata, 53
    I. quamoclit, 42
Iriartea, 43
Iridaceae, 54, 151
Iris
    I. ensata, 151
    I. hexagona, 151, 162
Isoetales, 106, 113
Isoetes, 111, 114
    I. lithophila, 111, 113
iteroparous, 249
Iva imbricata, 77

JABOWA, 256
Jacquinia ruscifolia, 31
Janzen-Connell hypothesis, 204, 242,
    264, 267, 280

jasmonic acid JA, 154, 162, 169
Juglandaceae, 7, 39, 120
Juglans nigra, 120
Juncaceae, 38, 384
Juncus, 38
    Juncus effusus, 383, 384
Juniperus, 26, 33, 46, 269
    J. sabina, 72
    J. virginiana, 126
juvenile, definition, 218

Kadsura
    K. coccinea, 140, 144
    K. scandens, 140
Kandelia candel, 51
Kennedia rubicunda, 46
Khaya, 348
Kingia australis, 27
KNOX homeobox genes, 158, 167
Kosteletzkya virginica, 70
Krameria, 97
Krameriaceae, 97

Labiatae, 46, 49
Laburnum vulgare, 42
Laccaria, 206
Lactuca sativa, 151, 253
Laguncularia, 49
    L. racemosa, 51
Lamiaceae, 151, 271, 272, 284,
    365
landscape, 279, 285
landscape patches, 307, 311, 312
    transformation, 323
Laportea canadensis, 50
LAR, see leaf area ratio
Larix, 297
    L. kaempferi, 203, 204, 207
    L. x eurolepis, 152
Lathraea clandestina, 92
Lathyrus, 23
    L. articulatus, 47
    L. nissolia, 47
    L. vernus, 241, 249, 250
Lauraceae, 26, 39, 97, 151
Lavandula stoechas, 151
leaf-area ratio (LAR), 63, 177, 178, 181,
    291
leaf nitrogen mass ratio (LNMR), 275
Lecythidaceae, 29, 42, 51
Lecythis zabucajo, 42
Leguminosae, 23, 25, 26, 27, 32, 33,
    34, 35, 36, 42, 43, 44, 46, 47, 49,
    50, 52, 67, 69, 74, 151, 187, 212,

233, 241, 249, 271, 284, 291, 298,
    299, 301, 319, 328, 348, 350
Lemna, 20, 21, 55
    L. minor, 30, 44, 151
Lemnaceae, 20, 21, 30, 44, 55, 151
Lennoa, 97
Lennoaceae, 97
Lens culinaris, 151
Lentibulariaceae, 42, 50, 54, 88
Lepidium, 27, 318
Lepidocarpon, 114
Lepidoceras, 92
Lessingia filaginifolia, 272
Leucanthemum vulgare, 286
Leucospermum glabrum, 152
Leymus mollis, 244
life cycle, 5, 11
life form, and light environments,
    285
life history traits, correlations
    among, 178, 227, 228
life span, 239
life table response experiment LTRE,
    249, 250
light responsiveness, 159, 186
lignotubers, 27, 31
Liliaceae, 27, 31, 32, 35, 47, 50, 52,
    54, 152, 198
Lilium superbum, 32
Limnanthaceae, 44
Limnanthes douglasii, 44
Limnocharis, 55
Limonia acidissima, 42
Limonium peregrinum, 151
Limosella australis, 5
Linaceae, 151
Linaria, 45
    L. cymbalaria, 43
Lindernia dubia, 4
Linum usitatissimum, 151
Liparis
    L. liliifolia, 84
    L. loeselii, 84
Lithospermum erythrorhizon, 151
litter, 64, 282, 365
    and seedling mortality, 243
Lodoicea maldivica, 22, 42, 240
logging, 334, 337
Lolium arundinaceum, 211
Lonas inodora, 43
Lonicera, 303
    L. japonica, 303
    L. maackii, 303
    L. morrowii, 303

Lonicera (cont.)
  L. tartarica, 298, 303
  L. x bella, 303
Loranthaceae, 54, 92, 94, 95
Lotus
  L. corniculatus, 212
  L. japonicus, 151
  L. scoparius, 271
Lucuma, 25, 43
Lupinus
  L. albus, 151
  L. sulphureus, 42
Luzuriagaceae, 20
Lychnis flos-cuculi, and genetic
    structure, 253
Lycium afrum, 43
Lycopersicon esculentum, 28
lycophytes, 113
Lycopodiales, 106, 113
Lycopodium, 111, 114
  L. cernuum, 125
  L. clavatum, 113, 114
Lythraceae, 38, 298
Lythrum salicaria, 38, 298

Macfadyena unguis-cati, 30
Macrolobium acaciifolium, 33
Macrozamia riedeli, 44
Magnolia, 132
Magnoliaceae, 132
Malacothamnus fasciculatus, 271
Malosma laurina, 271
Malphigiaceae, 48, 151
Malus x domestica, 151
Malvaceae, 7, 31, 43, 51, 53, 70, 151,
    271, 342
Mammillaria gaumeri, 39
Mangifera indica, 31, 51, 151
mangroves, 4, 19, 25, 33, 34, 68, 169
Manihot, 128
  M. esculenta, 128, 151
Manilkara kauki, 49
Marah, 28
Marchantia, 112
  M. domingensis, 112
  M. polymorpha, 110
Marchantiaceae, 110
masting, 258, 263, 266
maternal investment, 11, 56
Matricaria globifera, 36, 43
matrix models, 299
Medicago, 34, 74
  M. lupulina, 52, 212
  M. sativa, 151
  M. truncatula, 151

megasporangium, 103
Melastoma beccarianum, 298
Melastomataceae, 298, 299, 343,
    347
Meliaceae, 37, 49, 52, 348
Melicia excelsa, 348
Melilotus officinalis, 44
Melinis, and seedling inhibition, 347
Melocalamus compactiflorus, 53
Melocanna
  M. baccifera, 53
  M. bambusoides, 53
Menispermaceae, 42
Menispermum canadense, 42
Menyanthaceae, 37
meristems, loss, 31
Mertensia virginiana, 32
Mesembryanthemum nodiflorum, 318
Mesozoic, 113, 131, 147
  boundary with Cenozoic, 138
Messor, and seed predation, 318
Metrosideros polymorpha, 300
Miconia, 343
  M. argentea, 343
microbial communities, and
    disturbance, 341
microhabitat, 279
  and seedling dynamics, 280
Microloma, 42
microsites (*See* safe sites), 245, 247,
    248, 278, 280, 334, 341, 344, 347,
    352, 361, 367, 369, 370, 378
  limitation, 245
Mimosa
  M. pigra, 301
  M. pudica, 50, 151
Mimosaceae, 319
  and genetic structure, 253
Mimulus
  M. aurantiacus, 271
  M. lutea, 28
Mimusops balata, 45
Mirabilis
  M. californica, 271
  M. dichotoma, 44, 47
  M. linearis, 47
Misodendraceae, 92
mistletoes, 92
Mitella breweri, 47
Momordica charantia, 151
monilophytes, 107, 114
monocot (*See* monocotyledon), 8–11,
    19, 20–4, 26, 42, 49, 80, 84, 107,
    111, 120–2, 127–9, 135, 137, 138,
    142, 144, 149, 379, 383–6

monocotyledon (*See* monocot), 20, 22,
    23, 80, 111, 120
  functional aspects of seedlings, 20,
    22, 80, 112, 120
  woody, 22
monopodial growth habit, 132, 143
Monotropa, and mycorrhizae, 192
  M. uniflora, 98
Monotropoideae, 46, 92, 98, 192, 193
Montezuma speciosissima, 51
Mora
  M. megistoperma, 27, 32, 33, 36
  M. oleifera, 51
Moraceae, 44, 49, 52, 348
mortality, causes, 19, 241, 280, 282,
    283, 288, 289
  of invasive seedlings by natives,
    300
Morus latifolia, 52
Moscharia rosea, 48
Musa acuminata, 151
Musaceae, 151
myco-heterotrophic plants, 84, 92,
    98, 100, 193, 198, 240
mycorrhizae (*See* arbuscular
    mycorrhizae, ectomycorrhizae),
    39, 80, 84, 87, 91, 98, 99, 164,
    165, 170, 188, 191–3, 194, 244,
    264, 305, 316, 333, 336, 341, 355,
    363, 366, 367
  arbuscular AM, 191, 193, 194,
    366
  arbutoid, 192, 193
  and carnivorous plants, 91
  chlorophyllous orchid, 193
  dependency, 291
  and disturbance, 341
  ectomycorrhizae (EM), 191, 192,
    195
  and epiphytes, 87
  ericoid, 192, 193
  and facilitation, 264
  fungal specificity, 83
  interactions with protocorms, 83
  monotropoid, 192, 193
  of achlorophyllous orchid, 193
  and recruitment, 244
  types, 191, 193
mycorrhizal fungal communities,
    197, 366
  of AM fungi, 208
  differences between seedlings and
    adults, 197
mycorrhizal fungi, 19, 81, 83, 98,
    189, 190, 367

and carbon costs to hosts, 195
and colonization of land, 190
and continuum of
    mutualism/parasitism, 190,
    210
and drought, 189
and host specificity, 189, 196
and mineral uptake, 189, 190
and nurse logs, 363
and patchy distribution, 378
and pathogens, 189
and restoration, 363, 366
and seedling growth, 197, 208
shifts in composition, 83
mycorrhizal mutants, and
    phytohormones, 165
Myoporaceae, 67
Myoporum platicarpum, 67
Myrica, 212
    M. californica, 44
Myricaceae, 44
Myriophyllum variifolium, 5
Myristica hollrungii, 52
Myristicaceae, 52, 132
Myrmecoidia, 53
Myrsinaceae, 23, 31, 48, 49, 51, 69,
    151, 299
Myrtaceae, 48, 49, 64, 151, 193, 205,
    300, 301, 359
    and genetic structure, 253
    and tropical pastures, 347
NAC genes, 159
Najadaceae, 48
Najas marina, 48
NAR, 176, 178
Neolitsea sericea, 39
Neomarica gracilis, 54
Neophytum, 211
Nepenthaceae, 46, 49, 88
Nepenthes, 49, 88, 90
    N. khasiana, 46, 49
Nerine, 54
net assimilation rate NAR, 176,
    291
net primary production, and
    mycorrhizae, 195
neutral theory, 386
niche-construction, 263
Nicotiana
    N. attenuata, 152
    N. tabacum, 152
Nierembergia caerulea, 152
nitrate, interaction with hormones,
    163
nitrate reductase, induction, 86

nitrogen content, and growth rate,
    275
nitrogen-fixing bacteria, 209, 211
Noaea mucronata, 322, 323, 327
nodulation, and phytohormones, 164
nonmycorrhizal species, and
    disruption of mycorrhizal
    infection, 209
Nothofagus, 243
Nuphar, 141, 146
nurse plants, 39, 244, 280, 315, 319,
    322, 362, 378
nutrient acquisition, 90, 99
Nuytsia, 92
Nyctaginaceae, 44, 47, 52, 271
Nymphaea, 141, 146
    N. lotus, 47
Nymphaeaceae, 46, 47, 135
Nymphaeales, 135, 141, 144, 146,
    149
Nymphioides, 37
Nypa fruticans, 25, 53, 151

obligate seeding, and fire, 270
Ochroma pyramidale, 342
Odocoileus virginianus, 6
Oenothera, 305
Oidiodendron, 193
Oldenlandia affinis, 151
Olea europaea, 151
Oleaceae, 36, 43, 44, 151, 305
Onagraceae, 46, 305, 357
Ondatra zibethicus, 268
Ononis repens, 212
Ophioglossum, 115
    O. vulgatum, 116
optimality theory, 153
Opuntia labouretiana, 43
Orchidaceae, 27, 42, 44, 79, 80, 98,
    151, 240
    and mycorrhizae, 193
    and seed size, 240
Orchis simia, 84
Orobanchaceae, 27, 46, 92, 95, 151
Orobanche
    O. cumana, 151
    O. ramosa, 27, 151
Orontium aquaticum, 11
Orycteropus afer, and seed dispersal,
    315
Oryza sativa, 151, 211
    and genetic structure, 253
    and phytohormones, 168
Ostrya virginiana, 202
Ottelia alismoides, 121

Oxalidaceae, 50
Oxalis hirta, 50
Oxybaphus ovatus, 47

Pachystegia insignis var. *minor*, 52
Paeonia, 120
Palaquium amboinense, 49
Paleocene, 139
Paleozoic, 104, 107, 130, 131
Palmae (*See* Arecaceae), 22, 37, 39, 42,
    43, 44, 53, 75, 240
Pandanaceae, 19
Papaveraceae, 36
Papilionaceae, 23, 48
parasitic fungi, 190
parasitic plants, 27, 79, 91, 383
    and phytohormones, 165
Passiflora macrocarpa, 47
Passifloraceae, 47
pathogenic microbes, 209
pathogens, 65, 75, 242
Pectocarya recurvata, 318
Pedaliaceae, 45, 151
Pedicularis furbishiae, 250
Pelargonium peltatum, 151
Pellia, 109
    P. epiphylla, 110–12
Pelliciera, 34
    P. rhizophorae, 51
Pellicieraceae, 34, 51
pelotons, 83, 193
Peltandra virginica, 21, 28, 384
Pennisetum, and seedling inhibition,
    347
    P. glaucum, 151
Pentaclethra macroloba, 33
Pentoxylon, 141
Peperomia, 25
performance trade-offs, 342
perisperm, 141, 149
Permian, 106
Persea americana, 151
Petunia hybrida, 152
Peucedanum sativum, 42
Phacelia tanacetifolia, 48
Phalaris arundinacea, 28, 365,
    369
    and genetic structure, 253
Phanerozoic, 104
Pharbitis nil, 4, 151
Phaseolus, 127
    P. lunatus, 33, 151
    P. vulgaris, 35, 46, 151
phenotypic plasticity, 185
Philesiaceae, 20

Philodendron, 21
Phlox
  P. paniculata, 151
  P. setacea, 151
Phoenix
  P. dactylifera, 44
  P. reclinata, 151
Pholisma, 97
Phorodendron densum, 92
photoinhibition, 67, 71, 344
phototrophic, 157, 158
phototropins, 158
Phragmites australis, 151
Phyllocactus stenopetalus, 48
phyllomorph, 30
phylogeny, 7
  of embryophytes, 106
  and light environments, 285
physical damage, 73, 75, 179
Physoplexus comosa, 151
phytochrome, 156–8
  five forms, 157
phytohormones, 150, 153, 156, 161,
    164, 376
  classes, 153
  and herbivory, 167
  and mycorrhizae, 165
  and niche partitioning, 167
  and phenotypic range, 166
  and phytochrome, 156
  receptors, 153
  and restoration, 363
  and root growth, 163
  second messengers, 153
  and signal transduction, 154
  and symbioses, 164
  synergy and antagonism, 166
photosynthesis, 26, 34, 60, 127, 144,
    145, 158, 160–2, 173, 184, 186,
    240, 290, 344, 377
Picea, 119
  P. abies, 39
  P. engelmannii, 383
  P. glauca, 259
  P. omorika, and genetic structure,
    253
  P. sitchensis, 26, 202
Pilea pumila, 27, 303, 383, 384
Pinaceae, 26, 28, 36, 39, 42, 49,
    50, 117, 151, 192, 193, 196,
    197, 199, 202, 204, 207, 252,
    256, 257, 292, 297, 341, 350,
    357, 383
  and genetic structure, 253

Pinguicula, 88, 91
  P. alpina, 91
  P. villosa, 91
  P. vulgaris, 91
Pinus, 111, 115, 117, 119, 120, 195,
    205, 265, 269, 305, 365
  P. banksiana, 383
  and genetic structure, 252, 253
  P. caribaea, and forest restoration,
    350
  P. contorta, 126, 202, 206, 357
  P. densiflora, 197
  P. echinata, 28
  P. halepensis, 49
  P. koraiensis, 126
  P. lambertiana, 268
  P. merkusii, 49
  P. muricata, 205
  P. palustris, 269
  P. pinaster, and genetic structure,
    253
  P. pinea, 42, 151
  P. ponderosa, 257, 268
  P. radiata, 151
  P. rigida, 3, 27, 28
  P. serotina, 50
  P. strobus, 42, 201, 202, 256
  P. taeda, 50, 151, 169
Piper, 149, 343
Piperaceae, 25, 88, 343
Piperales, 120
Pisonia
  P. brunoniana, 52
  P. longirostris, 52
Pistia, 20
Pisum, 23, 127
  P. sativum, 43, 127, 151
  and phytohormones, 168
Pitcairnia, 87
  P. corallina, 11
  P. flammea, 85, 89
Pithecellobium
  P. pedicellare, and genetic structure,
    253
  P. racemosum, 51
Pitheoctenium crucigerum, 174
Pittosporaceae, 42, 49
Pittosporum
  P. crassifolium, 42
  P. erioloma, 42
  P. ferrugineum, 49
planerocotylar, 23
Plantaginaceae, 43, 194, 280, 317, 318
Plantago, 280

  P. callosa, 43
  P. coronopus, 317, 318
  P. lanceolata, 194
  P. patagonica, 318
plasticity, 28, 153, 158
Platysace cirrosa, 47
Plectilospermum, 120
Pleurophyllum hookeri, 19
Plumbaginaceae, 49, 51, 151
Plumeria alba, 49
Poa bulbosa, 319, 320
Poaceae, 20, 26, 28, 29, 44, 45, 50, 53,
    55, 73, 122, 151, 208, 211, 244,
    252, 254, 300, 301, 302, 304, 317,
    318, 319, 347, 356, 364, 365, 366,
    369
  and genetic structure, 253
Podocarpaceae, 42, 111
Podophyllum emodi, 47
Podostemaceae, 30, 44, 47, 48
Podostemum subulatus, 48
Polemoniaceae, 47
pollination, and seed size, 233
pollutants, and restoration, 362
polyamines, as phytohormones, 156
polyembryony, 30, 118, 119, 120
Polygonaceae, 5, 29, 30, 43, 47, 53,
    151, 271, 300, 328, 384
Polygonum, 7
  P. arifolium, 7, 383, 384
  P. bistorta, 25, 43, 47
  P. bistortoides, 25, 43
  P. hydropiper, 30
  P. perfoliatum, 300
  P. punctatum, 5, 383, 384
  P. sphaerostachyum, 43
  P. viviparum, 35
polymorphism, 29
polypeptide hormones, 154, 155
Polypleurum stylosum, 48
Polyradicion, 80
Pontederiaceae, 28, 48
Populus, 18, 285
  and phytohormones, 168
  P. deltoides, 152, 298
  P. nigra, 152
  P. simonii, 328
  P. tremuloides, 254, 383
Porella, 109
  P. bolanderi, 110, 112
Portulacaceae, 112, 120
Posidonia australis, 18
Posidoniaceae, 18
Potentilla reptans, 47

Pothos, 20
Primula
  P. denticulata, 42
  P. glaucescens, 151
  P. sinensis, 48
  P. veris, 250
Primulaceae, 38, 42, 45, 47, 48, 50, 151, 250
Proboscidea louisianica, 45
Proechimys semispinosus, and seed predation, 340
Prosopis, 320
  P. alpataco, 70
  P. argentina, 70
  P. flexuosa, 69
  P. glandulosa, 319
  P. juliflora, 328
Proteaceae, 33, 48, 151, 190, 233, 363
  and herbivory, 242
proteoid roots, 163, 170
Protium javanicum, 49
protocorm, 80, 81, 83, 85, 100, 114
Prunus, 23
  P. americana, 9, 23
  P. laurocerasus, 301
  P. persica, 151
  P. serotina, 243, 281
  P. virginiana, 9, 23
pseudorhizomatous growth, 142, 145
Pseudotsuga, 198, 205
  P. menziesii, 39, 151, 195, 196, 198–202, 204, 205
pseudovivipary, see also vivipary, 33, 35
Psilotum, 107, 108
Psittacanthus, 92, 94
Ptelea trifoliata, 36, 43
pteridophytes, 111
Pterospora, and mycorrhizae, 192
Pyracantha angustifolia, 244
Pyrus communis, 151
Pyxidanthera barbulata, 43

Quercus, 25, 41, 45, 46, 187, 205, 206, 244
  Q. alba, 6, 202, 261, 298
  Q. berberidifolia, and transient seed banks, 262
  Q. ellipsoides, 201
  Q. ilex, 25, 33, 233
  Q. kelloggii, 257
  Q. lobata, 357
  Q. macrocarpa, 201
  Q. montana, 201

Q. oleoides, 27
Q. petraea, 281
Q. prinus, and restoration, 261, 367
Q. robur, 151
Q. rubra, 72, 201, 202, 204, 260, 261, 367
Q. suber, 51
Q. turbinata, 46

radicle, 5, 8, 18, 32, 85, 92, 118
Ranunculaceae, 36, 42–4, 112, 240, 246, 280
Ranunculus, 280
  R. arvensis, 44
  R. fiscaria, 36
  R. hederaceus, 43
Raphanus sativus, 151
Rapistrum rugosum, 302
Rauvolfia serpentina, 151
Ravenea musicalis, 25, 53
recruitment, 239, 273, 342
  among annuals, 240
  and landscape patchiness, 312
  and population dynamics, 248
  among parasitic plants, 95
  failure, 349, 374
  coexistence among multiple strategies, 236
  strategies, 255
recruitment limitation, 241, 245, 251, 260, 264, 334, 373
regeneration, from detached cotyledons, 29
regeneration dynamics, 259–60
regeneration niche, 131, 237, 255, 260, 274, 379
region scale, 279
relative growth rate (RGR), 18, 63, 86, 128, 174, 176, 177, 220, 275, 283, 284, 291, 305
  decline, 176
  and herbivory, 242
  and seed size, 177–8, 231
  and seedling defense, 179, 181
  and tissue density, 231
reproductive lifespan, definition, 218
resource allocation theory, 62
resource use, limitations related to size, 6
resprouting, 231, 270
restoration, 10, 326, 352
RGR, see relative growth rate
Rhamnaceae, 272, 281
Rhamnus

R. cathartica, 281
  R. crocea, 262, 272
Rheum officinale, 43, 47
Rhipsalidopsis, 151
Rhizobiaceae, 211
Rhizobium, 244
rhizoids, 18, 81
Rhizophora, 34, 51, 69, 151
  R. mangle, 35, 162
Rhizophoraceae, 4, 34, 45, 51, 69, 151
Rhizopogon, 195, 197, 202, 205, 206
rhizosphere, 27, 199
Rhodomyrtus tomentosa, 48
Rhopalostylis sapida, 37
Rhus integrifolia, 271
Rhyncholacis macrocarpa, 44
Ribes, 271
Riccia, 108
Ricinus communis, 49, 151
Ripogonum scandens, 19, 53
risk reduction, 309
r-K spectrum, 235, 309
Robinia psuedoacacia, 151
roots
  types, 26
  contractile, 27
  dropper roots, 28
  exudates, 26
  proteoid (cluster), 163, 170
Rosa
  Rosa canina, 278
  R. rugosa, 280–2
Rosaceae, 9, 23, 47, 151, 243, 244, 277, 278, 280–2, 284, 299, 301, 330
Rostraria, 317
Rubia cordifolia, 44
Rubiaceae, 42, 44, 48, 52, 151, 185, 271, 272, 303, 347
Rubus, 299
  R. fruticosus, 151
ruderal paleoherb hypothesis, 133, 134
Rumex
  R. acetosa, 151
  R. palustris, 151
Ruschia spinosa, 309
Rutaceae, 23, 26, 30, 36, 42, 43, 49, 152

Sabina vulgaris, 319, 328
Saccharum, and seedling inhibition, 347
  S. officinarum, 211
  S. spontaneum, 347

safe sites, see also microsites, 6, 7, 38, 40, 245, 255, 280, 360, 379
  filters, 7, 8
  and restoration, 359
Sagittaria
  S. latifolia, 36, 47
  S. sagittifolia, 47
Salaciopsis ingifera, 53
Salicaceae, 4, 18, 152, 193, 195, 204, 205, 254, 263, 285, 298, 319, 328, 383
Salicornia
  S. dolichostachya, 55
  S. europaea, 55
salicylic acid SA, 154, 169
salinity, 68
Salix, 4, 195, 285
  S. nigra, 263
  S. psammophila, 319, 328
  S. reinii, 201, 203, 204, 207
Salsola, 317
  S. kali, 366
  S. paletzkiana, 328
salt deserts, 68
salt marshes, 68
Salvia
  S. apiana, 272
  S. lavandulifolia, and facilitating restoration, 365
  S. leucophylla, 272
  S. mellifera, 271
  S. miltiorrhiza, 151
Samanea saman, 50
Sambucus nigra, 278
Sansieveria, 20
Santalaceae, 92
Sapium sebiferum, 297, 302
Sapotaceae, 25, 37, 42, 43, 45, 49
Sarcandra, 141
  S. glabra, 143
Sarcodes, and mycorrhizae, 192
Sarcopoterium spinosum, 330
Sarracenia, 88, 90
  S. alata, 90
  S. purpurea, 88, 90
Sarraceniaceae, 49, 88
Saxifraga cernua, 35
Saxifragaceae, 35, 47
Schefflera morototoni, 347
Scheidea diffusa, 52
Schisandra, 144
  S. chinensis, 143, 144
  S. glabra, 144

Schisandraceae, 135, 142
Schismus, 317
  S. barbatus, 318
Schlumbergera, 151
Sciurus carolinensis, 6
Scrophularia aquatica, 38
Scrophulariaceae, 4, 5, 27, 28, 38, 43, 45, 53, 72, 92, 152, 250, 271, 282, 284
Scutellospora, and facilitating restoration, 366
seagrasses, 4, 35, 39
Sebacina, 193
Sebacinaceae, 84
Secale cereale, 151
Sechium edule, 53
sedges, 10, 36
seed
  dimorphic, 29
  heteromorphism, 30
  maternal reserves, 18, 24, 175, 332
  reserve utilization, 173
seed banks, 3, 142, 360, 364
  in deserts, 315
  dormant, 270
  dynamics, 5
  and forest lianas, 336
  persistent, 5, 230, 256
  and restoration, 359, 369
  and seed size, 5, 229
  and seedling banks, 382
  and soil disturbance, 337
  transient, 5, 262, 375
seed dormancy, 66, 118, 123, 125
  mechanisms, 125
  morphological, 118
  morphophysiological, 118
  physiological, 119
seed habit, selective advantages, 126
seed limitation, see also recruitment limitation and dispersal limitation, 246, 247, 349, 352, 381
  and disturbance, 334
  as a part of recruitment limitation, 245
  on large scales, 247
seed mass, see seed size
seed number/seedling survival tradeoff, see also seed size/seed number trade-off, 219, 233
seed predation, and restoration see also herbivory, 367

seed production, lifetime reproduction and seed number, 223
seed reserves, energy and nitrogen, 173, 176
seed size, 8, 20, 61, 65, 72, 76, 99, 132, 135, 136, 139, 145, 146, 217, 218, 275, 283, 288, 377
  across all species, 217
  and allocation patterns, 172, 177, 179, 181
  in basal angiosperms, 136
  and biomes, 287
  and cotyledon photosynthesis, 230
  evolution, 138–41
  and gap pioneers, 343
  and growth rates, 177, 231
  heritability, 233
  and initial seedling size, 234
  and life history traits, 227
  and lifetime reproduction, 223
  and nitrogen addition, 63
  and plant size, 223
  relationship to seedling mass, 176
  and RGR, 174, 176–8, 180
  and reserve dependency, 173–5, 179–82, 184, 187, 188
  and survival, including later stage, 220, 221, 222, 225, 234, 283
  variation within species, 233, 234
seed size/number trade-off, 220, 224, 247, 248
  within species, 234
seedling
  conservative, 7, 178, 179
  definition, 6, 17–18, 142, 218, 276
  and desert patch formation, 320
  development, 18
  dispersal by water, 37
  end of stage, 18
  epigeal, 8, 9, 18, 19, 23
  first leaves, 26
  fugitive, 7
  functional types, 9, 20, 23, 24
  generalists, 7
  growth patterns, 31
  herbivore defense, 26
  hypogeal, 9, 18, 19, 23
  interrupted growth, 31
  morphology, 23, 24
  opportunistic, 7, 178–81, 184, 186–8
  palatability, 33
  roots, 26–7

specialists, 7
storage, 20, 21, 27, 28, 36, 39, 40
stress tolerator, 7
surface features, 25
vulnerabilities, 12
dependence on mineral elements, 173
dependence on reserves, 173, 174
trade-off among defenses, 182
variation in morphology, 276
variation in size, 276
seedling banks, 36, 297, 375
types, 282, 381
seedling dormancy, 32, 160
and dehydrin proteins, 160
and phytohormones, 160
seedling equivalents, 33, 35, 181
seedling establishment, 5
and specialized seed beds, 363
seedling organs, longevity, 36
seedling size
and herbivory, 241
and light environments, 283
seedling stage, 17, 18, 38, 40
length under selection, 297
seedling survival (*See* mortality causes), 84, 86, 90, 92, 172, 178, 179, 181–5, 188, 209, 211
of orchids, 84
sources of mortality, 19
and growth rates, 178
and microhabitats, 290
and seed dormancy, 230
and seed size, 226
to reproductive maturity, 225, 227
turnover, 266
seedling traits, and adult traits, 232
seed-seedling conflicts, 40
Selaginella, 114
S. *kraussiana*, 114
S. *martensii*, 113, 114
Selaginellales, 106, 113
semelparous, 249
Senecio
S. johnstonii, 31
S. pulcher, 45
S. vulgaris, 48
shade adaptation, small seeds as, 140
shade tolerance, 185
Sherardia arvensis, 44
Shorea, 46
S. *leprosula*, 37
shrublands

and facilitating restoration, 364, 365
restoring Mediterranean scrub, 368
Sida carpinifolia, 31
Sideroxylon tomentosum, 43
Silene alba, 286
Silurian, 106, 107, 114
Simaroubaceae, 297
Simmondsia chinensis, 152
Simmondsiaceae, 152
skototropic movement, 158
SLA, see also specific leaf area, 178, 181, 184
Smilacaceae, 19, 53
Smith and Fretwell model, 218, 219, 233
Socratea, 43
S. *exorrhiza*, 75
soils, infertile, 175
Solanaceae, 28, 43, 152, 155, 272, 315
Solanum, 272
S. *khasianum*, 152
S. *lycopersicum*, 152, 155, 168
S. *tuberosum*, 152, 155
Sorghum bicolor, 26, 50, 151, 253
Spartina
S. *alterniflora*, 254, 300
S. *versicolor*, 55
spatial scale, 278
species richness, 310
specific leaf area SLA, 63, 178, 275, 285, 290, 291, 377
and herbivory, 242
specific leaf mass, 63
specific root length, 63
spermatophytes, 107
Sphagnaceae, 110
Sphagnum, 112, 362
S. *subsecundum*, 110, 112
Spirodela, 151
spores
dispersal, 206, 341
dispersal facilitating restoration, 367
dormant fungal spore bank, 205
of mycorrhizal fungi, 205
Stachys sieboldii, 151
Stellaria longipes, 151
stems, modified, 27
Sterculiaceae, 26, 152
Stipa, 319
S. *capensis*, 318
Stipagrostis, 319, 320
Stirlingia latifolia, 33

strategies, see also r-K continuum, successional species
conservative, 179, 180, 187
opportunistic, 179, 180, 186
stratification, 71, 368
Stratiotes, 46
S. *aloides*, 44
Streblus asper, 44, 49
Streptocarpus, 30, 47
S. *dunnii*, 42
S. *rexii*, 42
stresses, habitat, 6
Striga, 27, 95, 96
S. *asiatica*, 96
Stryphnodendron microstachyum, 348
Stylidiaceae, 36, 44
Stylidium adnatum, 36, 44
Suaeda, 317
S. *maritima*, 69
S. *monoica*, 151
succession, and restoration, 353, 357, 364
successional species, see also r-K continuum, strategies, 61
sugars, as phytohormone signals, 164
Suillus, 197, 202, 206
survival, see seedling survival
Swetia perennis, 233
Swietenia, 348
symbiotic fungus, 81
sympodial growth habit, 135, 142

Tabebuia rosea, 174
Tamaricaceae, 49, 298
Tamarix ramosissima, 298
Taraxacum hamatiforme, 29
Taxaceae, 9, 36, 42, 244
Taxodiaceae, 26, 28, 42
Taxodium distichum, 28
Taxus, 36
T. *baccata*, 9, 244
temperature stress (*See* thermal extremes), 308, 316, 329, 344, 365
Terminalia megalocarpa, 42
Tertiary, 113
Thalassia, 4
T. *hemprichii*, 55
Thallassodendron
T. *ciliatum*, 55
T. *pachyrhizum*, 55
Theobroma cacao, 152
Theophrastaceae, 31

thermal extremes (*See* temperature stress), 161, 308
thigmotropism, 93, 97
Thuja
  T. occidentalis, 341, 363
  T. plicata, 198, 199, 202
Thymelaea hirsuta, 323, 327, 330
Thymelaeaceae, 323
Tillandsia
  T. circinnata, 87
  T. dasylirifolia, 87
  T. paucifolia, 86, 87
Tillandsioideae, 85
Tipularia, 80
  T. discolor, 81, 84, 89
Tmesipteris, 107, 108
Tolumnia variegata, 84
Toona sinensis, 49
Torreya myristica, 9
Toxicodendron pubescens, 6
Tozzia, 96
tracheophytes, 106, 108, 109
trade-offs, 12
  between growth and survival, 178, 179, 181
Trapa natans, 29, 42
Trapaceae, 29, 42
Trema, 342
  T. micrantha, 343
Tricholoma matsutake, 197
Trifolium repens, 151, 212
Trillium grandiflorum, 32
Trimenia, 137, 142
  T. moorei, 143
  T. papuana, 142, 145
Trimeniaceae, 135
Triphysaria versicolor, 96, 152
Tristerix, 94
Tristichaceae, 30
Trithuria, 143
Triticum, 26, 50
  T. aestivum, 122, 151, 253
Triuridaceae, 98
Trollius ledebouri, 44
tropical forests, 172, 178, 179, 185
Tsuga, 205
  and advanced regeneration, 262
  T. canadensis, 201, 202, 257, 341
  T. heterophylla, 199, 201, 202

Tuber, 205
Tulasnella, 193
Tulipa, 27, 47
Tussilago farfara, 47
Typha, 37
  T. domingensis, 69, 70
  T. latifolia, 5, 152, 384
Typhaceae, 5, 37, 69, 152, 384

Ulmaceae, 31, 152
Ulmus
  U. effusa, 31
  U. glabra, 152
Ulocladium atrum, 304
unit leaf ratio ULR, 291
Urtica dioica, 50
Urticaceae, 27, 50, 303, 347, 384
Utricularia, 50, 88, 89, 90
  U. geminiscapa, 42
  U. nelumbifolia, 54
  U. reniformis, 54
  U. radiata, 92
  U. striatula, 89
  U. subulata, 90

Valeriana glechomifolia, 152
Valerianaceae, 152
Vallisneria, 48
  V. americana, 151
Verbascum, 4
Verbena officinalis, 43
Verbenaceae, 26, 43
Veronica peregrina, 282
Vicia, 23
  V. faba, 151
Vigna radiata, 151
Viola palustris, 47
Violaceae, 47
Viscaceae, 92, 95
viscin, 94
Vismia, and tropical pastures, 347
Vitaceae, 152
Vitellaria paradoxa, 42
Vitex pubescens, 26
Vitis vinifera, 152
vivipary, see also cryptovivipary, 4, 25, 33, 34, 68, 69, 169

Vriesea
  V. geniculata, 87
  V. hieroglyphica, 86
  V. sanguinolenta, 86, 87
  V. scalaris, 85

Washingtonia filifera, 44
waterlogging, 252
Welwitschia, 118
  W. mirabilis, 31, 36, 152
Welwitschiaceae, 31, 152
wetlands, and restoration, 358
Winteraceae, 132
Wolffia, 31
woody magnoliid hypothesis, 132, 134
wound response, 162

Xanthium strumarium, 4, 6
Xanthorrhoea, and excluding grazers during restoration, 367
  X. australis, 27
  X. gracilis, 367
  X. preisii, 367
Xanthorrhoeaceae, 27, 367
Xanthosoma, 21
  X. sagittifolium, 151

Yucca, 316, 320

Zamia, 108, 115, 118
  Z. pumila, 117
Zamiaceae, 44
Zannichelliaceae, 38
Zantedeschia aethiopica, 151
Zanthoxylum stenophyllum, 152
Zea mays, 151, 168
  and phytohormones, 168
  mutants and phytohormone function, 168
Zilla myagroides, 42
Zinnia elegans, 151
Zizania, 28
zonation, 70
Zostera, 35
Zosteraceae, 20, 35
zosterophyllophytes, 106
Zygomycota, 193
Zygophyllaceae, 318

Printed in the United States
by Baker & Taylor Publisher Services